Numerical Methods in Geomechanics

NATO ADVANCED STUDY INSTITUTES SERIES

Proceedings of the Advanced Study Institute Programme, which aims at the dissemination of advanced knowledge and the formation of contacts among scientists from different countries

The series is published by an international board of publishers in conjunction with NATO Scientific Affairs Division

A	Life Sciences	Plenum Publishing Corporation
B	Physics	London and New York
C	Mathematical and Physical Sciences	D. Reidel Publishing Company Dordrecht, Boston and London
D	Behavioural and Social Sciences	Martinus Nijhoff Publishers
E	Engineering and Materials Sciences	The Hague, London and Boston
F	Computer and Systems Sciences	Springer Verlag Heidelberg
G	Ecological Sciences	

Series C – Mathematical and Physical Sciences

Volume 92 – Numerical Methods in Geomechanics

Numerical Methods in Geomechanics

Proceedings of the NATO Advanced Study Institute, University of Minho, Braga, Portugal, held at Vimeiro, August 24 - September 4, 1981

edited by

J. B. MARTINS
School of Engineering, University of Minho, Braga, Portugal

D. Reidel Publishing Company

Dordrecht : Holland / Boston : U.S.A. / London : England

Published in cooperation with NATO Scientific Affairs Division

Library of Congress Cataloging in Publication Data

CIP

NATO Advanced Study Institute (1981 : Vimeiro, Lisbon, Portugal)
Numerical methods in geomechanics.

(NATO advanced study institutes series. Series C, Mathematical and physical sciences ; v. 92)
Includes bibliographical references and index.
1. Soil mechanics–Mathematics–Congresses. 2. Rock mechanics–Mathematics–Congresses. 3. Numerical calculations–Congresses.
I. Martins, J. B. (J. Barreiros) II. Title. III. Series.
TA710.A1N37 1981 624.1'513 82-13263
ISBN 90-277-1461-4

Published by D. Reidel Publishing Company
P.O. Box 17, 3300 AA Dordrecht, Holland

Sold and distributed in the U.S.A. and Canada
by Kluwer Boston Inc.,
190 Old Derby Street, Hingham, MA 02043, U.S.A.

In all other countries, sold and distributed
by Kluwer Academic Publishers Group,
P.O. Box 322, 3300 AH Dordrecht, Holland

D. Reidel Publishing Company is a member of the Kluwer Group

Printed in The Netherlands

CONTENTS

Preface ix
List of Participants xi
Lecture Program xv

CHAPTER I - FUNDAMENTALS 1

1 Field Equations
E.R. Arantes e Oliveira 3

2 Basic Formulation of Static and Dynamic Behaviour of Soil and Other Porous Media
O.C. Zienkiewicz 39

3 Generalized Plasticity and Some Models for Geomechanics
O.C. Zienkiewicz 57

4 Constitutive Equations for Soil Media
Jean H. Prevost 79

5 Constitutive Modeling and Soil-Structure Interaction
C.S. Desai 103

6 Rock Mass Behaviour and Its Mathematical Modelling
K. Kovari 145

CHAPTER II - FINITE ELEMENTS TECHNIQUES 165

1 Numerical Analysis of Shallow Foundations
I.M. Smith 167

2 Numerical Analysis of Deep Foundations
I.M. Smith 187

3 Finite Element Analysis of Coupled Deformation and Fluid Flow in Porous Media
Ranbir S. Sandhu 203

4 Finite Elements and Slope Stability
D.J. Naylor 229

CHAPTER III - MATHEMATICAL PROGRAMMING METHODS 245

1 Plastic Analysis in Geomechanics by Mathematical Programming
John Munro 247

2 Optimization Methods for Parametric Identification of Geotechnical Systems
G. Maier and G. Gioda 273

3 Embankments and Slopes by Mathematical Programming
J.B. Martins 305

CHAPTER IV - BOUNDARY ELEMENT TECHNIQUES 335

1 Fundamentals of Boundary Elements
C.A. Brebbia 337

2 Boundary Elements in Groundwater Flow Problems
L.C. Wrobel and C.A. Brebbia 355

3 Boundary Element Formulation to Solve No-Tension Problems in Geomechanics
W.S. Venturini and C.A. Brebbia 375

4 Elastoplastic Analysis Using Boundary Elements
J.C.F. Telles and C.A. Brebbia 389

CHAPTER V - SEISMIC PROBLEMS 407

1 Seismic Site and Soil-Structure Interaction Analysis
John Lysmer 409

2 Liquefaction of Sandy Soils
José Folque 455

CHAPTER VI - STATISTICAL METHODS 471

- Statistical and Probabilistic Methods in Geomechanics
A. Haldar 473

APENDICES 505

LECTURE
- Liquefaction of Soils During Earthquakes
I.M. Idriss 507

RESEARCH PRESENTATIONS* IN PANELS
1 Dynamic, Cyclic and Transient Aspects in Soil Engineering
M.P. Luong 531

2 Buried Structures-Surrounding Soil Interaction
M.P. Luong 541

3 Bearing Capacity and Shape Factors for Shallow Foundations
M.P. Luong 547

4 Coefficient of Potential Rupture for Earth Structures
M.P. Luong 557

NAME INDEX 567

SUBJECT INDEX 573

* written papers

PREFACE

This volume contains the edited texts of the lectures and written discussions prepared for the NATO sponsored Advanced Study Institute (ASI) on "Numerical Methods in Geomechanics" promoted by the University of Minho, Braga, Portugal and held at Vimeiro Hotel (Torres Vedras) - August 24 to September 4, 1981.

The object of the NATO-ASI has been to gather, as lecturers, a wide range of scientists and engineers specialized on the above branch of knwoledgement in order to cover the whole variety of approaches for the solution of the set of problems concerning Geomechanics. In this way we could make available to the students, most of them researchers in this field, a large number of concepts, views and techniques for the solution of the problems they are concerned with. At the same time it was possible to disclose relationships between apparently different techniques for the solution of the same problem. For example some plasticity problems may be solved either by the usual finite elements technique or by mathematical programming or by the boundary elements method. Therefore, a link exists amongst those methods. However, there has been in the past lack of effort to spring out these links, since there is a tendency for the researchers to stress the method they are working with and to forget about other methods, at least in what concerns their fundamentals.

Bringing together scientists from distinct schools of thought, this ASI has provided a forum for an interchange of views and of experiences amongst the lecturers and panelists themselves. In this way a stimulating and rewarding experience was given to each participant, mainly to the students, most of them preparing their doctoral theses.

Therefore, we hope these proceedings will be a way of reviving the "spirit of Vimeiro" giving an opportunity to the students to apply some important new concepts put forward at this ASI, and to the lecturers and panelists to try, in their new experiments, to apply each others concepts and methods in order to make a correct comparison and to find the reasons for agreement or disagreement between results.

J. B. Martins (ed.), Numerical Methods in Geomechanics, ix–x.

We hope further that the Proceedings will give to all professionals an up-to-date and broad view of the methods available to solve geomechanical problems and a forward looking for future developments.

In fact many lecturers have brought to this Advanced Institute a number of quite new ideas for the solution of geomechanical problems which are particularly difficult, since they concern materials made by Nature and not by Man, with a most complex constitution and dealing with rather intrincated geometric shapes.

So, we wish to express our appreciation to those lecturers who put so great interest in this ASI.

In organizing the Chapters of these proceedings we have found some difficulties due to the fact that some lecturers have written their lectures in one final paper covering a somewhat large range of subjects. Therefore, the chapters do not follow exactly the order of the lectures and probably some people may feel that a better arrangement could be done.

We are greatfull to my Colleague Carlos Bernardo from U.M. who from the very begining was a full-time organizer of this Advanced Institute and to the staff of the University of Minho whose dedication allowed to overcome the many difficulties that always arrive in organizations such as this one.

Thanks are also due to the Vice-Chancelor of the U.M. for the facilities made available for the Institute.

Finally we extend our deep appreciation to the NATO Division of Scientific Affairs whose generous financial support made possible this NATO-ASI and the publication of these Proceedings.

January 1982

J.B. Martins

LIST OF PARTICIPANTS

Abascal, J.D.
Escuela Superior de Ingenieros Industriales
Sevilla - 12, Spain

Alarcón, E.
Escuela Politecnica Superior de Ingenieros Industriales
Madrid - 6, Spain

Anand, S.C.
Department of Civil Engineering
Clemson University
Clemson, S. Carolina, USA 29631

Barros, M.H.F.
Departamento de Engenharia Civil
Universidade de Coimbra
3000 Coimbra, Portugal

Barroso, M.J.G.
Laboratório Nacional de Engenharia Civil
1700 Lisboa, Portugal

Béliveau, J.G.
Department of Civil Engineering
Université de Sherbrooke
Sherbrooke, Québec, Canada J1K 2R1

Bénitez, F.G.
Escuela Politecnica Superior de Ingenieros Industriales
Madrid - 6, Spain

Bolle, A.
Institut du Génie Civil - Bât C 2
Université de Liège
B - 4000 Liège, Belgium

Borm, G.
Lersthul fur Felsmechanik
Universitat Fridericiana zu Karlsruhe
D - 7500 Karlsruhe 1, West Germany

Brebbia, C.A.
Department of Civil Engineering
The University
Southampton, England SO9 5NH

Carballo, F.A.
Escuela Superior de Ingenieros Industriales
Sevilla - 12, Spain

Cavilhas, J.L.A.
Hidrotécnica Portuguesa
Ed HP Guiné, Prior Velho
Lisboa, Portugal

Cerejeira, J.M.G.
OMAN - Centro de Estudos e Projectos
R. Eugénio de Castro, 8 - 8º Esqº
Almada, Portugal

Chapel, F.M.
Laboratoire de Mécanique des Sols
Ecole Centrale des Arts et Manufactures
92290 Chatenay - Malabry, France

Civelek, M.B.
Department of Civil Engineering
Bogaziçi University
P.K. 2 Bebek, Istanbul, Turkey

Cividini, A.
Ist. di Scienza e Tecnica delle Construzioni
Politecnico
20133 Milano, Italy

Cormeau, I.
Consultant Control Data Belgium N.V. - S.A.
Leopold III Lan 111
1130 Brussels, Belgium

Correia, A.G.
Instituto Superior Técnico
1000 Lisboa, Portugal

Cozzarelli, F.A.
Department of Civil Engineering
State University of New York at Buffalo
Buffalo, N.Y., USA 14214

Cruz, J.M.L.
Hidrotécnica Portuguesa
Ed HP Guiné, Prior Velho
Lisboa, Portugal

Cunha, A.J.V.P.
Laboratório Nacional de Engenharia Civil
1700 Lisboa, Portugal

Davies, T.G.
Department of Civil Engineering
State University of New York at Buffalo
Buffalo, N.Y., USA 14214

Desai, C.S.
Department of Civil Engineering
Virginia Polytechnic Institute and State University
Blacksburg, Virginia, USA 24061

Doblaré, M.
Escuela Politecnica Superior de Ingenieros Industriales
Madrid - 6, Spain

Durgunoglu, T.
Earthquake Engineering Research Institute
Bogaziçi University
P.K. 2 Bebek, Istanbul, Turkey

Ferreira, A.M.S.
Geotecnia
Universidade Nova de Lisboa
1800 Lisboa, Portugal

Figueiras, J.A.
University College of Swansea
Swansea, England SA2 8PP

Fonseca, A.M.A.
CECUP - Centro de Engenharia Civil
Universidade do Porto
R. dos Bragas
4000 Porto, Portugal

Gaspar, J.E.T.
Cubiertas Y MZOV S.A.
Ayala 42
Madrid 1, Spain

Georgiou, P.
Computational Mechanics Centre
Southampton, Englang SO1 0AA

Gioda, G.
Ist. di Scienza e Tecnica delle Construzioni
Politecnico
20133 Milano, Italy

Gonzalez, J.M.V.
Instituto Superior Técnico
1000 Lisboa, Portugal

Grossmann, N.
Laboratório Nacional de Engenharia Civil
1700 Lisboa, Portugal

Gumusoglu, M.C.
Department of Engineering Geology and Rock Mechanics
Tesvikiye, Istanbul, Turkey

Gussmann, P.
Institut fur Grundbau und Bodenmechanik
Pfaffernwaldring 35
7 Stuttgart 80, West Germany

Haahr, F.
Soil Mechanics and Foundation Engineering
Technical University of Denmark
Maglebjergvej 1
DK - 2800 Lingby, Denmark

Haldar, A.
Georgia Institute of Technology
Atlanta, Georgia, USA 30332

Halonen, O.
Technical Research Centre of Finland
Geotechnical Laboratory
Lampomiehenkuja 2 B
SF - 02150, Espoo 15, Finland

Heuze, F.E.
Lawrence Livermore National Lab.
University of California
Livermore, Calif., USA 94550

Idriss, I.M.
Woodward-Clyde Consultants
Three Embarcadero Center, S. 700
S. Francisco, Calif., USA 94111

Jaudenes, M.D.R.
Cubiertas Y MZOV S.A.
Ayala 42
Madrid 1, Spain

Katzir, M.
Foundation and Soil Ltd
44 Revivim St, Box 10107
Tel-Aviv 69274, Israel

Kausel, E.A.M.
Department of Civil Engineering
Massachusetts Institute of Technology
Cambridge, Mass., USA 02139

Kelly, W.E.
Department of Civil Environmental Engineering
University of Rhode Island
Kingston, Rhode Island, USA 02881

Kovari, K.
Institut fur Strassen-, Eisenbahn- und Felsbau
ETH - Honggerberg
CH-8093 Zurich, Switzerland

Lemoine, M.E.
Koninklijke/Shell Exploratie en Produktie Laboratorium
Volmerlaan 6
2288 GD Rijswijk, Holland

Loret, B.
Laboratoire de Mécanique des Solides
Ecole Polytechnique
91128 Palaiseau, France

Luong, M.P.
Laboratoire de Mécanique des Solides
Ecole Polytechnique
91128 Palaiseau, France

Lysmer, L.
Department of Civil Engineering
University of California
Berkeley, Calif., USA 94720

Maier, G.
Ist. di Scienza e Tecnica delle Construzioni
Politecnico
20133 Milano, Italy

Marques, J.C.
University of Swansea
Swansea SA2 8PP, England

Martins, J.B.
Unidade Pedagógica de Engenharia
Universidade do Minho
4700 Braga, Portugal

Matos, A.B.C.
F.E.U.P.
Universidade do Porto
4000 Porto, Portugal

Matos, A.C.
F.E.U.P.
Universidade do Porto
4000 Porto, Portugal

Mohiuddin, G.M.
Gesellschaft fur Reaktorsicherheit
Postfach 10 16 50
5000 Koln 1, West Germany

Morancho, A.L.
Departamento de Geotecnia
Universidad Politecnica de Barcelona
Barcelona 17, Spain

Mota, C.A.M.
Construções Técnicas SARL
Av. 24 de Julho 24
1200 Lisboa, Portugal

Munro, J.
Department of Civil Engineering
Imperial College
London SW7 2BU, England

Natau, O.
Lehrstuhl fur Felsmechanick
Universitat Fridericiana zu Karlsruhe
D-7500 Karlsruhe, West Germany

Naylor, D.J.
University College of Swansea
Swansea SA2 8PP, England

Neves, E.M.
Laboratório Nacional de Engenharia Civil
1700 Lisboa, Portugal

Neves, M.L.S.O.
Construções Técnicas SARL
Av. 24 de Julho 24
1200 Lisboa, Portugal

Nordal, S.
Institut fur Geoteknikk og Fundamenteringslaere
N-7034 Trondheim-NTH, Norway

Oliveira, E.R.A.
Universidade Técnica de Lisboa
1100 Lisboa, Portugal

Ottosen, N.S.
Riso National Laboratory
Postbox 49
DK-4000 Roskilde, Denmark

Paixão, J.C.L.
F.E.U.P.
Universidade do Porto
4000 Porto, Portugal

Prater, E.G.
Institut fur Strassen-, Eisenbahn- und Felsbau
ETH - Honggerberg
CH-8093 Zurich, Switzerland

Prevost, J.H.
Princeton University
Princeton, New Jersey, USA 08544

Quaresma, C.F.
Departamento de Engenharia Civil
Universidade de Coimbra
3000 Coimbra, Portugal

Quiroga, A.S.
Departamento Analisis de las Estruturas
Universidad de Santander
Santander, Spain

Ratigan, J.L.
Geotechnical Engineering
University of California
Berkeley, Calif., USA 94720

Reis, E.B.
F.E.U.P.
Universidade do Porto
4000 Porto, Portugal

Rodrigues, F.M.C.P.
Laboratório Nacional de Engenharia Civil
1700 Lisboa, Portugal

Rucker, W.
Bundesanstalt fur Materialprufung
Unter den Eichen 87
1000 Berlin 45, West Germany

Sadgorski, W.
Bayer Landesamt fur Wasserwirtschaft
Dienstgebande Torring Str. 2
8000 Munchen 80, West Germany

Sandhu, R.S.
Department of Civil Engineering
The Ohio State University
Columbus, Ohio, USA 43210

Shaw, R.P.
Department of Engineering Science
University of New York at Buffalo
New York, USA 14214

Silva, A.J.
University of Rhode Island
Narragansett
Rhode Island, USA 02882

Silva, M.A.G.
Gabinete da Área de Sines
R. Artilharia 1
1200 Lisboa, Portugal

Smith, I.M.
Simon Engineering Laboratories
University of Manchester
Manchester M13 9PL, England

Soeiro, F.A.
Bureau d'Etudes et de Recherches Geotechniques Engineering
6 Rue des Colonels Renard
75017 Paris, France

Sousa, L.M.R.
Laboratório Nacional de Engenharia Civil
1700 Lisboa, Portugal

Swoboda, G.
Universitat Innsbruck
Technikerstrasse 13
A - 6020 Innsbruck, Austria

Taylor, B.B.
Department of Civil Engineering
University of Waterloo
Ontario, Canada N2L 3GL

Veiga, E.V.
Universidade do Minho
4700 Braga, Portugal

Vilar, C.
Universidade do Minho
4700 Braga, Portugal

Wolfe, W.E.
The Ohio State University
2070 Neil Avenue
Columbus, Ohio, USA 43210

Wynendaele, H.
Katholieke Universiteit te Leuven
Celestijnenlaan 200 A
3030 Haverlee, Belgium

Zacas, M.
Geomechaniki S.A.
25 Bairaktari Str.
Athens (607), Greece

Zeman, A.J.
Shore Processes Section
Hydraulic Research Division
National Water Research Institute
P.O. Box 5050
Burlington, Ontario, Canada

Zienkiewicz, O.C.
Applied Science Building
University College of Wales
Swansea SA2 8PP, England

DAY/HOUR	9.00 - 9.50	9.00 - 10.40	(a)	11.00-11.50	11.50-12.40	(b)	14.30-15.20	15.20-16.10	(a)	16.30 - 17.20	17.20 - 18.10
24/8	A. e Oliveira	A. e Oliveira	"	A. e Oliveira	Desai	"	Desai	Desai	"	Prevost	Prevost
25	Zienckiewicz	Zienckiewicz	"	Zienckiewicz	Zienckiewicz	"	(*)Constitutive Laws for Soils (Statics)		"	(*)Constitutive Laws For Soils (Dynamics)	
26	Kovari	Kovari	"	Borm	Natau	"	(*)Rock Masses Behaviour		"	(*)Underground Openings	
27	Smith	Smith	"	Naylor	Martins	"	(*) Foundations		"	(*)Embankments and Slopes	
28	Brebbia	Brebbia	"	Brebbia	Brebbia	"	(*) Boundary Element Methods		"	(*) Boundary Element Methods	
29	Desai	Desai	"	(*) Soil Structure Interaction (Statics)		"	**SOCIAL PROGRAMME**				
30	**SOCIAL PROGRAMME**										
31	Munro	Munro	"	Maier	Maier	"	(*) Identification by Mathem. Programming		"	(*)Plastic Analysis by Mathem. Programming	
1/9	Sandhu	Sandhu	"	Idriss	Idriss	"	(*) Seepage and Consolidation		"	(*) Liquefaction	
2	Lysmer	Lysmer	"	Lysmer	Idriss	"	Idriss		"	(*) Seismic Analysis	
3	(*) Seismic Analysis			Soil - Structure Interaction (Dynamics)		"	Soil - Structure Interaction (Dynamics)		"		
4	(*)Statistical and Probab. Meth. in Geomechanics		"	(*)Statistical and Probab. Meth. in Geomechanics		"	**DEPARTURE**				

(a) Coffe-break. (b) Lunch. (*) Discussion Panel

NATO A S I "NUMERICAL METHODS IN GEOMECHANICS" 24TH AUGUST TO 4TH SEPTEMBER 1981

STRUCTURE OF THE DISCUSSION PANELS

SUBJECTS	PANELISTS	
	LECTURERS	OTHER MEMBERS
Constitutive Laws for Soils (Statics)	C.Desai, Blacksburg, U.S.A. J.Prevost, Princeton, U.S.A.	F.A.Cozarelli, Buffalo, U.S.A.
Constitutive Laws for Soils (Dynamics)	O.C.Zienckiewicz, Swansea, U.K.	M.P.Luong, Palaiseau, France
Foundations	I.M.Smith, Manchester, U.K.	M.P.Luong, Palaiseau, France T.Davies, Buffalo, U.S.A.
Rock Masses Behaviour	K.Kovari, Zurich, Switzerland	F.E.Heuze, Livermore, Calif.,U.S.A. T.Durgunoglu, Istanbul, Turkey
Underground Openings	G.Borm, Karlrsruhe, Germany O.Natau, Karlrsruhe, Germany	G.Swoboda, Innsbruck, Austria F.E.Heuze, Livermore, Calif.,U.S.A. S.C.Anand, Clemson, S.Carol.,U.S.A. T.Durgunoglu, Istanbul, Turkey
Seepage and Consolidation	R.S.Sandhu, Columbus, U.S.A.	G.Gioda, Milano, Italy
Identification by Mathem. Programming	G.Maier, Milano, Italy	G.Gioda, Milano, Italy
Plastic Analysis by Mathem. Programming	J.Munro, London, U.K.	G.Gioda, Milano, Italy
Boundary Element Methods	C.Brebbia, Southampton, U.K.	R.P.Shaw, Buffalo, U.S.A. H.Winendaele, Haverlee, Belgium E.Alarcon, Madrid, Spain J.Dominguez, Sevilla, Spain P.Georgiou, Southampton, U.K.
Soil-Structure Interaction (Statics)	C.Desai, Blacksburg, U.S.A.	M.P.Luong, Palaiseau, France M.B.Civelek, Istanbul, Turkey S.C.Anand, Clemson, S.Carol.,U.S.A.
Soil-Structure Interaction (Dynamics)	J.Lysmer, Berkeley, U.S.A.	Jean-Guy Béliveau, Québec, Canada E.Kausel, Massachusetts, U.S.A. M.B.Civelek, Istanbul, Turkey
Seismic Analysis	J.Lysmer, Berkeley, U.S.A.	I.M.Idriss, S.Franc., Calif.,U.S.A.
Soil Liquefaction	I.M.Idriss, S.F.,Calif., U.S.A.	J.Folque, Lisbon, Portugal
Statistical and Probabilistic Methods in Geomechanics		A.Haldar, Atlanta, Georgia, U.S.A.

CHAPTER I

FUNDAMENTALS

FIELD EQUATIONS

E. R. Arantes e Oliveira

Technical University (IST) - Lisbon

ABSTRACT

The paper consists of four parts: an overture on second order tensors, a chapter on Kinematics of deformable bodies, a chapter on stresses and balance equations and a summary of field equations in Soil Mechanics. The criterion has been followed of defining first the basic concepts and establishing the exact non-linear equations of Continuum Mechanics, and then introducing the simplifying assumptions.

INTRODUCTION

Geomechanics is, or is wished to be, a part of Mechanics.

It is not frequent however to start a exposition on Geomechanics from the beginning, i.e., from a presentation of the basic concepts and of the general laws of Continuum Mechanics.

It is even very often believed that rigour in Geomechanics is not consistent with the magnitude of the errors committed in predicting what is measured from what is computed.

We believe however that, although such errors are unavoidable, because Nature cannot easily be simulated, there is no reason for adding new errors just by failing in Mathematics or, what is the same, as Trusdell points out in his remarkable Introduction to [1], by failing in Logics.

For this reason, and also because, in many fields, the trend to more complex problems has led to a need for more exact equations, we chose to start by defining the basic concepts and establishing the exact non-linear equations of Continuum Mechanics, and then proceed by introducing the simplifying assumptions aimed to provide, when the nature of the problem consents, more economical solutions.

J. B. Martins (ed.), Numerical Methods in Geomechanics, 3–37.

1. SECOND ORDER TENSORS

1.1. Tensors and matrices

A second order tensor $\underset{\approx}{A}$ on a N-dimensional vector space L^N is a linear transformation which assigns to a vector $\underset{\approx}{u} \in L^N$ another vector $\underset{\approx}{v} \in L^N$. We write

$$\underset{\approx}{v} = \underset{\approx}{A}\,\underset{\approx}{u} \quad (1.1)$$

If a basis $\{\underset{\approx}{i}_k\}_{k=1}^{N}$ is chosen for L^N and

$$\underset{\approx}{u} = u_k\,\underset{\approx}{i}_k \quad (1.2)$$

$$\underset{\approx}{v} = v_k\,\underset{\approx}{i}_k \quad (1.3)$$

$$\underset{\approx}{A} = A_{kl}\,\underset{\approx}{i}_k \otimes \underset{\approx}{i}_l \quad (1.4)$$

we write

$$v_i = A_{ij}\,u_j \,. \quad (1.5)$$

The second-order tensor $\underset{\approx}{A}$ can thus be identified with the Nth order square matrix $\underset{\sim}{A} = [A_{ij}]$ of its components. Such identification supposes however that a basis has been settled. Only once this is done, as it will be assumed in the sequel, we can indifferently write $\underset{\approx}{A}$ or $\underset{\sim}{A}$ and operations on tensors are isomorphic of operations on matrices.

The Cartesian decomposition of $\underset{\sim}{A}$ is defined as

$$\underset{\sim}{A} = \underset{\sim}{A}_s + \underset{\sim}{A}_a \quad (1.6)$$

where

$$\underset{\sim}{A}_s = \tfrac{1}{2}(\underset{\sim}{A} + \underset{\sim}{A}^T) \quad (1.7)$$

is a symmetric tensor and

$$\underset{\sim}{A}_a = \tfrac{1}{2}(\underset{\sim}{A} - \underset{\sim}{A}^T) \quad (1.8)$$

a skew-symmetric tensor.

A further decomposition for tensor $\underset{\sim}{A}_s$ into spherical and deviatoric tensors is

$$\underset{\sim}{A}_s = \tfrac{1}{3}\,A_{ii}\,\underset{\sim}{I} + \underset{\sim}{A}_d \quad (1.9)$$

where $\underset{\sim}{A}_d$ is a traceless tensor.

Another kind of decomposition results from the so-called polar decomposition theorem which states that every invertible matrix possesses invariant right and left decompositions

$$\underset{\sim}{F} = \underset{\sim}{R}\,\underset{\sim}{U} \qquad \underset{\sim}{F} = \underset{\sim}{V}\,\underset{\sim}{R} \quad (1.10\text{-}11)$$

where $\underset{\sim}{R}$ is an orthogonal tensor and $\underset{\sim}{U}$ and $\underset{\sim}{V}$ are positive definite matrices defined by

$$\underset{\sim}{U}^2 = \underset{\sim}{F}^T \underset{\sim}{F} \qquad\qquad \underset{\sim}{V}^2 = \underset{\sim}{F}\,\underset{\sim}{F}^T \tag{1.12-13}$$

and interconnected by

$$\underset{\sim}{V} = \underset{\sim}{R}\,\underset{\sim}{U}\,\underset{\sim}{R}^T. \tag{1.14}$$

$\underset{\sim}{U}$ and $\underset{\sim}{V}$ are similar matrices in the sense that their proper values are equal.
Assuming that an inner-product is defined on L^N, we require now the basis vectors $\underset{\approx}{i}_k$ to be orthonormal, i.e., to satisfy the condition

$$\underset{\approx}{i}_k \cdot \underset{\approx}{i}_l = \delta_{kl}\,. \tag{1.15}$$

The components of the tensor referred to such basis are called cartesian.
The long tradition of confounding a tensor with the matrix of its components is responsible for the pratice of calling the tensor itself a cartesian tensor. Such practice has however not really more meaning that calling a vector a cartesian vector if it is referred to a cartesian basis.
If a change of basis is made to another orthonormal basis $\{\underset{\approx}{i}_k^*\}_{k=1}^N$, the matrix $\underset{\sim}{T} = [t_{km}]$ such that

$$\underset{\approx}{i}_k = t_{km}\,\underset{\approx}{i}_m^* \tag{1.16}$$

is an orthogonal matrix, i.e., a matrix characterized by

$$\underset{\sim}{T}\,\underset{\sim}{T}^T = \underset{\sim}{T}^T\,\underset{\sim}{T} = \underset{\sim}{I} \tag{1.17}$$

so that $|\underset{\sim}{T}| = \pm 1$.
Introduction of (1.11) into (1.2) and (1.4) yields

$$\underset{\approx}{u} = u_k\, t_{km}\,\underset{\approx}{i}_m^* \tag{1.18}$$

$$\underset{\approx}{A} = A_{kl}\, t_{km}\, t_{ln}\,\underset{\approx}{i}_m^* \otimes \underset{\approx}{i}_n^*\,, \tag{1.19}$$

so that the components of $\underset{\approx}{u}$ and $\underset{\approx}{A}$ with respect to the new basis are

$$u_m^* = u_k\, t_{km} \tag{1.20}$$

$$A_{mn}^* = A_{kl}\, t_{km}\, t_{ln} \tag{1.21}$$

or, using the matrix notation,

$$\underset{\sim}{u}^* = \underset{\sim}{T}^T \underset{\sim}{u} \qquad \text{or} \qquad \underset{\sim}{u} = \underset{\sim}{T}\,\underset{\sim}{u}^*\,, \tag{1.22-23}$$

$$\underset{\sim}{A}^* = \underset{\sim}{T}^T \underset{\sim}{A}\,\underset{\sim}{T} \qquad \text{or} \qquad \underset{\sim}{A} = \underset{\sim}{T}\,\underset{\sim}{A}^*\,\underset{\sim}{T}^T. \tag{1.24-25}$$

By virtue of (1.24), symmetry does not depend on the basis, so that, if $\underset{\sim}{A}$ is symmetric, tensor $\underset{\approx}{A}$ itself can be called symmetric.

1.2. Principal values and principal vectors

The N principal or proper values a_n of a second order tensor with respect to a certain basis are the principal or proper values of the corresponding matrix $\underset{\sim}{A}$, i.e., the roots of the equation

$$|\underset{\sim}{A} - a \underset{\sim}{I}| = 0 , \tag{1.26}$$

where $\underset{\sim}{I}$ is the unit tensor.
By virtue of (1.24),

$$[\underset{\sim}{A}^* - a \underset{\sim}{I}] = \underset{\sim}{T}^T [\underset{\sim}{A} - a \underset{\sim}{I}] \underset{\sim}{T} , \tag{1.27}$$

so that

$$|\underset{\sim}{A}^* - a \underset{\sim}{I}| = |\underset{\sim}{A} - a \underset{\sim}{I}| . \tag{1.28}$$

The roots of the equation (1.26) are thus the roots of the equation $|\underset{\sim}{A}^* - a\underset{\sim}{I}| = 0$, and the invariance can thus be stated of the principal values of $\underset{\approx}{A}$ with respect to changes from orthonormal bases to orthonormal bases. Only such changes will be considered in the sequel.
Expanding (1.26), there results

$$a^3 - I_1 a^2 + I_2 a - I_3 = 0 , \tag{1.29}$$

where

$$I_1 = A_{mm} \tag{1.30}$$

$$I_2 = 1/2 (A_{mm} A_{nn} - A_{mn} A_{nm}) \tag{1.31}$$

$$I_3 = |A| \tag{1.32}$$

are the so-called principal invariants of $\underset{\approx}{A}$, i.e., a set of independent invariants in terms of which any invariant of $\underset{\approx}{A}$ can be expressed.
As, by virtue of (1.29), the principal values of $\underset{\approx}{A}$ can be expressed in terms of the principal invariants, these last, and therefore any invariant of $\underset{\approx}{A}$, can be expressed in terms of the principal values.
If $\underset{\approx}{A}$ is symmetric, an orthonormal basis exists such that components $A^*_{mn} = 0$ whenever $m \neq n$. The directions of the vectors of such basis are called the principal directions.
In order to demonstrate this statement, we prove first that the components $A'_{\underline{mm}}$ (an underlined index means that the repeated index convention must not be applied) referred to such basis are the principal values a_m.

Indeed, (1.24-25) imply

$$\underset{\sim}{A}\,\underset{\sim}{T} = \underset{\sim}{T}\,\underset{\sim}{A}^* \tag{1.33}$$

or

$$A_{kl}\, t_{ln} = t_{km}\, A^*_{mn} \;. \tag{1.34}$$

Making

$$A^*_{mn} = A_{\underline{n}}\, \delta_{m\underline{n}}\;, \tag{1.35}$$

there results

$$A_{kl}\, t_{ln} = A_{\underline{n}}\, t_{k\underline{n}} \tag{1.36}$$

or

$$[A_{kl} - A_{\underline{n}}\, \delta_{kl}]\, t_{l\underline{n}} = 0 \tag{1.37}$$

which may be rewritten as

$$[\underset{\sim}{A} - A_{\underline{n}}\, \underset{\sim}{I}]\, \underset{\sim}{t}_{\underline{n}} = \underset{\sim}{0}\,, \tag{1.38}$$

where $\underset{\sim}{t}_n$ denotes the nth column of matrix $\underset{\sim}{T}$.
The existence of non-trivial solutions requires the condition

$$|\underset{\sim}{A} - A_n\, \underset{\sim}{I}| = 0 \tag{1.39}$$

to be satisfied, which shows that A_n coincide with a_n.
Vectors $\underset{\sim}{t}_n$ associated to the principal values of $\underset{\sim}{A}$ define the principal directions.
The second step in the demonstration consists in proving that, to different principal values a_m and a_n , there correspond orthogonal principal directions.
(1.38) implies indeed

$$\underset{\sim}{A}\,\underset{\sim}{t}_n = a_{\underline{n}}\,\underset{\sim}{t}_{\underline{n}} \qquad\qquad \underset{\sim}{A}\,\underset{\sim}{t}_m = a_{\underline{m}}\,\underset{\sim}{t}_{\underline{m}} \tag{1.40-41}$$

and, therefore,

$$\underset{\sim}{t}_m^T\,\underset{\sim}{A}\,\underset{\sim}{t}_n = a_{\underline{n}}\,\underset{\sim}{t}_m^T\,\underset{\sim}{t}_{\underline{n}} \;,\qquad \underset{\sim}{t}_n^T\,\underset{\sim}{A}\,\underset{\sim}{t}_m = a_{\underline{m}}\,\underset{\sim}{t}_n^T\,\underset{\sim}{t}_{\underline{m}}\,. \tag{1.42-43}$$

But, by virtue of the symmetry of $\underset{\sim}{A}$, (1.43) yields, by transposition,

$$\underset{\sim}{t}_m^T\,\underset{\sim}{A}\,\underset{\sim}{t}_n = a_{\underline{m}}\,\underset{\sim}{t}_{\underline{m}}^T\,\underset{\sim}{t}_n\,, \tag{1.44}$$

so that, subtracting (1.42) from (1.44), we obtain

$$(a_{\underline{m}} - a_{\underline{n}})\,\underset{\sim}{t}_{\underline{m}}^T\,\underset{\sim}{t}_{\underline{n}} = 0. \tag{1.45}$$

Therefore, if a_m and a_n have different values, $\underset{\sim}{t}_m^T\,\underset{\sim}{t}_n$ must be equal to zero, i.e., the two vectors must be orthogonal.
The third and last step in the demonstration that a second-order symmetric tensor can be diagonalized consists in proving that all its principal values are real. The proof is straightforward if we

take the conjugates of both sides in (1.40). There results

$$\bar{\underset{\sim}{A}}\,\bar{\underset{\sim}{t}}_n = \bar{a}_n\,\bar{\underset{\sim}{t}}_n . \tag{1.46}$$

But, as $\underset{\sim}{A}$ is a real matrix, $\underset{\sim}{A}=\bar{\underset{\sim}{A}}$. Therefore,

$$\underset{\sim}{A}\,\bar{\underset{\sim}{t}}_n = \bar{a}_n\,\bar{\underset{\sim}{t}}_n . \tag{1.47}$$

From the comparison between (1.40) and (1.47), there results that a_n and $\bar{a}_n$ are both principal values of $\underset{\sim}{A}$. Thus, if they are distinct, the corresponding principal values, $\underset{\sim}{t}_n$ and $\bar{\underset{\sim}{t}}_n$, must be orthogonal.
However

$$\underset{\sim}{t}_{\underline{n}}\cdot\bar{\underset{\sim}{t}}_{\underline{n}} = t_{k\underline{n}}\,t_{k\underline{n}} = \sum_{k=1}^{N} t_{kn}^2 > 0 \tag{1.48}$$

so that $\underset{\sim}{t}_n$ and $\bar{\underset{\sim}{t}}_n$ cannot be orthogonal and, thus, a_n and $\bar{a}_n$ cannot be distinct. This means that $a_n=\bar{a}_n$ and, therefore, that a_n must be real.

1.3. Some notations

Matrix, vector and tensor notations will be used in the sequel. As already stated, (1.1) and (1.5) have the same meaning.
A dot will denote the internal product of two vectors, i.e.,

$$\underset{\sim}{u}\cdot\underset{\sim}{v} = \underset{\sim}{u}^T\underset{\sim}{v} . \tag{1.49}$$

The same notation will be used if, instead of vectors, we have matrices. Therefore,

$$\underset{\sim}{A}\cdot\underset{\sim}{B} = \underset{\sim}{A}^T\underset{\sim}{B} = [A_{ij}\cdot B_{ik}] \tag{1.50}$$

and

$$\underset{\sim}{A}\cdot\underset{\sim}{u} = \{A_{ij}\cdot v_i\} \tag{1.51}$$

We define the square of a matrix by

$$\underset{\sim}{A}^2 = \underset{\sim}{A}^T\cdot\underset{\sim}{A} = \underset{\sim}{A}\cdot\underset{\sim}{A} , \tag{1.52}$$

the trace of $\underset{\sim}{A}$ by

$$tr\,\underset{\sim}{A} = A_{ii} \tag{1.53}$$

and the magnitude of $\underset{\sim}{A}$ by

$$\|A\| = [tr\,\underset{\sim}{A}^T\underset{\sim}{A}]^{1/2} . \tag{1.54}$$

If $\underset{\sim}{A}$ is a tensor function of $\underset{\sim}{x}$, the divergence of $\underset{\sim}{A}$ is defined as

$$div\,\underset{\sim}{A} = \underset{\sim}{A}\cdot\underset{\sim}{\nabla} = \{A_{ij,i}\} . \tag{1.55}$$

The gradient and curl of a vector $\underset{\sim}{v}$ are, on the other hand,

$$grad\ \underset{\sim}{v} = \underset{\sim}{\nabla}\underset{\sim}{v}^T = [v_{i,j}] \quad , \quad curl\ \underset{\sim}{v} = \underset{\sim}{\nabla}\wedge\underset{\sim}{v} = \{e_{ijk}v_{k,j}\} \tag{1.56-57}$$

where

$$\underset{\sim}{\nabla} = \{\nabla_i\} = \left\{\frac{\partial}{\partial x_i}\right\} \tag{1.58}$$

and e_{ijk} is the permutation symbol.
As orthonormal bases only are to be considered, no distinction between covariant and contravariant components will be necessary.

2. KINEMATICS OF DEFORMABLE BODIES

2.1. Preliminaries

Kinematics is the study of the geometry of motion and deformation independently of their causes.
A material body B is a three-dimensional set, the elements of which may be called particles, P .
A one-to-one correspondence can be established between each particle of the set and a point in $\mathbb{R}^3$.
The subsets of $\mathbb{R}^3$ occupied by the body at any time t are called the configurations of the body. The origin of time is chosen in such a way that the configuration corresponding to $t=0$ is a special known configuration which we call the reference configuration.
We write therefore

$$P = P(X_1, X_2, X_3) = P(\underset{\sim}{X}) \tag{2.1}$$

and call X_i the material co-ordinates of the particles.
The co-ordinates x_i of the particle at a given configuration are called spatial coordinates.
Each configuration can be identified by a function $\underset{\sim}{\chi}$ such that

$$\underset{\sim}{x} = \underset{\sim}{\chi}(\underset{\sim}{X}) \ . \tag{2.2}$$

A motion of B is a one-parameter family of configurations of the body. The parameter is the time, t . We write

$$\underset{\sim}{x} = \underset{\sim}{\chi}(\underset{\sim}{X}, t) \ . \tag{2.3}$$

The displacement, velocity and acceleration vectors are defined by

$$\underset{\sim}{u} = \underset{\sim}{x} - \underset{\sim}{X} = \underset{\sim}{\chi}(\underset{\sim}{X}, t) - \underset{\sim}{X} \tag{2.4}$$

$$\underset{\sim}{v} = \dot{\underset{\sim}{x}} = \dot{\underset{\sim}{u}} = \frac{D}{Dt}\underset{\sim}{\chi}(\underset{\sim}{X}, t) \tag{2.5}$$

$$\underset{\sim}{a} = \ddot{\underset{\sim}{x}} = \ddot{\underset{\sim}{u}} = \frac{D^2}{Dt^2} \underset{\sim}{\chi}(\underset{\sim}{X}, t) . \tag{2.6}$$

The Jacobian

$$J = \left| \frac{\partial x_i}{\partial X_j} \right| \tag{2.7}$$

is assumed to be positive, i.e.,

$$J > 0 , \tag{2.8}$$

so that function $\underset{\sim}{\chi}$ has an inverse, $\underset{\sim}{\chi}^{-1}$. We can write therefore

$$\underset{\sim}{X} = \underset{\sim}{\chi}^{-1}(\underset{\sim}{x}, t) \tag{2.9}$$

Two kinds of descriptions are thus possible for mechanical phenomena: either the material dexcription, which uses $\underset{\sim}{X}$ and t as independent variables, or the spatial description which uses $\underset{\sim}{x}$ and t. Each description can be translated into the other with the help of (2.3) and (2.9).

2.2. Deformations and strains: the material description

The deformation gradient $\underset{\sim}{F}(\underset{\sim}{X}, t)$ is the tensor with components

$$F_{ij}(\underset{\sim}{X}, t) = \chi_{i.j} \tag{2.10}$$

where the dot denotes differentiation with respect to the material co-ordinates.
Co-ordinates X_i are supposed Cartesian and orthogonal at the reference configuration and the orthogonal unit vectors $\underset{\sim}{i}_i$ are tangent to the corresponding co-ordinate lines.
As the body deforms, the co-ordinates X_i of each particle are kept the same, i.e., the X_i co-ordinate lines follow the deformation of the body and thus cease to be straight and orthogonal.
We denote by $\underset{\sim}{G}_j$ the vectors

$$\underset{\sim}{G}_j(\underset{\sim}{X}, t) = \underset{\sim}{\chi}_{.j}(\underset{\sim}{X}, t) \tag{2.11}$$

whose components with respect to the unit vectors $\underset{\sim}{i}_i$ are the components F_{ij} of the deformation gradient. Vectors $\underset{\sim}{G}_j$ are tangent to the co-ordinate lines X_j and thus coincide with vectors $\underset{\sim}{i}_j$ in the reference configuration, i.e., for $t = 0$.
As, by virtue of (2.4),

$$F_{ij} = u_{i,j} + \delta_{ij} , \tag{2.12}$$

there results

$$\underset{\sim}{G}_j = (u_{i.j} + u_{j.i}) \underset{\sim}{i}_i . \tag{2.13}$$

The Green's deformation tensor is the symmetric tensor

$$\underset{\sim}{G} = \underset{\sim}{F}^T \underset{\sim}{F} \tag{2.14}$$

with components

$$G_{ij} = F_{mi} F_{mj} = \underset{\sim}{G}_i \cdot \underset{\sim}{G}_j = (\delta_{mi} + u_{m,i})(\delta_{mj} + u_{m,j}) =$$
$$= \delta_{ij} + u_{i,j} + u_{j,i} + u_{m,i} u_{m,j}. \tag{2.15}$$

The Green's strain tensor is the symmetric tensor.

$$\underset{\sim}{\gamma} = \tfrac{1}{2} (\underset{\sim}{G} - \underset{\sim}{I}) \tag{2.16}$$

with components

$$\gamma_{ij} = \tfrac{1}{2} (G_{ij} - \delta_{ij}) = \tfrac{1}{2}(u_{i,j} + u_{j,i} + u_{m,i} u_{m,j}). \tag{2.17}$$

As both deformation and strain tensors are symmetric, each of them has only six independent components. This means that, in (2.17), there are six independent equations.
As, on the other hand, the displacement vector has only three components, the six equations (2.17) have no solution for arbitrary precribed strain components, unless such components satisfy a set of integrability conditions which are called the compatibility equations.
These equations result from the condition that the space remains Euclidean after the deformation, i.e., from setting equal to zero the Riemann-Christoffel tensor.

$$R_{klmn} = \tfrac{1}{2} (G_{kn,lm} + G_{lm,kn} - G_{km,ln} - G_{ln,km}) +$$
$$+ G_{rs} (\Gamma_{knr} \Gamma_{lms} - \Gamma_{kmr} \Gamma_{lns}). \tag{2.18}$$

The entities

$$\Gamma_{ijk} = \tfrac{1}{2} (G_{ik,j} + G_{jk,i} - G_{ij,k}) \tag{2.19}$$

are the Christoffel symbols of the first kind.

2.3. The meaning of the strain and deformation components

A vector with components dX_j in the reference configuration becomes a vector with components

$$dx_i = \chi_{i,j}\, dX_j. \tag{2.20}$$

in the current configuration. We can write therefore

$$d\underset{\sim}{x} = \underset{\sim}{F}\, d\underset{\sim}{X} \tag{2.21}$$

and state that the deformation gradient is a tensor which transforms vectors in the reference configuration into vectors in the current configuration.

The strech of the vector is measured by

$$\Lambda = \frac{ds}{dS} = \frac{\sqrt{(d\underset{\sim}{X}^T \underset{\sim}{F}^T \underset{\sim}{F} d\underset{\sim}{X})}}{dS} \tag{2.22}$$

or

$$\Lambda = \frac{\sqrt{(d\underset{\sim}{X}^T \underset{\sim}{G} d\underset{\sim}{X})}}{dS} \tag{2.23}$$

where ds and dS are the lenghts of vectors $d\underset{\sim}{x}$ and $d\underset{\sim}{X}$.
If $\underset{\sim}{F}$ is an orthogonal tensor, $\underset{\sim}{G}$ becomes the unit tensor $\underset{\sim}{I}$ and the strectch equal to unity, so that no deformation will occur.
If the vector $d\underset{\sim}{X}$ is parallel to the unit vector $\underset{\sim}{i}_k$, the stretch becomes

$$\Lambda_k = \sqrt{(G_{kk})} \tag{2.24}$$

which means that G_{kk} is the square of the stretch of any segment which, before the deformation, is parallel to $\underset{\sim}{i}_k$.
On the other hand,

$$ds^2 - dS^2 = \Lambda^2 dS^2 - dS^2 = d\underset{\sim}{X}^T \underset{\sim}{G} d\underset{\sim}{X} - d\underset{\sim}{X}^T d\underset{\sim}{X} = d\underset{\sim}{X}^T (\underset{\sim}{G} - \underset{\sim}{I}) d\underset{\sim}{X}, \tag{2.25}$$

so that, by virtue of (2.16),

$$ds^2 - dS^2 = 2\, d\underset{\sim}{X}^T \underset{\sim}{\gamma}\, d\underset{\sim}{X}. \tag{2.26}$$

No deformation will occur if $\underset{\sim}{\gamma} = \underset{\sim}{0}$.
Equation (2.26) gives a meaning to γ_{kk}. Indeed, for segments parallel to $\underset{\sim}{i}_k$ before the deformation,

$$ds_k^2 - dS_k^2 = 2\gamma_{kk}\, dX_k\, dX_k \tag{2.27}$$

and, therefore,

$$\gamma_{kk} = \tfrac{1}{2}(\Lambda_k^2 - 1). \tag{2.28}$$

We now define the shear between two initially orthogonal vectors, $d\underset{\sim}{X}_1$ and $d\underset{\sim}{X}_2$, as

$$\Gamma(d\underset{\sim}{X}_1, d\underset{\sim}{X}_2) = \frac{\pi}{2} - \theta(d\underset{\sim}{x}_1, d\underset{\sim}{x}_2), \tag{2.29}$$

where $\theta(d\underset{\sim}{x}_1, d\underset{\sim}{x}_2)$ denotes the angle between $d\underset{\sim}{x}_1$ and $d\underset{\sim}{x}_2$.
Now,

$$\sin \Gamma = \cos \theta = \frac{d\underset{\sim}{x}_1 \cdot d\underset{\sim}{x}_2}{ds_1\, ds_2} = \frac{d\underset{\sim}{X}_1^T \underset{\sim}{F}^T \underset{\sim}{F}\, d\underset{\sim}{X}_2}{ds_1\, ds_2} \tag{2.30}$$

and, therefore,

$$\sin \Gamma = \frac{d\underset{\sim}{X}_1^T \underset{\sim}{G}\, d\underset{\sim}{X}_2}{ds_1\, ds_2} \tag{2.31}$$

Assuming that $d\underset{\sim}{X}_1$ and $d\underset{\sim}{X}_2$ are paralell to $\underset{\sim}{i}_1$ and $\underset{\sim}{i}_2$, there results

$$\sin \Gamma_{12} = \frac{G_{12}}{\sqrt{G_{11} G_{22}}}, \tag{2.32}$$

which gives a meaning to G_{12} and also to

$$\gamma_{12} = \tfrac{1}{2} \sin \Gamma_{12} \sqrt{G_{11} G_{22}}\ . \tag{2.33}$$

In what concerns the element of volume, we can write

$$dV = dX_1\, dX_2\, dX_3 \tag{2.34}$$

in the reference configuration.
After deformation, the edge $dX_k \underset{\sim}{i}_k$ of the element becomes $dX_k \underset{\sim}{G}_k$. Thus, the volume of the same element becomes, after the deformation,

$$dv = |\underset{\sim}{G}_1 \cdot \underset{\sim}{G}_2 \wedge \underset{\sim}{G}_3|\, dX_1\, dX_2\, dX_3 = |\underset{\sim}{F}|\, dV. \tag{2.35}$$

Therefore,

$$\frac{dv}{dV} = |\underset{\sim}{F}| = J = \sqrt{(|\underset{\sim}{F}^T \underset{\sim}{F}|)} = \sqrt{(G)}\ , \tag{2.36}$$

where G is the determinant of $\underset{\sim}{G}$.
By virtue of (2.16),

$$G = 8\gamma_3 + 4\gamma_2 + 2\gamma_1 + 1, \tag{2.37}$$

where γ_1, γ_2 and γ_3 are the invariants of tensor $\underset{\sim}{\gamma}$.
Let us remember now the polar decomposition of $\underset{\sim}{F}$ expressed by (1.10). Matrix $\underset{\sim}{R}$, which is orthogonal, implies no deformation and corresponds thus to a rigid body motion.

On the other hand, introducing (1.12) into (2.22), (2.31) and (2.36), these equations become

$$\Lambda = \frac{\sqrt{(d\underset{\sim}{X}^T \underset{\sim}{U}^2 d\underset{\sim}{X})}}{dS} \tag{2.38}$$

$$\sin \Gamma = \frac{d\underset{\sim}{X}_1^T \underset{\sim}{U}^2 d\underset{\sim}{X}_2}{\Lambda_1 \Lambda_2 dS_1 dS_2} \tag{2.39}$$

$$\frac{dv}{dV} = |\underset{\sim}{U}|. \tag{2.40}$$

This means that the deformation exclusively depends on $\underset{\sim}{U}$. Equation

$$d\underset{\sim}{x} = \underset{\sim}{R}\underset{\sim}{U}\, d\underset{\sim}{X}, \tag{2.41}$$

which results from introducing (1.10) into (2.21), can thus be interpreted in the sense that, when motion occurs, vector $d\underset{\sim}{X}$ is first stretched and then rotated.
The right polar decomposition of $F(P,t)$ means thus indeed a decomposition of the motion into a deformation and a rigid body rotation of each particle P.

2.4. Small strains

In case the strains γ_{kk} are very small, equation (2.28) shows that the value of Λ_k is very near to unity.
Then, if we make

$$\Lambda_k = 1 + e_k, \tag{2.42}$$

e_k is also very small, so that γ_{kk} becomes

$$\gamma_{kk} = e_k + \tfrac{1}{2} e_k^2 \simeq e_k. \tag{2.43}$$

Now,

$$e_k = \Lambda_k - 1 = \frac{dx_k - dX_k}{dX_k} \tag{2.44}$$

i.e., e_k is the linear extension of the vector $dX_k \underset{\sim}{i}_k$. The strain γ_{kk} acquires thus the meaning of the linear extension of a segment initially parallel to $\underset{\sim}{i}_k$.
In what concerns γ_{12}, as the values of G_{11} and G_{22} become very near to unity, and Γ_{12} becomes very small, (2.33) yields

$$\gamma_{12} \simeq \tfrac{1}{2}\Gamma_{12}, \tag{2.45}$$

i.e., the strain components γ_{ij}, where $i \neq j$, take the meaning of half the shear between two segments initially parallel to $\underset{\sim}{i}_i$ and $\underset{\sim}{i}_j$.
Finally, neglecting γ_2 and γ_3 as compared to γ_1, (2.36) and (2.37) yield

$$J = \frac{dv}{dV} = \sqrt{(1+2\gamma_1)} \simeq 1 + \gamma_1 . \tag{2.46}$$

Therefore,

$$\gamma_1 \simeq \frac{dv - dV}{dV} = \Delta , \tag{2.47}$$

i.e., the trace of the strain tensor becomes the volume extension, Δ.

2.5. Small displacements and geometrical linearity

Assume now that all the first derivatives of the displacements are very small as compared to unity, or, what is the same, that the difference between the displacements of two arbitrary points in the body are very small as compared with the distance between such points. Such is the small displacement assumption.
If the first derivatives of the displacements are very small, we can neglect products of such derivatives. This means that the expressions (2.17) for the strains can be approximated by

$$\gamma_{ij} \simeq \varepsilon_{ij} = \tfrac{1}{2} (u_{i,j} + u_{j,i}), \tag{2.48}$$

where $\underset{\sim}{\varepsilon}$ is called the infinitesimal strain tensor. In other words, the strain - displacement equations become linear.
The small displacement assumption implies the small strain assumption but the latter does not imply the first. Indeed, the strains can be small and the rotations be large.
However, if the strains and the rotations are both small, then all the first derivatives of the displacements are also small and the small displacement assumption is valid.
The small displacement assumption is the foundation of geometrical linearity which supposes that:

a) the co-ordinates of the points of the body after the deformation are very near the co-ordinates of the same points after the deformation, so that the equilibrium equations can be written for the undeformed body;
b) the strain-displacement equations are linear.

If the small displacement assumption is made, a rotation of a particle P of the body can be expressed by

$$d\underset{\sim}{u} = d\underset{\sim}{x} - d\underset{\sim}{X} = \underset{\sim}{\Theta} \wedge d\underset{\sim}{X} . \tag{2.49}$$

(2.49) can indeed be written under the form

$$dx_i = (e_{ijk}\,\Theta_j + \delta_{ik})\,dX_k . \tag{2.50}$$

The deformation gradient components are therefore

$$F_{ik} = e_{ijk}\,\Theta_j + \delta_{ik} , \tag{2.51}$$

so that

$$G_{kl} = F_{ik}F_{il} = (e_{ijk}\,\Theta_j + \delta_{ik})(e_{iml}\Theta_m + \delta_{il}) =$$

$$= e_{ijk}\,e_{iml}\,\Theta_j\,\Theta_m + (e_{ljk} + e_{kjl})\Theta_j + \delta_{kl} . \tag{2.52}$$

Considering that $e_{ljk} = -e_{kjl}$ and neglecting, by virtue of the small displacement assumption, the products $\Theta_j\,\Theta_m$, there results

$$G_{kl} = \delta_{kl} , \tag{2.53}$$

which means that (2.49) indeed represents a small rotation. Comparing (2.51) with (2.12), there results

$$u_{i,k} = e_{ijk}\,\Theta_j \tag{2.54}$$

and, therefore,

$$\Theta = 1/2\ \text{Curl}\ \underset{\sim}{u} , \tag{2.55}$$

where the fact that the differential operator is denoted with capital letters means that the derivatives are to be determined with respect to the material co-ordinates.
The components of the rotation vector $\underset{\sim}{\Theta}$ have thus the same values as the components

$$\theta_{ij} = 1/2\,(u_{i,j} - u_{j,i}) \tag{2.56}$$

of a new tensor called the <u>infinitesimal rotation tensor.</u>
Attention is called to the fact that

$$u_{i,j} = \varepsilon_{ij} + \theta_{ij} , \tag{2.57}$$

so that the infinitesimal strain and rotation tensors represent the Cartesian decomposition of the displacement gradient, i.e., of the tensor with components $u_{i,j}$.
On the other hand, by virtue of the small displacements assumption, (2.26) can be rewritten as

$$ds^2 - dS^2 \simeq 2\varepsilon_{ij}\, dX_i\, dX_j \tag{2.58}$$

or, as $\theta_{ij}\, dX_i\, dX_j$ (because $\underset{\sim}{\theta}$ is skew-symmetric),

$$ds^2 - dS^2 \simeq 2u_{i,j}\, dX_i\, dX_j\,. \tag{2.59}$$

This equation, which is valid only in case os small displacements, is to be compared with equation (2.70). It remains to write the simplified form of the compatibility equations,

$$\gamma_{kn,lm} + \gamma_{lm,kn} - \gamma_{km,ln} - \gamma_{ln,km} = 0\,, \tag{2.60}$$

which results directly from (2.48), or from neglecting the second--order terms in (2.18).

2.5. Velocities and deformation rates: the spatial description

In fluid mechanics, the rates of change of the different kinematic magnitudes are generally more important than the quantities themselves.
If the spatial description, which is usual in fluid mechanics, is adopted, i.e., if the magnitudes are expressed in terms of $\underset{\sim}{x}$ and t , their derivatives with respect to time must be determined considering that $\underset{\sim}{x}$ also depends on time. Therefore, for any function $f(\underset{\sim}{x}, t)$,

$$\frac{D}{Dt} f(\underset{\sim}{x}, t) = \frac{\partial f}{\partial t} + \frac{\partial f}{\partial x_j}\frac{Dx_j}{Dt} = \frac{\partial f}{\partial t} + \frac{\partial f}{\partial x_j} v_j \tag{2.61}$$

where D denotes differentiation with respect to time.
Considering (2.61), the acceleration components are given by

$$a_i = \frac{D}{Dt} v_i(\underset{\sim}{x}, t) = \frac{\partial v_i}{\partial t} + L_{ij} v_j\,, \tag{2.62}$$

where, denoting by a comma the derivatives with respect to the spatial co-ordinates x_i ,

$$L_{ij} = \frac{\partial v_i}{\partial x_j} = v_{i,j}\,. \tag{2.63}$$

The tensor

$$\underset{\sim}{L} = L_{ij}\, \underset{\sim}{i}_i \otimes \underset{\sim}{i}_j \tag{2.64}$$

is called the velocity gradient tensor.
$\underset{\sim}{L}$ can be decomposed into its symmetric and skew-symmetric parts, i.e., we can make

$$\underset{\sim}{L} = \underset{\sim}{d} + \underset{\sim}{\omega}\,, \tag{2.65}$$

where, of course,

$$\underset{\sim}{d} = \tfrac{1}{2}(\underset{\sim}{L} + \underset{\sim}{L}^T) \quad \text{and} \quad \underset{\sim}{\omega} = \tfrac{1}{2}(\underset{\sim}{L} - \underset{\sim}{L}^T). \tag{2.66-67}$$

$\underset{\sim}{d}$ is called the rate of deformation tensor and $\underset{\sim}{\omega}$ the spin tensor.
Let us justify that tensor $\underset{\sim}{d}$ with components

$$d_{ij} = \tfrac{1}{2}(v_{i,j} + v_{j,i}) \tag{2.68}$$

be called the rate of deformation tensor.
By expressing the derivative $\frac{d}{dt}(ds^2 - dS^2)$ in terms of $\underset{\sim}{d}$, we obtain indeed

$$\frac{D}{Dt}(ds^2 - dS^2) = \frac{D}{Dt}(ds^2) = \frac{D}{Dt}(dx_i\, dx_i) =$$

$$= 2\, dx_i \frac{D}{Dt}(dx_i) = 2\, dx_i \frac{D}{Dt}\left(\frac{\partial x_i}{\partial X_j}\right) dX_j =$$

$$= 2\, dx_i \frac{\partial v_i}{\partial X_j}\, dX_j = 2\, dx_i \frac{\partial v_i}{\partial x_k}\frac{\partial x_k}{\partial X_j}\, dX_j =$$

$$= 2\frac{\partial v_i}{\partial x_k}\, dx_i\, dx_k \tag{2.69}$$

and, therefore,

$$\frac{D}{Dt}(ds^2 - dS^2) = 2 v_{i,j}\, dx_i\, dx_j\,. \tag{2.70}$$

Equation (2.70) is to be compared with equation (2.59). It is remarked however that equation (2.59) is valid only in case of small displacements, while (2.70) is valid in any case.
From (2.70) there finally results

$$\dot{\overline{ds^2 - dS^2}} = 2\, d_{ij}\, dx_i\, dx_j. \tag{2.71}$$

from which it can be concluded that tensor $\underset{\sim}{d}$ is a good measure for the rate of deformation.
Tensor $\underset{\sim}{d}$ vanishes indeed if and only if the configuration of the body does not change with time. The same does not happen with tensor $\underset{\sim}{L}$, which does not necessarily vanish if the configuration does not change with time.
It has, on the other hand, the merit of introducing a great simplicity in the spatial description of the motion.
Suppose, e.g., that the rate of change of volume is to be determined, and let us consider that the element that, at time t, has edges $dx_i \cdot \underset{\sim}{i}_i$ becomes an element with edges $dx_i \cdot \underset{\sim}{i}_i + v_{j,i} \, dx_i \, \underset{\sim}{i}_j \, Dt$, at time $t + Dt$, so that, neglecting higher order infinitesimal terms,

$$D(dv) = v_{i,i} \; dv \, Dt \tag{2.72}$$

or, as $d_{ii} = v_{i,i}$,

$$\frac{D}{Dt}(dv) = d_{ii} \; dv. \tag{2.73}$$

The rate of change of volume is thus

$$\dot{\overline{dv}} = d_{ii} \; dv \,, \tag{2.74}$$

where d_{ii} is the first invariant of $\underset{\sim}{d}$.
Equation (2.74) is to be compared with (2.47), which is valid only in the case of small strains.
From (2.72) an expression can be derived for the time derivative of the magnitude J defined by (2.7).
Indeed, by virtue of (2.36),

$$dv = J \, dV. \tag{2.75}$$

Therefore,

$$\dot{J} \, dV = \dot{\overline{dv}} = \operatorname{div} \underset{\sim}{v} \; dv = \operatorname{div} \underset{\sim}{v} \, J \, dV \tag{2.76}$$

and, thus,

$$\dot{J} = J \operatorname{div} \underset{\sim}{v} \,. \tag{2.77}$$

In what concerns the spin tensor, $\underset{\sim}{\omega}$, with components

$$\omega_{ij} = \tfrac{1}{2} \left(v_{i,j} - v_{j,i} \right), \tag{2.78}$$

let us remark that it implies no rate of deformation. Indeed, as $\underset{\sim}{\omega}$ is skew-symmetric, $\omega_{ij} \cdot dx_i \cdot dx_j = 0$

The vector $\underset{\sim}{\Omega}$ with components

$$\Omega_k = e_{ijk} \; v_{i,j} \; , \tag{2.79}$$

i.e., such that

$$\underset{\sim}{\Omega} = \mathrm{curl} \; \underset{\sim}{v} \tag{2.80}$$

is called the <u>vorticity vector</u>.
The spin tensor components ω_{ij} are of course expressed in terms of the components of the vorticity vector by

$$\omega_{ij} = -\tfrac{1}{2} \, e_{ijk} \, \Omega_k \, . \tag{2.81}$$

3. STRESSES AND BALANCE PRINCIPLES

3.1. Basic assumptions concerning stresses

Classical Continuum Mechanics is associated to a set of fundamental assumptions, the first of which is that the actions exerted on one another by the different parts of a body are <u>contact actions which can be assimilated to forces</u> distributed on the boundaries of each part.
According to this assumption, if a body B acted by external forces is cut into two parts, B_1 and B_2, by a surface S which intersects it, the state of each part is not modified if the remaining part is replaced by forces distributed on S.
Let us consider now B_1 acted by the external forces which act on B_1 as a part of B, and by the forces distributed on S, which represent the action of B_2 on B_1.
Let $d\underset{\sim}{F}$ be the force exerted on an element $dS \subset S$, with area da, centered on an arbitrary point P localized on S.
The second fundamental assumption is that when dS tends to P, the $\lim_{dS \to P} \frac{dF}{da}$ exists and, provided dS is kept on S, is independent of the way in which dS tends to P.
We call such limit the traction vector at P and denote it by $\underset{\sim}{t}$, i.e., we make $\underset{\sim}{t}(P) = \lim_{dS \to P} \frac{d\underset{\sim}{F}}{da}$.
A third fundamental assumption is that $d\underset{\sim}{F}$, i.e., the force exerted on dS, does not change if the surface S is replaced by another surface containing dS.
As dS can be so small as we wish, there results from this assumption that the traction vector at P is the same for any two surfaces S and S' tangent to each other at P, i.e., which are normal to the same unit vector, $\underset{\sim}{n}$, at P. We write therefore $\underset{\sim}{t} = \underset{\sim}{t}(P, \underset{\sim}{n})$, where $\underset{\sim}{n}$ is supposed to be directed towards the outside of B_1.
If dS is considered a part of B_2, the force exerted on dS is $-d\underset{\sim}{F}$, by virtue of the action and reaction principle, and,

therefore, the traction vector becomes $-\underset{\sim}{t}$. The normal vector is directed towards the outside of B_2 and thus becomes $-\underset{\sim}{n}$. We may write then, as a consequence of the action and reaction principle $\underset{\sim}{t}(P,\underset{\sim}{n}) = -\underset{\sim}{t}(P,-\underset{\sim}{n})$.

The Cauchy's stress principle, a consequence of Euler's first law of motion, establishes that $\underset{\sim}{t}(P,\underset{\sim}{n}) = \underset{\sim}{T}(P).\underset{\sim}{n}$ for an arbitrary $\underset{\sim}{n}$, $\underset{\sim}{T}$ representing a tensor the values of whose components do not depend on $\underset{\sim}{n}$.

In order to prove this statement, we start by considering three distinct planes, Π_1, Π_2 and Π_3, intersecting at P, and three unit vectors, $\underset{\sim}{n}_1$, $\underset{\sim}{n}_2$ and $\underset{\sim}{n}_3$, normal to them.

We can always write

$$[\underset{\sim}{t}_1 \ \ \underset{\sim}{t}_2 \ \ \underset{\sim}{t}_3] = \underset{\sim}{T} \cdot [\underset{\sim}{n}_1 \ \underset{\sim}{n}_2 \ \underset{\sim}{n}_3] \ , \tag{3.1}$$

where

$$\underset{\sim}{T}^T = [\underset{\sim}{t}_1 \ \underset{\sim}{t}_2 \ \underset{\sim}{t}_3] \, [\underset{\sim}{n}_1 \ \underset{\sim}{n}_2 \ \underset{\sim}{n}_3]^{-1} ,$$

and, therefore, either

$$\underset{\sim}{t}_i = \underset{\sim}{T}\,\underset{\sim}{n} \qquad \text{or} \qquad -\underset{\sim}{t}_i = \underset{\sim}{T}(-\underset{\sim}{n}) . \tag{3.2-2'}$$

It remains to prove that

$$\underset{\sim}{t} = \underset{\sim}{T} \cdot \underset{\sim}{n} \tag{3.3}$$

for an arbitrary $\underset{\sim}{n}$, i.e., for a unit vector $\underset{\sim}{n}$ not necessarily parallel to any of the vectors $\underset{\sim}{n}_i$.

In order to do so, we consider the tetrahedron τ obtained by intersecting the three planes Π_1, Π_2 and Π_3 by another plane not containing P, normal to another unit vector, $\underset{\sim}{n}$, distinct from $\underset{\sim}{n}_1$, $\underset{\sim}{n}_2$ and $\underset{\sim}{n}_3$.

As P is a fixed point, i.e., a point with given co-ordinates, the divergence theorem permits to write

$$\int_{\partial\tau} \underset{\sim}{T}(P) \cdot \underset{\sim}{n} \, d\partial\tau = \int_\tau \operatorname{div} \underset{\sim}{T}(P) \, d\tau \tag{3.4}$$

Assume now the tetrahedron to be so small that the tractions can be supposed constant on each face, and consider that, according to (3.2-2'), the values of the tractions on the faces belonging to Π_1, Π_2 and Π_3 are respectively $\underset{\sim}{T}.\underset{\sim}{n}_1$, $\underset{\sim}{T}.\underset{\sim}{n}_2$ and $\underset{\sim}{T}.\underset{\sim}{n}_3$.

There results, by virtue of (3.4) and of Euler's first law of motion (force equilibrium), that the traction on the face normal to $\underset{\sim}{n}$ is $\underset{\sim}{T}.\underset{\sim}{n}$.

An important remark is that we have been reasoning on the body acted by external forces, i.e., not on the reference configuration, but on a current configuration B_χ.

3.2. Stress tensors

We call Cauchy stress tensor $\underset{\sim}{T}$ the second order tensor which transforms, according to (3.3), the unit normal $\underset{\sim}{n}$ into the traction vector $\underset{\sim}{t}$ at a given point P.
Denoting by $\underset{\sim}{N}$ and dA the external normal to dS at P and the area of dS in the reference configuration, we make now

$$\underset{\sim}{\sigma} = \lim_{dS \to P} \frac{d\underset{\sim}{F}}{dA} \tag{3.5}$$

and define the first Piolla-Kirchhoff tensor $\underset{\sim}{\Sigma}$ as the second order tensor which transforms $\underset{\sim}{N}$ into $\underset{\sim}{\sigma}$.
Therefore,

$$\underset{\sim}{\sigma} = \underset{\sim}{\Sigma} \cdot \underset{\sim}{N}. \tag{3.6}$$

Now, the following formula,

$$\underset{\sim}{n}\, da = J(\underset{\sim}{F}^{-1})^T \underset{\sim}{N}\, dA, \tag{3.7}$$

called Nanson's formula, can be demonstrated, so that, as $\underset{\sim}{\sigma}\, dA = \underset{\sim}{t}\, da = d\underset{\sim}{F}$ the following relation can be established between $\underset{\sim}{\Sigma}$ and $\underset{\sim}{T}$:

$$\underset{\sim}{\Sigma} = J\, \underset{\sim}{F}^{-1} \underset{\sim}{T}. \tag{3.8}$$

The fact that $\underset{\sim}{T}$ and $\underset{\sim}{\Sigma}$ cannot be both symmetric, leads to the definition of still another stress tensor, the second Piolla-Kirchhof tensor,

$$\underset{\sim}{\Sigma}^{\circ} = \underset{\sim}{\Sigma}\,(\underset{\sim}{F}^{-1})^T = J\,\underset{\sim}{F}^{-1}\underset{\sim}{T}\,(\underset{\sim}{F}^{-1})^T \tag{3.9}$$

which is symmetric if and only if $\underset{\sim}{T}$ is symmetric.
The meaning of the stress components T_{ij} and Σ_{ij} can be established just by writing equations (3.4) and (3.6) in component form, i.e.,

$$t_i = T_{ij}\, n_i, \tag{3.10}$$

$$\sigma_i = \Sigma_{ij} \cdot N_i. \tag{3.11}$$

T_{ij} and Σ_{ij} are thus the jth components of tractions $\underset{\sim}{t}$ and $\underset{\sim}{\sigma}$ respectively, applied to a face whose normal unit vector is $\underset{\sim}{i}_i$.

3.3. Non-mechanical energy vectors

Let h denote the density of influx of non-mechanical power across the boundary of the body in the configuration B_χ (see [3]). Let $-\underset{\sim}{h}$ denote the vector which transforms the normal unit vector $\underset{\sim}{n}$ into h, i.e., let

$$h = -\underset{\sim}{h} \cdot \underset{\sim}{n} . \tag{3.12}$$

Let now H denote the density of influx of non-mechanical power across the boundary of the body in the reference configuration and let $-\underset{\sim}{H}$ denote the vector which transforms the normal unit vector $\underset{\sim}{N}$ into H, i.e., let

$$H = -\underset{\sim}{H} \cdot \underset{\sim}{N} . \tag{3.13}$$

(3.12) corresponds to (3.4) and (3.13) to (3.6).
As

$$h\,da = H\,dA , \tag{3.14}$$

the same reasoning as in Section 3.2 leads to

$$\underset{\sim}{H}^T = J \underset{\sim}{F}^{-1} \underset{\sim}{h}^T . \tag{3.15}$$

The components h_i and H_i represent of course the outward flux of energy through a face whose normal unit vector is $\underset{\sim}{i}_i$.

3.4. The balance of mass

The mass of the body is

$$M = \int_{B_\chi} \rho \, dv . \tag{3.16}$$

As the conservation of mass requires $\rho\,dv$ to be an invariant and, by virtue of (2.36),

$$dv = J\,dV , \tag{3.17}$$

ρJ is also an invariant, i.e.,

$$\rho J = P , \tag{3.18}$$

where P is the value of the mass density in the reference configuration.
Therefore,

$$\frac{D}{Dt}(\rho J) = 0 \tag{3.19}$$

and, thus,

$$\dot{\varrho} J + \varrho \frac{DJ}{Dt} = 0 . \tag{3.20}$$

By virtue of (2.77), (3.20) yields

$$\dot{\varrho} + \varrho \operatorname{div} \underset{\sim}{v} = 0 \tag{3.21}$$

or, what is the same,

$$\frac{\partial \varrho}{\partial t} + \operatorname{div} (\varrho \underset{\sim}{v}) = 0 . \tag{3.22}$$

3.5 - The balance of momenta

Let $\underset{\sim}{f}$ denote the resultant of the external forces acting on the body and let us distinguish between body forces $\underset{\sim}{f}_b$ and contact forces $\underset{\sim}{f}_c$, making

$$\underset{\sim}{f} = \underset{\sim}{f}_b + \underset{\sim}{f}_c . \tag{3.23}$$

Let us introduce now the force densities $\underset{\sim}{b}$ and $\underset{\sim}{t}$ defined by

$$\underset{\sim}{f}_b = \int_{B'} \underset{\sim}{b} \, dm = \int_{v'} \underset{\sim}{b} \varrho \, dv \tag{3.24}$$

and

$$\underset{\sim}{f}_c = \int_{\partial v'} \underset{\sim}{t} \, da , \tag{3.25}$$

where B' is any part of the body with volume v' in the current configuration.
The resultant couple is

$$\underset{\sim}{M} = \int_{v'} \varrho \, \underset{\sim}{x} \wedge \underset{\sim}{b} \, dv + \int_{\partial v'} \underset{\sim}{x} \wedge \underset{\sim}{t} \, da \tag{3.26}$$

Euler's laws of motion require that

$$\frac{D}{Dt}\left(\int_{v'} \varrho \underset{\sim}{v} \, dv\right) = \int_{v'} \varrho \, \underset{\sim}{b} \, dv + \int_{\partial v'} \underset{\sim}{t} \, da \tag{3.27}$$

$$\frac{D}{Dt}\left(\int_{v'} \underset{\sim}{x} \wedge \underset{\sim}{v} \varrho \, dv\right) = \int_{v'} \underset{\sim}{x} \wedge \underset{\sim}{b} \varrho \, dv + \int_{\partial v'} \underset{\sim}{x} \wedge \underset{\sim}{t} \, da , \tag{3.28}$$

where the left-hand sides respectively represent the momentum and the moment of momentum of B'.

3.6. The balance of energy

We define the kinetic energy

$$K = \int_{B'} \tfrac{1}{2}\, \underset{\sim}{v} \cdot \underset{\sim}{v} \; dm \tag{3.29}$$

and the internal energy

$$E = \int_{B'} e \; dm \,, \tag{3.30}$$

where e is the energy density associated with heat, deformation, or some other form of energy.
We also define the mechanical power

$$W = \int_{B'} \underset{\sim}{b} \cdot \underset{\sim}{v} \; dm + \int_{\partial B'} \underset{\sim}{t} \cdot \underset{\sim}{v} \; da \tag{3.31}$$

and the non-mechanical power

$$Q = \int_{B'} r \, dm + \int_{\partial B'} h \, da \,, \tag{3.32}$$

where r is the density of sources on non-mechanical power and h, as defined in Section 3.3, is the density of influx of non-mechanical power across the boundary of the body.
Energy balance requires

$$\frac{D}{Dt}\left(\int_{B'} (\tfrac{1}{2}\underset{\sim}{v} \cdot \underset{\sim}{v} + e)\, dm\right) = \int_{B'} (\underset{\sim}{b} \cdot \underset{\sim}{v} + r)\, dm + \int_{\partial B'} (\underset{\sim}{t} \cdot \underset{\sim}{v} + h)\, da \,, \tag{3.33}$$

i.e., that the rate of increase of energy in B' equals the rate of supply.

3.7. Local equations in the spatial description

Local field equations are to be derived from the global equations (3.27), (3.28) and (3.33).
Considering that, by virtue of the invariance of $\rho \, dv$,

$$\frac{D}{Dt} \int_{v'} \underset{\sim}{v} \rho \, dv = \int_{v'} \dot{\underset{\sim}{v}} \rho \, dv \tag{3.34}$$

and introducing (3.4) into (3.27), we obtain

$$\int_{v'} \dot{\underset{\sim}{v}} \rho \, dv = \int_{v'} \rho \underset{\sim}{b} \, dv + \int_{\partial v'} \underset{\sim}{T} \cdot \underset{\sim}{n} \, da \tag{3.35}$$

or, remarking that

$$\int_{\partial v'} \underset{\sim}{T} \cdot \underset{\sim}{n} \, da = \int_{v'} \underset{\sim}{T} \cdot \underset{\sim}{\nabla} \, dv = \int_{v'} \operatorname{div} \underset{\sim}{T} \, dv \tag{3.36}$$

the equation

$$\int_{V'} \dot{\underset{\sim}{v}} \rho \, dv = \int_{V'} \underset{\sim}{b} \rho \, dv + \int_{V'} \operatorname{div} \underset{\sim}{T} \, dv \, , \tag{3.37}$$

which, being valid for any $B' \subset B$, implies

$$\operatorname{div} \underset{\sim}{T} + \rho \underset{\sim}{b} = \rho \dot{\underset{\sim}{v}} \, . \tag{3.38}$$

Considering that

$$\frac{D}{Dt} \int_{V'} \underset{\sim}{x} \wedge \underset{\sim}{v} \, \rho \, dv = \int_{V'} \overline{\underset{\sim}{x} \wedge \underset{\sim}{v}}^{\,\cdot} \, \rho \, dv, \tag{3.39}$$

and introducing (3.4) into (3.28), we obtain

$$\int_{V'} \overline{\underset{\sim}{x} \wedge \underset{\sim}{v}}^{\,\cdot} \, \rho \, dv = \int_{V'} \underset{\sim}{x} \wedge \underset{\sim}{b} \, \rho \, dv + \int_{V'} \operatorname{div}(\underset{\sim}{x} \wedge \underset{\sim}{T}) \, dv \tag{3.40}$$

or

$$\operatorname{div}(\underset{\sim}{x} \wedge \underset{\sim}{T}) + \rho \, \underset{\sim}{x} \wedge \underset{\sim}{b} = \rho \, \overline{\underset{\sim}{x} \wedge \underset{\sim}{v}}^{\,\cdot} \, . \tag{3.41}$$

Now,

$$\operatorname{div}(\underset{\sim}{x} \wedge \underset{\sim}{T}) = \underset{\sim}{x} \wedge \operatorname{div} \underset{\sim}{T} + (\operatorname{grad} \underset{\sim}{x}) \wedge \underset{\sim}{T} \tag{3.42}$$

and

$$\overline{\underset{\sim}{x} \wedge \underset{\sim}{v}}^{\,\cdot} = \underset{\sim}{v} \wedge \underset{\sim}{v} + \underset{\sim}{x} \wedge \dot{\underset{\sim}{v}} = \underset{\sim}{x} \wedge \dot{\underset{\sim}{v}} \, , \tag{3.43}$$

so that, introducing (3.38), (3.41) yields

$$(\operatorname{grad} \underset{\sim}{x}) \wedge \underset{\sim}{T} = \underset{\sim}{0} \, . \tag{3.44}$$

As

$$(\operatorname{grad} \underset{\sim}{x}) \wedge \underset{\sim}{T} = \{ e_{ijk} \, x_{m,j} \, T_{m,k} \} = \{ e_{ijk} \, \delta_{mj} \, T_{mk} \} = \{ e_{ijk} \, T_{jk} \} \tag{3.45}$$

there follows

$$\underset{\sim}{T} = \underset{\sim}{T}^{T} . \tag{3.46}$$

$\underset{\sim}{T}$ is thus a symmetric tensor.
From equation (3.33) the following local equation,

$$\tfrac{1}{2}\rho\,\dot{\overline{\mathbf{v}\cdot\mathbf{v}}} + \rho\,\dot{e} = \rho\,\mathbf{b}\cdot\mathbf{v} + \rho r + \operatorname{div}(\mathbf{v}\cdot\mathbf{T}) - \operatorname{div}\mathbf{h}, \tag{3.47}$$

results which yields

$$\rho\mathbf{v}\cdot\dot{\mathbf{v}} + \rho\dot{e} = \rho\mathbf{b}\cdot\mathbf{v} + \rho r + \mathbf{v}\cdot\operatorname{div}\mathbf{T} + \operatorname{tr}(\mathbf{T}\mathbf{L}) - \operatorname{div}\mathbf{h} \tag{3.48}$$

or, using (3.36) and remarking that, by virtue of the symmetry of $\mathbf{T}$,

$$\operatorname{tr}(\mathbf{T}\mathbf{L}) = \operatorname{tr}(\mathbf{T}\mathbf{D}) = T_{ij}\, d_{ij}, \tag{3.49}$$

the following equation results

$$\rho\dot{e} + \operatorname{div}\mathbf{h} - T_{ij}\, d_{ij} = \rho r, \tag{3.50}$$

which represents a generalization of the principle of virtual work (see [3]).

3.8. Local equations in the material description

The global equation (3.27) can be translated into

$$\frac{D}{Dt}\int_{V'} P\mathbf{v}\, dV = \int_{V'} P\mathbf{b}\, dV + \int_{\partial V'} \boldsymbol{\sigma}\, dA, \tag{3.51}$$

from which the equation

$$\operatorname{Div}\boldsymbol{\Sigma} + P\mathbf{b} = P\dot{\mathbf{v}} \tag{3.52}$$

results which is the material form of (3.38).
Considering (3.8), equation (3.46) yields

$$\boldsymbol{\Sigma}^T = \mathbf{F}\boldsymbol{\Sigma}(\mathbf{F}^{-1})^T, \tag{3.53}$$

so that $\boldsymbol{\Sigma}$ is not a symmetric tensor.
Conversion of (3.33) into

$$\frac{D}{Dt}\int_{B'} (\tfrac{1}{2}\mathbf{v}\cdot\mathbf{v} + e)\, dm = \int_{B'} (\mathbf{b}\cdot\mathbf{v} + r)\, dm = \int_{\partial V'} (\boldsymbol{\sigma}\cdot\mathbf{v} + H)\, dA \tag{3.54}$$

finally yields the material local form of the energy equation:

$$P\dot{e} + \operatorname{Div}\mathbf{H} - \tfrac{1}{2}\operatorname{tr}(\boldsymbol{\Sigma}^{\circ}\dot{\mathbf{C}}) = Pr, \tag{3.55}$$

where $\boldsymbol{\Sigma}^{\circ}$ is the second Piolla-Kirchhoff stress tensor.

3.9. Material frame-indiference

Let us assume that the frame of reference moves together with an observer who is supposed to be associated with it.
The displacements, velocities and accelerations measured with respect to the new frame are of course different from those which were associated to the old one.
The principle of material frame-indiference or of material objectivity, states however that, for the observer who moves with the frame, the response of the body does not change.
More precisely, if the change of the frame of reference consists in a superimposed rigid motion defined by

$$\underset{\sim}{x}^* = \underset{\sim}{Q}^*(t)\,\underset{\sim}{x} + \underset{\sim}{c}^*(t), \tag{3.56}$$

where $\underset{\sim}{Q}^*$ is an orthogonal matrix (the new frame is still Cartesian and orthogonal), the scalars, vectors and tensors which caracterize the state of the body are invariant.
Therefore, any scalar keeps its value and, if $\underset{\sim}{a}$ is one of such vectors, or $\underset{\sim}{A}$ any of such tensors, the corresponding component matrices must transform according to the following laws

$$\underset{\sim}{a}^* = \underset{\sim}{Q}^*\underset{\sim}{a} \qquad\qquad \underset{\sim}{A}^* = \underset{\sim}{Q}^*\underset{\sim}{A}\,\underset{\sim}{Q}^{*T} \tag{3.57-58}$$

Important consequences can be drawn which concern the constitutive equations, but the point which interests us now is that the momentum and moment of momentum balance equations (3.27) and (3.28) can be derived from the energy balance equation (3.33) with the help of the material objectivity principle.
In order to derive the balance of momentum equation, let us make $\underset{\sim}{Q}^* = \underset{\sim}{I}$ and $\underset{\sim}{c}^* = \underset{\sim}{c}\,t$ where $\underset{\sim}{c}$ is a constant, i.e.,

$$\underset{\sim}{x}^* = \underset{\sim}{x} + \underset{\sim}{c}\,t \tag{3.59}$$

and, therefore,

$$\underset{\sim}{v}^* = \underset{\sim}{v} + \underset{\sim}{c}\,. \tag{3.60}$$

The energy equation for the new frame is

$$\frac{D}{Dt}\int_{B'}(1/2\ \underset{\sim}{v}^*\cdot\underset{\sim}{v}^* + e^*)\,dm = \int_{B'}(\underset{\sim}{b}^*\cdot\underset{\sim}{v}^* + r^*)\,dm + \int_{\partial B'}(\underset{\sim}{t}^*\cdot\underset{\sim}{v}^* + h^*)\,da \tag{3.61}$$

Considering (3.60) and the scalars invariance, and subtracting (3.33) from (3.61), there results the equation

$$\frac{D}{Dt}\int_{B'}(\underset{\sim}{v}\cdot\underset{\sim}{c} + 1/2\,\underset{\sim}{c}^T\underset{\sim}{c})\,dm = \int_{B'}\underset{\sim}{b}\cdot\underset{\sim}{c}\,dm + \int_{\partial B'}\underset{\sim}{t}\cdot\underset{\sim}{c}\,da \tag{3.62}$$

In order that this equation may hold for any $\underset{\sim}{c}$, the balance of momentum equation (3.27) must be satisfied (remember that $\frac{D}{Dt}(\underset{\sim}{c}^T\underset{\sim}{c})=0$). The balance of moment of momemtum equation, can in turn be derived by making

$$\underset{\sim}{Q}^* = \underset{\sim}{Q}\,\underset{\sim}{t} , \tag{3.63}$$

where $\underset{\sim}{Q}$ is a constant matrix, and $\underset{\sim}{c}^* = \underset{\sim}{0}$.
Therefore,

$$\underset{\sim}{x}^* = \underset{\sim}{Q}^*\underset{\sim}{x} \qquad \text{and} \qquad \underset{\sim}{v}^* = \underset{\sim}{Q}^*\underset{\sim}{v} + \dot{\underset{\sim}{Q}}^*\underset{\sim}{x} , \tag{3.64.65}$$

so that

$$\frac{D}{Dt}\int_{B'} \tfrac{1}{2}\, \underset{\sim}{v}^* \cdot \underset{\sim}{v}^* dm = \frac{D}{Dt}\int_{B'} \tfrac{1}{2}(\underset{\sim}{Q}^*\underset{\sim}{v} + \dot{\underset{\sim}{Q}}^*\underset{\sim}{x})^T(\underset{\sim}{Q}^*\underset{\sim}{v} + \dot{\underset{\sim}{Q}}^*\underset{\sim}{x})\, dm . \tag{3.66}$$

Now, $\underset{\sim}{Q}^{*T}\underset{\sim}{Q}^*$ and, thus,

$$\dot{\underset{\sim}{Q}}^{*T}\underset{\sim}{Q}^* + \underset{\sim}{Q}^{*T}\dot{\underset{\sim}{Q}}^* = \underset{\sim}{0} . \tag{3.67}$$

This means that

$$\dot{\underset{\sim}{Q}}^T\underset{\sim}{Q}^* = -\underset{\sim}{Q}^{*T}\dot{\underset{\sim}{Q}}^* , \tag{3.68}$$

i.e., that the matrix

$$\underset{\sim}{S} = \dot{\underset{\sim}{Q}}^{*T}\underset{\sim}{Q}^* \tag{3.69}$$

is a skew-symmetric matrix.
Differentiating (3.67) with respect to time, we obtain

$$\ddot{\underset{\sim}{Q}}^{*T}\underset{\sim}{Q}^* + 2\,\dot{\underset{\sim}{Q}}^{*T}\dot{\underset{\sim}{Q}}^* + \underset{\sim}{Q}^{*T}\ddot{\underset{\sim}{Q}}^* = \underset{\sim}{0} \tag{3.70}$$

or, as $\ddot{\underset{\sim}{Q}}^* = \underset{\sim}{0}$,

$$\dot{\underset{\sim}{Q}}^{*T}\dot{\underset{\sim}{Q}}^* = \underset{\sim}{0} . \tag{3.71}$$

Equation (3.66) can thus be transformed into

$$\frac{D}{Dt}\int_{B'} \tfrac{1}{2}\, \underset{\sim}{v}^* \cdot \underset{\sim}{v}^* dm = \frac{D}{Dt}\int_{B'} (\tfrac{1}{2}\, \underset{\sim}{v}^T\underset{\sim}{v} + \underset{\sim}{x}^T\underset{\sim}{S}\,\underset{\sim}{v})\, dm . \tag{3.72}$$

On the other hand,

$$\int_{B'} \underset{\sim}{b}^* \cdot \underset{\sim}{v}^* dm = \int_{B'} \underset{\sim}{b}^T\underset{\sim}{Q}^{*T}(\underset{\sim}{Q}^*\underset{\sim}{v} + \dot{\underset{\sim}{Q}}^*\underset{\sim}{x})\, dm = \int_{B'} (\underset{\sim}{b}^T\underset{\sim}{v} + \underset{\sim}{x}^T\underset{\sim}{S}\,\underset{\sim}{b})\, dm \tag{3.73}$$

$$\int_{\partial B'} \underset{\sim}{t}^{*} \cdot \underset{\sim}{v}^{*} da = \int_{\partial B'} \underset{\sim}{t}^{T} \underset{\sim}{Q}^{*T} (\underset{\sim}{Q}^{*} \underset{\sim}{v} + \dot{\underset{\sim}{Q}}^{*} \underset{\sim}{x}) da = \int_{\partial B'} (\underset{\sim}{t}^{T} \underset{\sim}{v} + \underset{\sim}{x}^{T} \underset{\sim}{S} \underset{\sim}{t}) da. \quad (3.74)$$

Therefore, subtracting (3.33) from (3.61), we obtain

$$\frac{D}{Dt} \int_{B'} \underset{\sim}{x}^{T} \underset{\sim}{S} \underset{\sim}{v} \, dm = \int_{B'} \underset{\sim}{x}^{T} \underset{\sim}{S} \underset{\sim}{b} \, dm + \int_{\partial B'} \underset{\sim}{x}^{T} \underset{\sim}{S} \underset{\sim}{t} \, da \quad (3.75)$$

Now, the fact that $\underset{\sim}{S}$ is a skew-symmetric matrix allows us to make

$$S_{jk} = s_i \, e_{ijk} \, , \quad (3.76)$$

so that

$$\underset{\sim}{x}^{T} \underset{\sim}{S} \underset{\sim}{v} = S_{jk} x_j v_k = e_{ijk} s_i x_j v_k = \underset{\sim}{s} \cdot \underset{\sim}{x} \wedge \underset{\sim}{v} \quad (3.77)$$

As $\ddot{\underset{\sim}{S}} = \underset{\sim}{0}$ i.e., as $\dot{\underset{\sim}{S}}$ is independent of time, vector $\underset{\sim}{s}$ is also independent of time.

Introducing (3.77) into (3.75), there results

$$\frac{D}{Dt} \int_{B'} \underset{\sim}{s} \cdot \underset{\sim}{x} \wedge \underset{\sim}{v} \, dm = \int_{B'} \underset{\sim}{s} \cdot \underset{\sim}{x} \wedge \underset{\sim}{b} \, dm + \int_{\partial B'} \underset{\sim}{s} \cdot \underset{\sim}{x} \wedge \underset{\sim}{t} \, da, \quad (3.78)$$

or

$$\underset{\sim}{s} \cdot \frac{D}{Dt} \int_{B'} \underset{\sim}{x} \wedge \underset{\sim}{v} \, dm = \underset{\sim}{s} \cdot \int_{B'} \underset{\sim}{x} \wedge \underset{\sim}{b} \, dm + \underset{\sim}{s} \cdot \int_{\partial B'} \underset{\sim}{x} \wedge \underset{\sim}{t} \, da \, , \quad (3.79)$$

which, as $\underset{\sim}{s}$ is arbitrary, finally yields the moment of momentum balance equation (3.28).

4. SUMMARY OF FIELD EQUATIONS IN SOIL MECHANICS

4.1. Field equations for solids

Within the frame of the local description, the equilibrium field equations are equations (3.38) and (3.46), which, in component form, and denoting by ρ_s the mass density of the solid, become

$$T_{ij,i} + \rho_s b_j = \rho_s \dot{v}_j \quad (4.1)$$

and

$$T_{ij} = T_{ji} \, . \quad (4.2)$$

The comma denotes, as usual, differenciation with respect to co-ordinates x_i.

Assuming the small displacement assumption to be a valid approximation, equations (2.48) provide the strain-displacement equations

$$\gamma_{ij} = 1/2\,(u_{i,j} + u_{j,i}) . \tag{4.3}$$

Under the same assumption, the distinction disappears between Cauchy's and Piolla-Kirchhoff's stresses and between spatial and material co-ordinates.
Equation (3.50) provides, if the non-mechanical energy terms are dropped, i.e., if the assumption is made of only mechanical energy exchanges being significant,

$$\rho_s \dot{e} = T_{ij} \cdot d_{ij} \tag{4.4}$$

where e becomes the energy density associated with deformation. No restrictions on the displacement magnitudes are made for the validity of equation (4.4). If, however, the small displacement assumption is adopted,

$$d_{ij} \simeq \dot{\gamma}_{ij} , \tag{4.5}$$

so that equation (4.4) may be written as

$$\rho_s De = T_{ij} \cdot D\gamma_{ij} , \tag{4.6}$$

where De and $D\gamma_{ij}$ represent variations of e and γ_{ij} corresponding to the same infinitesimal time increment Dt. From (4.6) there results, assuming that $\rho_s e$ is expressed as an explicit function of the strains alone,

$$T_{ij} = \frac{\partial(\rho_s e)}{\partial \gamma_{ij}} \tag{4.7}$$

which represents the most general form of the constitutive equations for an elastic material, i.e., for a material such that an internal energy density function exists which exclusively depends on the strains.
Equations (4.1), (4.3) and (4.7) are 15 in number. The number of unknowns is also 15: 6 stresses, 6 strains and 3 displacements.

4.2. Field equations for fluids

Neglecting the internal viscosity of the fluid, and assuming the pressure p to be positive in case of compression, the stress tensor components become

$$T_{ij} = -p\,\delta_{ij} . \tag{4.8}$$

Introducing in (3.38), there results the equilibrium equation

$$-p_{,i} + \rho_f \, b_i = \rho_f \, \dot{v}_i \, . \tag{4.9}$$

Equation (2.68) provides the rate of deformation-velocity equations

$$d_{ij} = \tfrac{1}{2} \left(v_{i,j} + v_{j,i} \right) . \tag{4.10}$$

To these equations, the continuity equation (3.21) may be added which yields

$$\dot{\rho}_f + \rho_f \, v_{i,i} = 0 . \tag{4.11}$$

Supposing the fluid to be compressible, we can assume, as a first approximation,

$$\frac{D(dv)}{dv} = - \frac{Dp}{K_f} \; , \tag{4.12}$$

where K_f is the bulk modulus of the fluid and the minus sign stands for the fact the an increase in pressure implies a decrease in volume.

Dividing both sides by Dt, (4.12) yields

$$\frac{\dot{\overline{dv}}}{dv} = - \frac{\dot{p}}{K_f} \; . \tag{4.13}$$

Considering now that $\rho_f \, dv$ is as invariant, there results

$$\dot{\rho}_f \, dv + \rho_f \, \dot{\overline{dv}} = 0 \; , \tag{4.14}$$

which, combined with (4.13), yields

$$\frac{\dot{\rho}_f}{\rho_f} = - \frac{\dot{p}}{K_f} \; . \tag{4.15}$$

Introducing (4.15) into (4.11), the equation becomes

$$v_{i,i} = - \frac{\dot{p}}{K_f} \; . \tag{4.16}$$

In case of incompressibility, we obtain

$$v_{i,i} = 0 \; . \tag{4.17}$$

Equations (4.9), (4.11) and (4.16) are 5 equations which contain 5 unknowns: the velocities, the pressure and the fluid density. In case of incompressibility, ρ_f ceases to be unknown and (4.9) and (4.11) are replaced by (4.17).

4.3. Effective and neutral stresses

According to the usual idealization, soils are composed by grains or particles, which, bound together by molecular forces, constitute a porous media.
The voids of the solid skeleton, i.e., the pores, are supposed to be filled with water and air bubbles, i.e., by a (generally)compressible fluid which is usually assumed to be free of tangential stresses.
The stress field in the fluid is thus hydrostatic, i.e., with components $-p\delta_{ik}$, where p is the pressure.
Grains or grain-size portions of the fluid will be considered in the sequel. Although the size of such portions of matter are so large, if compared to the size of molecules, that the basic assumptions of Continuum Mechanics remain valid, they will be termed microscopic solid or fluid elements and the average values of the significant magnitudes determined on such elements will be called microscopic average values.
The analysis will require reasoning also on portions of soil much larger than the grains themselves, in order that such portions can be considered as macroscopically homogeneous, and yet so small that the average magnitudes which characterize the behaviour or the response of the soil may be considered as constant throughout. These will be termed macroscopic soil elements and the average values determined on such portions will be called macroscopic average values.
As the soil is loaded, the loads are resisted both by the solid and fluid phases.
The grains which compose the solid skeleton are thus stressed and the stresses transmited from one grain to another through contact surfaces which, although not reduced to points, as it would happen if the grains were spherical, and so remained after the deformation, are however very small.
Reasoning now on a macroscopic element detached from the soil,let us decompose the total tractions acting on such element into two terms, according to fig. 1 and to the equation

$$\underset{\sim}{t} = -p\underset{\sim}{n} + \underset{\sim}{t}' , \tag{4.18}$$

where $\underset{\sim}{n}$ represents the outward normal to the element surface and p the fluid pressure, which is supposed to be constant throughout the element.

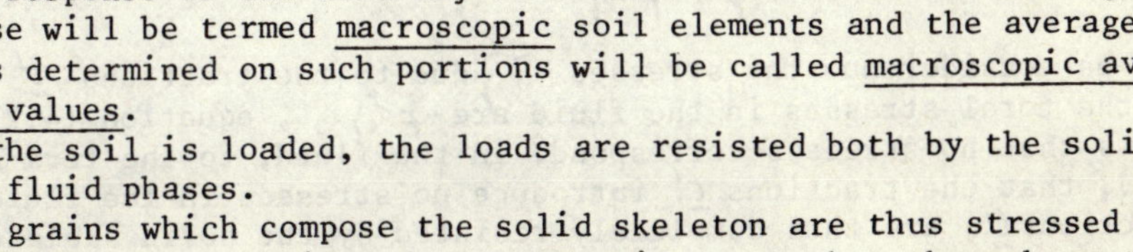

Fig. 1

The first term corresponds thus to compressive normal tractions of magnitude p , the second to the remaining tractions.
Each of these introduces a stress field into the macroscopic element and it is clear that the sum of such fields reproduces the total stress field due to the total tractions. We write then

$$T_{ij} = T^{\circ}_{ij} + T'_{ij} \tag{4.19}$$

where T°_{ij} corresponds to $-p\underset{\sim}{n}$ and T'_{ij} to $\underset{\sim}{t}'$.
Tractions $-p\underset{\sim}{n}$ can of course be equilibrated by a stress field with components $-p\delta_{ij}$.
If each grain is assumed homogeneous, it is clear, on the other hand, that such stress field is indeed the one which the tractions $-p\underset{\sim}{n}$ really introduce both in the solid and fluid phases of the body.
Therefore,

$$T^{\circ}_{ij} = -p\delta_{ij} , \tag{4.20}$$

so that, introducing (4.20) into (4.19), we obtain

$$T_{ij} = -p\delta_{ij} + T'_{ij} . \tag{4.21}$$

Let us consider now the stresses T'_{ij} due to the tractions $\underset{\sim}{t}'$.
As the total stresses in the fluid are $-p\delta_{ij}$, equation (4.21) shows that no stresses correspond, in the fluid, to the term T'_{ij}, i.e., that the tractions $\underset{\sim}{t}'$ introduce no stresses in the fluid. Tractions $\underset{\sim}{t}'$ are thus completely resisted by the solid skeleton. In other words, stresses T'_{ij} are the stresses which would be installed in the solid skeleton if the fluid did not exist.
The fact that the stresses are transmitted from grain to grain through very small contact surfaces implies however that high stress concentrations and comparatively large strains are introduced in the grains by such tractions.
For this reason, we usually neglect the much smaller deformation due to the stresses $-p\delta_{ij}$, which are called the <u>neutral stresses</u>, and suppose the stresses T'_{ij} to control the resistance and the deformation of the solid skeleton from which depend both the resistance and the deformation of the soil. This is the reason why these stresses are called the <u>effective stresses</u>.
An essential point to mention however is that, while the neutral stresses $-p\delta_{ij}$, are really, i.e., microscopically, introduced by the tractions $-p\underset{\sim}{n}$, the effective stresses T'_{ij} are <u>not</u> the microscopic stresses, i.e., the stresses that are really introduced in the grains by the tractions $\underset{\sim}{t}'$.
The effective stresses T'_{ij} are indeed always determined under the assumption that the solid skeleton is a continuous three-dimensional body whose mechanical behaviour is governed by the soil (macroscopic) constitute equations.

Only from a macroscopic point of view it is thus possible to tell that they are introduced in the soil by the tractions $\underset{\sim}{t}'$.

4.4. The coupled equations

It turns out from the preceeding Section that the soil must be considered as a two-phase continuum.
The external loads are equilibrated by the total stresses which, according to equation (4.21), are decomposed into two terms, the neutral stresses and the effective stresses. Only the latter are suppose to produce displacements and strains.
Such displacement and strains are interconnected by the effective strain-displacement equations (4.3).
Two kinds of equilibrium equations will now be established.
The first concern the equilibrium of a macroscopic soil element and may be called the macroscopic equilibrium equations.
The second refer to the fluid phase and are thus microscopic in character.
No equilibrium equations are needed for the solid phase, since the equilibrium of the whole and the equilibrium of one of its parts imply the equilibrium of the remaining part.
Let us consider first the equilibrium of a macroscopic soil element occupying a three-dimensional domain τ .
Denoting by $\underset{\sim}{v}_f$ the absolute velocity of the fluid, and by

$$\underset{\sim}{w}_f = \underset{\sim}{v}_f - \dot{\underset{\sim}{u}} \tag{4.22}$$

its relative velocity with respect to the soil skeleton, the momentum becomes

$$\int_\tau \rho \underset{\sim}{v}\, dv = \int_\tau \rho \dot{\underset{\sim}{u}}\, dv + \int_{\tau^\circ} \rho_f \underset{\sim}{w}_f\, dv , \tag{4.23}$$

where τ° denotes the part of τ occupied by the pores.
Introducing the porosity,

$$n = \frac{\tau^\circ}{\tau} , \tag{4.24}$$

denoting by $\overline{\underset{\sim}{w}}_f$ the microscopic average value of $\underset{\sim}{w}_f$, denoting by $\bar{\rho}$ the macroscopic average value of ρ on τ, and assuming $\dot{\underset{\sim}{u}}$ and $\overline{\underset{\sim}{w}}_f$ to be constant on τ , equation (4.23) becomes

$$\int_\tau \rho \underset{\sim}{v}\, dv = \bar{\rho}\, \dot{\underset{\sim}{u}}\, \tau + \rho_f\, \overline{\underset{\sim}{w}}_f\, \tau^\circ = (\bar{\rho}\, \dot{\underset{\sim}{u}} + \rho_f \overline{\underset{\sim}{w}})\tau = \int_\tau (\bar{\rho} \dot{\underset{\sim}{u}} + \rho_f \overline{\underset{\sim}{w}})\, dv , \tag{4.25}$$

where

$$\overline{\underset{\sim}{w}} = n\, \overline{\underset{\sim}{w}}_f \tag{4.26}$$

represents the macroscopic average value of the relative velocity $\underset{\sim}{w}$.

The term $\frac{D}{Dt}\int_{v}\varrho \underset{\sim}{v}\, dv$ must thus be replaced by $\frac{D}{Dt}\int_{v}(\bar{\varrho}\underset{\sim}{\dot{u}} + \varrho_f \bar{\underset{\sim}{w}})\, dv$ in the momentum balance equation (3.27), so that the equilibrium equations referring to the two-phase medium results from (3.38) just by replacing $\varrho \underset{\sim}{v}$ by $\bar{\varrho}\underset{\sim}{\dot{u}} + \varrho_f \bar{\underset{\sim}{w}}$. If it is admitted, on the other hand, that the only body forces acting on the soil are the gravity forces, with density components $\bar{\varrho} g_i$, where magnitudes g_i are the components of the gravity acceleration $\underset{\sim}{g}$, the macroscopic equilibrium equations become

$$T_{ij,i} + \bar{\varrho} g_j = \bar{\varrho}\ddot{u}_j + \varrho_f \dot{\bar{w}}_j \quad (j=1,2,3). \tag{4.27}$$

The second set of equilibrium equations govern the motion of the fluid itself through the porous medium.
Such equation can be derived from (4.9) if the microscopic average value of $\underset{\sim}{w}_f$, i.e., $\bar{\underset{\sim}{w}}_f$, is considered as the true value of the relative velocity, and if the body force density components become

$$b_j = g_j - \frac{g}{k}\bar{w}_j \,, \tag{4.28}$$

where k is the <u>permeability coefficient</u>, g the magnitude of the gravity acceleration and $\bar{w}_j$ the jth component of the macroscopic relative velocity, $\bar{\underset{\sim}{w}}$.
The second term in (4.21) represents a viscous flow-resistant force and results from Darcy's law, which states that, in the absence of inertial forces, the relative velocity of a fluid flowing through a porous medium is proportional to the gradient of the pressure.
Equation (4.9) yields therefore

$$-p_{,j} + \varrho_f g_j = \varrho_f \ddot{u}_j + \varrho_f \frac{\dot{\bar{w}}_j}{n} + \frac{\varrho_f g}{k}\bar{w}_j \tag{4.29}$$

which represents the missing equilibrium equations.
A supplementary equation is the microscopic continuity equation of the fluid phase which can be derived from (4.17) just by making

$$\underset{\sim}{v} = \underset{\sim}{\dot{u}} + \bar{\underset{\sim}{w}}_f = \underset{\sim}{\dot{u}} + \frac{\bar{\underset{\sim}{w}}}{n} \,. \tag{4.30}$$

We obtain

$$\dot{u}_{i,i} + \frac{\bar{w}_{i,i}}{n} = -\frac{\dot{p}}{K_f} \,. \tag{4.31}$$

The 6 equations (4.3), 6 equations (4.19), 3 equations (4.27), 3 equations (4.29) and the equation (4.31) are, in the whole, 19 equations involving 25 unknowns: the displacements, u_i, the relative velocities, w_i, the strains γ_{ij}, the total stresses T_{ij}, the effective stresses T'_{ij} and the pressure p. The 6 missing equations are the constitutive equations of the solid skeleton, whose treatment in the frame of this course is trusted to other lecturers.

References

[1] - Truesdell, C. and Toupin, R. "The Classical Field Theories", in S. Flügge (ed.) Encyclopedia of Physics, vol.III,Pt.1, Springer Verlag OHG, Berlin, 1960.

*[2] - Truesdell, C. and Noll, W. - "The Nonlinear Field Theories of Mechanics", in S. Flügge (ed.), Encyclopedia of Physics, vol. III, Pt.3, Springer Verlag OHG Berlin, 1965.

[3] - Pister, K.S. - "Continuum Mechanics. A Brief Review", in J.T. Oden and E.R. de Arantes e Oliveira (ed.), Lectures on Finite Elements in Continuum Mechanics, The University of Alabama in Huntsville Press, Huntsville, 1973.

[4] - Oden, J.T. - "The Finite Element Method in Fluid Mechanics", in J.T. Oden and E.R. de Arantes e Oliveira (ed.), Lectures on Finite Elements in Continuum Mechanics, U.A.H. Press, Huntsville, 1973.

[5] - Oden, J.T. and Reddy, J.N. - "Variational Methods in Theoretical Mechanics", Springer Verlag, Berlin, 1976.

[6] - Arantes e Oliveira, E.R. - "Advanced Theory of Structures", lectures in the CEB International Course on Structural Concrete, Laboratório Nacional de Engenharia Civil,Lisbon,1973.

[7] - Biot, M.A. - "Theory of Three-Dimensional Consolidation",J. Appl. Phys. 12, pp. 155-164, 1941.

[8] - Zienkiewicz, O.C., Chang, C.T. and Betess, P. - "Drained, Undrained, Consolidating and Dynamic Behaviour Assumptions in Soils", Géotechnique 30, Nº 4, pp. 385-395,1980.

[9] - Aronofsky, J.S. - "Flow Through Porous Media", in W. Flügge (ed.) Handbook of Engineering Mechanics, Mc Graw-Hill, New York, 1962.

In all that concerns Continuum Mechanics, the present notes were mainly inspired in [3] and, of course, in [1] and [2]. In what concerns the applications to Soil Mechanics, Biot [7] is quoted as a pioneer but the complete equations were drawn from [8]. Paper [9] was included as a reference text for Darcy's law.

Acknowledgement

The support of the Instituto Nacional de Investigação Científica (INIC) is gratefully acknowledged.
Thanks are also due to Mrs. Cristina Bragança who kindly taped my unreadable manuscript.

BASIC FORMULATION OF STATIC AND DYNAMIC BEHAVIOUR OF SOIL AND OTHER POROUS MEDIA

O.C. Zienkiewicz

Professor, Department of Civil Engineering, University of Wales, Swansea, U. K.

Notation: We shall use throughout a 'vectorial' tensor notation for convenience in subsequent finite element manipulations. Thus strain and stress increments are written as

$$(d\underset{\sim}{\varepsilon})^T = (d\,\varepsilon_x,\ d\varepsilon_y\ ,\ d\varepsilon_z\ ,\ d\gamma_{xy}\ ,\ d\gamma_{yz}\ ,\ d\gamma_{zx}\ )$$

$$(d\underset{\sim}{\sigma})^T = (d\sigma_x\ ,\ d\sigma_y\ ,\ d\sigma_z\ ,\ d\tau_{xy}\ ,\ d\tau_{yz}\ ,\ d\tau_{zx}\ )$$

and the equivalent of Kroneker delta is

$$\underset{\sim}{m}^T = (1,\ 1,\ 1,\ 0,\ 0,\ 0)$$

Symbols occurring are defined in text.

INTRODUCTION

The soil or indeed any other porous medium such as rock or concrete shall be considered as a two phase medium composed of a

J. B. Martins (ed.), Numerical Methods in Geomechanics, 39–55.

"solid skeleton" and an interstitial fluid. Fig. 1. We shall be thus concerned with the saturated material such as certainty occurs in foundations of offshore structures, dams, etc. Under unsaturated conditions a third phase enters the considerations and additional complexity results in the treatment of which drastic assumptions have to be additionally introduced.

In the treatment of the problem we shall consider in turn
(1) Constitutive relations
(2) Dynamic and kinematic relations
in terms of which the problem can be formulated.

This will be followed by an introduction of
(3) Various approximations and discussion of their limits of validity. Here we shall propose a formulation which embraces most practical engineering situations and consider such special situations as the consolidation problem (where dynamic terms are normally omitted) and the undrained behaviour which represents a convenient limiting case.

In Part II of these notes we shall consider the finite element discretization and time dependent solution of various classes of soil problems.

PART I

1. GENERAL FORM OF CONSTITUTIVE RELATIONS FOR POROUS , TWO PHASE ,MEDIA

It is convenient in such materials to separate the total stress tensor $\underset{\sim}{\sigma}$ into the 'effective' stress $\underset{\sim}{\sigma}'$ and a hydrostatic component p equal to pore pressure

$$\underset{\sim}{\sigma} = \underset{\sim}{\sigma}' - \underset{\sim}{m}\, p \tag{1}$$

(tension +ve convention maintained for case of adaptation to standard programs).

If the displacements of the solid skeleton are $\underset{\sim}{u}$ the strain increments $d\underset{\sim}{\varepsilon}$ are given as

$$d\underset{\sim}{\varepsilon} = \underset{\sim}{L}\, d\underset{\sim}{u} \tag{2}$$

where $\underset{\sim}{L}$ is the linear, small strain, operator.

A uniform increase of pore pressure dp (with $d\,\underset{\sim}{\sigma}' = 0$) can only result in a uniform compression of the matrix of the solid, giving a volumetric strain change

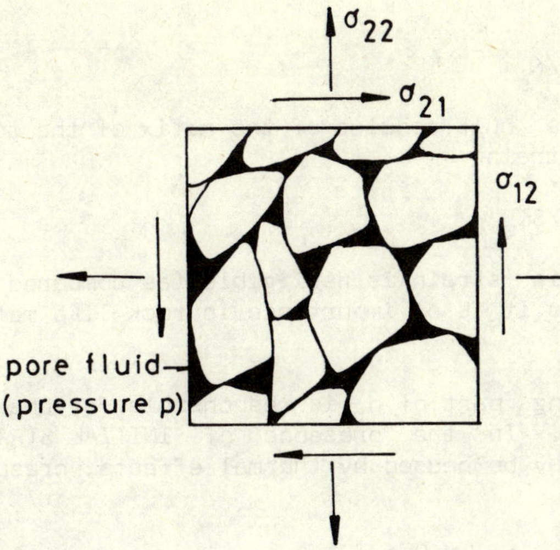

Fig. 1 Total stress $\underset{\sim}{\sigma}$ and pore pressure p in an element of a porous material

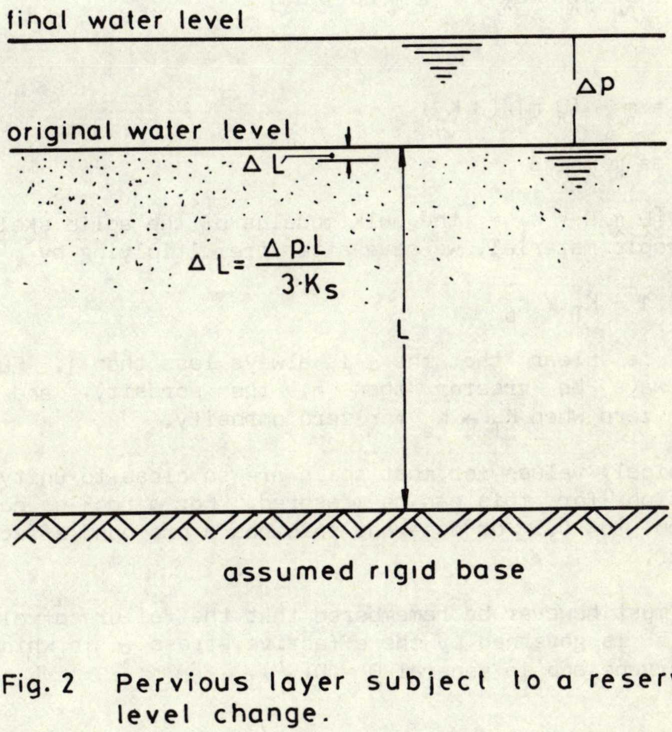

Fig. 2 Pervious layer subject to a reservoir level change.

$$d\varepsilon^{p}_{vol} = - dp/K_s \tag{3}$$

where K_s is the bulk modulus of the matix of the solid. This gives a uniform strain

$$d\underset{\sim}{\varepsilon}^{p} = - \underset{\sim}{m}\, dp/3K_s \tag{4}$$

Although this strain is negligible (as compared with total strain) for soils it is of importance in rock like materials and will be retained.

The remaining part of $d\underset{\sim}{\varepsilon}$ is responsible for changes of the effective stress. In the presence of INITIAL STRAIN changes $d\underset{\sim}{\varepsilon}^{o}$ - such as may be caused by thermal effects, creep, etc. we can write

$$d\underset{\sim}{\sigma}' = \underset{\sim}{D}(d\underset{\sim}{\varepsilon} - d\underset{\sim}{\varepsilon}^{o} - d\underset{\sim}{\varepsilon}^{p}) \tag{5}$$

where $\underset{\sim}{D}$ is a tangent matrix (dependent on strain/stress and history in a general non linear case).

Using (1) (4) and (5) we can write quite generally that

$$d\underset{\sim}{\sigma} = \underset{\sim}{D}\,(d\underset{\sim}{\varepsilon} - d\underset{\sim}{\varepsilon}^{o}) - \alpha\, \underset{\sim}{m}\, dp + d\underset{\sim}{\sigma}_R \tag{6}$$

where

$$\alpha\, \underset{\sim}{m} = \underset{\sim}{m} - (\underset{\sim}{D}\, \underset{\sim}{m})/(3\, K_s)$$

However as $\underset{\sim}{m}^T \underset{\sim}{m} \equiv 3$

and $\underset{\sim}{m}^T \underset{\sim}{D}\, \underset{\sim}{m}/9 = K_T =$ the bulk modulus of the solid skeleton for an isotropic material, we have after premultiplying by $\underset{\sim}{m}^T$

$$\alpha = 1 - K_T\, /\, K_s \tag{7}$$

It is clear that the α is always less than 1. Further it must always be greater than n, the porosity, and can only approach zero when $K_T \rightarrow K_s$ for zero porosity.

Typical values for most soils are so close to unity that no distinction for this can be measured. For a rock or concrete α can drop to 0.5 or less and this has a profound effect on the behaviour.

It must however be remembered that the failure of all porous materials is governed by the effective stress $\underset{\sim}{\sigma}'$ in which α does not intervene and in general $\underset{\sim}{D} = \underset{\sim}{D}(\underset{\sim}{\sigma}')$.

The last term of equation (6) is negligible for small displacement and if $\|\underset{\sim}{\sigma}\| << \|\underset{\sim}{D}\|$. This term accounts for rotation of the stress without straining and is thus a function of $\underset{\sim}{\sigma}$ and $d\underset{\sim}{u}$. In large displacement and strain analysis the inclusion of this term is of importance.

To complete the constitutive relations it is necessary to determine the change of the fluid volume stored per unit volume of the solid skeleton $d\Theta$.

This change in storage is due to four courses.

(a) change due to total strain with mtrix assumed incompressible

$$\underset{\sim}{m}^T d\underset{\sim}{\varepsilon}$$

(b) change due to compression of the matrix by dp. If n is the porosity this is simply

$$(1 - n)\, dp/K_s$$

(c) change due to compression of the fluid with a bulk modulus K_f

$$n\, dp/K_f$$

(d) change due to compression of solid grains related to change of effective stress $d\underset{\sim}{\sigma}'$. Assuming the average change of mean effective stress i.e. $1/3\, \underset{\sim}{m}^T d\underset{\sim}{\sigma}'$ is representive of the matrix stress this component is

$$-\,1/3\, \underset{\sim}{m}^T d\underset{\sim}{\sigma}'/K_s$$

or using (1) and (6)

$$-\,1/3\, \underset{\sim}{m}^T(\underset{\sim}{D}\,(d\underset{\sim}{\varepsilon} - d\underset{\sim}{\varepsilon}^o) + (1 - \alpha)\, \underset{\sim}{m}dp)/K_s$$

Addition of all the components, use of relation (7) and some manipulation yields

$$d\,\Theta = \alpha\, \underset{\sim}{m}^T d\underset{\sim}{\varepsilon} + (1/Q)\, dp + (\underset{\sim}{m}^T \underset{\sim}{D}\, d\underset{\sim}{\varepsilon}^o)/(3\, K_s) \qquad (8)$$

where

$$\frac{1}{Q} = \left[\frac{n}{K_f} + \frac{1 - n}{K_s} - (1 - \alpha)\frac{1}{K_s}\right] \qquad (9)$$

NOTE: Equations (6) and (8) have been originally derived by Biot in a different manner and the author is grateful to Dr. Bruce Simon of University of Arizona, Tucson, for pointing out this identity of the two forms. Equations (7) and (9) give a physically meaningful interpretation of Biot's parameters and Q in terms of microstructure.

The basic equations (6) and (8) determine the changes of total stress and stored fluid volume when changes of $\underset{\sim}{\varepsilon}$ and p are imposed.

Immediately we can obtain the behaviour of the material when UNDRAINED CONDITIONS are specified. Now $d\,\Theta \equiv 0$ and from equations (8) and (6) (neglecting $d\underset{\sim}{\varepsilon}^{o}$)

$$dp = -\,Q\,\alpha\,\underset{\sim}{m}^{T}\,d\underset{\sim}{\varepsilon} \tag{10}$$

$$d\underset{\sim}{\sigma} = \underset{\sim}{D}\,d\underset{\sim}{\varepsilon} - \alpha\,\underset{\sim}{m}\,dp = (\underset{\sim}{D} + \alpha^{2}\,\underset{\sim}{m}\,Q\,\underset{\sim}{m}^{T})d\underset{\sim}{\varepsilon} \equiv \bar{\underset{\sim}{D}}\,d\underset{\sim}{\varepsilon} \tag{11}$$

in which $\bar{\underset{\sim}{D}} = \underset{\sim}{D} + \alpha^{2}\,\underset{\sim}{m}\,Q\,\underset{\sim}{m}^{T}$ gives the relation between DRAINED AND UNDRAINED TANGENT MATRICES which is unique. It is of interest to consider the change of pore pressure as a fraction of the mean stress $-\underset{\sim}{m}^{T}d\underset{\sim}{\sigma}/3$ due to a unit volumetric strain $d\underset{\sim}{\varepsilon} = \underset{\sim}{m}/3$. Now

$$B = -\frac{3\,dp}{\underset{\sim}{m}^{T}\,d\underset{\sim}{\sigma}} = \frac{Q\,\alpha\,\underset{\sim}{m}^{T}\underset{\sim}{m}}{\underset{\sim}{m}^{T}\,(\underset{\sim}{D} + \alpha^{2}\,\underset{\sim}{m}\,Q\,\underset{\sim}{m}^{T})\,\underset{\sim}{m}/3}$$

$$= \frac{3\,Q\,\alpha}{3\,K_{T} + 3\,\alpha^{2}\,Q} = \frac{\alpha}{K_{T}/Q + \alpha^{2}} \tag{12}$$

It is immediately observed that in soils where $K_{s} >> K_{T}$, $B \rightarrow \alpha \rightarrow 1$ but that this is not the case in rock like materials.

B is the well known soil parameter whose values range from 0.3 to 0.5 for concretes to 1 for sands and clays. (viz. Lamb and Whitman ,pp.395)

2. DYNAMIC AND KINEMATIC RELATIONS

The constitutive relation (6) or (8) and the strain rate/displacement rate relation (2) here to be supplemented by addition equation before a solution can be attempted.

Before doing so it is necessary to introduce an addition variable $\underset{\sim}{w}$ which is the DISPLACEMENT OF THE PORE FLUID RELATIVE TO THE SOIL SKELETON. This is measured in terms of total quantity of fluid displaced divided by the GROSS area . Thus the actual mean displacement in the pores is $\underset{\sim}{w}/n$ where n is the porosity.

The force of the additional relation is that of overall equilibrium of a unit soil skeleton + fluid volume . This relation is explicity :

$$\text{in x direction} \qquad \frac{\partial\sigma_x}{\partial x} \quad \frac{\partial\tau_{xy}}{\partial y} + \frac{\partial\tau_{xz}}{\partial z} + \rho g_x = \rho\, \ddot{u}_x + \rho_f\, \ddot{w}_x \qquad (13)$$

etc. for other directions, or simply

$$\underset{\sim}{L}^T\underset{\sim}{\sigma} + \rho\, \underset{\sim}{g} = \rho\, \underset{\sim}{\ddot{u}} + \rho\; \underset{\sim}{\ddot{w}} \qquad (14)$$

where $\underset{\sim}{L}$ is the same operator as that defining strain rate in eq.(2). In above

ρ = density of the total fluid/solid mixture

ρ_f = density of fluid

$\underset{\sim}{g}$ = vector of gravity accerelation components

Observe that in the last term porosity does not enter as the volume of fluid is equal to n whcih acceleration relative to solid is $\underset{\sim}{\ddot{w}}/n$.

The second equation is that of EQUILIBRIUM OF FLUID PHASE ALONE. Here the main DRIVING force is the pressure gradient and the main resisting (viscous) force is proportional to velociety of fluid (DARCY relationship). In addition acceleration forces occur. Thus we have

$$-\underset{\sim}{\nabla}p + \rho_f\, \underset{\sim}{g} = \rho_f\underset{\sim}{\ddot{u}} + \rho_f\, \underset{\sim}{\ddot{w}}/n + (\rho_f g/k)\, \underset{\sim}{\dot{w}} \qquad (15)$$

In above k is the well known permeability coefficient (for anisotropic media we have a permeability matrix $\underset{\sim}{k}$ and use $\underset{\sim}{k}^{-1}$ in place of 1/k). The notation $\underset{\sim}{\nabla}$ is

$$\underset{\sim}{\nabla} = \left(\frac{\partial}{\partial x}\, ,\, \frac{\partial}{\partial y}\, ,\, \frac{\partial}{\partial z}\right)^T \qquad (16)$$

The final relationship is a kinematic one relating the fluid inflow to the storage θ . Hence simply

$$\underset{\sim}{\nabla}^T\underset{\sim}{\dot{w}} + \dot{\theta} = 0 \qquad (17)$$

where $\underset{\sim}{\nabla}^T\underset{\sim}{\dot{w}} \equiv \text{div}\, \underset{\sim}{\dot{w}}$

3. SUMMARY OF ESSENTIAL EQUATIONS

3.1 Full $\underset{\sim}{u}/\underset{\sim}{w}$ formulation

The full soil problem is now posed by equations (2), (6), (8), (14), (15) and (17) summerised below

$$
\begin{aligned}
& d\underset{\sim}{\varepsilon} = \underset{\sim}{L}\, d\underset{\sim}{u} && \text{(a)} \\
& d\underset{\sim}{\sigma} = \underset{\sim}{D}\,(d\underset{\sim}{\varepsilon} - d\underset{\sim}{\varepsilon}^{o}) - \alpha\, \underset{\sim}{m}\, dp + (d\underset{\sim}{\sigma}_R) && \text{(b)} \\
& d\Theta = \alpha\, \underset{\sim}{m}^T d\underset{\sim}{\varepsilon} + (1/Q)\, dp + \underset{\sim}{m}^T \underset{\sim}{D}\, d\underset{\sim}{\varepsilon}^{o}/(3\, K_s) && \text{(c)} \\
& \underset{\sim}{L}^T \underset{\sim}{\sigma} + \rho\, \underset{\sim}{g} = \rho\, \ddot{\underset{\sim}{u}} + \rho\, \ddot{\underset{\sim}{w}} && \text{(d)} \quad (18) \\
& -\underset{\sim}{\nabla} p + \rho_f\, \underset{\sim}{g} = \rho_f\, \ddot{\underset{\sim}{u}} + \rho_f\, \ddot{\underset{\sim}{w}}/n + (\rho_f\, g/k)\, \dot{\underset{\sim}{w}} && \text{(e)} \\
& \underset{\sim}{\nabla}^T \dot{\underset{\sim}{w}} + \dot{\Theta} = 0 && \text{(f)}
\end{aligned}
$$

This supplemented by boundary and initial condition such as

$\underset{\sim}{u} = \hat{\underset{\sim}{u}}$ on Γ_u (prescribed displacements)

$\underset{\sim}{t} = \hat{\underset{\sim}{t}}$ on Γ_t (prescribed tractons)

$k\,(\partial p/\partial n) = \hat{q}$ on Γ_q (prescribed flow)

$p = \hat{p}$ on Γ_p (prescribed pressure)

can be solved incrementally for most problems.

In this six equations for those six variables $\underset{\sim}{u}$, $\underset{\sim}{w}$, $\underset{\sim}{\varepsilon}$, $\underset{\sim}{\sigma}$, p and Θ the first two are PRIMARY and solution can be obtained in terms of $\underset{\sim}{u}$ and $\underset{\sim}{w}$ with all other variables eliminated without any differentation. This full formulation of a dynamic case will lead to a large finite element systems as will be seen later .

We shall now consider possible approximations.

3.2 Displacement - pressure ($\underset{\sim}{u}$/p) formulation (Zienkiewicz)

If the motion of fluid ($\dot{\underset{\sim}{w}}$) is relatively small, terms insolving $\ddot{\underset{\sim}{w}}$ can be omitted. Now $\dot{\underset{\sim}{w}}$ can be easily eliminated from 18(e) and (f) giving the system

$$
\begin{aligned}
& d\underset{\sim}{\varepsilon} = \underset{\sim}{L}\, d\underset{\sim}{u} && \text{(a)} \\
& d\,\underset{\sim}{\sigma} = \underset{\sim}{D}\,(d\underset{\sim}{\varepsilon} - d\underset{\sim}{\varepsilon}^{o}) - \alpha\, \underset{\sim}{m}\, dp + (d\underset{\sim}{\sigma}_R) && \text{(b)}
\end{aligned}
$$

$$d\Theta = \alpha \underset{\sim}{m}^T \, d\underset{\sim}{\varepsilon} + (1/Q)\, dp + \underset{\sim}{m}^T \underset{\sim}{D} \, d\underset{\sim}{\varepsilon}^o/(3 K_s) \qquad (c) \qquad (19)$$

$$\underset{\sim}{L}^T \underset{\sim}{\sigma} + \rho \, \underset{\sim}{g} = \rho \, \ddot{\underset{\sim}{u}} \qquad (d)$$

$$- \underset{\sim}{\nabla}^T \bar{k} \, \underset{\sim}{\nabla} p + \underset{\sim}{\nabla}^T \bar{k} \rho_f \, \underset{\sim}{g} - \underset{\sim}{\nabla}^T \rho_f \, \bar{k} \, \ddot{\underset{\sim}{u}} + \dot{\Theta} = 0 \qquad (e)$$

where $\underset{\sim}{u}$ an p are priamary variables and $\bar{k} = k/(\rho_f \, g)$

3.3 Slow quasi-static consolidation phenomena- $\underset{\sim}{u}$/p form

If the motion is so slow that all dynamic force terms can be neglected (i.e., $\ddot{\underset{\sim}{u}}$ terms omitted) equations (19) give the well known coupled consolidation equation frequency solved by numerical method. This particular case is EMBEDDED in the full dynamic form of eqs.(19) and as the general solution of (19) is no more difficult than the solution of the quasi-static case we shall not discuss it seperately. It is desirable that dynamic programs capable of solving (19) should have facility for continuing the solution in the solw quasi-static domain. This enables problem of earthquake and post earthquake reaction to be dealt with simultaneously.

3.4 Undrained behaviour

If the permeability k tends to be zero no motions of the fluid relative to the solid skeleton can occur and thus $\underset{\sim}{w} = \dot{\underset{\sim}{w}} = \ddot{\underset{\sim}{w}} = 0$.

From eq.(18f) we find that $\dot{\Theta} = 0$ or $d\Theta = 0$ at all times.

Equation (18e) is indentically satisfying and equation (18c) allows a relation between $d\underset{\sim}{\varepsilon}$ and dp to be established (viz. eq.(10)). The system becomes

$$d\underset{\sim}{\varepsilon} = \underset{\sim}{L} \, d\underset{\sim}{u}$$

$$d\underset{\sim}{\sigma} = \underset{\sim}{D} \, d\underset{\sim}{\varepsilon} - (\underset{\sim}{I} - (\alpha \underset{\sim}{m} Q \underset{\sim}{m}^T)/(3 K_s)) \underset{\sim}{D} \, d\underset{\sim}{\varepsilon}^o \qquad (20)$$

$$\underset{\sim}{L}^T \underset{\sim}{\sigma} + \rho \, \underset{\sim}{g} = \rho \, \ddot{\underset{\sim}{u}}$$

This is indentical (except for the initial strain term) with the dynamic equations governing the behaviour of a single phase medium. The primary variable is now $\underset{\sim}{u}$ and solution procedures are simpler.

3.5 Static (limiting) solution

Undrained static solution

If the permeability is so small that eqs.(20) are applicable in solution the undrained static solution is simply one resulting from solution of eqs.(20) with $\ddot{u} = 0$. Such solutions are of importance in the initial analysis of foundation on clay etc. But it must be remembered that even here these represent but an intermediate effect before consolidation transients set in. With undrained solution it is impossible to require that p = 0 on any boundaries - such as a physical condition is always present we simply neglecte a boundary layer effect.

Drained static solution

With damping inherent in the full dynamic equations (18) or their approximations (19) we shall expect an asymptotic state to be reached at some LARGE TIME when

$$\dot{u} = \ddot{u} = 0$$

$$\ddot{w} = 0$$

but $u \neq 0$ and $\dot{w} \neq 0$ and $\dot{p} = \dot{\varepsilon} = \dot{\Theta} = 0$

The equation (19e) is now decoupled giving the usual steady Darcy flow equation,

$$-\nabla^T \bar{k} \nabla p + \nabla^T \bar{k} \rho_f g = 0 \tag{21}$$

and this can be solved for p. The solid phase equation is

$$L^T \sigma + \rho g = 0$$

$$d\varepsilon = L du \tag{22}$$

and $\sigma = \int D (d\varepsilon - d\varepsilon^o) - \alpha m p$

These are again a single phase static equations with $\alpha m p$ playing the role of an "initial stress".

3.6 Limits of validity of various assumptions

To determine the limits of validity of the various assumptions the full dynamic equations (18) were linearised and solved analytically for a one dimensional problem of an elastic layer subject to a periodic force applied at the surface (Fig.2). The pressure condition on the surface is that of full drainage (Zienkiewicz and Bettess, 1980).

It is found that the main parameters affecting the problem can be put into two non-dimensional groups

Zone I B=Z=C Slow phenomena ($\ddot{w}$ and $\ddot{u}$ can be neglected)

Zone II B=Z≠C Moderate speed ($\ddot{w}$ can be neglected)

Zone III B≠Z≠C Fast phenomena ($\ddot{w}$ can not be neglected) only full Biot equation valid

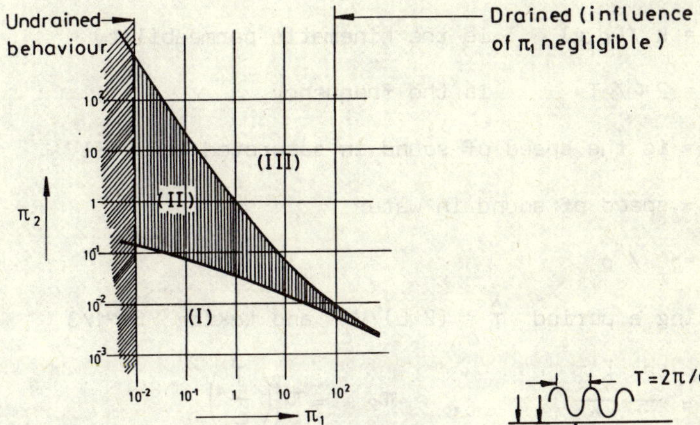

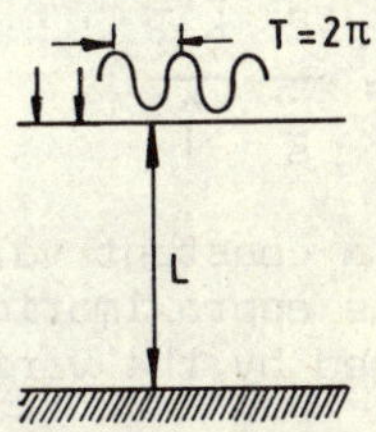

$$\pi_1 = \frac{\hat{k}\rho V_c^2}{\omega L^2} = \frac{2\rho T}{\pi\ \hat{T}^2}$$

$$\pi = \frac{\omega^2 L^2}{V_c^2} = \pi^2 \left(\frac{\hat{T}}{T}\right)^2$$

$\hat{k} = k/\rho_f g$ k-kinematic permeability

$\hat{T} = 2L/V_c$ $V_c^2 = (D + K_f/n)/\rho$

$= \beta K_f/\rho_f n$

$\approx K_f/\rho_f$ (speed of sound in water)

$\beta = \rho_f/\rho$ $n \approx 0.33$ $\beta \approx 0.33$

Fig. 3

Zones of applicability of various assumption

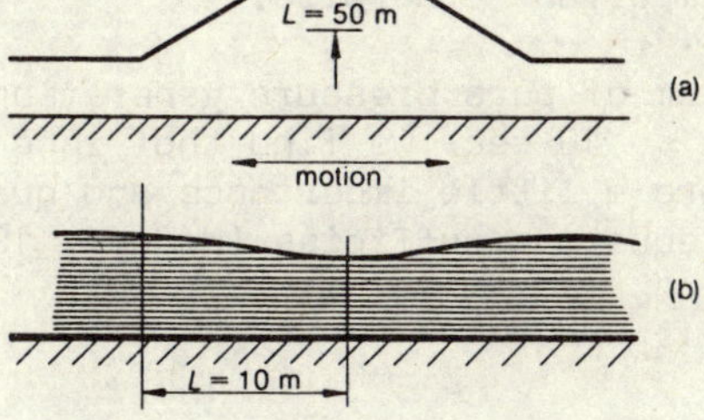

Fig. 4. Some extrapolation of results. Typical *L* values for (a) an earth dam earthquake motion, and (b) wave pressure of Madsen (1978)

$$\pi_1 = \frac{\bar{k}\, V_c^2}{(g\beta)\,\omega\, L^2} \quad , \quad \pi_2 = \frac{\omega^2 L^2}{V_c^2}$$

where

$\bar{k} = k/(\rho_f g)$ is the kinematic permeability

$\omega = 2\pi/T$ is the frequency

V_c^2= is the speed of sound in saturated material

$\simeq$ speed of sound in water

$\beta = \rho_f / \rho$

Introducing a period $\hat{T} = (2\,L)/V_c$ and taking $\beta \simeq 1/3$

$$\pi_1 \simeq \frac{2}{g}\,\frac{\bar{k}\,T}{\hat{T}^2} \quad , \quad \pi_2 = \pi_1^2\left(\frac{\hat{T}}{T}\right)^2$$

For a constant value of $n \simeq 1/3$, the results show that a reasonable approximation (excluding surface boundary layers) can be obtained by the various approximations in the manner shown in Fig.3.

This result obtained for a simple case of a single layer can be used to estimate the type of approximation required for analysis of real situations if L is taken as a typical dimension of a drainage distance and an approximate periodically of interest is specified . In Fig.4 we indicate how such a typical dimension can be estimated and the reader can readily find that for instance:

In a study of earthquake response of a dam (or foundation) with L = 50 m and longest period of circle 5 sec. undrained behaviour can only be assumed of $k < 10^{-4}$ m/sec. with drainage effects being important otherwise.

In a problem of pore pressure generation in a sea bad with L = 10 m and T $\simeq$ 10 sec. we find that zone 1 is reached where dynamic effect are a little importance and quasi-static analysis by consolidation equation sufficies (Madsen, 1978).

PART II

INTRODUCTION

In PART I we have presented the various forms of full dynamic , coupled soil-fluid equations and found that the approximate form of eqs.(19) (repeated below) is sufficient to describe all but the very high frequency phenomena. We shall show how this system can be discretized in space and solved for transient problems in the time domain.

5. FINITE ELEMENT DISCRETIZATION OF FLUID DYNAMIC AND STATIC SOIL/FLUID EQUATION

In the form of equation given below we

(a) eliminate $d\underset{\sim}{\varepsilon}$ and $d\Theta$, and

(b) separate the determination of stress changes into that of effective and total stress. This is convenient as invariably the non linearity of D depends on effective stress.

Thus we have

$$\underset{\sim}{L}^T\underset{\sim}{\sigma} + \rho\,\underset{\sim}{g} = \rho\,\ddot{\underset{\sim}{u}} \qquad \text{(a)}$$

$$d\underset{\sim}{\sigma}' = \underset{\sim}{D}\,(\underset{\sim}{L}\,d\underset{\sim}{u} - d\underset{\sim}{\varepsilon}^{o}) + (1-\alpha)\underset{\sim}{m}\,dp + \underset{\sim}{R}\,d\underset{\sim}{u}\,\underset{\sim}{\sigma}' \qquad \text{(b)} \qquad (23)$$

$$d\underset{\sim}{\sigma} = d\underset{\sim}{\sigma}' - \underset{\sim}{m}\,dp \qquad \text{(c)}$$

$$-\underset{\sim}{\nabla}^T\bar{k}\,\underset{\sim}{\nabla}p + \underset{\sim}{\nabla}^T\,\bar{k}\,\rho_f\underset{\sim}{g} - \underset{\sim}{\nabla}^T\,\rho_f\bar{k}\,\ddot{\underset{\sim}{u}} + \alpha\,\underset{\sim}{m}^T\,\underset{\sim}{L}\dot{\underset{\sim}{u}} + \qquad \text{(d)}$$

$$(1/Q)\,\dot{p} + (\underset{\sim}{m}^T\,\underset{\sim}{D}\dot{\underset{\sim}{\varepsilon}}^{o})/(3\,K_s) = 0$$

In above $\underset{\sim}{R}$ is a matrix defining suitably the rate of rotation in terms of displacements.

To discretize the above system we take suitable shape functions expression

$$\underset{\sim}{u} = \underset{\sim}{N}\bar{\underset{\sim}{u}} \qquad\qquad \underset{\sim}{p} = \bar{\underset{\sim}{N}}\bar{\underset{\sim}{p}} \qquad (24)$$

and form use Galerkin weighting. Details can be found elwsewhere (Zienkiewicz 1963, 71, 77).

The first equation (23a) yields the 'standard' equilibrium equation

$$\int_{\Omega} \underset{\sim}{B}^{T} \underset{\sim}{\sigma}\, d\Omega + \underset{\approx}{M}\ddot{\bar{\underset{\sim}{u}}} + \underset{\sim}{f} = 0 \tag{25a}$$

$$\int_{\Omega} \underset{\sim}{B}^{T} \underset{\sim}{\sigma}'\, d\Omega + \underset{\approx}{M}\ddot{\bar{\underset{\sim}{u}}} - \underset{\approx}{Q}\bar{\underset{\sim}{p}} + \underset{\sim}{f} = 0$$

where

$\underset{\sim}{B} = \underset{\approx}{L}\underset{\sim}{N}$ is the strain matrix

$\underset{\sim}{M} = \int_{\Omega} \underset{\sim}{N}^{T} \rho\, \underset{\sim}{N}\, d\Omega$ is the mass matrix

$$\underset{\sim}{Q} = \int_{\Omega} \alpha\, \underset{\sim}{B}^{T}\, \underset{\sim}{m}\, \bar{\underset{\sim}{N}}\, d\Omega$$

and $\underset{\sim}{f} = \int_{\Omega} \underset{\sim}{N}^{T} \rho\, \underset{\sim}{g}\, d\Omega + \int_{\Gamma} \underset{\sim}{N}^{T}\, \bar{\underset{\sim}{t}}\, d\Gamma$

The auxiliary equation (23b/c) are simply written in a discretize form as

$$d\bar{\underset{\sim}{\sigma}}' = \underset{\sim}{D}\,(\underset{\sim}{B}\, d\bar{\underset{\sim}{u}} - d\underset{\sim}{\varepsilon}) + (\underset{\approx}{RN})\, d\bar{\underset{\sim}{u}}\, \underset{\sim}{\sigma}' \tag{25b}$$

Note that $d\underset{\sim}{\sigma}' = d\bar{\underset{\sim}{\sigma}}' + (1-\alpha)\underset{\sim}{m}\, dp$ and can be evaluated if justified

Finally the coupled seepage equation (23d) becomes

$$\underset{\approx}{H}\bar{\underset{\sim}{p}} + \underset{\approx}{S}\dot{\bar{\underset{\sim}{p}}} + \underset{\sim}{Q}^{T}\dot{\bar{\underset{\sim}{u}}} - \hat{\underset{\approx}{M}}\ddot{\bar{\underset{\sim}{u}}} + \underset{\sim}{q} = 0 \tag{25c}$$

where

$$\underset{\sim}{H} = \int_{\Omega} (\underset{\approx}{\nabla \bar{N}})^{T}\, \bar{k}\, (\underset{\approx}{\nabla \bar{N}})\, d\Omega$$

$$\underset{\sim}{S} = \int_{\Omega} \bar{\underset{\sim}{N}}^{T} \frac{1}{Q} \bar{\underset{\sim}{N}}\, d\Omega$$

$$\hat{\underset{\sim}{M}} = \int_{\Omega} \bar{\underset{\sim}{N}}^{T} \underset{\sim}{\nabla}^{T} (\rho_{f}\, \bar{k}\, \underset{\sim}{N})\, d\Omega$$

and $\underset{\sim}{q}$ involves initial strain term.

In general the system of ordinary differential equations (25) is highly non linear as

(1) $\underset{\sim}{D}$ depends on $\underset{\sim}{\sigma}'$ and previous strain history.

(2) The problem (and element) domains Ω and Ω_e are current i.e., are changing with time as deformation proceeds.

(3) Matrices $\underset{\sim}{B}$ and $(\underset{\sim}{RN})$ depend on current geometry.

and finally

(4) The permeability matrix $\underset{\sim}{H}$ can change if k is dependent on deformation as it often is.

In step by step solution all such non linearities can be accomodated without undue complication.

If the problem is transient eqs. (25) will provide solution of both dynamic and static consolidation behaviour.

Undrained behaviour can be obtained by noting that $k \to 0$, $\underset{\sim}{H} \to 0$ and $\underset{\sim}{\hat{M}} \to 0$ thus permitting determination of dp interms of $d\underset{\sim}{u}$. The equations are valid for steady static drained behaviour and uncoupled then.

It is of interest to write down the linear form of eq.(25) . Now the auxiliary stress increment equation can be integrated and the system can be written as

$$\begin{bmatrix} \underset{\sim}{M} & 0 \\ -\underset{\sim}{\hat{M}} & 0 \end{bmatrix} \begin{bmatrix} \ddot{\underset{\sim}{\bar{u}}} \\ \ddot{\underset{\sim}{\bar{p}}} \end{bmatrix} + \begin{bmatrix} \underset{\sim}{0} & \underset{\sim}{0} \\ \underset{\sim}{Q}^T & \underset{\sim}{S} \end{bmatrix} \begin{bmatrix} \dot{\underset{\sim}{\bar{u}}} \\ \dot{\underset{\sim}{\bar{p}}} \end{bmatrix} + \begin{bmatrix} \underset{\sim}{K} & -\underset{\sim}{Q} \\ \underset{\sim}{0} & \underset{\sim}{H} \end{bmatrix} \begin{bmatrix} \underset{\sim}{\bar{u}} \\ \underset{\sim}{\bar{p}} \end{bmatrix} + \begin{bmatrix} \underset{\sim}{f} \\ \underset{\sim}{q} \end{bmatrix} = 0 \qquad (26)$$

where $\underset{\sim}{f}$ and q incorporate any initial stress effects. We note that above equation system is non symmetric and is not conveniently treated by combined solution methods.

5. TIME STEPPING - STAGGERED PROCEDURES

As we have just noted that even the linear coupled system is not conveniently structured for a direct simultaneous solution. Hence staggered solution process (viz Park and Felippa, 1980) are conventionally used.

In such STAGGERED SOLUTIONS we use a standad time stepping algorithm for solving the first equation system (eq.(25a/b)) treating the pore pressure system as known and thus putting in effects as the known force, i.e.

$$\int_\Omega \underset{\sim}{B}^T \underset{\sim}{\sigma}' \, d\Omega + \underset{\approx}{M}\ddot{\underset{\sim}{u}} + \underset{\sim}{\bar{f}} = 0 \qquad (27a)$$

$$d\underset{\sim}{\sigma}' = \underset{\sim}{D}\,(\underset{\sim}{B}\, d\underset{\sim}{u} - d\underset{\sim}{\varepsilon}^o) + (\underset{\approx}{RN})\underset{\sim}{\sigma}'$$

becomes a straightforeward standard dynamic equation of second order if

$$\bar{f} = f - Q\,\bar{p} \tag{27b}$$

is taken as known. This can be solved for a time step using Newmark or similiar procedures

Once the computation of increments Δu is accomplished solution for the first order system

$$H\bar{p} + S\dot{\bar{p}} + \bar{q} = 0 \tag{27c}$$

can be found with q now known as

$$\bar{q} = q + Q^T\dot{\bar{u}} - \hat{M}\ddot{\bar{u}} \tag{27d}$$

In the examples of dynamic analysis of dams and other structures we have used extensively EXPLICIT (central difference) procedures for solving (27a) and IMPLICIT (backward difference) procedures for (27d).

Clearly other procedures can be used and it is important to derive schemes in which the time step can be varied by several orders of magnitude if rapid (earthquake) phenomena proceeded a slow consolidation phase.

REFERENCES

Biot, M. A.; Theory of three-dimensional consolidation, J. of Appl. Phys., 12, pp. 155-164, 1941

Biot, M. A.; Mechanics of deformation and acoustic propagation in porous media, J. Appl. Phys., 33, pp. 1483-98, 1960

Madsen, O. S. ; Wave induced pore pressure and effective stresses in a porous bed. Geotechnique , 28, pp.377-393

Park, K. C. and Felippa, C. A.; Partitioned transient analysis procedures for coupled field problems - accuracy analysis. - to be published

Terzhagi, K.; Erdboumechanik, Franz Deuticke, Vienna, 1925

William, T. and Whitman, R. V.; Soil Mechanics. John Wiley, 1969.

Zienkiewicz, O. C. ; The Finite Element Method., third edition, McGraw Hill, 1977

Zienkiewicz, O. C., Humpheson, C. and Lewis, R. W.; A unified approach to soil mechanics problems, Chapter 4 of Finite Elements in Geomechanics, pp. 151-78, ed. G. Gudehus, J. Wiley, 1977

Zienkiewicz, O. C. and Bettess, P.; Soil and other porous material under transient , dynamic condition, general formulation and validity of various simplified assumptions. in Soils under Cyclic and Transient Loading, eds. Panade, G.N. and Zienkiewicz, O. C., John Wiley, to be published

Zienkiewicz, O. C., Leung, K. H., Hinton, E. and Chang, C. T.; Earthdam analysis for earthquakes. Proceeding Conference on 'Design of Dams to resist Earthquakes' Inst. Civ. Eng. London, Oct., 1-2, 1980

GENERALIZED PLASTICITY AND SOME MODELS FOR GEOMECHANICS

O. C. Zienkiewicz

Professor, Department of Civil Engineering, University of Wales, Swansea, U.K.

1. UNIAXIAL MODELS

Consider the nonlinear behaviour of a uniaxial sample under the imposition of a stress which increases uniformly. If the material is insensitive to time a unique curve will be registered, Fig. 1. If on unloading the same curve is followed we can describe the behaviour as ELASTIC with nonlinear characteristics (hyper-elastic). One way of describing such a model would be to write a tangent relation

$$d\sigma = D d\varepsilon \tag{1}$$

where

$$D = D(\sigma) \text{ or } D = D(\varepsilon) \tag{2}$$

is a "tangent modulus".

J. B. Martins (ed.), Numerical Methods in Geomechanics, 57–78.

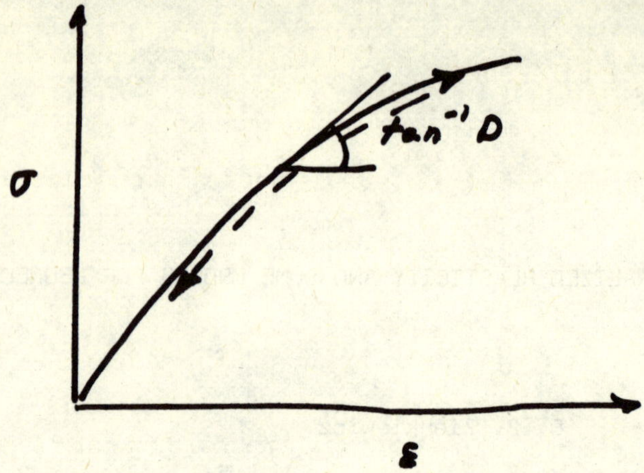

Fig. 1.
Non-linear elasticity

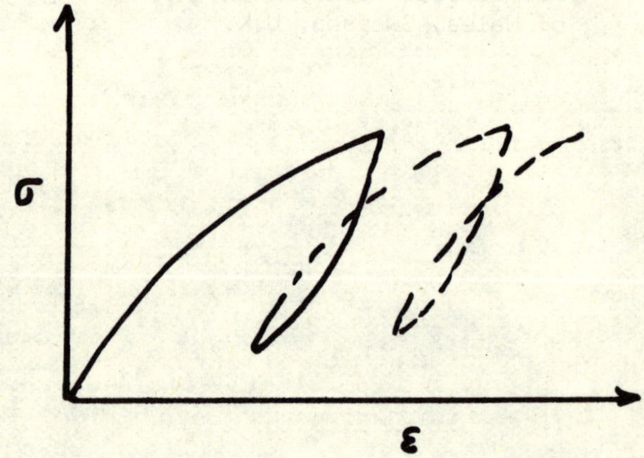

Fig. 2.
Rcal material behaviour

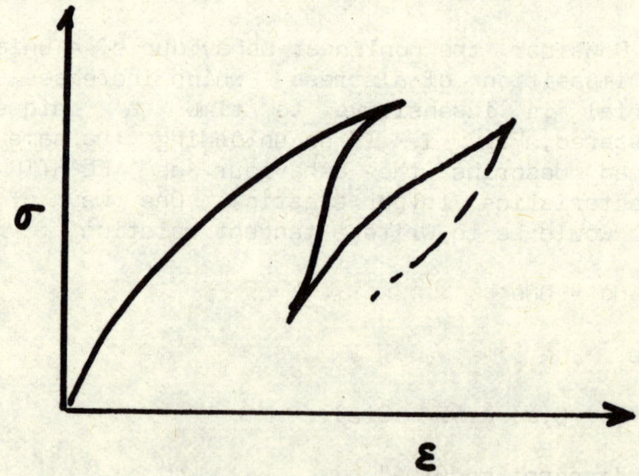

Fig. 3.
Model where D=D(state) cy Endochronic

Real materials show generally different paths on unloading and again on subsequent reloading (Fig.2). Such a material, if again time insensitive, will show permanent deformation on unloading and quite generally will be called plastic.

Again we could characterise the material by a tangent relation (1) but now D is not uniquely defined by the strain (or stress) state of the material. Two essential bits of information are needed for definition of D

(a) parameter (or parameters) describing history of previous loading (say κ)

(b) specification whether next $d\sigma$ (or $d\varepsilon$) involves 'loading' or unloading.

If only history parameters are used then at any point of the stress path only one value of D can exist but branching is still possible as shown in Fig. 3. A particular theory of plasticity - the ENDOCHRONIC THEORY is based on such a model.

For a more general description it is obviously necessary to use both the history dependence and a specification of loading direction.

Thus a general plastic model will be characterized by a relationship $d\sigma = D d\varepsilon$

where D for any state has two values D_l and D_u (load/unload moduli) both in general dependant on previous history.

Special cases of plasticity postulate the existence of an elastic zone where $D_l = D_u = D_e$ (elastic).

Limit of elastic zone is known as YIELD point.

In "ideal" plasticity (elasto-plasticity)

$D_l = 0$; $D_u = D_e$ (elastic)

at yield point viz. Fig. 4a.

In "strain hardening" plasticity

$D_l > 0$; $D_u = D_e$

at yield point viz. Fig. 4b.

In "strain softening" plasticity

$D_l < 0$; $D_u = D_e$ viz. Fig. 4c.

(Note here 'loading' is best defined in terms of strain).

2. GENERALISED PLASTICITY IN MULTIAXIAL STRESS

In a multiaxial stress/strain state the incremental relations become

$$d\underset{\sim}{\sigma} = \underset{\sim}{D}\, d\underset{\sim}{\varepsilon} \tag{3}$$

where $\underset{\sim}{\sigma}$ and $\underset{\sim}{\varepsilon}$ are appropriate tensors. It is convenient for our purpose however to continue matrix/vector notation listing appropriate components.

Once again we can define (1) nonlinear elastic
(2) endochronic
and (3) generalized plastic

models which are characterized by;

(1) $\underset{\sim}{D}$ being a function of the state of stress
(2) $\underset{\sim}{D}$ being a function of the state of stress and in addition of historic variable κ
(3) as (2) but defining at each point loading and unloading directions with appropriate $\underset{\sim}{D}_l$ and $\underset{\sim}{D}_u$

In practical problems with only increasing loading, nonlinear elastic behaviour can often be assumed and identical results to those given by more elaborate theories are obtained. In uniaxial behaviour this is simple to do as direction of "loading" or "unloading" is self evident. This is not the case under multiaxial stress and load/unload directions need to be clearly specified. (Disastrous results have been obtained in soil mechanics by use of nonlinear elastic models due to this cause - and use of such models should be avoided).

With the drawback of "endochronic" models mentioned we will proceed to describe a very general model of generalized plasticity.

In this model the state of stress $\underset{\sim}{\sigma}$ and the 'historic' parameters κ define a direction in the stress space $\underset{\sim}{n}_f$ which is associated with "loading" and "unloading". $\underset{\sim}{n}_f$ is defined as a unit vector in the stress space and if $d\underset{\sim}{\sigma}$ is the direction of a stress change then the projection $\underset{\sim}{D} = \underset{\sim}{D}_l$; $\underset{\sim}{n}_f^T \; d\underset{\sim}{\sigma} > 0$ is loading, $\underset{\sim}{D} = \underset{\sim}{D}_u$, $\underset{\sim}{n}_f^T \, d\underset{\sim}{\sigma} < 0$ is unloading.

The plane NN normal to $\underset{\sim}{n}_f$ defines a neutral direction. The form of the matrices $\underset{\sim}{D}_l$ and $\underset{\sim}{D}_u$ is related by the continuity conditions (Mroz et al 1980). This requires (by obvious physical argument that for given stress changes the same strain change should be observed) that

$$\underset{\sim}{D}_u^{-1} = \underset{\sim}{D}_l^{-1} \quad \text{when } \underset{\sim}{n}_f^T \, d\underset{\sim}{\sigma} = 0 \tag{5}$$

The simplest relation guaranteeing above is written as

$$d\underset{\sim}{\varepsilon} = \underset{\sim}{D}_l^{-1} \, d\underset{\sim}{\sigma} = (\underset{\sim}{D}_u^{-1} + \underset{\sim}{n}_q \, \underset{\sim}{n}_f^T /K) d\underset{\sim}{\sigma} \tag{6}$$

The equivalent relation between moduli is

$$\underset{\sim}{D}_l^{-1} = (\underset{\sim}{D}_u^{-1} + \underset{\sim}{n}_q \, \underset{\sim}{n}_f^T /K) \tag{7}$$

in above $\underset{\sim}{n}_q$ is a completely arbitrary unit direction vector in stress space and K a scalar known as HARDENING MODULUS.

Note (1): This is particularly convenient if $\underset{\sim}{D}_u = \underset{\sim}{D}_e$ i.e. unloading operation is elastic. It is of course possible not to have any elastic regions and to write

$$\underset{\sim}{D}_l^{-1} = (\underset{\sim}{D}_o^{-1} + \underset{\sim}{n}_{ql} \, \underset{\sim}{n}_f^T /K) \tag{8}$$

$$\underset{\sim}{D}^{-1} = (\underset{\sim}{D}_o^{-1} + \underset{\sim}{n}_{qu} \, \underset{\sim}{n}_f^T /K) \tag{9}$$

where $\underset{\sim}{D}_o$ is a matrix dependent only on state and $\underset{\sim}{n}_{ql}$ may not be the same as $\underset{\sim}{n}_{qu}$.

Note (2): Definition of loading suggested by Dafalias is $\underset{\sim}{n}_f^T \, d\underset{\sim}{\sigma}/K \geq 0$.

This gives a positive quantity even if negative K (or strain softening) occurs. However ambiguity with elastic unloading still occurs. In such cases it is necessary to use strain control and check $\underset{\sim}{n}_f^T \, d\underset{\sim}{\varepsilon}$. In F.E.M. analysis $\Delta\underset{\sim}{\varepsilon}$ is generally computed at start of an increment and this determines whether the increment is to be treated as plastic or elastic throughout all subsequent iterations in that increment.

Equation (6) may be interpreted as

$$d\underset{\sim}{\varepsilon} = d\underset{\sim}{\varepsilon}^e + d\underset{\sim}{\varepsilon}^p \qquad \text{for} \qquad \underset{\sim}{n}_f^T \, d\underset{\sim}{\sigma} \geq 0 \tag{10}$$

where plastic strain increment is

$$d\underset{\sim}{\varepsilon}_p = (\underset{\sim}{n}_q \, \underset{\sim}{n}_f^T /K) \, d\underset{\sim}{\sigma} \tag{11}$$

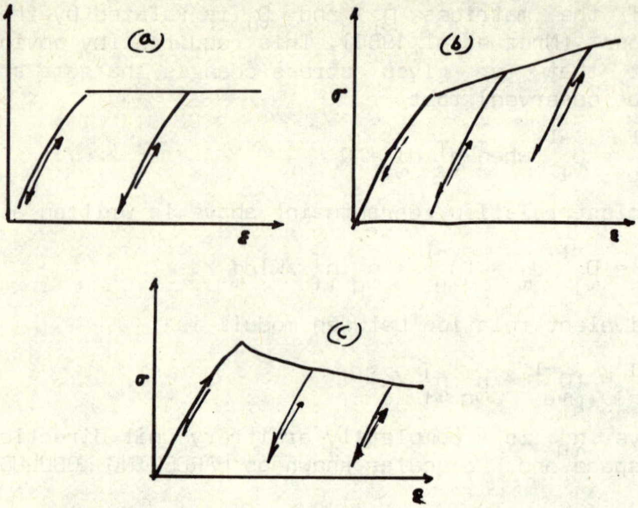

Fig. 4. Classical plastic behaviour (with elasticity)
(a) ideal
(b) hardening (c) softening

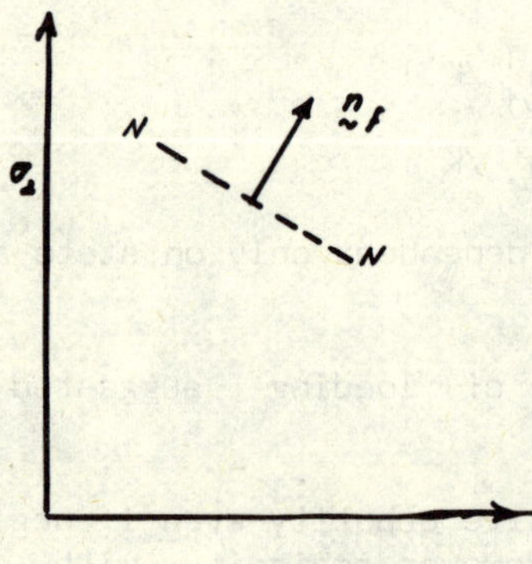

Fig. 5. Stress Space - loading direction specified at each point.

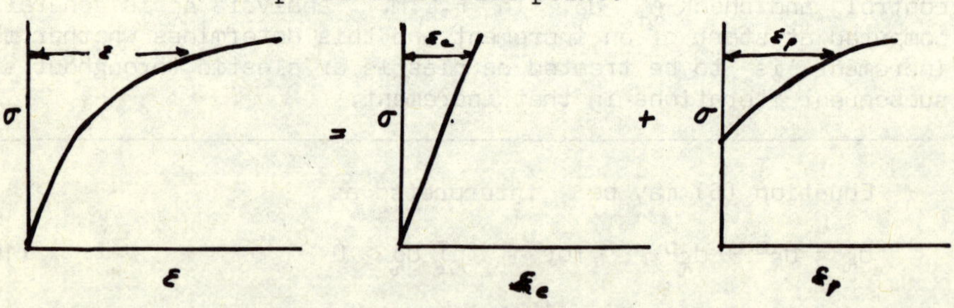

Fig. 6. Separation of total stress into 'elastic' and plastic components.

In uniaxial stress consider $n_f = n_q = 1$ and K is the slope of $\sigma - \varepsilon$ curve (Fig. 6).

K can be zero (ideal plasticity)
positive (hardening plasticity)
or negative (softening plasticity)

To obtain inverse relation we write

$$d\underset{\sim}{\sigma} = \underset{\sim}{D}_1 \, d\underset{\sim}{\varepsilon} \tag{12}$$

and

$$\underset{\sim}{D}_1 = \left[\underset{\sim}{D}_u - \frac{\underset{\sim}{D}_u \underset{\sim}{n}_q \underset{\sim}{n}_f^T \underset{\sim}{D}_u}{K + \underset{\sim}{n}_f^T \underset{\sim}{D}_u \underset{\sim}{n}_q} \right] \tag{13}$$

When $\underset{\sim}{D}_u$ is symmetric and $\underset{\sim}{n}_f = \underset{\sim}{n}_q$ we have associated flow and $\underset{\sim}{D}_1$ is symmetric. $\underset{\sim}{n}_f \neq \underset{\sim}{n}_q$ is called non associated flow.

NOTE: Derivation of expressions (13) is not trivial and is included for completeness. Starting from (6)

$$d\underset{\sim}{\varepsilon} = (\underset{\sim}{D}_u^{-1} + \underset{\sim}{n}_q \underset{\sim}{n}_f^T /K) \, d\underset{\sim}{\sigma} \tag{a}$$

if $\quad \underset{\sim}{n}_f^T d\underset{\sim}{\sigma}/K \equiv \lambda$ (b)

we have $\quad d\underset{\sim}{\varepsilon} = \underset{\sim}{D}_u^{-1} \, d\underset{\sim}{\sigma} + \underset{\sim}{n}_q \, \lambda$ (c)

multiplying by $\underset{\sim}{n}_f^T \underset{\sim}{D}_u$

$$\underset{\sim}{n}_f^T \underset{\sim}{D}_u d\underset{\sim}{\varepsilon} = \underset{\sim}{n}_f^T d\underset{\sim}{\sigma} + \underset{\sim}{n}_f^T \underset{\sim}{D}_u \underset{\sim}{n}_q \, \lambda \tag{d}$$

or $\quad \underset{\sim}{n}_f^T \underset{\sim}{D}_u d\underset{\sim}{\varepsilon} = \lambda K + \underset{\sim}{n}_f^T \underset{\sim}{D}_u \underset{\sim}{n}_q \, \lambda$ (e)

and

$$\lambda = \frac{\underset{\sim}{n}_f^T \underset{\sim}{D}_u d\underset{\sim}{\varepsilon}}{K + \underset{\sim}{n}_f^T \underset{\sim}{D}_u \underset{\sim}{n}_q} \tag{f}$$

substituting into (c)

$$d\underset{\sim}{\varepsilon} = \underset{\sim}{D}_u^{-1} \, d\underset{\sim}{\sigma} + \frac{\underset{\sim}{n}_q \underset{\sim}{n}_f^T \underset{\sim}{D}_u d\underset{\sim}{\varepsilon}}{K + \underset{\sim}{n}_f^T \underset{\sim}{D}_u \underset{\sim}{n}_q} \tag{g}$$

or multiplying by $\underset{\sim}{D}_u$

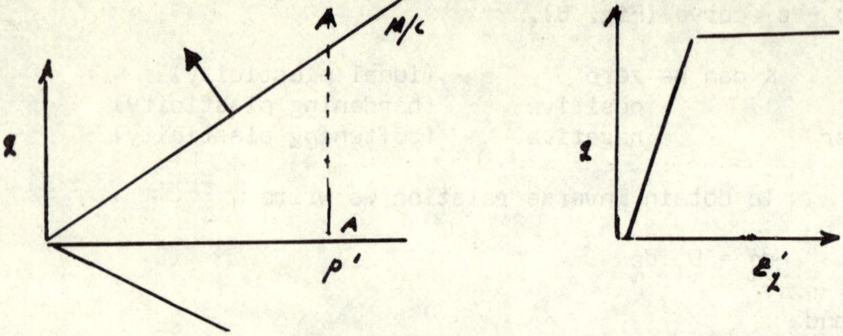

Fig. 7. Model A. Ideal plasticity - Associated

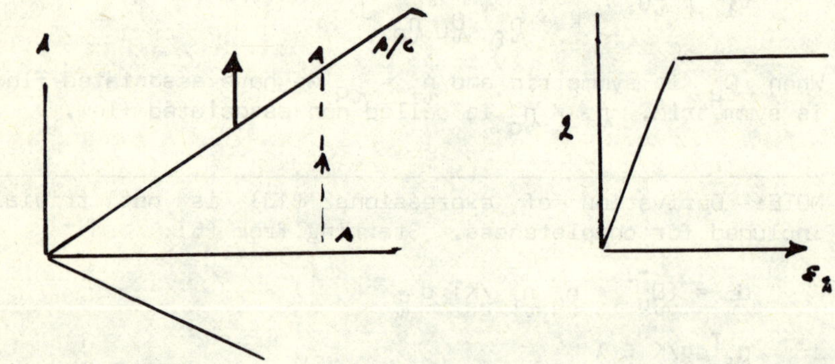

Fig. 8. Model B. Ideal plasticity - Nonassociated

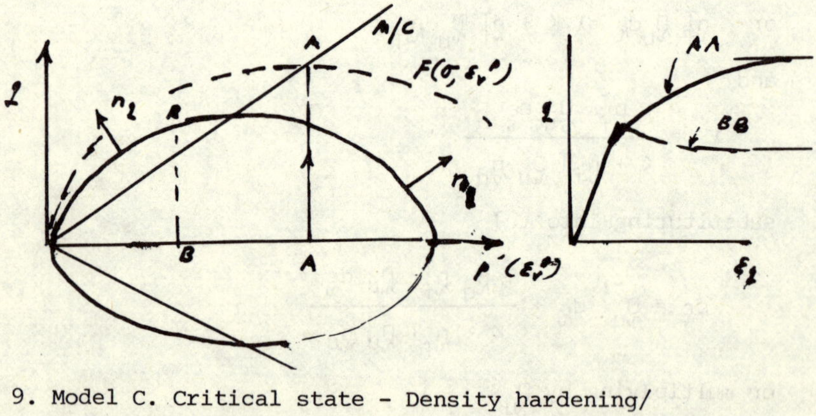

Fig. 9. Model C. Critical state - Density hardening/
softening plasticity.

$$d\underset{\sim}{\sigma} = (\underset{\sim}{D}_u - \frac{\underset{\sim}{D}_u \underset{\sim}{n}_q \underset{\sim}{n}_f^T \underset{\sim}{D}_u}{K + \underset{\sim}{n}_f^T \underset{\sim}{D}_u \underset{\sim}{n}_q}) d\underset{\sim}{\varepsilon} \qquad \text{(h)}$$

Q.E.D.

This is the standard elasto-plastic matrix (viz. Zienkiewicz et al 1968).

Note that the definition of a plastic law is complete if for any stress state we can specify

1. the 'historical' parameters
2. the direction $\underset{\sim}{n}_f$ of LOADING
3. the direction $\underset{\sim}{n}_q$ of flow
4. the 'elastic' (or unloading) modulus $\underset{\sim}{D}_u$

and

5. the plastic modulus K

Obviously many possible ways for such specification exist and these include CLASSICAL PLASTICITY, BOUNDING SURFACE PLASTICITY, and many as yet untried variants.

So called hypo-elasticity (Truesdel) is a particular case of above relations.

3. CLASSICAL PLASTICITY

(1) Here direction $\underset{\sim}{n}_f$ is defined by a yield surface

$$F(\underset{\sim}{\sigma}, \kappa) = 0 \qquad (14)$$

or $$\underset{\sim}{n}_f = \frac{\partial F}{\partial \underset{\sim}{\sigma}} \Bigg/ \left[\left(\frac{\partial F}{\partial \underset{\sim}{\sigma}}\right)^T \left(\frac{\partial F}{\partial \underset{\sim}{\sigma}}\right)\right]^{\frac{1}{2}} \quad \text{(unit vector)} \qquad (15)$$

i.e. by a normal to the surface F.

(2) The direction of flow $\underset{\sim}{n}_q$ is defined by normals to surfaces $Q(\underset{\sim}{\sigma})$ (plastic potential) as

$$\underset{\sim}{n}_q = \frac{\partial Q}{\partial \underset{\sim}{\sigma}} \Bigg/ \left[\left(\frac{\partial Q}{\partial \underset{\sim}{\sigma}}\right)^T \left(\frac{\partial Q}{\partial \underset{\sim}{\sigma}}\right)\right]^{\frac{1}{2}} \qquad (16)$$

(3) The hardening parameters κ are defined as functions of $\underset{\sim}{\varepsilon}^p$ (including history of $\underset{\sim}{\varepsilon}^p$).

(4) The hardening modulus K follows from above definitions. It is postulated that for $\underset{\sim}{n}_f^T \, d\underset{\sim}{\sigma} \geq 0$ the stress point remains on yield surface F. Thus we must have (consistency condition)

$$dF = \frac{\partial F}{\partial \underset{\sim}{\sigma}} d\underset{\sim}{\sigma} + \frac{\partial F}{\partial \kappa} \frac{\partial \kappa}{\partial \underset{\sim}{\varepsilon}^p} d\underset{\sim}{\varepsilon}^p = 0 \tag{17}$$

putting eq.(11)

$$d\underset{\sim}{\varepsilon}^p = \underset{\sim}{n}_q \, \underset{\sim}{n}_f^T \, d\underset{\sim}{\sigma}/K \tag{18}$$

and dividing by $\left[\left(\frac{\partial F}{\partial \underset{\sim}{\sigma}}\right)^T\left(\frac{\partial F}{\partial \underset{\sim}{\sigma}}\right)\right]^{\frac{1}{2}}$ given

$$(1 + \beta\, /K)\, \underset{\sim}{n}_f^T \, d\underset{\sim}{\sigma} = 0 \; ; \; \beta = \frac{\frac{\partial F}{\partial \kappa}\frac{\partial \kappa}{\partial \underset{\sim}{\varepsilon}^p}}{\left[\left(\frac{\partial F}{\partial \underset{\sim}{\sigma}}\right)^T\left(\frac{\partial F}{\partial \underset{\sim}{\sigma}}\right)\right]^{\frac{1}{2}}} \underset{\sim}{n}_q \tag{19}$$

and as $\underset{\sim}{n}_f^T \, d\underset{\sim}{\sigma} \geq 0$

$$K = -\beta \tag{20}$$

In soil mechanics classical plasticity models include many useful models all of which must show continuing deformation when stress reaches Mohr Coulomb envelope in principal stress space $M(\sigma) = 0$. In diagrams given below variants of such models are graphically described in invariant space σ_m and $\bar{\sigma}$ (first and second stress invariants) corresponding to triaxial loading. Notation conventionally used $\sigma_m \equiv -p$, $\bar{\sigma} = q/\sqrt{3}$. All refer of course , to effective stress behaviour here.

Model A. Ideal associated plasticity (Fig 7).

$$K = 0$$

$$F(\underset{\sim}{\sigma}) = M(\underset{\sim}{\sigma}) = 0$$

$$Q(\underset{\sim}{\sigma}) = F(\underset{\sim}{\sigma})$$

Model B. Ideal non-associated plasticity (Fig.8).

$$K = 0$$

$$F(\underset{\sim}{\sigma}) = M(\underset{\sim}{\sigma}) = 0$$

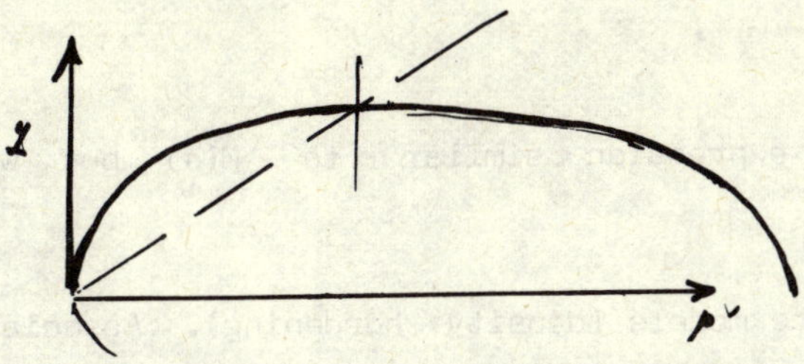

Fig. 10. Model C. With non symmetric surface

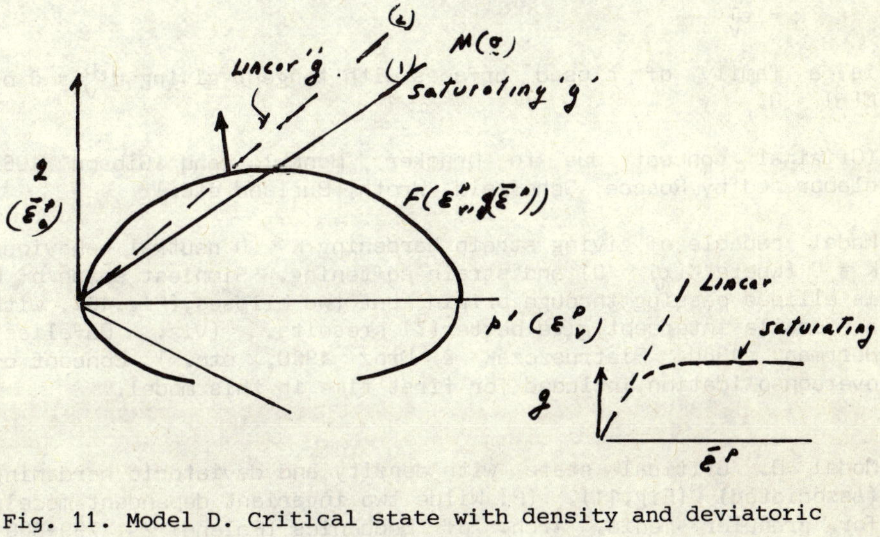

Fig. 11. Model D. Critical state with density and deviatoric hardening.

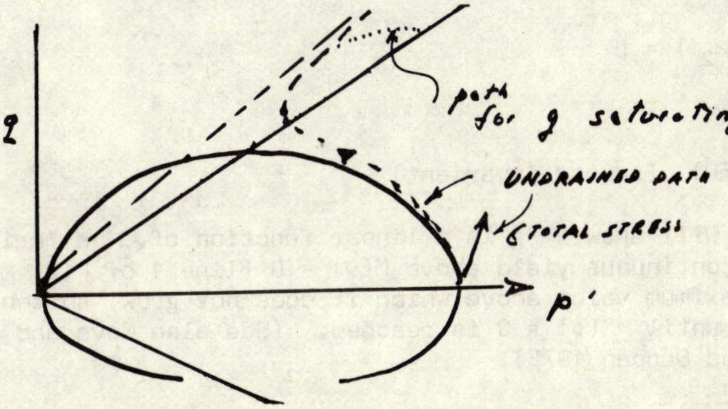

Fig. 12. Model D. Typical undrained stress paths.

$$Q(\underset{\sim}{\sigma}) = \bar{M}(\underset{\sim}{\sigma})$$

$\bar{M}(\underset{\sim}{\sigma})$ defined by expression similar to $M(\underset{\sim}{\sigma})$ but with different Φ.

Model C. Critical state models (density hardening). Associated behaviour (Fig.9).

$$F(\underset{\sim}{\sigma},\kappa) = 0$$

$$\kappa = \varepsilon^p_v$$

is a family of closed surfaces with tangent giving $d\varepsilon^p_v = 0$ on $M(\underset{\sim}{\sigma}) = 0$.

(Original concept due to Drucker , Henkel and Gibson 1957 elaborated by Roscoe, Schofield, Wroth, Burland etc.)

Model capable of giving strain hardening $K > 0$ neutral behaviour $K = 0$ (where $M(\underset{\sim}{\sigma}) = 0$) and strain softening. Simplest shape of F is ellipse passing through origin but two elipses,(Fig.10), with a tensile intercept give better(?) results. (viz. Dafalias, Herrman, 1980, Pietruszczak & Mroz 1980, etc.) Concept of overconsolidation included for first time in this model.

Model D. Critical state with density and deviatoric hardening (associated) (Fig.11). (P. Wilde two invariant dependent models for granular media. Arch. of Mechanics (Poland) 29, 799-809 1977).

$$F(\sigma,\kappa_1,\kappa_2) = 0$$

$$\kappa_1 = \varepsilon^p_v$$

$$\kappa_2 = g(\bar{\varepsilon}^p) \quad \text{(second invariant)}$$

This model will show if g is a linear function of κ_2 a "criticl state" or continuous yield above $M(\underset{\sim}{\sigma}) = 0$ (line 1 of Fig. 10). If g has a maximum value above which it does not grow, softening will occur until $M(\underset{\sim}{\sigma}) = 0$ is reached. (See also Nova and Wood 1979, Lade and Duncan 1975).

This model is apparently a good performer for granular materials - and is a refinement of the standard critical state, model to which it reduces with an elimination of one parameter. For undrained behaviour a completely different behaviour characteristic of sands is noted for this model, Fig. 11.

4. 'BOUNDING SURFACE' MODELS

The classical plasticity models described in previous section are incapable of representing simply the nonlinear phenomena observed on repeated load and unload cycles viz. Fig. 2. Each cycle in the elastic domain of a classical yield surface is completely reversible and thus no permanent strains (such as the well known densification of sands under repeated load) are recorded. Such difficulties of modelling led to development of alternative plasticity presentations. Mroz (1967) introduced a multisurface model, Dafalis and Popov (1975), Krieg (1975), formalized a simple interpolation theory for determination of moduli and Mroz, Norris, Zienkiewicz (1978, 79, 1980) developed a usable theory for soil mechanics problems. Many variants today exist for such models.

The basic idea of bounding surface plasticity is

(1) A bounding surface

$$B(\tilde{\underset{\sim}{\sigma}},\kappa) = 0$$

is defined in the stress space in a manner identical to that of a yield surface in classical plasticity.

On this surface the vectors $\tilde{\underset{\sim}{n}}_f$ and $\tilde{\underset{\sim}{n}}_g$ as well as the plastic moduli K are defined by classical plasticity rules.

(2) A rule is established for associating the stress point $\underset{\sim}{\sigma}$ (with its previous history) which is within the bounding surface and a point $\tilde{\underset{\sim}{\sigma}}$ on the bounding surface.

(3) A set of rules is established for obtaining the values of $\underset{\sim}{n}_f$, $\underset{\sim}{n}_q$ and K at point $\underset{\sim}{\sigma}$ from the corresponding values $\tilde{\underset{\sim}{n}}_f$, $\tilde{\underset{\sim}{n}}_q$ and $\tilde{K}$ at $\tilde{\underset{\sim}{\sigma}}$.

The rules of (2) and (3) can be of varying complexity and many models currently used are in this category. All allow plastic deformation to be continued within the bounding surface and in some no purely elastic domain occurs.

Some typical models of this class are now illustrated - all showing a cyclic densification under drained conditions (or cyclic pressure rise in undrained behaviour).

In all the models described here the critical state ellipse is used on the bounding surface with the corresponding $\tilde{\underset{\sim}{n}}_f = \tilde{\underset{\sim}{n}}_q$ and $\tilde{K}$.

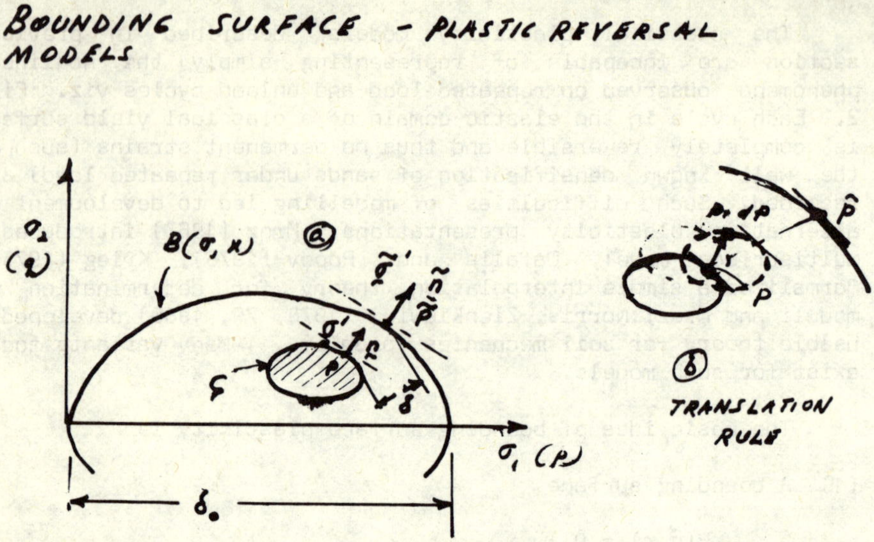

Fig. 13. Model I. Two surface M-N-Z model.

Fig. 14. Cyclic undrained behaviour of model I.

Model I. Two surface anisotropic hardening model. (Mroz, Norris, Zienkiewicz, 1978/79).

(1) A moveable ellipse C, Fig. 13, exists with $\dot{\sigma}$ always on or inside this. This ellipse is proportional in size and shape to B.

(2) All strains inside this ellipse are elastic.

(3) For a point on ellipse (P) with σ the associated point on B i.e. $\tilde{P}$ (or $\tilde{\sigma}$) is one with same tangents $n_f = \tilde{n}_f$ and $\tilde{K}$ is there determined by critical state conditions.

(4) The value of K at σ (P) is obtained by an interpolation rule depending on δ, the distance $P\tilde{P}$, (so that as $\delta \to 0$ $K \to \tilde{K}$). A usable rule is:

$$K = \tilde{K} + (\hat{K} - \tilde{K})(\delta/\delta_o)^{\gamma}$$

where $\hat{K}$ is a large positive value, δ_o is some large distance such as diameter of ellipse, B.

An alternative rule used by Dafalias & Herrmann (1980) is

$$K = \tilde{K} + \hat{K}\,(\delta/\delta_o - \delta)^{m}$$

with $\hat{K}$ increasing toward the p axis. This will show, as indicated in experiments bigger plastic flow in L.H.S. of bounding ellipse.

(5) The plastic movement at the stress point determines the moements of the ellipse - its centre moving along PP instantaneously and with the new stress point P + dP being on it, viz. Fig. 13b.

This model will give plastic strains during cycling and a limited densification (pressure rise). Bounding surface dependent only on $\kappa = \varepsilon_v$ expands with cycling.

Typical stress path for undrained cyclic load is shown in Fig. 14.

Note that this model can also be interpreted as a model of kinematic plasticity with C being now a yield surface. However answers will be the same -- and indeed above is a simpler way of postulating kinematic hardening behaviour.

Model II. The bounding surface model with degradation.

As model I with degradation now introduced. In this model the bounding surface is $B_N = B_N(\sigma, \kappa, h)$ where h is some measure of accumulated deviatoric plastic strain

$$h = \int (de^p_{ij} \cdot de^p_{ij})^{\frac{1}{2}}$$

Now cyclic load is capable of producing liquefaction providing the size of the bounding surface decreases.

An alternative degradation mode may be introduced to the value of K (or in another way, into the interpolation formula for K) to show softening behaviour after many cycles (viz. Mroz, Norris, Zienkiewicz (1981)).

Model III. Zero-elastic region model.

A simplification of model I in which the elastic domain C becomes zero. Now n_f depends only on direction of $d\sigma$, the stress increment. This model is simpler in formulation and has almost same performance as Model I with one less parameter.

Model IV. Dafalias and Herrmann 1980.

This model (Fig. 15) is very simple and defines a unique n_f (or $\tilde{\sigma}$) correspondence. Here the associted point $\tilde{P}$ is obtained by a simple 'radial' extension of line OP of Fig. 15. Similar interpolation as for model I used to obtain K. This model gives pressure build up and liquefaction as well as larger hysterysis loops if suitable K softening formula used. This model is incapable of simulating anisotropy. As in Model I critical surface expands on cycling. Unloading is strictly elastic.

Model V(a). Infinite surface model.

None of the previous models exhibit what may be termed a fading memory behaviour in which cycles of bigger amplitude eraze the memory of previous events while those of a smaller amplitude 'remember' previous larger amplitudes. This behaviour is often observed in real soils.

An infinite surface model (INS) was introduced by Mroz et al in 1980.

Its behaviour can be described in stages shown in Fig. 17. The rule is now that from every stress reversal point on a surface of same geometric shape as the bounding surface (homologous surface) is drawn tangent to the surface at which the

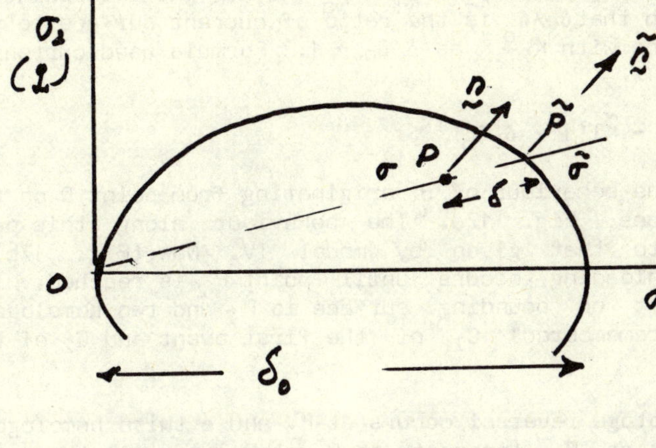

Fig. 15. Model IV (D-H) Radial rule model

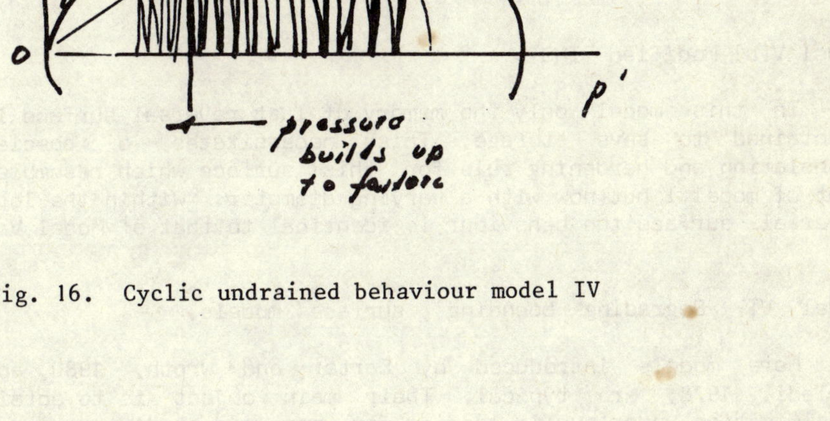

Fig. 16. Cyclic undrained behaviour model IV

reversal occurs. The stress point σ lies on this surface and the associated point $\tilde{\sigma}$ on the bounding surface is one with the same tangent. This determines $\tilde{n}_f = n_f = n_g$ and $\tilde{K}$. The interpolation formula is such that δ / δ_0 is the ratio of current surface to the bounding surface with $K \to \tilde{K}$ as $\delta/\delta_0 \to 1$. Formula used currently is

$$K = \tilde{K} + (\hat{K} - \tilde{K})(1 - \delta/\delta_0)^{\gamma}$$

(1) Consider the behaviour of σ originating from point 0 on the bounding surfaces, Fig. 17a. The behaviour along this path is identical to that given by model IV. Now (Fig. 17b) a reversal or unloading occurs until point P_2 is reached. Its associated point on bounding surface is $\tilde{P}_2$ and two homologous surfaces are "remembered" C_1 of the first event and C_2 of the second.

(3) At this stage reversal occurs at P_2 and a third homologous surface develops at P_2 (tangent to C_2 at P_2) and all three are "remembered" (Fig. 17c).

(4) If loading continues to a point P_3 on C_3 the memory of this is obliterated and if it reaches P_3 beyond the surface C_2 this is forgotten and the current surface (C_3) reverts to form (but not size) of (1) as shown in Fig. 17a.

The rule is simple in principle but in some loading histories may require a large number of the homologous reversal surfaces to be remembered.

This model gives a versatile performance but without degradation does not allow liquefaction to occur in undrained conditions (performance similar to Model I). Degradation as in Model II improves performance.

Model V(b) Modified INS

In this model only the memory of last reversal surface is maintained to save storage. This necessitates a special translation and hardening rule for this surface which resembles that of model I but now with a varying diameter. Within the last reversal surface the behaviour is identical to that of Model Va.

Model VI. Degrading bounding surface models.

Here models introduced by Carter and Wroth, 1980, and Balladi, 1978, are typical. Their main object is to obtain densification under cyclic loading for undrained conditions and a

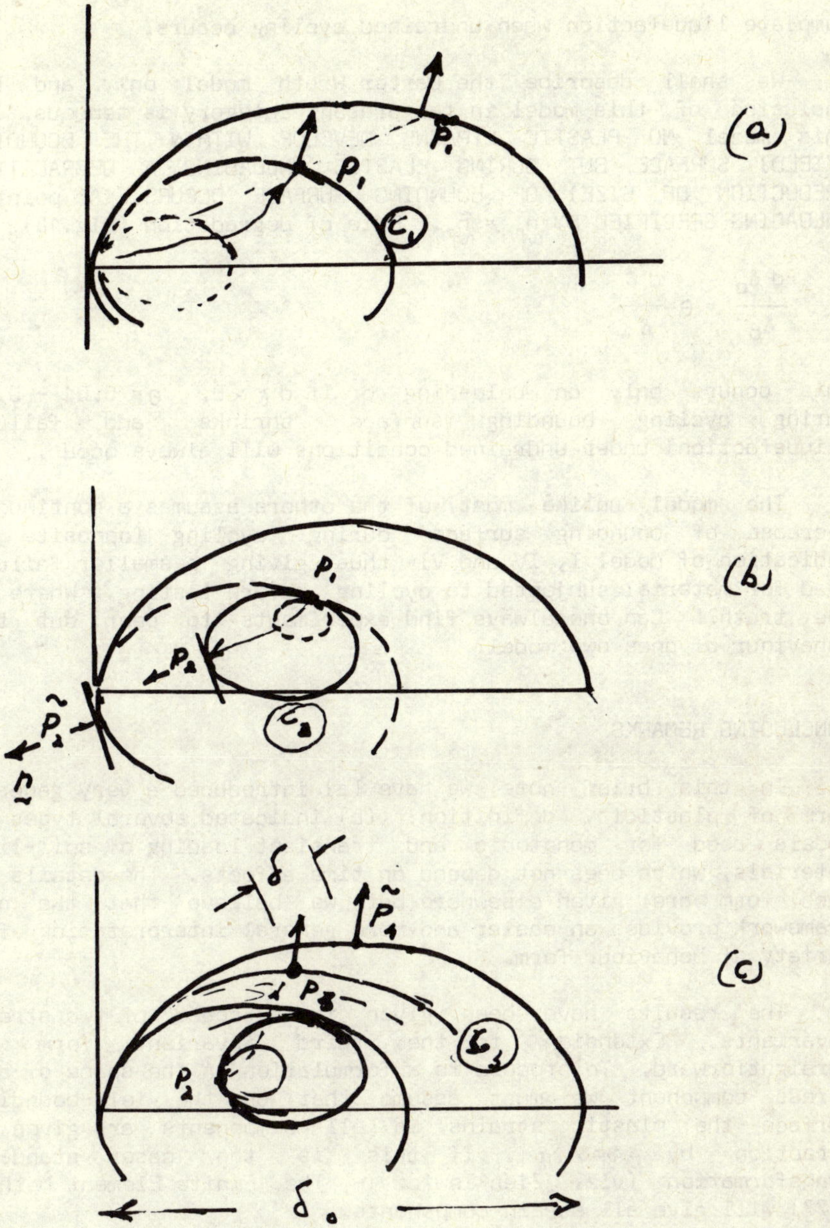

Fig. 17. Model V (a) Infinite surface (INS) M-N-Z
Model and hardening rule (Correspondence rule)

complete liquefaction when undrained cycling occurs.

We shall describe the Carter Wroth model only and the inclusion of this model in the present category is tenuous. In this model NO PLASTIC STRAINS DEVELOP WITHIN THE BOUNDING (YIELD) SURFACE BUT DURING ELASTIC UNLOADING A DEGRADATION (REDUCTION OF SIZE) OF BOUNDING SURFACE OCCURS. At point P UNLOADING SPECIFIED BY $\underset{\sim}{n}_f = \tilde{\underset{\sim}{n}}_f$. Rule of degradation (Fig.18)

$$\frac{d\,\delta_o}{\delta_o} = \Theta\,\frac{d\,\delta}{\delta}$$

this occurs only on unloading or if $d\,\delta < 0$. $\theta = 0.01 - 0.1$. During cycling bounding surface shrinks and failure (liquefaction) under undrained conditions will always occur.

The model unlike most of the others assumes a continuous decrease of bounding surface during cycling (opposite to indication of Model I, IV and V) thus giving a smaller failure load for material subjected to cycling before testing. Where is the truth?! Can one always find experiments to bear out the behaviour of ones own model?

CONCLUDING REMARKS

In this brief note we have (a) introduced a very general form of plasticity definition; (b) indicated several types of models used for monotonic and transient loading of soil-like materials, which does not depend on time effects. The details of each form are given elsewhere but we believe that the new framework provides an easier and more general interpretation of a variety of behaviour form.

The results have been given in a space of two stress invariants. Extension to the third invariant form is straightforward. To proceed to a formulation in the space of all stress component we must assume that on the yield-bounding surface the plastic strains in all components are given in direction by $\partial F / \partial \underset{\sim}{\sigma}$. If this is the case standard transformation (viz. Zienkiewicz - The Finite Element Method 1977) will give all strain components.

Generalization of present models to the viscoplatic domain to include time effects is discussed by Mroz et al. 1981.

REFERENCES

Bazant, Z; Endochronic in elasticity and incremental plasticity, Int. J. Sol. Struct., 14, pp.691-714, 1978

Carter, J. P., Booker, J. R. and Worth, C. P.; A critical state soil for cyclic loading. in Soils under Cyclic and Transient Loading, ed. Panade, G.N. and Zienkiewicz, O.C., John Wiley, to be published

Dafalias, Y. F. and Popov, E. P.; A model of non-linearly hardening materials for complex loadings, Acta Mechanica, 21, pp. 173-192, 1975

Dafalias, Y. F; On cyclic and anisotropic plasticity: i) A general model including material behaviour under stress reversals, ii) anisotropic hardening for initially orthotropic materials, Ph. D. Thesis, University of Califoria, Berkeley, 1975

Dafalias, Y. F. and Herrmann, L. R.; A bounding surface soil plasticity model. Int. Symp. on Soils under Cyclic and Transient Loading, Swansea, U.K. 1980.

Drucker, D. C., Gibson, R. E. and Henkel, D. J.; Soil mechanics and work hardening theories of plasticity, Trans. A. S. C. E., 122 : 388 : 346, 1957

Krieg, R. D.; A practical two surface plasticity theory, J. Appl. Mech., E42, pp.641-646, 1975

Lade, P. V. and Duncan, J. M.; Elastoplastic stress-strain theory for cohesionless soil, J. Geotech. Eng. Div., Proc. A.S.C.E., 101, GT10, pp.1037-1053, 1975

Mroz, Z.; On the description of anisotropic work- hardenking, J. Mech. Phys. Solids, 15, pp.163-175, 1967

Mroz, Z., Norris, V. A. and Zienkiewicz, O. C.; An anisotropic hardening model for soils and its application to cyclic loading, Int. J. Num. Anal. Meth. Geom., 2, pp.203-221, 1978

Mroz, Z., Norris, V. A. and Zienkiewicz, O. C.; Application of an anisotropic hardening model in the analysis of elasto-plastic deformation of soils, Geotechnique, 29, pp. 1-34, 1979

Mroz, Z., Norris, V. A. and Zienkiewicz, O. C.; Simulation of behaviour of soils under cyclic loading by using a more general hardening rule, Research report, Departmant of Civil Engineering, University of Wales, Swansea, 1980

Mroz, Z., Norris, V. A. and Zienkiewicz, O. C.; An anisotropic critical state model for soils subjected to cyclic loading , to be published in Geotechnique.

Nova, R. and Wood, D. M.; A constitutive model for sand in triaxial compression, Int. J. Num. Anal. Meth. Geom., 3, pp. 255-278, 1979

Pietruszczak, St. and Mroz, Z.; Description of mechanical behaviour of anisotropically consolidated clays, Proc. Euromech . Coll., 'Anisotropy in Mechanics', Grenbole, Noordhoff Int. publ, 1979

Roscoe, K. H. and Burland, J. B.; On the generalised stress-strain behaviour of 'wet' clay, in the 'Engineering Plasticity , ed. Heyman,J. and Leckie, F.A., Cambridge Univ. Press, pp. 535-609, 1968

Schofield, A. and Worth, P.; Critical State Soil Mechanics, McGraw Hill, London, 1968

Wilde, P.; Two invariant depending models of granular media, Arch. Mech. steo., 29, pp.799-809, 1977

Zienkiewicz, O. C., Valliappan, S. and King, I. P.; Elastoplastic solutions of engineering problems, initial tress, finite element approach, Int. J. Num. Meth. Eng., pp. 75-100, 1969

Zienkiewicz, O. C.; The Finite Element Method, third edition,1977

CONSTITUTIVE EQUATIONS FOR SOIL MEDIA

Jean H. Prevost

Princeton University, Princeton, New Jersey, U.S.A.

ABSTRACT

A general analytical model which describes the nonlinear, anisotropic, elastoplastic, stress-strain-strength properties of the soil skeleton when subjected to complicated three-dimensional loading paths is presented. A brief summary of the model's basic principle is included and the constitutive equations are provided. It is shown that the model parameters required to characterize the behavior of any given soil can be derived entirely from the results of conventional soil tests. The procedure can be easily programmed. The model's accuracy is evaluated by applying it to represent the behavior of a cohesive soil under undrained loading conditions for nonproportional stress paths. Implementation of the proposed formulation for solution of boundary value problems is discussed.

INTRODUCTION

Soil consists of an assemblage of particles with different sizes and shapes which form a skeleton whose voids are filled with various fluids. The stresses carried by the soil skeleton are conventionally termed "effective stresses" [23] in the soil mechanics literature, and those in the fluids are called "pore-fluid pressures". It is observed experimentally that the stress-strain behavior of the soil skeleton is strongly nonlinear, anisotropic and hysteretic. In order to relate the changes in effective stresses carried by the soil skeleton to the skeleton rate of deformation, a general analytical model which describes the nonlinear, anisotropic, elastoplastic, stress and strain dependent,

J. B. Martins (ed.), Numerical Methods in Geomechanics, 79–101.

stress-strain-strength properties of the soil skeleton when subjected to complicated three-dimensional, and in particular to cyclic loading paths [6] is presented. A brief summary of the model's basic principles [8] is included and the constitutive equations are provided. It is shown that the model parameters required to characterize the behavior of any given soil can be derived entirely from the results of conventional soil tests. The procedure can be easily programmed and the listing of a simple computer code is reported in [20]. The model's accuracy is evaluated by applying it to represent the behavior of a cohesive soil under undrained loading conditions for nonproportional stress paths. Implementation of the proposed formulation for solution of boundary value problems is discussed.

CONSTITUTIVE EQUATIONS

The constitutive equations for the solid skeleton are written in one of the following forms:

$$\underset{\sim}{C} : \underset{\sim}{d} = \begin{cases} \dot{\underset{\sim}{\sigma}}' & \text{small deformations} \\ \overset{\nabla}{\underset{\sim}{\sigma}}' + \underset{\sim}{\sigma}' \operatorname{div} \underset{\sim}{v} & \text{finite deformations} \end{cases} \tag{1}$$

in which $\underset{\sim}{\sigma}'$ = effective Cauchy stress tensor; $\underset{\sim}{v}$ = (spatial) velocity of solid phase; $\underset{\sim}{d}$ = rate of deformation tensor for the solid phase (= symmetric part of the spatial solid velocity gradient); a dot denotes the material derivative; and $\overset{\nabla}{\underset{\sim}{\sigma}}'$ = Jaumann derivative [3], viz.

$$\overset{\nabla}{\underset{\sim}{\sigma}}' = \dot{\underset{\sim}{\sigma}}' + \underset{\sim}{\sigma}' \cdot \underset{\sim}{w} - \underset{\sim}{w} \cdot \underset{\sim}{\sigma}' \tag{2}$$

where $\underset{\sim}{w}$ = spin tensor for the solid phase (= skew-symmetric part of the spatial solid velocity gradient). In Eq. 1, C_{abcd} is an (objective) tensor valued function of, possibly, $\underset{\sim}{\sigma}'$ and the deformation gradients. Many nonlinear material models of interest can be put in the above form (e.g., all nonlinear elastic materials, and many elasto-plastic materials). The finite deformation form of the constitutive equation above was first proposed by Hill [1] within the context of plasticity theory.

For soil media, the form of the $\underset{\sim}{C}$ tensor is given as follows [8],

$$\underset{\sim}{C} = \underset{\sim}{E} - \frac{1}{|\underset{\sim}{Q}|^2 H' + \underset{\sim}{Q}:\underset{\sim}{E}:\underset{\sim}{P}} (\underset{\sim}{E}:\underset{\sim}{P})(\underset{\sim}{Q}:\underset{\sim}{E}) \quad (3)$$

in which H' is the plastic modulus; $\underset{\sim}{P}$ and $\underset{\sim}{Q}$ are symmetric second-order tensors, such that $\underset{\sim}{P}$ gives the direction of plastic deformations and $\underset{\sim}{Q}$ the outer normal to the active yield surface; and $\underset{\sim}{E}$ is the fourth-order tensor of elastic moduli, assumed isotropic, viz.

$$E_{abcd} = \Lambda\delta_{ab}\delta_{cd} + G(\delta_{ac}\delta_{bd} + \delta_{ad}\delta_{bc}) \quad (4)$$

where Λ, G = Lame's constant; δ_{ab} = Kzonecker delta. It is convenient to decompose $\underset{\sim}{P}$ and $\underset{\sim}{Q}$ into their deviatoric and volumetric components, and in the following

$$\underset{\sim}{P} = \underset{\sim}{P}' + P''\underset{\sim}{1} \qquad \underset{\sim}{Q} = \underset{\sim}{Q}' + Q''\underset{\sim}{1} \quad (5)$$

where

$$3P'' = \text{trace } \underset{\sim}{P} \qquad 3Q'' = \text{trace } \underset{\sim}{Q} \quad (6)$$

The yield function is selected of the following form:

$$f = \frac{3}{2}(\underset{\sim}{s}-\underset{\sim}{\alpha}):(\underset{\sim}{s}-\underset{\sim}{\alpha}) + C^2(p'-\beta)^2 - k^2 = 0 \quad (7)$$

where $\underset{\sim}{s} = \underset{\sim}{\sigma}' - p'\underset{\sim}{1}$ = deviatoric stress tensor; $p' = \frac{1}{3}$ trace $\underset{\sim}{\sigma}'$ = mean stress; $\underset{\sim}{\alpha}$ and β are the coordinates of the center of the yield surface in the deviatoric stress subspace and along the hydrostatic stress axis, respectively; k = size of the yield surface; and C = material parameter called the yield surface axis ratio. When $C = 0$, Eq. 7 reduces to the one used in [6,7]. From Eq. 7,

$$\underset{\sim}{Q} = 3(\underset{\sim}{s}-\underset{\sim}{\alpha}) + \frac{2}{3}C^2(p'-\beta)\underset{\sim}{1} \quad (8)$$

where

$$\underset{\sim}{Q}' = 3(\underset{\sim}{s}-\underset{\sim}{\alpha}) \qquad 3Q'' = 2C^2(p'-\beta) \quad (9)$$

and

$$|\underset{\sim}{Q}|^2 = 6k^2 - 6C^2(1 - \frac{2}{9}C^2)(p'-\beta)^2 \quad (10)$$

The plastic potential is selected such that, in agreement with experimental observations, the plastic rate of deformation vector remains normal to the projection of the yield surface onto the deviatoric stress subspace, i.e.,

$$\underset{\sim}{P}' = \underset{\sim}{Q}' \tag{11a}$$

and

$$\text{trace } \underset{\sim}{P} = \text{trace } \underset{\sim}{Q} + \bar{A} \tag{11b}$$

where $\bar{A}$ is a material function given by

$$\bar{A} = \begin{cases} 81\sqrt{6}\ \dfrac{J_3'}{\{\underset{\sim}{Q}':\underset{\sim}{Q}'\}}\ A & \text{if } \{\underset{\sim}{Q}':\underset{\sim}{Q}'\} \neq 0 \\ 0 & \text{if } \{\underset{\sim}{Q}':\underset{\sim}{Q}'\} = 0 \end{cases} \tag{12}$$

where

$$J_3' = \frac{1}{3}(s_{ab} - \alpha_{ab})(s_{bc} - \alpha_{bc})(s_{ca} - \alpha_{ca}) \tag{13}$$

and A is a material parameter which measures the departure from an associative flow rule. When $A = 0$, $\underset{\sim}{P} = \underset{\sim}{Q}$ and consequently the $\underset{\sim}{C}$ tensor possesses the major symmetry and leads to a symmetric tangent stiffness. On the other hand, when $A \neq 0$, $\underset{\sim}{P} \neq \underset{\sim}{Q}$ and $\underset{\sim}{C}$ does not possess the major symmetry. Eq. 12 allows for a continuously varying function for trace $\underset{\sim}{P}$ along the yield surface.

In order to allow for the adjustment of the plastic hardening rule to any kind of experimental data, for example data obtained from axial or simple shear soil tests, a collection of nested yield surfaces is used. Much pros and cons have been said about multi-surface versus two-surface plasticity theories and a critical assessment of their relative merits and shortcomings is reported in [16]. A plastic modulus is associated with each of the yield surfaces, and

$$H' = h' + \frac{\text{trace } \underset{\sim}{Q}}{\{3\underset{\sim}{Q}:\underset{\sim}{Q}\}^{1/2}}\ B' \tag{14}$$

where h' = plastic shear modulus; $(h' + B')$ and $(h' - B')$ = plastic bulk moduli to be mobilized in consolidation tests upon loading and unloading, respectively. The projections of the yield surfaces onto the deviatoric stress subspace thus define regions of constant plastic shear moduli [6,7].

The yield surfaces' initial positions and sizes reflect the past stress-strain history of the soil skeleton, and in particular their initial positions are a direct expression of the material's "memory" of its past loading history. Because the $\underset{\sim}{\alpha}$'s are not necessarily all equal to zero, the yielding of the material is anisotropic. Direction is therefore of great importance and the physical reference axes (x,y,z) are fixed with respect to the material element and specified to coincide with the reference axes of consolidation. For a soil element whose anisotropy initially exhibits rotational symmetry about the y-axis, $\alpha_x = \alpha_z = -\alpha_y/2$ and Eq. 7 simplifies to

$$[(\sigma'_y - \sigma'_x) - \alpha]^2 + c^2(p' - \beta)^2 - k^2 = 0 \qquad (15)$$

in which $\alpha = 3\alpha_y/2$. The yield surfaces then plot as ellipses in the axisymmetric stress plane $(\sigma'_x = \sigma'_z)$ as shown in Fig. 1. Points C and E on the outermost yield surface define the critical state conditions (i.e., H' = 0) for axial compression and extension loading conditions, respectively [2]. It is assumed that the slopes of the critical state lines OC and OE remain constant during yielding.

The yield surfaces are allowed to change in size as well as to be translated by the stress point. Their associated plastic moduli are also allowed to vary and in general both k and H' are functions of the plastic strain history. They are conveniently taken as functions of invariant measures of the amount of plastic volumetric strains and/or plastic shear distortions, respectively [6,8].

Complete specification of the model parameters requires the determination of

(i) the initial positions and sizes of the yield surfaces together with their associated plastic moduli;

(ii) their size and/or plastic modulus changes as loading proceeds, and finally,

(iii) the elastic shear G and bulk B $(=\Lambda + 2G/3)$ moduli.

The soil's anisotropy originally develops during its deposition and subsequent consolidation which, in most practical cases, occurs under no lateral deformations. In the following, the y-axis is vertical and coincides with the direction of consolidation, the horizontal xz-plane is thus a plane of material's isotropy and the material's anisotropy initially exhibits rotational symmetry about the vertical y-axis. The model parameters required

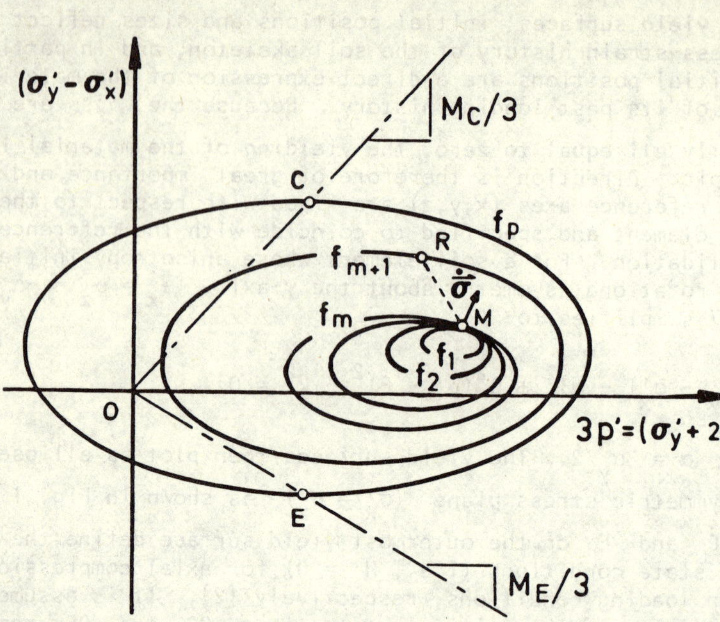

Figure 1.

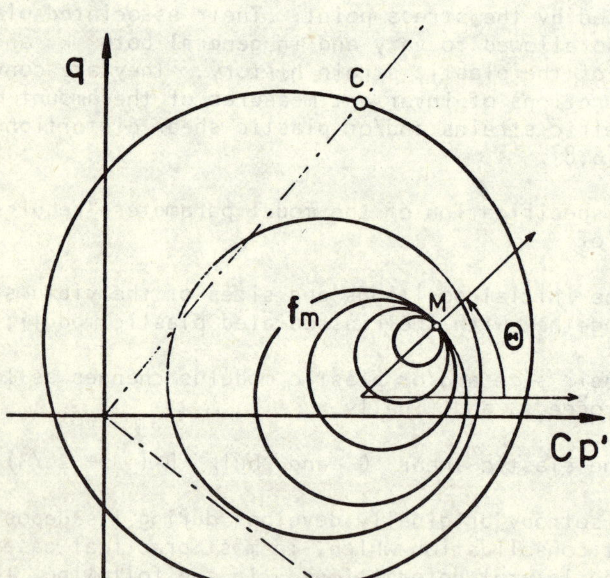

Figure 2.

to characterize the behavior of any given soil can then be derived entirely from the results of conventional monotonic axial and cyclic strain-controlled simple shear soil tests [6]. This is explained and further discussed in the following.

DETERMINATION OF MODEL PARAMETERS

In order to follow common usage in soil mechanics, compressive stresses and strains are considered positive in the following. All stresses are effective stresses unless otherwise specified.

As explained previously, for a material which initially exhibits cross-anisotropy about the vertical y-axis, the initial position in stress space of the yield surfaces is defined by the sole determination of the two parameters $\alpha^{(m)}$ and $\beta^{(m)}$ $(m = 1,\ldots,p)$ and Eq. 7 simplified to

$$[q - \alpha^{(m)}]^2 + C^2[p' - \beta^{(m)}]^2 - [k^{(m)}]^2 = 0 \tag{16}$$

for axial loading conditions (i.e., $\sigma'_x = \sigma'_z$ and $\tau_{xy} = \tau_{yz} = \tau_{zx} = 0$), where $q = (\sigma'_y - \sigma'_x)$. The yield surfaces then plot as circles in the q versus Cp' plane (referred to as the axial stress plane hereafter) as shown in Fig. 2. When the stress point reaches the yield surface f_m,

$$q = \alpha^{(m)} + k^{(m)} \sin\theta \tag{17a}$$

$$p' = \beta^{(m)} + \frac{k^{(m)}}{C} \cos\theta \tag{17b}$$

where θ is defined in Fig. 2, and Eq. 3 simplifies to:

$$\frac{\dot{\bar{\varepsilon}}}{\dot{q}} = \frac{1}{2G} + \frac{1}{H'_m} \frac{\sin\theta(\sin\theta + C\gamma\cos\theta)}{\sin^2\theta + \frac{2}{9}C^2\cos^2\theta} \tag{18a}$$

$$\frac{\dot{\varepsilon}_v}{\dot{p}'} = \frac{1}{B} + \frac{1}{H'_m} \frac{(2C\cos\theta + A_m\sqrt{6}\sin\theta)\frac{1}{3\gamma}(\sin\theta + C\gamma\cos\theta)}{\sin^2\theta + \frac{2}{9}C^2\cos^2\theta} \tag{18b}$$

in which

$$\dot{\varepsilon}_v = \text{trace}\ \underset{\sim}{d} = \dot{\varepsilon}_y + 2\dot{\varepsilon}_x \tag{19}$$

$$\dot{\bar{\varepsilon}} = (\dot{\varepsilon}_y - \dot{\varepsilon}_x) \tag{20}$$

$$\gamma = \frac{\dot{p}'}{\dot{q}} \tag{21}$$

G = elastic shear modulus, B = elastic bulk modulus. The dependence of the model parameters upon the effective mean normal stress and volumetric strain are assumed to be of the following forms:

$$x = x_1 \left(\frac{p'}{p_1'}\right)^n \tag{22}$$

and

$$y = y_1 \exp(\lambda \varepsilon_v) \tag{23}$$

respectively, where $x = B, G, h_m', B_m'$ and $y = \alpha^{(m)}, \beta^{(m)}, k^{(m)}$; n is an experimental parameter ($n = 0.5$ for most cohesionless soils [21], and $n = 1$ for most cohesive soils; p_1' = initial effective mean normal stress (i.e., at $\varepsilon_v = 0$); $\varepsilon_v = (v_1 - v)/v_1$ where v = current volume and v_1 = initial volume (i.e., at $p' = p_1'$ and $q = q_1$) of the soil specimen.

It is assumed that when the soil is in a normally consolidated state, the consolidation soil test results plot (1) as a straight line parallel to the projections of critical state lines on the ε_v versus ln p' plane [22], and (2) as a straight line in the q versus p' plane. The parameter λ is then simply determined from the results of K_0-consolidation soil test results, viz (from Eqs. 17 and 23).

$$\lambda = \frac{1}{p_K'} \frac{\dot{p}_K'}{\dot{\varepsilon}_v^K} \tag{24}$$

where the subscript/superscript K refers to K_0-loading conditions.

(i) Interpretation of Monotonic Drained Axial Compression and Extension Soil Test Results

Let θ_C and θ_E denote the values of θ (Eq. 17) when the stress point reaches the yield surface f_m in axial compression and extension loading conditions. Combining Eqs. 17-23, one finds that:

$$\frac{1}{\tan\theta_C} - \frac{1}{\tan\theta_E} = \frac{1}{2C}\left[3\gamma_C \frac{x_C}{y_C} - 3\gamma_E \frac{x_E}{y_E}\right] \tag{25}$$

$$\cos\theta_C - \cos\theta_E = R_{CE}(\sin\theta_C - \sin\theta_E) \tag{26}$$

in which

$$R_{CE} = \frac{p'_C - p'_E \exp\left[\lambda\left(\varepsilon_v^C - \varepsilon_v^E\right)\right]}{q_C - q_E \exp\left[\lambda\left(\varepsilon_v^C - \varepsilon_v^E\right)\right]} \tag{27}$$

$$\frac{1}{x_C} = \left(\frac{p'_C}{p'_1}\right)^n \frac{\dot{\bar{\varepsilon}}}{\dot{q}} - \frac{1}{2G_1} \tag{28}$$

$$\frac{1}{y_C} = \left(\frac{p'_C}{p'_1}\right)^n \frac{\dot{\varepsilon}_v}{\dot{p}} - \frac{1}{B_1} \tag{29}$$

and similarly for x_E and y_E, where the subscripts C and E refer to axial compression and extension loading conditions, respectively.

The smooth experimental stress-strain curves obtained in axial tests are now to be approximated by linear segments along which the tangent (or secant) modulus is constant. Evidently, the degree of accuracy achieved by such a representation of the experimental curves is directly dependent upon the number of linear segments used. The model parameters associated with the yield surface f_m are determined by the condition that the slopes $\dot{q}/\dot{\bar{\varepsilon}}$ are to be the same in axial compression and extension tests when the stress point has reached the yield surface f_m [6-8,11]. The corresponding values θ_C and θ_E are determined by combining Eqs. 25 and 26. For that purpose, note that Eq. 26 is more conveniently rewritten as

$$\left(\frac{1}{\tan\theta_C} + \frac{1}{\tan\theta_E}\right) + \frac{2R_{CE}}{1-R_{CE}^2}\left(\frac{1}{\tan\theta_C}\,\frac{1}{\tan\theta_E} - 1\right) = 0 \tag{30}$$

Once θ_C and θ_E have been determined, the model parameters associated with f_m are simply obtained by combining Eqs. 17 and 18, viz.

$$B'_m = \frac{1}{(DC-DE)}\left[\frac{x_C \sin\theta_C TC}{JC} - \frac{x_E \sin\theta_C TE}{JE}\right] \tag{31}$$

$$h'_m = \frac{x_C \sin\theta_C TC}{JC} - DCB'_m \tag{32}$$

$$A_m\sqrt{6} = 3\gamma_C \frac{x_C}{y_C} - \frac{2C}{\tan\theta_C} \tag{33}$$

where

$$JC = \sin^2\theta_C + \frac{2}{9}C^2\cos^2\theta_C$$

$$JE = \sin^2\theta_E + \frac{2}{9}C^2\cos^2\theta_E \tag{34}$$

$$TC = \sin\theta_C + C\gamma_C\cos\theta_C$$

$$TE = \sin\theta_E + C\gamma_E\cos\theta_E \tag{35}$$

$$DC = \frac{\sqrt{2}}{3}\,\frac{C\cos\theta_C}{\sqrt{JC}}$$

$$DE = \frac{\sqrt{2}}{3}\,\frac{C\cos\theta_E}{\sqrt{JE}} \tag{36}$$

and

$$k_1^{(m)} = \frac{q_C\exp(-\lambda\varepsilon_v^C) - q_E\exp(-\lambda\varepsilon_v^E)}{\sin\theta_C - \sin\theta_E} \tag{37}$$

$$\alpha_1^{(m)} = q_C\exp(-\lambda\varepsilon_v^C) - k_1^{(m)}\sin\theta_C \tag{38}$$

$$\beta_1^{(m)} = p'_C\exp(-\lambda\varepsilon_v^C) - \frac{k_1^{(m)}}{C}\cos\theta_C \tag{39}$$

The yield surface f_1 is chosen as a degenerate yield surface of size $k_1^{(1)} = 0$. Further, in order to get a smooth transition from the elastic into the plastic regime $H'_1 = \infty$, so that the material behavior inside f_2 is elastic. The elastic shear, G_1, and bulk, B_1, moduli are then determined from the steepest slopes observed at the origin of the plots q versus $\bar{\varepsilon}$, and

p' versus ε_v of the axial compression/extension soil test results, respectively.

At the critical state, $x_C = x_E = 0$, and the stress point is on the outermost yield surface f_p. From Eq. 32, it is apparent that $h'_p = 0$ in that case.

(ii) Interpretation of Monotonic Undrained Axial Compression and Extension Soil Test Results

In undrained tests, $\dot{\varepsilon}_v = 0$, and (from Eq. 29) $y_C = y_E = -B_1$ in that case. The model parameters associated with the yield surface f_m are again determined by the condition that the slopes $\dot{q}/\dot{\bar{\varepsilon}}$ are to be the same in axial compression and extension tests when the stress point has reached the yield surface f_m. As previously, the corresponding values θ_C and θ_E are determined from Eqs. 25 and 26, in which

$$R_{CE} = C \frac{p'_C - p'_E}{q_C - q_E} \quad . \tag{40}$$

Knowing θ_C and θ_E, the model parameters associated with f_m are computed using Eqs. 31-39, in which $\varepsilon_v^C = \varepsilon_v^E = 0$.

(iii) Interpretation of Simple Shear Soil Test Results

In simple shear soil tests, $\dot{\varepsilon}_x = \dot{\varepsilon}_y = \dot{\varepsilon}_z = 0$. The necessary algebra for the determination of the model parameters is considerably simplified in that case if the elastic contributions to the normal strains is neglected. Eqs. 31-39 then yield: $\sigma'_x = \sigma'_z$,

$$\frac{\dot{\gamma}_{xy}}{\dot{\tau}_{xy}} = \frac{1}{G} + \frac{2}{h'_m} \tag{41}$$

$$\alpha^{(m)} = (\sigma'_y - \sigma'_x) \tag{42}$$

$$k^{(m)} = \sqrt{3}\,\tau_{xy} \tag{43}$$

$$\beta^{(m)} = \sigma'_y - \frac{2}{3}(\sigma'_y - \sigma'_x) \tag{44}$$

The model parameters associated with the yield surface f_m are then simply determined from the above equations and a piecewise linear representation of the shear stress-strain curves obtained in a simple shear test. Note that the sole use of simple shear test results does not allow the determination of the parameters λ, B'_m and A_m. On the other hand, it is apparent from Eqs. 41 and 43 that the degradation of the mechanical properties of the material under cyclic shear loading conditions, i.e.

$$k^{(m)}(\bar{e}) \qquad h'_m(\bar{e}) \tag{45}$$

with

$$\bar{e} = \int \left\{ \frac{2}{3}\, d' : d' \right\} \qquad \underset{\sim}{d}' = \underset{\sim}{d} - \frac{1}{3}(\text{trace } \underset{\sim}{d})\underset{\sim}{\delta} \tag{46}$$

where the integration is carried along the strain path, are most conveniently determined from the results of cyclic strain-controlled simple shear tests

$(\bar{e} = \int \frac{1}{\sqrt{3}} |\dot{\gamma}|$ in that case). This is explained and further discussed in [6].

The procedures described above for the determination of the model parameters may be easily programmed, and a simple computer code has been written for that purpose. A listing of that code is reported in [20].

MODEL EVALUATION

The soil data to be used in this section are part of the ones which were collected by the organizing committee of the NSF/NSERC North American Workshop on plasticity and generalized stress-strain applications in soil engineering held May 28-30, 1980 at McGill University, Montreal, Canada. Laboratory axial test data on a laboratory-prepared kaolinite clay had then been transmitted to the author. Predictions about the constitutive behavior of the soil subjected to loading stress paths not identified in the data had been requested by the organizing committee. This section describes the test results, their analysis, and compares the model predictions with observed behavior in the tests.

Laboratory Prepared Kaolinite Clay

The experimental tests had been conducted on cylindrical samples in a torsional shear testing device. All samples had

TABLE 1

Kaolinite Clay σ'_{vc} = 40 psi Model Parameters

G_1 = 3121 psi; B_1 = 8323 psi; n = 1.0; λ = 15.6; C = $3/\sqrt{2}$

m	$\alpha^{(m)}$ (psi)	$\beta^{(m)}$ (psi)	$k^{(m)}$ (psi)	h'_m (psi)	B'_m (psi)	A_m
2	7.819	38.29	19.93	4476.0	-2942.0	-0.336
3	8.698	38.63	20.89	2110.0	-949.7	-0.308
4	10.660	38.83	22.89	2098.0	-1008.0	-0.320
5	12.620	38.95	24.87	1651.0	-885.7	-0.303
6	15.530	39.51	27.92	1131.0	-505.2	-0.333
7	17.480	39.86	29.97	995.3	-444.0	-0.353
8	18.540	39.93	33.19	558.1	-271.3	-0.434
9	14.570	37.77	40.38	216.5	-152.5	-0.595
10	9.350	38.26	49.10	0.0	0.0	-0.484

KAOLINITE CLAY - UNDRAINED AXIAL TESTS

AXIAL SHEAR PLOT

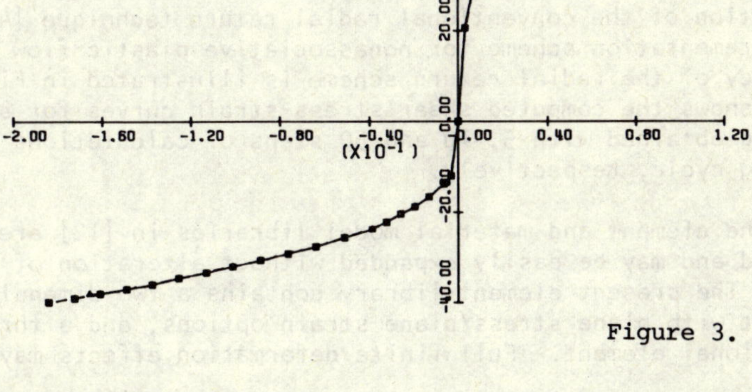

Figure 3.

first been K_0-consolidated with a cell pressure of 58 psi and a backpressure of 18 psi, and then left to rebound to an equal-all-around cell pressure of 58 psi, and a back pressure of 18 psi (in other words, the excess axial load necessary for K_0- consolidation was then released). All the tests were stress-controlled and performed under constant volume conditions (i.e., undrained).

Figures 3 and 4 show in dashed-lines the experimental results obtained in conventional monotonic axial compression/extension soil tests, and in solid-lines the design curves used to determine the model parameters for that clay. Note that some data points close to failure have been ignored when selecting the design curves because these do not seem consistent with the rest of the data. This inconsistency may be due to experimental difficulties in capturing failure states in stress-controlled testing devices.

Table 1 gives the model parameter values for the Kaolinite clay. Figures 5-12 show the model predictions for shear tests in which the major principal stress is inclined at $\theta = 15°$ and $45°$, respectively, to the vertical axis of the soil specimen. Figures 5-12 also show a comparison between predicted (solid lines) and observed (dashed-lines) behavior of the soil in these tests. Note that all the model predictions agree well with the experimental test results.

COMPUTATIONAL ASPECTS

The soil model described above has been coded and incorporated into a finite-element program [12] for analysis of boundary value problems of interest in soil mechanics. Use of elastic-plastic equations of the above type in analysis of boundary value problems requires that an efficient and accurate numerical method of integrating the equations be available. Substantial efforts have thus been devoted to finding a computational method with the best balance of accuracy and computational speed. The integration methods presently used in the stress routine are an adaptation of the conventional radial return technique [4] and a subincrementation scheme for nonassociative plastic flow rules. Accuracy of the radial return scheme is illustrated in Figure 13 which shows the computed shear stress-strain curves for shear loading obtained with 5, 10 and 50 steps of calculations per loading cycle, respectively.

The element and material model libraries in [12] are modularized and may be easily expanded without alteration of the main code. The present element library contains a two-dimensional element with plane stress/plane strain options, and a three-dimensional element. Full finite deformation effects may be

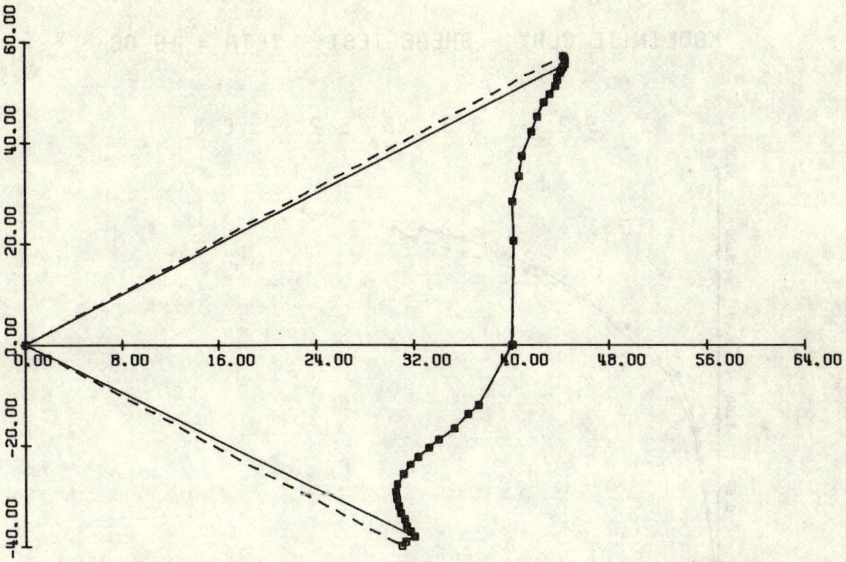

Figure 4.

KAOLINITE CLAY - SHEAR TEST - TETA = 15 DG
S 1 - S 2 VS E 1 - E 2

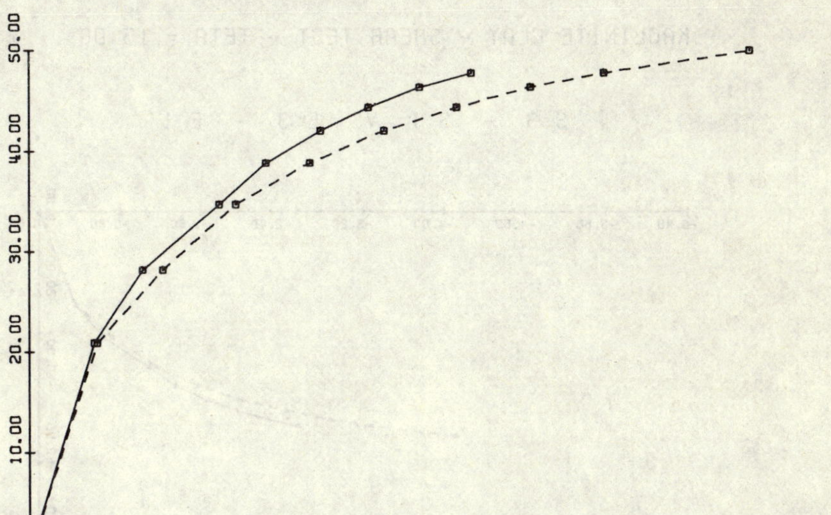

Figure 5.

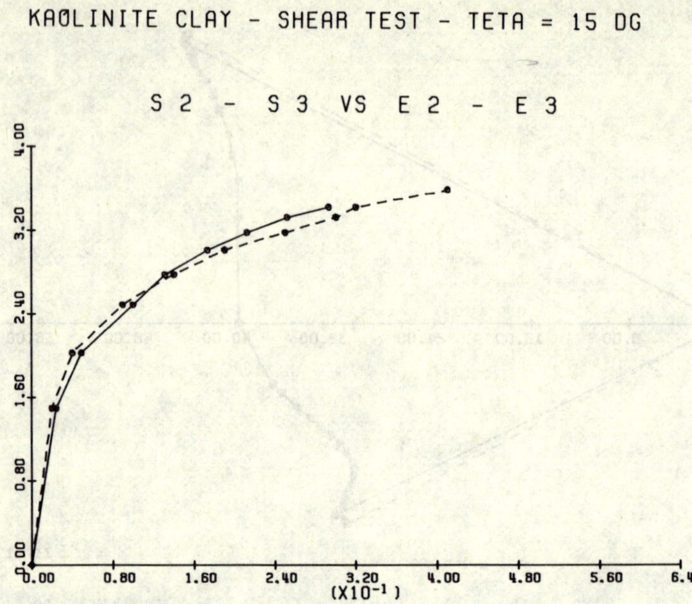

Figure 6.

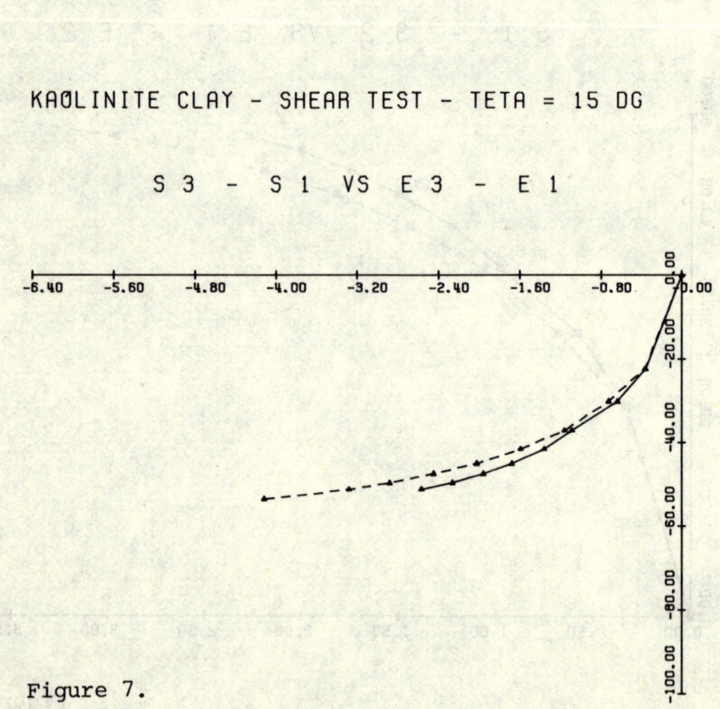

Figure 7.

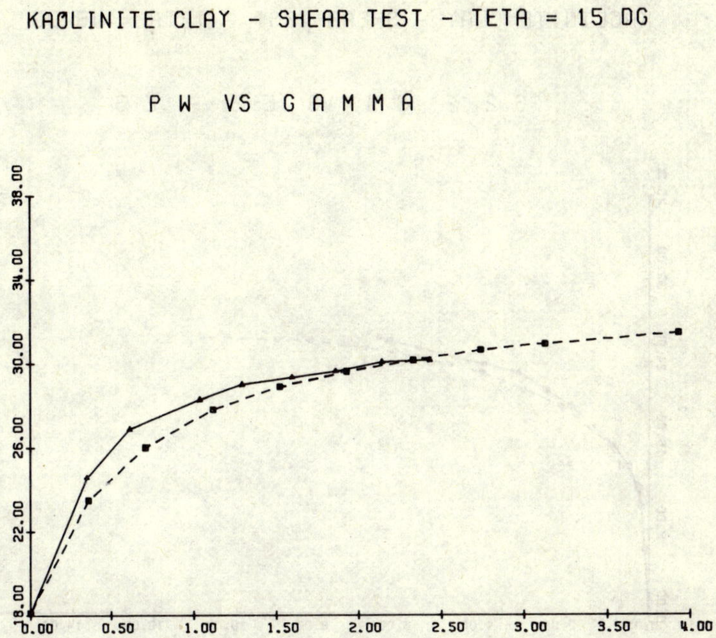

Figure 8.

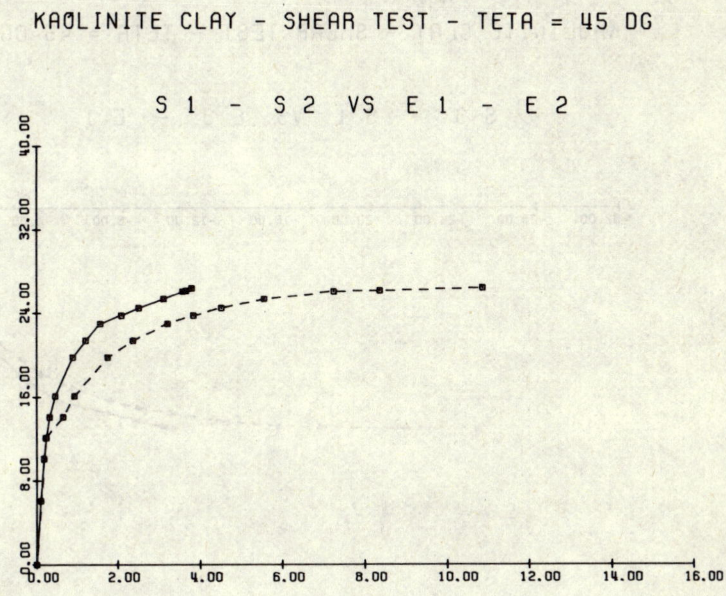

Figure 9.

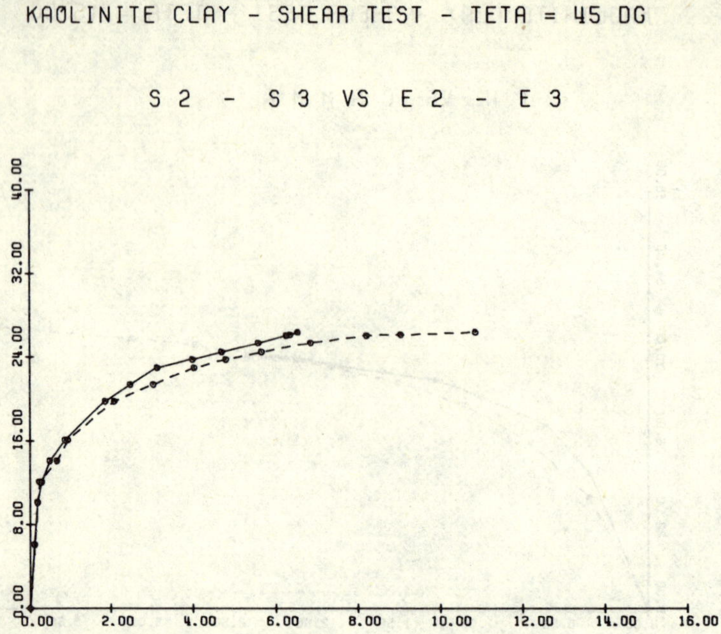

Figure 10.

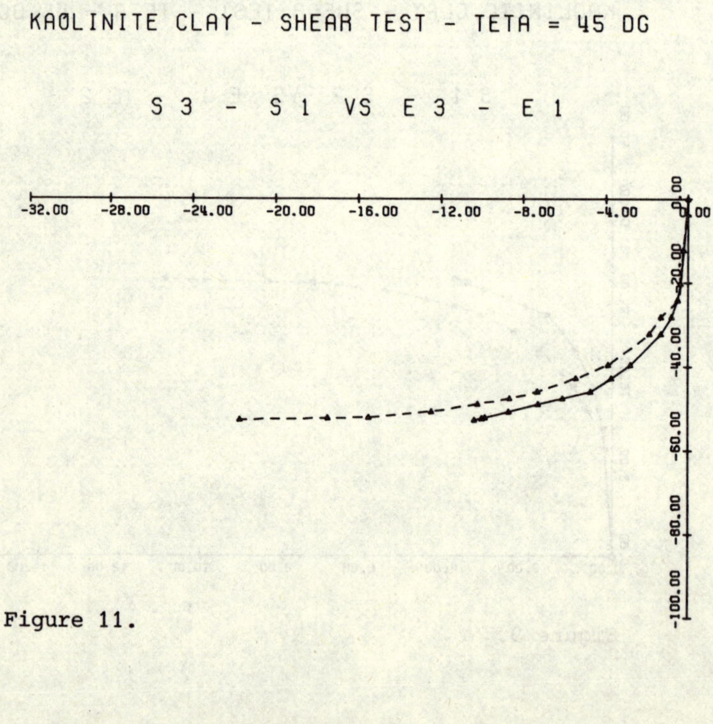

Figure 11.

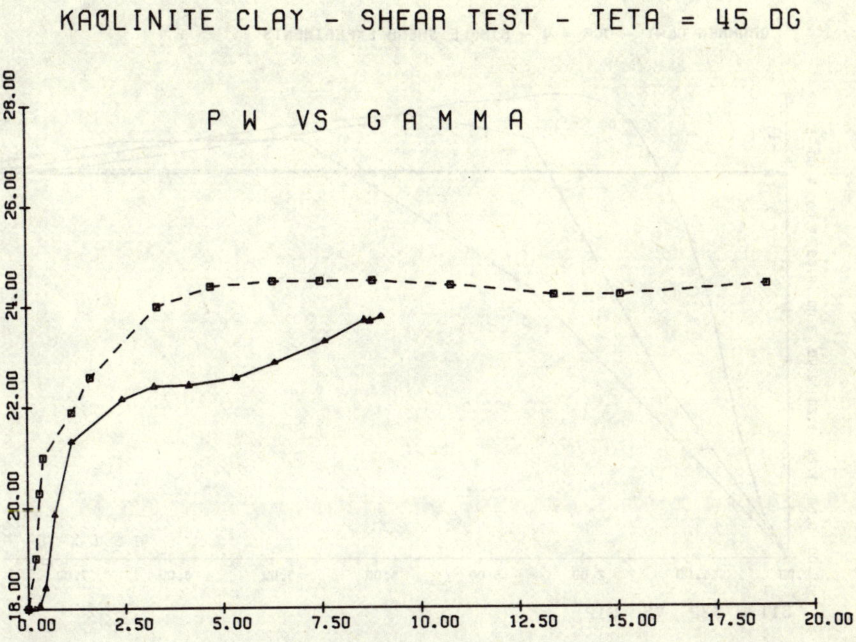

Figure 12.

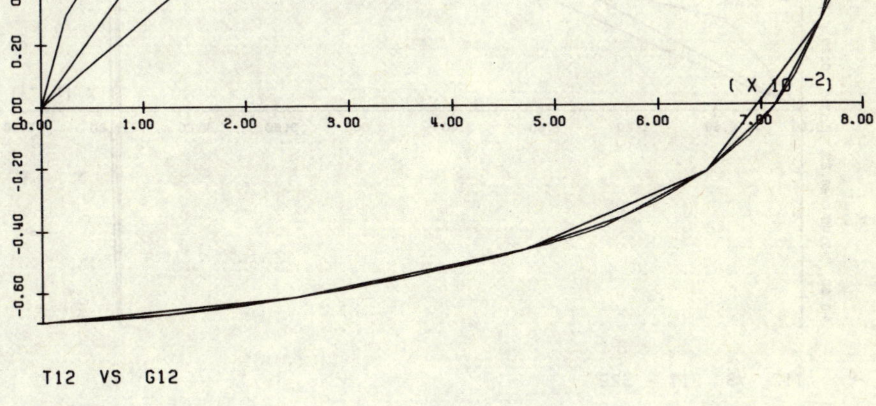

Figure 13.

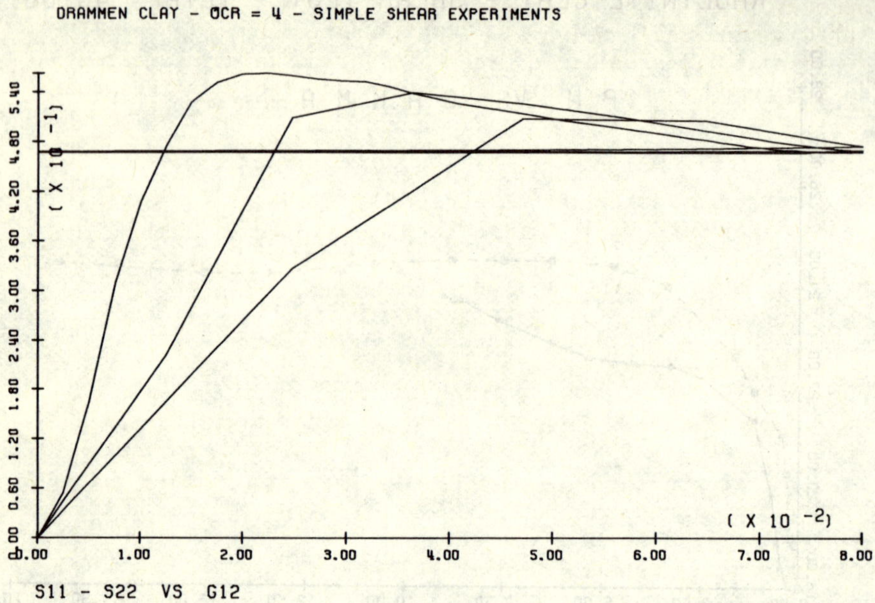

Figure 14.

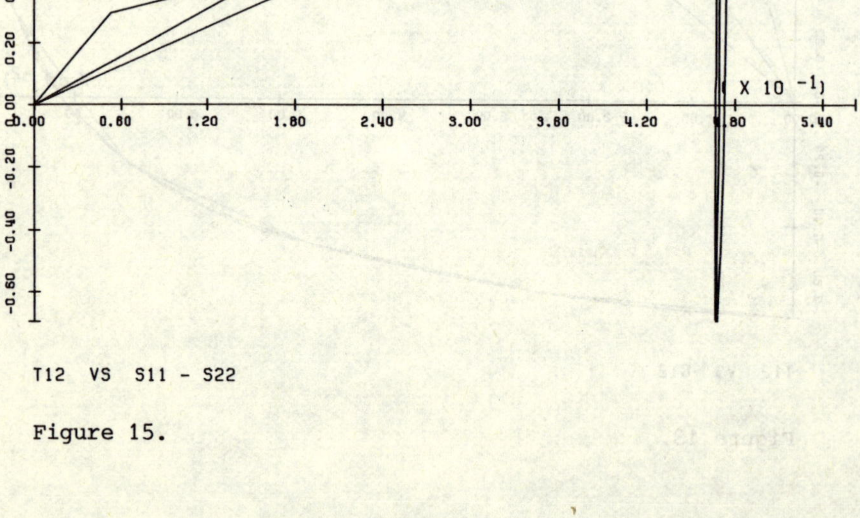

Figure 15.

accounted for. A contact element is also available for two and three-dimensional analysis. The present material model library contains a linear elastic model and various elasto-plastic and soil models. Some features which are available in the program are

. Both symmetric and nonsymmetric matrix equation solvers.

. Reduced/selective integration procedures for effective treatment of incompressibility constraints [5].

. Pesudo-static and dynamic analysis capabilities.

. Coupled field equations for analysis of transient phenomena in saturated soils.

Time integration in pseudo-static analysis is achieved by using a backward difference scheme together with an incremental predictor-corrector type algorithm as reported in [13,19]. Time integration in dynamic analysis is achieved by using an implicit-explicit predictor-multicorrector scheme developed by Hughes and co-workers as reported in [17,18].

Accuracy and versatility of the proposed procedure in applications of interest in soil mechanics are reported in [9,10,13, 14,15,17-19].

SUMMARY AND CONCLUSIONS

A general analytical model which describes the nonlinear, anisotropic, elastoplastic, stress-strain-strength properties of the soil skeleton when subjected to complicated three-dimensional loading paths is proposed. A brief summary of the model's basic principle is included and the constitutive equations are provided. It is shown that the model parameters required to characterize the behavior of any given soil can be derived entirely from the results of conventional soil tests. The procedure can be easily programmed and the listing of a simple computer code is reported in [20]. The model's accuracy is evaluated by applying it to represent the behavior of a cohesive soil under undrained loading conditions for non-proportional stress paths.

ACKNOWLEDGMENTS

Computer time was provided by Princeton University Computer Center.

REFERENCES

1. Hill, R., "A General Theory of Uniqueness and Stability in Elastic-Plastic Solids," J.Mech.Phys.Solids, Vol. 6, 1958, pp. 236-249.

2. Hvorslev, M. J., "Uber die Festigkeitseigenschaften gestorler bindiger boden," Ingeniozvidenskabelige Skrifter, A. no. 45, Copenhagen, 1937.

3. Jaumann, G., Gzundlagen der Bewegungolehze, Leipzig, 1903.

4. Krieg, R. D., and Krieg, D. B., "Accuracies of Numerical Solution Methods for the Elastic-Perfectly Plastic Model," J.of Pressure Vessel Technology, ASME, Vol. 99, No. 4, 1977 pp. 510-515.

5. Malkus, D. S., and Hughes, T. J. R., "Mixed Finite Element Methods - Reduced and Selective Integration Techniques: A Unification of Concepts," Computer Methods in Applied Mechanics and Engineering, Vol. 15, No. 1, 1978, pp. 63-81

6. Prevost, J. H., "Mathematical Modeling of Monotonic and Cyclic Undrained Clay Behavior," Int.J.Num.Analyt.Methods in Geomechanics, Vol. 1, No. 2, 1977, pp. 196-216.

7. Prevost, J. H., "Anisotropic Undrained Stress-Strain Behavior of Clays," J.Geotech.Eng.Div., ASCE, Vol. 104, No. GT8, 1978, pp. 1075-1090.

8. Prevost, J. H., "Plasticity Theory for Soil Stress-Strain Behavior," J.Eng.Mech.Div., ASCE, Vol. 104, No. EM5, 1978, pp. 1177-1194.

9. Prevost, J. H. and Hughes, T. J. R., "Analysis of Gravity Offshore Structure Foundations Subjected to Cyclic Wave Forces," Proceedings, 1978 Offshore Technology Conference, Houston, Texas, Vol. 1 (May 1978), pp. 1809-1818.

10. Prevost, J. H., and Hughes, T. J. R., "Finite Element Solution of Boundary Value Problems in Soil Mechanics," Proceedings, International Symposium on Soils under Cyclic and Transient Loading, Swansea, U.K., (January 1980), pp. 263-276.

11. Prevost, J. H., "Constitutive Theory for Soil," Proceedings, NSF/NSERC North American Workshop on Plasticity and Generalized Stress-Strain Applications in Soil Engineering, Montreal, Canada, May 1980.

12. Prevost, J. H., "DYNA-FLOW: A Nonlinear Transient Finite Element Analysis Program," Department of Civil Engineering, Report 81-SM-1, Princeton University (January 1981).

13. Prevost, J. H., "Consolidation of Anelastic Porous Media," Journal of the Engineering Mechanics Division, ASCE, Vol. 107, No. EM1, Proc. Paper 16042 (February 1981), pp. 169-186.

14. Prevost, J. H., Cuny, B., Hughes, T. J. R. and Scott, R. F., "Foundations of Offshore Gravity Structures: Analysis," Journal of the Geotechnical Engineering Division, ASCE, Vol. 107, No. GT2, Proc. Paper 16045, (February 1981), pp. 143-165.

15. Prevost, J. H. and Hughes, T. J. R., "Finite Element Solution of Elastic-Plastic BVP," Journal of Applied Mechanics, ASME, Vol. 48, No. 1, Paper No. 81-APM-17 (March 1981), pp. 69-84.

16. Prevost, J. H., "Two-Surface vs. Multi-Surface Plasticity Theories," International Journal for Numerical and Analytical Methods in Geomechanics, (1981) (to appear).

17. Prevost, J. H., "Nonlinear Transient Phenomena in Saturated Porous Media," Computer Methods in Applied Mechanics and Engineering (1981) (to appear).

18. Prevost, J. H., "Nonlinear Transient Phenomena in Elastic-Plastic Solids," Journal of the Engineering Mechanics Division, ASCE (April 1981)

19. Prevost, J. H., "Localization of Deformations in Elastic-Plastic Solids," Journal of the Engineering Mechanics Division, ASCE (April 1981)

20. Prevost, J. H. and Wakugawa, J. M., "Constitutive Equation for Pressure Sensitive Soils," International Journal for Numerical and Analytical Methods in Geomechanics, (April 1981)

21. Richard, R. E., Woods, R. D. and Hall, J. R., Vibrations of Soils and Foundations, Prentice-Hall, Inc., Englewood Cliffs, New Jersey, 1970.

22. Roscoe, K. H. and Burland, J. B., "On the Generalized Stress-Strain Behavior of 'Wet' Clay," in Engineering Plasticity, Heymand and Leckis, eds., Cambridge University Press, Cambridge, England, 1968, pp. 535-609.

23. Terzaghi, K., Theoretical Soil Mechanics, Wiley, New York, 1943.

CONSTITUTIVE MODELING AND SOIL-STRUCTURE INTERACTION

C. S. Desai*

Professor, Department of Civil Engineering and Engineering Mechanics, University of Arizona, Tucson, Arizona

ABSTRACT

The purpose of these lecture notes is to discuss the topic of constitutive modelling for geological materials and interfaces, and their importance in problems involving soil-structure interaction. The presentation is covered in three parts. The first part discusses some commonly used constitutive models for solids with particular attention to laboratory testing and determination of constitutive parameters. The second part covers the subject of modelling for interfaces and determination of parameters from laboratory tests. Most of this presentation covers static loading, however, in the case of interfaces, consideration is also given to cyclic testing. The last part covers a number of applications involving solution of several boundary value problems. Here attention is given to the topic of verification of constitutive models, and selection of the most appropriate model for a given problem. Although relevant references have been cited at appropriate locations, a majority of these notes include recent research investigations performed by the author and his coworkers.

I. MODELING FOR SOLIDS

Recent times have seen a significant activity and attention toward this important area. This is evident from a large number of publications, and symposia, workshops and conferences. Details and comprehensive reviews are available in various publications, and hence will not be presented here; only brief descriptions of the models used herein will be included.

*Formerly at Department of Civil Engineering, Virginia Polytechnic Institute and State University, Blacksburg, Virginia.

J. B. Martins (ed.), Numerical Methods in Geomechanics, 103–143.

Philosophy

Since there is no unique model yet developed that is valid for all geologic materials under all conditions, the philosophy used constitutes five steps (1, 2) which are depicted schematically in Fig. 1. The procedure involves adoption of a number of appropriate models, definition of parameters and their determination, implementation and verification phases.

Models such as variable moduli (VM) (3), von Mises, Drucker-Prager (DP) (4), Critical Stage-Cambridge (CAM) (5) and Hardening-Cap (CAP) (6) are implemented in two- and three-dimensional finite element procedures. For a given boundary value problem (BVP), material parameters for these models are determined from laboratory test data. Then numerical solutions with these models are compared with closed-form and/or experimental observations for the BVP. The model(s) that provide the optimum correlation with the closed-form and/or observed data is then recommended for the given problem.

It may be noted that only a limited number of models are considered herein. Some recent models (7-11) including those proposed by the author and his coworkers (12-14) are not included in these notes; one of the reasons being that the available information on their implementation is considered to be somewhat limited at this time. Also, some of them will be presented by other lecturers in this seminar.

Determination of Parameters

Cylindrical or conventional triaxial, consolidation and direct shear tests have been some of the common devices used for finding parameters for stress-strain or constitutive laws. Although these tests can provide useful data for some practical problems, they are usually restricted to a few stress paths, and loading conditions. In order to simulate general three-dimensional states of stress, it can be appropriate to employ truly triaxial or multiaxial testing devices. Two truly triaxial devices (15, 16) used herein are shown in Figs. 2 and 3. The first (Fig. 2) has a capacity of 200 psi (1380 kPa) and permits both drained and undrained testing of softer materials. The second, Fig. 3, has a capacity of 20,000 psi (1380 MPa) and permits undrained tests. It is used for materials with relatively high strength such as ballast, coal, rock and concrete.

In various research investigations, the foregoing devices have been used for detailed testing of materials such as artificial soil, Ottawa sand, Agricultural Soil and glass beads; and rock, ballast and wood. In each case comprehensive series

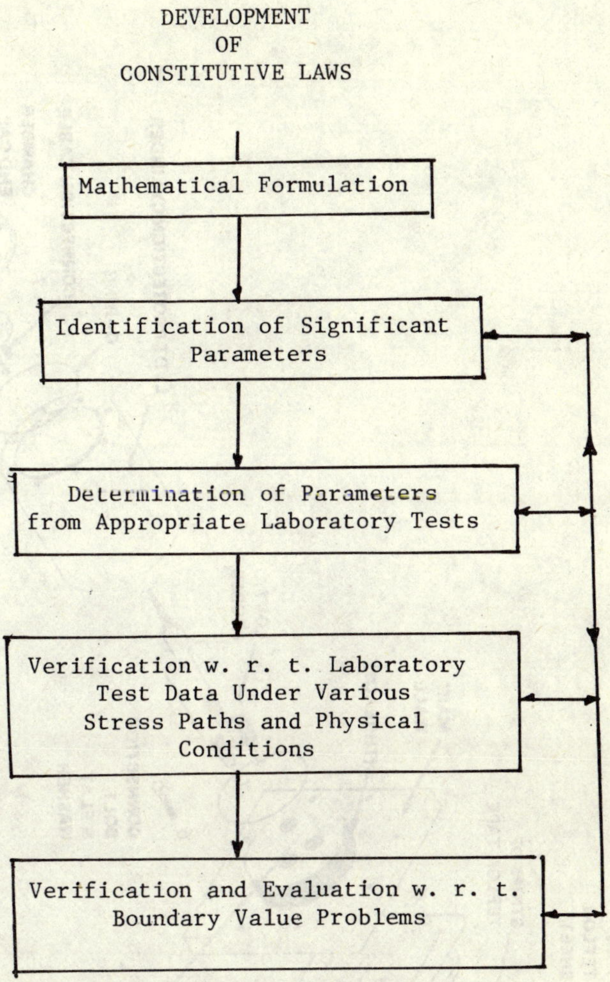

Fig. 1 Steps in Development of Constitutive Law(s)

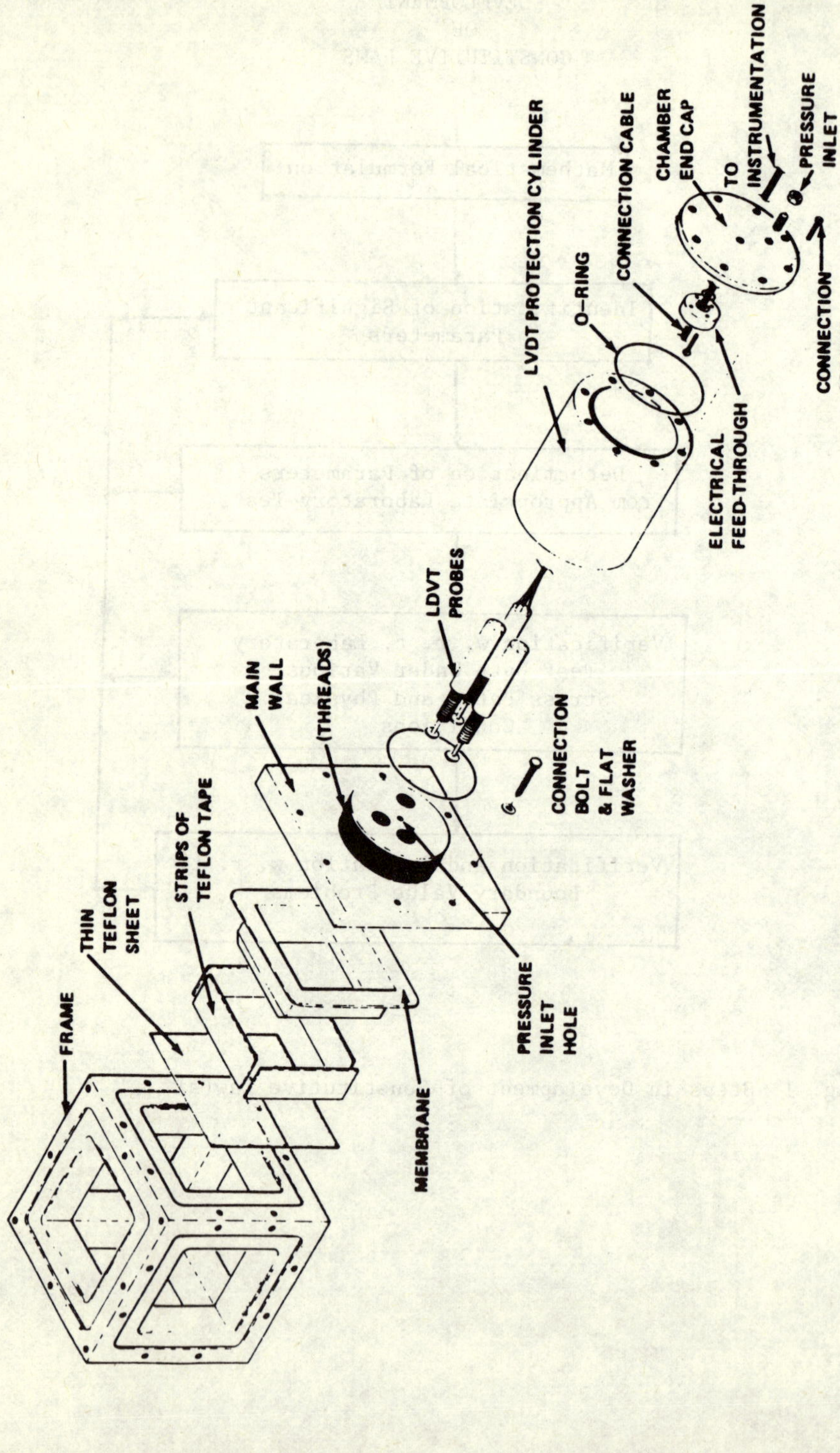

Figure 2 Schematic Details of Multiaxial Device : 200 psi

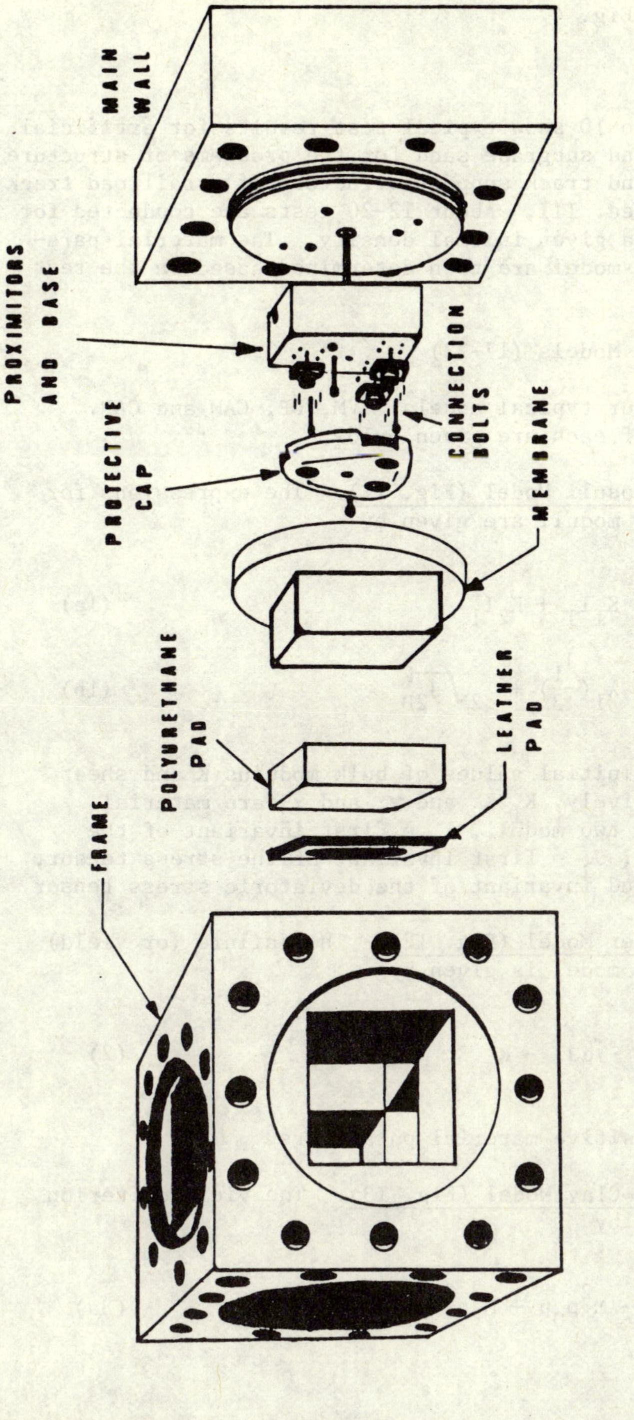

Figure 3 Schematic Details of Multiaxial Device : 20,000 psi

of tests have been performed with different initial densities, and stress paths, Fig. 4.

Typical Results

Figures 5 to 10 show typical test results for artificial soil, subballast and subgrade sand for the problems of structure pushed in ground and track support structure of a railroad track bed described in Sec. III. About 12-20 tests are conducted for each material for a given initial density. The material parameters for a given model are then determined based on the test results.

Details of Typical Models (17-23)

Consider four typical models: VM, DP, CAM and CAP. Brief statements of each are given below.

Variable Moduli Model (Fig. 11). The expressions for the bulk and shear moduli are given by

$$K = K_0 + K_1 I_1 + K_2 I_1^2 \tag{1a}$$

$$G = G_0 + \lambda_1 \left(\frac{J_1}{3}\right) + \lambda_2 \sqrt{J_{2D}} \tag{1b}$$

where K_o and G_o = initial values of bulk modulus K and shear modulus G, respectively, K_1 K_2 and γ_1 and γ_2 are material parameters for the two moduli, I_1 = first invariant of the strain tensor, ε_{ij}, J_1 = first invariant of the stress tensor, σ_{ij} and J_{2D} = second invariant of the deviatoric stress tensor.

Drucker-Prager Model (Fig. 12). The failure (or yield) criterion for this model is given by

$$F = \sqrt{J_{2D}} - \alpha J_1 - k \tag{2}$$

where α and k = positive material parameters.

Modified Cam-Clay Model (Fig. 13). The yield criterion used here is given by

$$F_c = q^2 - M^2 p_0 p + M^2 p^2 = 0 \tag{3a}$$

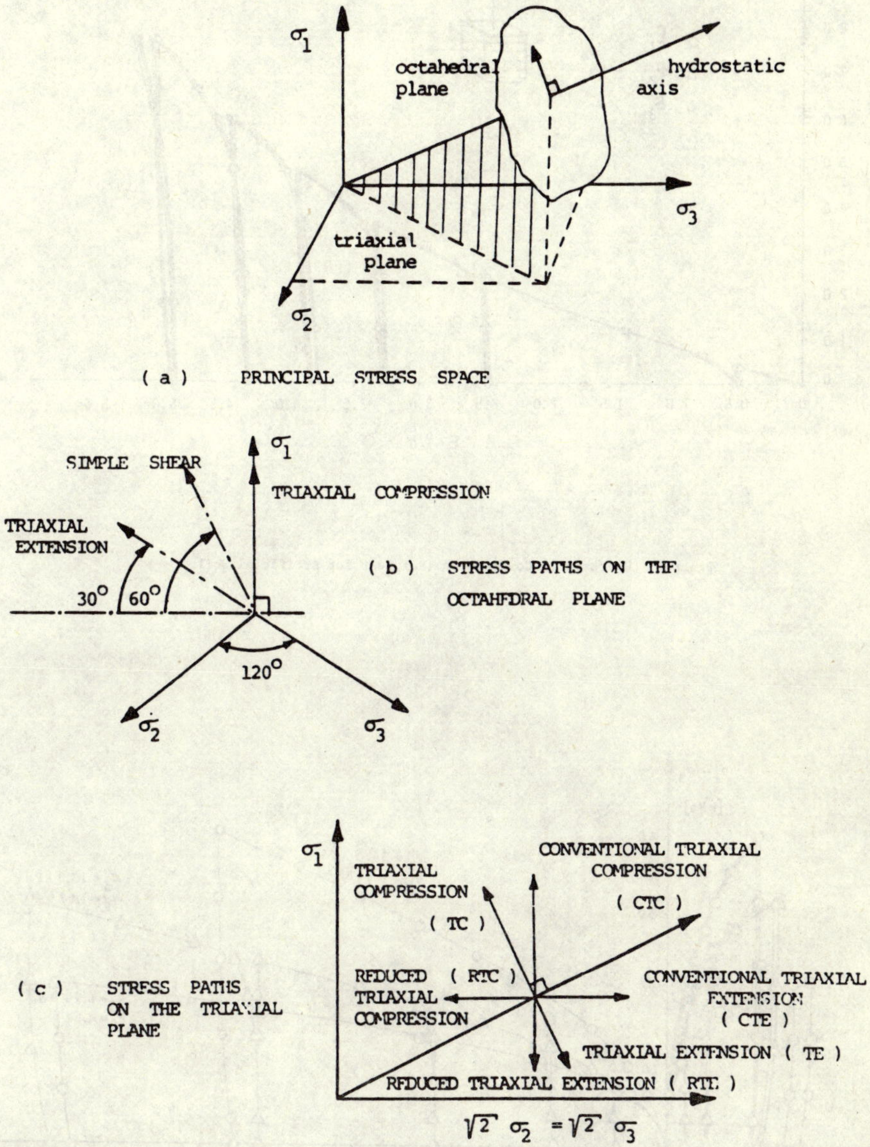

Figure 4 Commonly Used Stress Paths for Multiaxial Testing

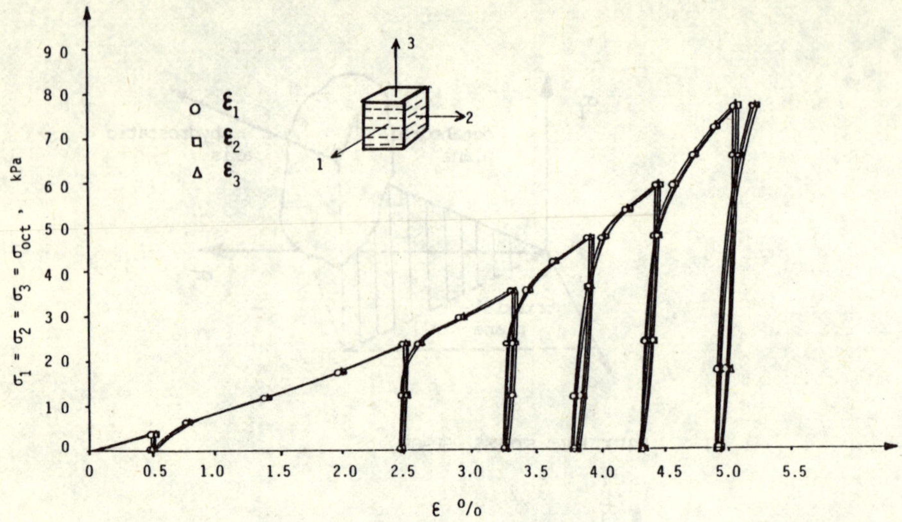

Figure 5 Hydrostatic Compression (HC) Test for Artificial Soil

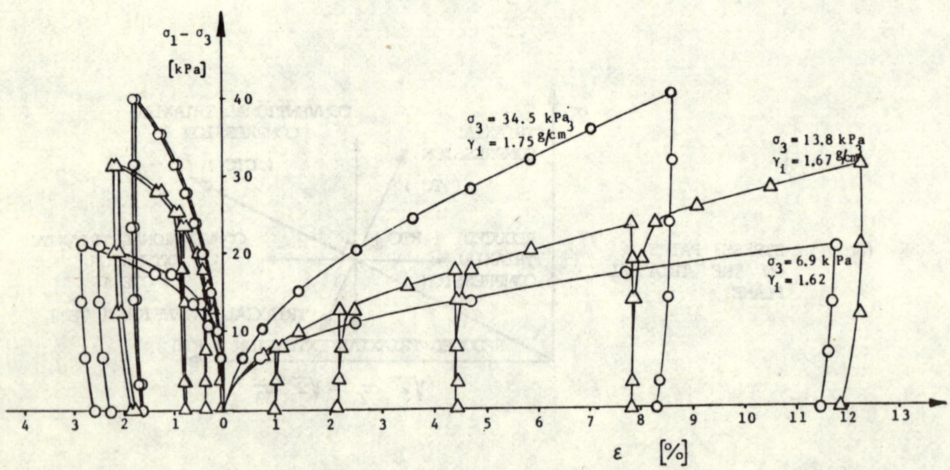

Figure 6 Conventional Triaxial Compression (CTC) Test for Artificial Soil

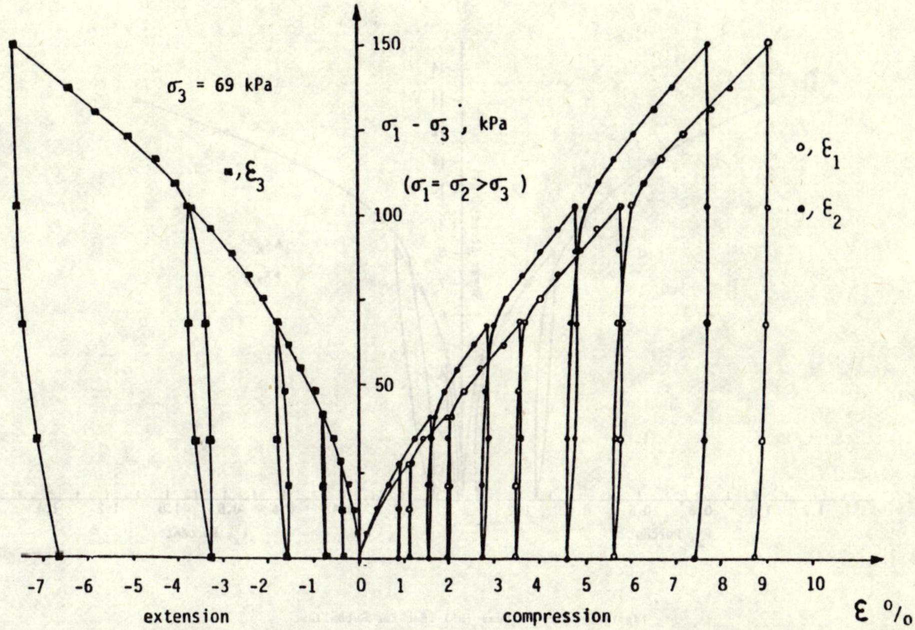

Figure 7 Conventional Triaxial Extension (CTE) Test for Artificial Soil

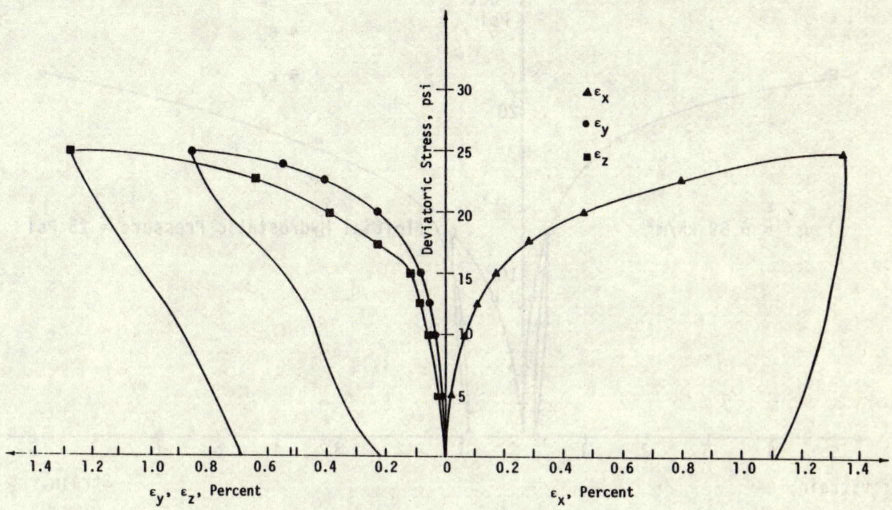

Figure 8 Reduced Triaxial Compression (RTC) Test for Subballast

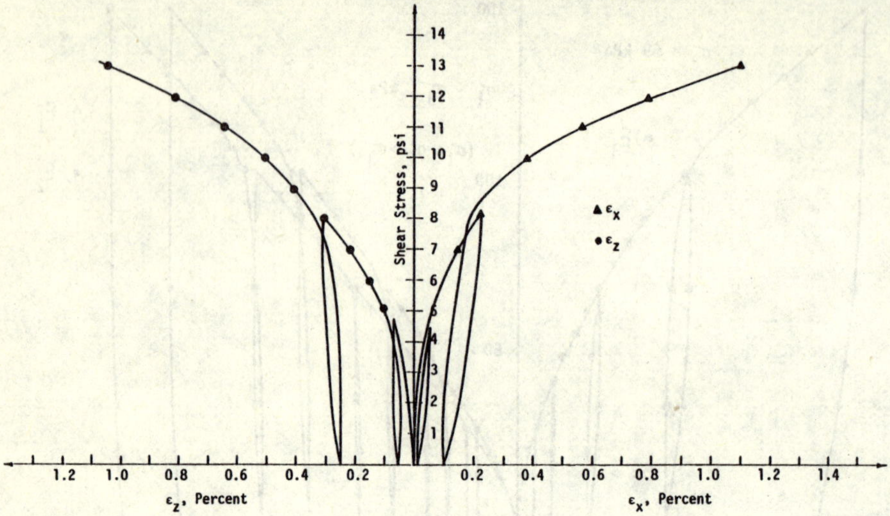

Figure 9 Simple Shear (SS) Test for Subballast

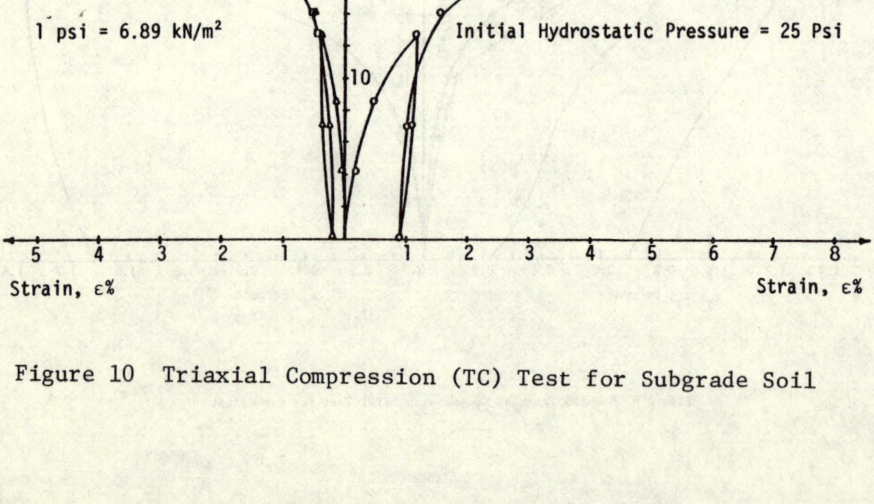

Figure 10 Triaxial Compression (TC) Test for Subgrade Soil

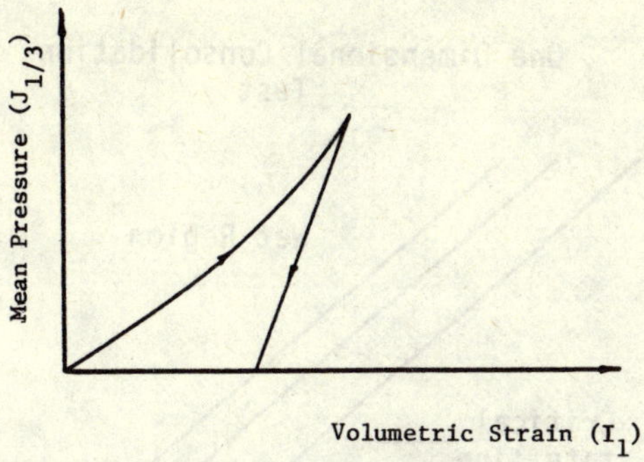

(a) Hydrostatic Test Data

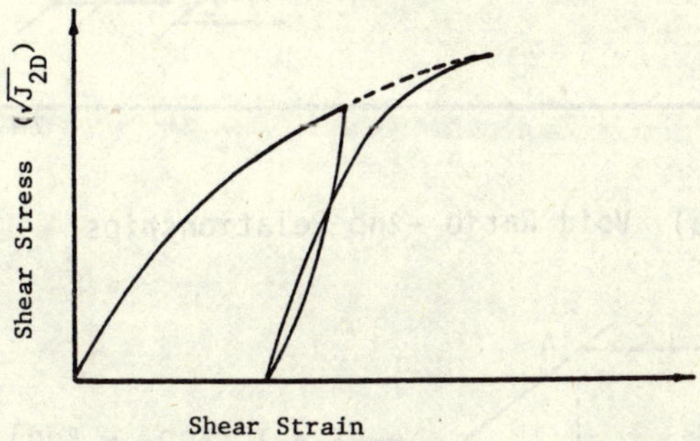

(b) Shear Test Data

Figure 11 Representation of Variable Moduli Model

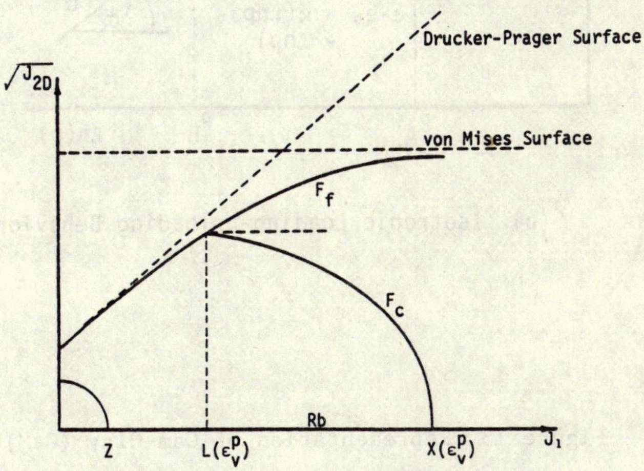

Figure 12 Representation of Drucker-Prager and CAP Models

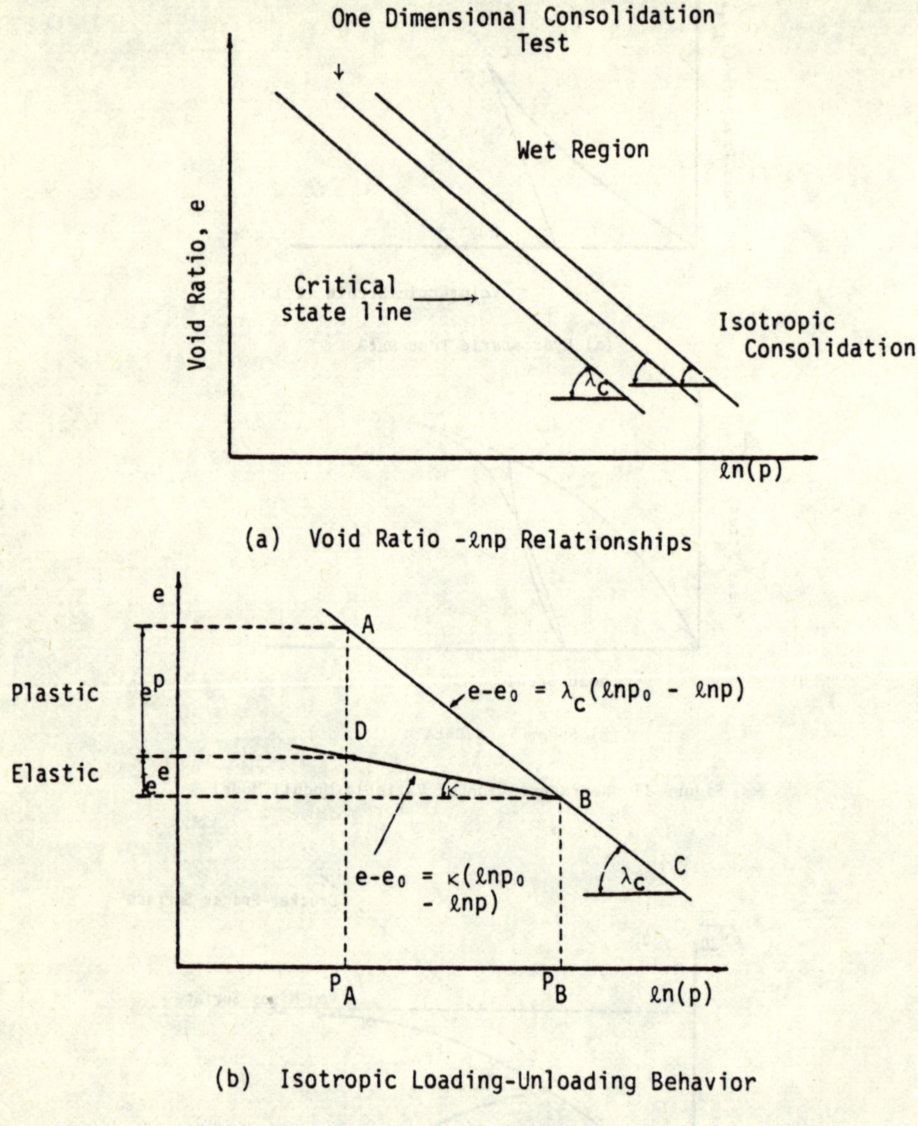

Figure 13 Representation of Cam-Clay (CAM) Model

where $q = \sigma_1 - \sigma_3 = \sqrt{3J_{2D}}$, σ_1 and σ_2 = principal stresses, $p = J_1/3$ mean pressure, p_o reference pressure or hardening parameter, and M = slope of the critical state line on the q-p plot. The plastic volumetric strain dv^p is expressed as

$$dv^p = \frac{\lambda_c - \kappa}{1+e_0} \frac{dp_0}{p_0} \tag{3b}$$

where e_0 = initial void ratio, λ_c = slope of the consolidation or HC test curve and κ = slope of swelling curve.

<u>Cap Model (Fig. 12)</u>. In this model, the failure envelope F_f and the yield caps, F_c, are expressed in a modified form as

$$F_f = \sqrt{J_{2D}} \quad [\alpha + \theta J_1 - \gamma e^{-\beta J_1}] \tag{4a}$$

$$F_c = [\frac{J_1 - L(\varepsilon_v^p)}{X(\varepsilon_v^p) - L(\varepsilon_v^p)}]^2$$

$$- [\frac{\sqrt{J_{2D}}}{\alpha - \gamma e^{-\beta LC\varepsilon_v^p}}]^2 - 1 = 0 \tag{4b}$$

where α, β, θ = material parameters, and $X(\varepsilon_v^p)$ and $L(\varepsilon_v^p)$ define geometrical properties of the yield cap, Fig. 12, and $X(\varepsilon_v^p)$ is given by

$$X(\varepsilon_v^p) = L(\varepsilon_v^p) + R(\alpha + \theta J_1 - \gamma e^{-\beta L(\varepsilon_v^p)}] \tag{4c}$$

where R = ratio of the major to minor axes of the ellipse. The HC test provides the relation between volumetric plastic strain ε_v^p and X or L as

$$\varepsilon_v^p = W(1 - e^{-DX}) \tag{4d}$$

where W and D = material parameters.

The laboratory test data is used to find the parameters for the models. For the DP Model the parameters α and k can be often determined from CTC tests only. It may be noted that one can find different values of the parameters if tests results under other and additional stress paths are also used; this comment will apply to other models also. The parameters for the VM Model can be found by using test data under CTC and HC stress paths. Although the parameters for the CAM and CAP models can be found from CTC and HC tests, it is advisable to use data from a greater number of tests under different stress paths.

In the case of the track support structure problem described subsequently, linear, variable moduli, CAP and CAM (Modified Cam-Clay) models were used for the concrete tie, ballast, subballast and sand respectively. Table 1 gives the parameters determined for the materials in field section at Transportation Test Center, Pueblo, Colorado (20-23). Simularly, parameters have been determined for other materials considered in the problems presented herein.

II. MODELING FOR INTERFACES

Behavior of interfaces between structural and geologic media pose different problems as compared to that of solid soil or structural materials. Under loading, there can occur relative motions during which the interface may remain intact, slide, separate (debond) and/or close (rebonding); Figure 14 shows a schematic representation of these nodes.

A number of different models have been proposed by various investigators (24-28) and others. Comprehensive reviews on this subject are provided in Refs. 29 to 31.

The foregoing models have been applied to various soil-structure interaction problems (29). Here material parameters have been usually obtained on the basis of static direct shear tests.

Relative merits and disadvantages of these models have been discussed by Desai (31). One of the major difficulties with these interface models arises while adopting material parameters such as normal and shear stiffness. Although the (nonlinear) variation of the latter can be found from direct shear tests until sliding or separation occurs, adoption of normal stiffness has essentially remained empirical. A value of the normal stiffness is chosen (arbitrarily) for no-sliding and contact modes, and then it is arbitrarily reduced to a small value during separation which is often identified by tensile normal stress. This approach can cause computational difficulties in many soil-structure interaction problems.

Recently a rather "new" approach has been attempted by various investigators (20, 22, 23, 31-38); the details of some basic aspects with respect to cyclic loading are given in

Table 1: Material Properties for Nonlinear UMTA Section

Rail (steel): Linear Elastic	Tie Concrete: Linear Elastic
$E = 30\text{x}10^6$ psi ($207.0\text{x}10^6$ (kN/m^2)	$E = 4.2\text{x}10^6$ psi ($29\text{x}10^6$ (kN/m^2)
$\nu = 0.3$	$\nu = 0.20$
$I = 71.4$ in^4 (2971.9 cm^4)	
$A = 11.65$ in^2 (75.16 cm^2)	

Subballast: Cap Model

$E = 20{,}000$ psi ($138\text{x}10^3$ kN/m^2)

$D = 0.308\text{x}10^{-4}$ psi^{-1} ($0.446\text{x}10^{-4}$ (kN/m^{-1})

$\nu = 0.40$, $\alpha = 26.0$ psi (179.4 kN/m^2)

$\gamma_m = 0.0833$ lb/in^3 (0.00230 Kg/cm^3)

$\gamma = 21.5$ psi (148.35 kN/m^2), $R = 1.24$, $\beta = 0.02$, $W = 0.035$

Ballast: Variable Moduli Model

$K_o = 4\text{x}10^3$ psi (27.6 kN/m^2)

$G_o = 1.7414\text{x}10^3$ psi (11.83 kN/m^2)

$K_1 = -5.4917\text{x}19^5$ psi ($37.89\text{x}10^5$ kN/m^2)

$\gamma_1 = 2.5593$

$K_2 = 2.536\text{x}10^7$ psi ($17.50\text{x}10^7$ kN/m^2)

$\gamma_2 = -60.172$

$\gamma_m = 0.0613$ lb/in^3 (0.00169 Kg/cm^3)

Subgrade (Sand): Modified Cam Clay Model

$M = 1.24$

$E_o = 12{,}000$ psi ($82.8\text{x}10^3$ kN/m^2)

$\lambda_c = 0.014$, $\nu = 0.28$, $\kappa = 0.0024$, $\gamma_m = 0.081$ lb/in^3 (0.00224 Kg/cm^3)

$e_o = 0.34$

Here γ_m = density of material.

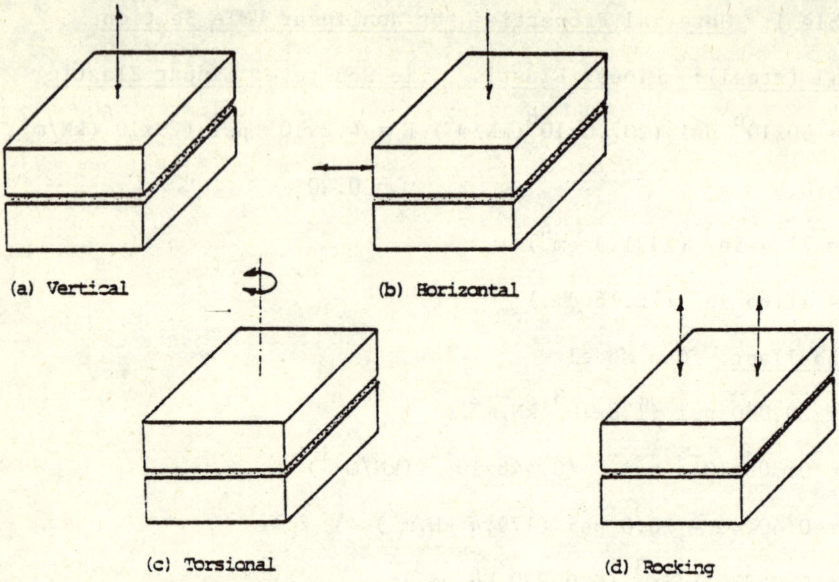

Figure 14 Modes of Deformation at Interface

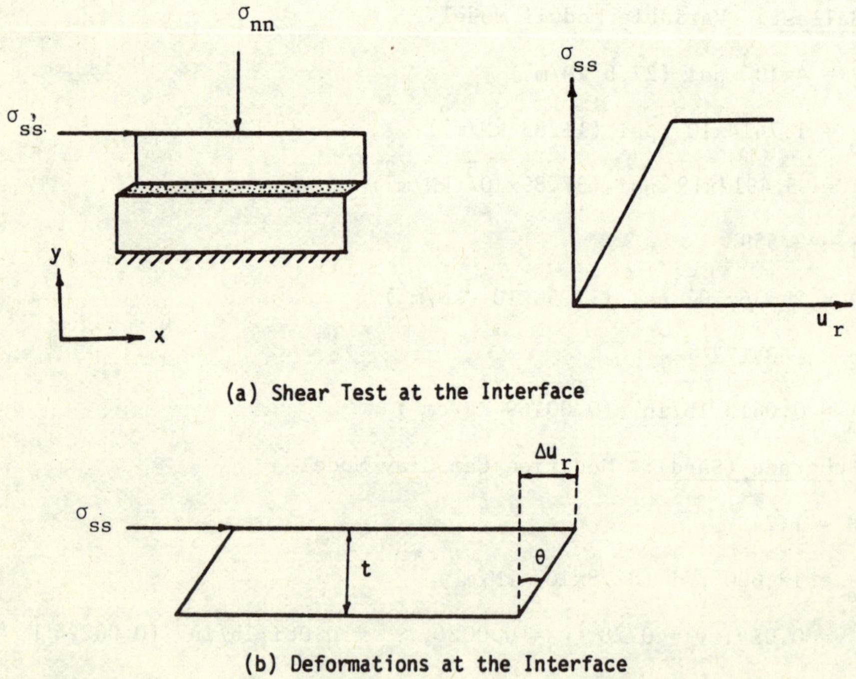

Figure 15 'Thin' Interface Element

Ref. 31. In this concept the interface is assigned a small finite thickness (Fig. 14) and is called 'thin' interface. The normal behavior is defined by using the same parameters as the soil or structure or some average (nonlinear) values of the adjoining soil and structural elements. Thus the normal stiffness is made dependent upon the state of stress in the interface and/or surrounding elements.

For plane strain conditions, the constitutive matrix of the interface for nonlinear elastic approach can be written as

$$[C_i] = \begin{bmatrix} A & B & 0 \\ & A & 0 \\ \text{symm.} & & G_i \end{bmatrix} \tag{5a}$$

where $A = \dfrac{\bar{E}(1-\bar{\nu})}{(1+\bar{\nu})(1-2\bar{\nu})}$ and $B = \dfrac{\bar{E}\,\bar{\nu}}{(1+\bar{\nu})(1-2\bar{\nu})}$

Here $\bar{E}$ and $\bar{\nu}$ are the (average) elastic modulus and Poisson's ratio, respectively and G_i is the shear modulus of the interface. As depicted in Fig. 15, its values is found from

$$G_i = \frac{\Delta\sigma_{ss}}{\Delta u_r/t} \tag{5b}$$

where σ_{ss} = shear stress and Δu_r= relative motion in direct shear or other relevant test results.

The foregoing approach has been implemented in a number of two- and three-dimensional finite element procedures for soil-structure interaction problems. Figures 16 and 17 show the solid and thin interface elements used.

Testing

Testing devices for interfaces are not as advanced as those for the solid elements. A review of available and previously used static direct, ring and annular shear devices is presented by Desai (30,31). Here brief details are presented for a new static and cyclic multi-degree-of-freedom (CYMDOF) shear device for interfaces (39).

The CYMDOF device will permit testing of interfaces under both translational (horizontal and vertical) and rotational

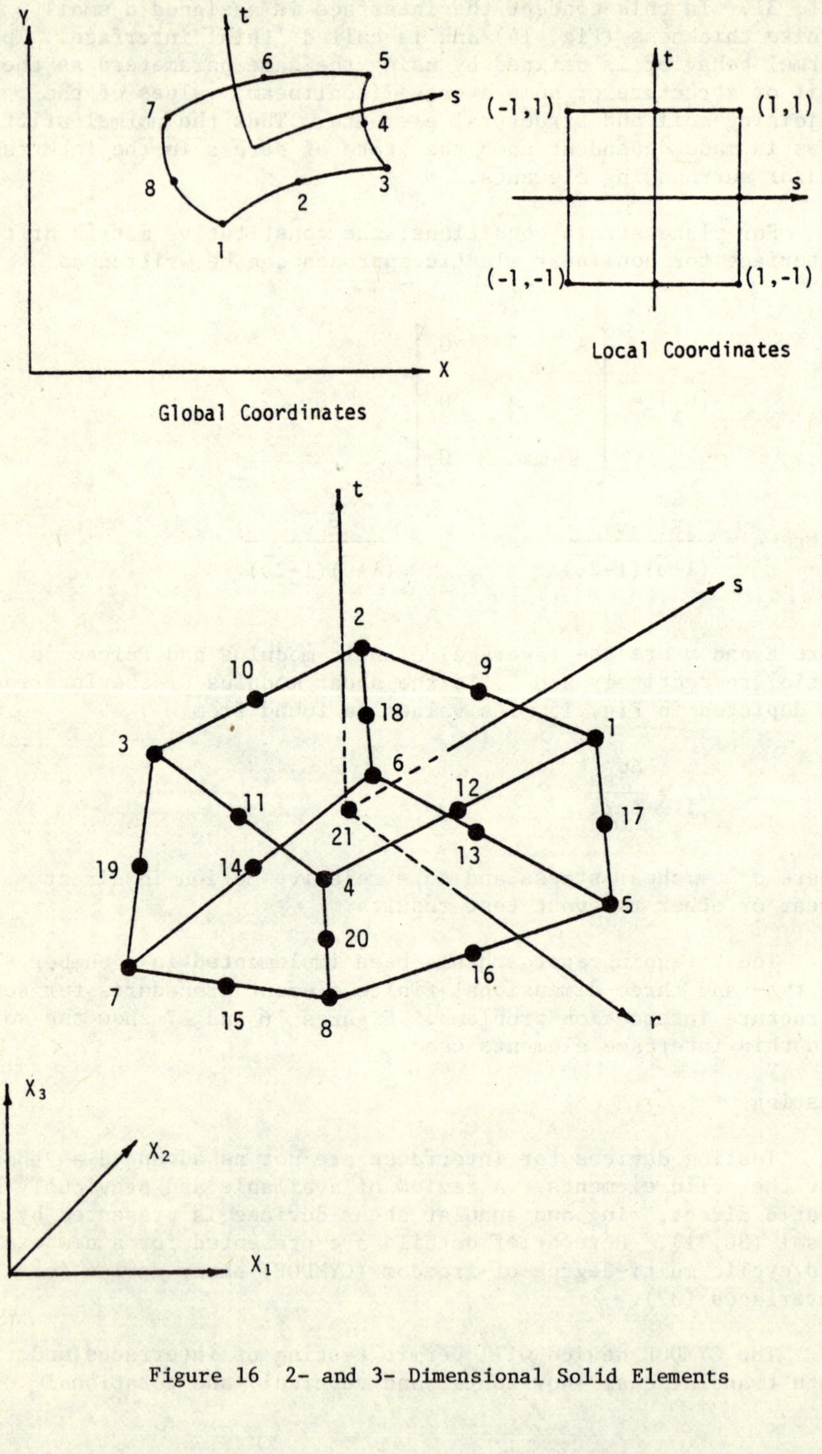

Figure 16 2- and 3- Dimensional Solid Elements

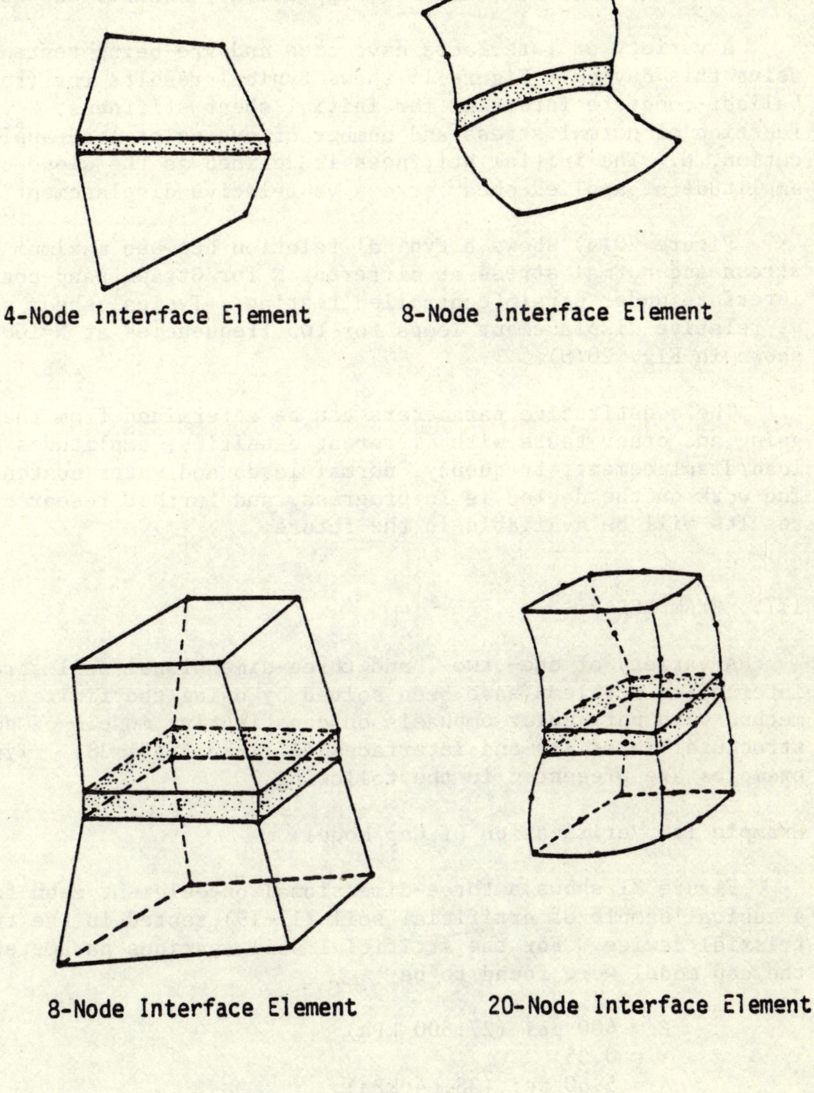

Figure 17 2- and 3- Dimensional Thin Interface Elements

motions (torsional and rocking). Figure 18(a) shows a schematic diagram of the devices, and Fig. 18(b) shows some of the functions for stress (force) and strain (displacement) controlled tests.

A variety of interfaces have been and are being tested by using this device. Figure 19 shows typical results for (railroad) ballast-concrete interface for initial shear stiffness, K_{si}, as function of normal stress and number of cycles of load application, N. The initial stiffness is defined as the slope of the amplitude of applied shear stress vs relative displacement curve.

Figure 20(a) shows a typical relation between maximum shear stress and normal stress at different N for Ottawa sand-concrete interface under strain controlled testing. Typical shear stress vs relative displacement loops for two frequencies at N=100 are shown in Fig. 20(b).

The constitutive parameters can be determined from the foregoing and other tests with different densities, amplitudes of load/displacement, frequency, normal loads and water content. The work on the device is in progress, and further research results will be available in the future.

III. EXAMPLES

A variety of one-,two-, and three-dimensional soil-structure interaction problems have been solved by using the finite element method with particular emphasis on constitutive models of soils, structural materials and interfaces (1,17-23,32,35-38). Typical examples are presented in the following.

Example 1: Verification of Cap Model

Figure 21 shows a three-dimensional one-element mesh for a cubical sample of artificial soil (17-19) tested in the truly triaxial device. For the artificial soil, various parameters for the cap model were found to be

$E = 400$ psi (27,600 kPa)
$\nu = 0.35$
$A = 5.60$ psi (38.64 kPa),
$\beta = 0.11$
$B = 0.062$ psi^{-1} (0.009 kPa^{-1})
$C = 5.60$ psi (38.64 kPa)
$R = 2.0$
$W = 0.18$
$D = 0.05$ psi^{-1} (0.0072 kPa^{-1})

These parameters were found on the basis of a number of tests conducted under various stress pahts, Fig. 4.

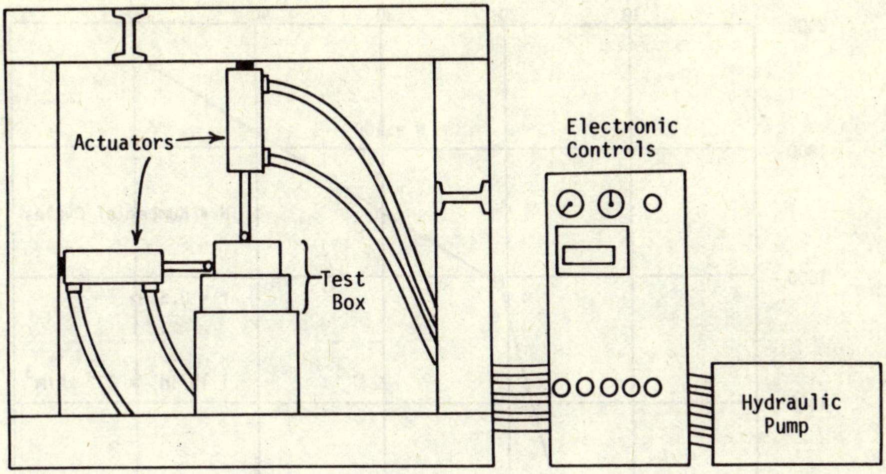

Figure 18(a) Schematic of CYMDOF Device

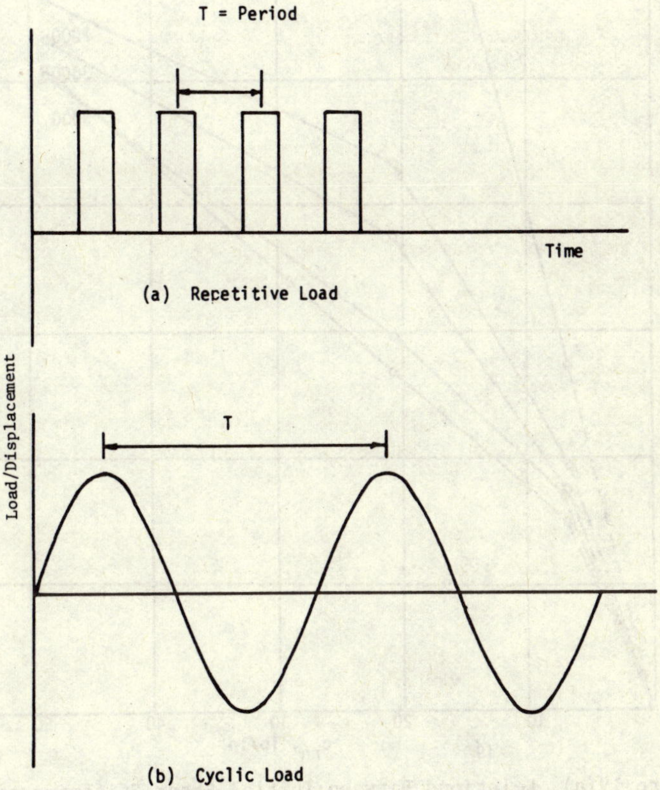

Figure 18(b) Typical Loading Functions

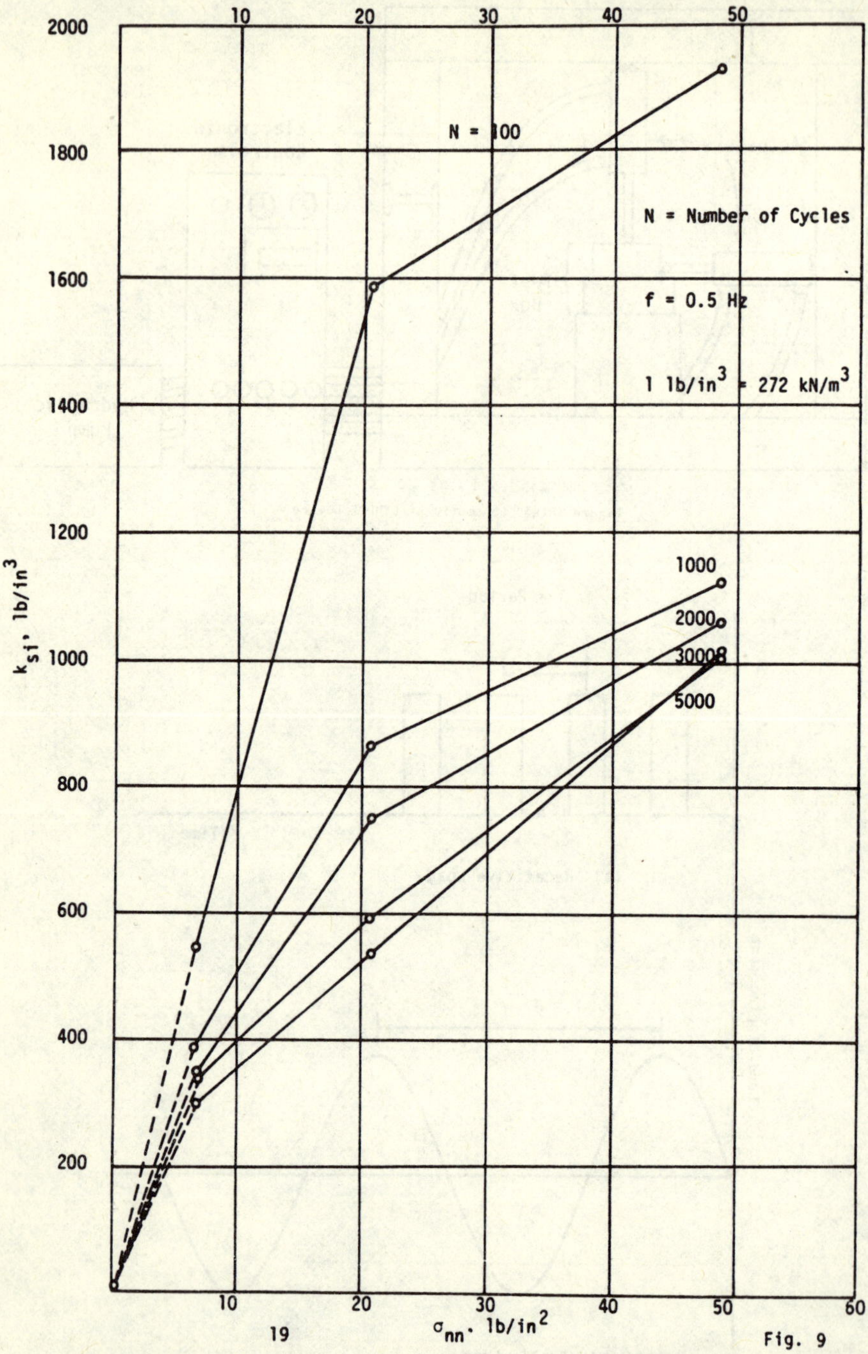

Figure 19(a) Relations Between Initial Shear Stiffness and Normal Normal Stress, σ_{nn}

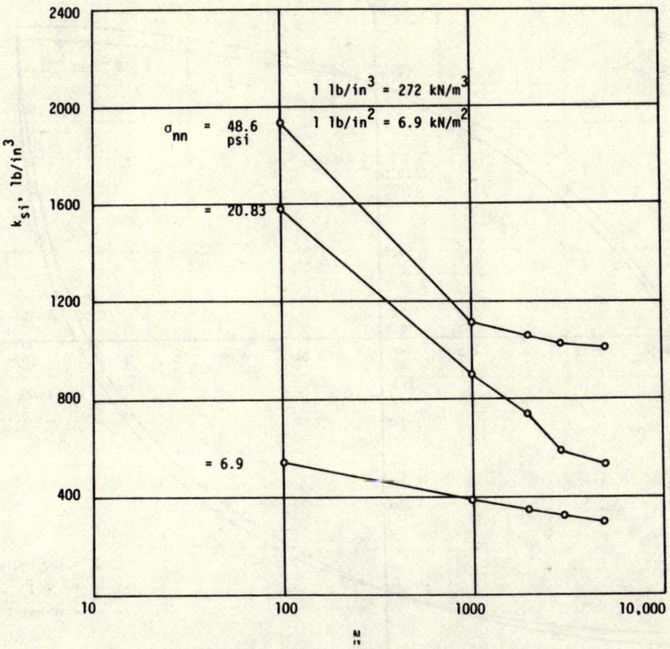

Figure 19(b) k_{si} vs N for Ballast-Concrete Interface

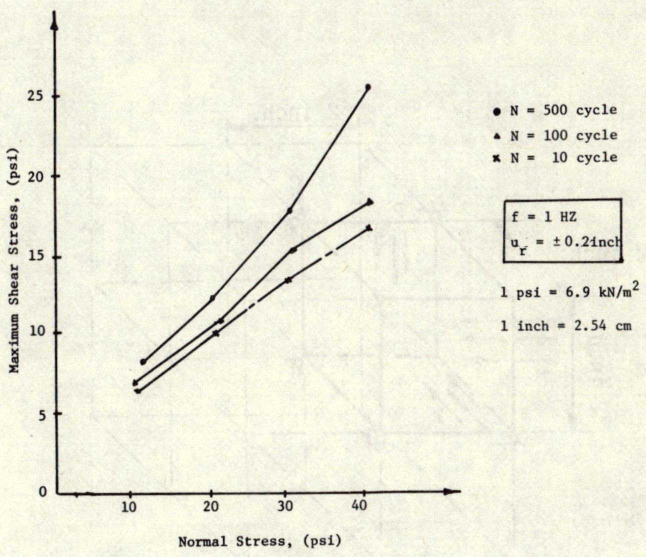

Figure 20(a) Relation Between Maximum Shear Stress and Normal Stress: Concrete-Sand Interface

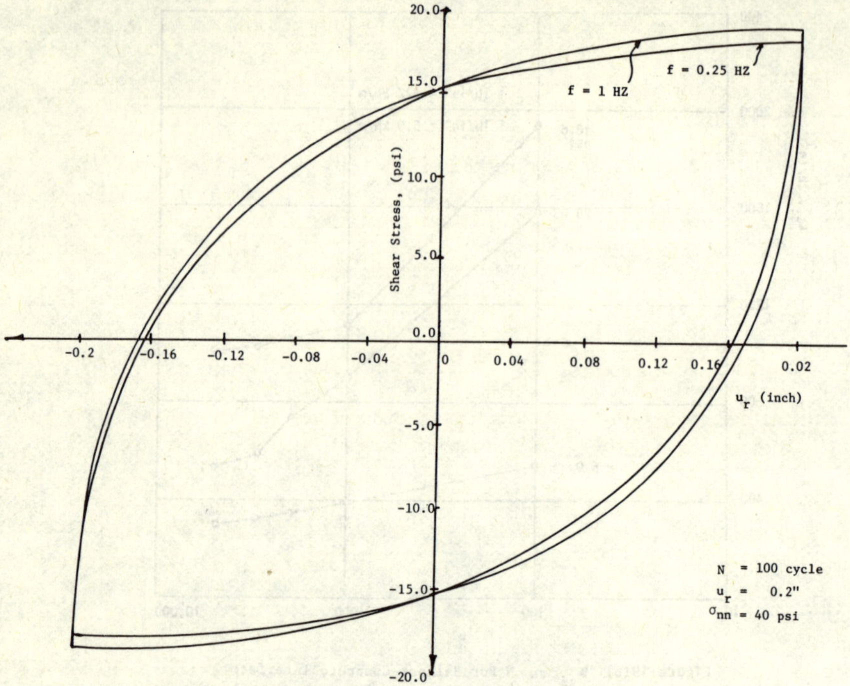

Figure 20(b) Shear Stress vs Relative Displacement

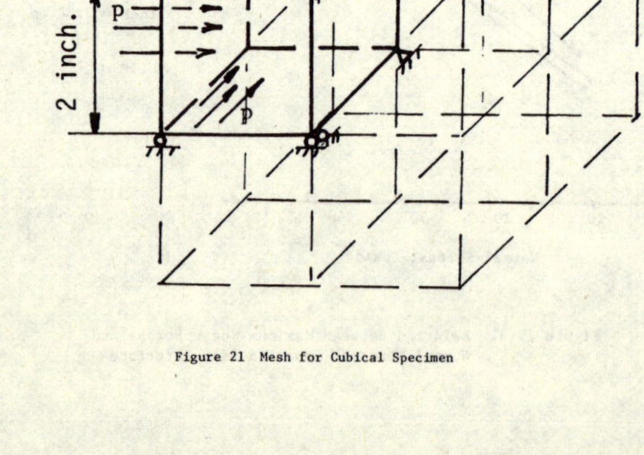

Figure 21 Mesh for Cubical Specimen

Loadings consistent with these paths were simulated on the discretized mesh for the HC, CTC, CTE and SS test. The comparisons between predictions and observations are shown in Figs. 22(a) to (d). There is excellent agreement between the measurements and predictions for the HC and CTC tests, whereas there is poor agreement between the two for the CTE and SS tests. It may be noted that no general constitutive low(s) is yet available that can simulate all stress paths. At the same time, use of such a law can yield satisfactory results for overall stress-deformation behavior of some (relevant) boundary value problems (see subsequent examples).

Example 2: Comparison on Various Models for Footing Problem (17-19)

Figure 23 shows a three-dimensional mesh for a laboratory footing made of plastic resting on a bed of the artificial soil. The load was applied in increments and the displacements at the center of the footing were measured. Photographs were taken of one side of the box in order to record its deformed shapes.

Figure 24 shows comparisons between predicted load-displacement behavior using Drucker-Prager, CAM and CAP models and the observations. The cap model that allows for stress path effects and appropriate volume change behavior yields the best correlation. The D-P models fail to predict the behavior except for the ultimate load. Thus, for the artificial soil, the cap model is considered to be a suitable model.

Example 3: Three-Dimensional Analysis of Structure (Tool) Pushed in Soil (17-19)

Figure 25 shows the finite element mesh with three-dimensional soil (solid) and interface elements. The short structure was pushed in a bed of the artificial soil filled in a special soil-bin test facility. Tests were conducted with the tool of different dimensions, and inclinations to the horizontal. Measurements were made for the loads and moments, and displacements, and photographic records were obtained for displacement patterns.

Figure 26 and 27 show typical comparison between predictions and observations for tool angles of inclinations of 90 and 45 with the horizontal. The cap model was used to simulate the behavior of the artificial soil, and a (conventional) interface element was used to simulate the interaction effects (18,19).

The results indicate that it is necessary to employ a constitutive model that allows for stress path effects and

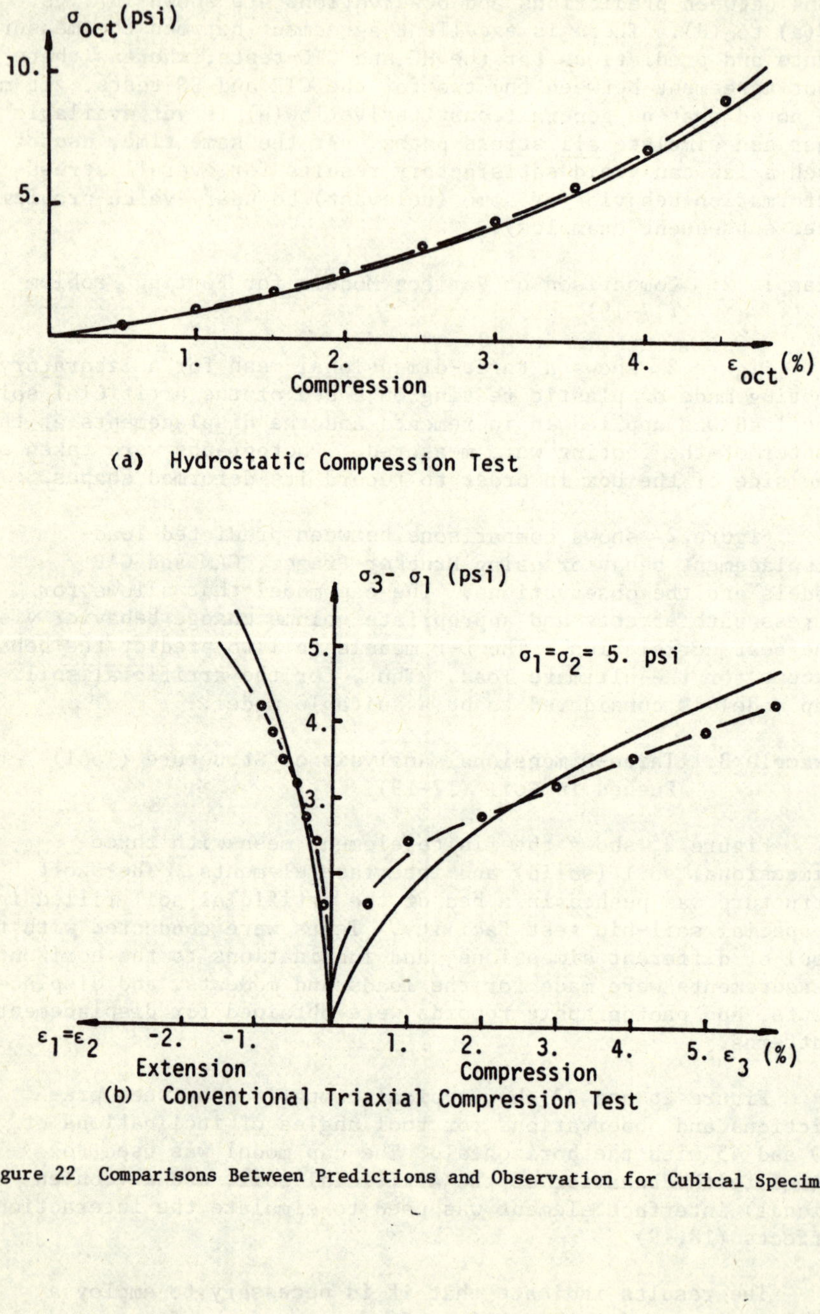

Figure 22 Comparisons Between Predictions and Observation for Cubical Specimen

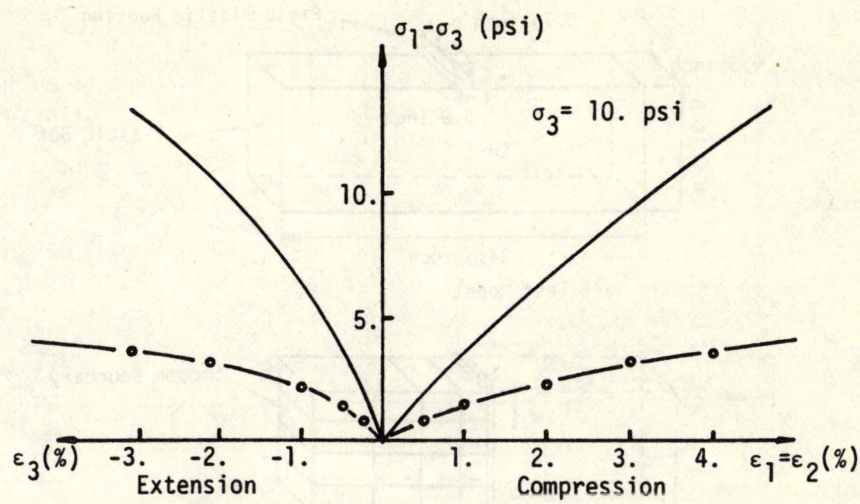

(c) Conventional Triaxial Extension Test

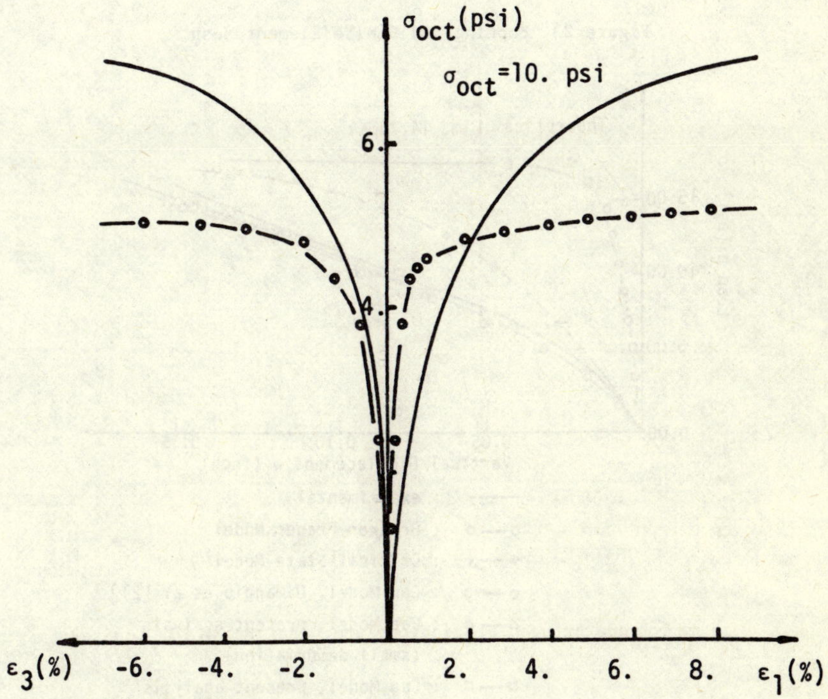

(d) Simple Shear Test

Figure 22 (cont'd) Comparisons Between Predictions and Observations for Cubical Specimen

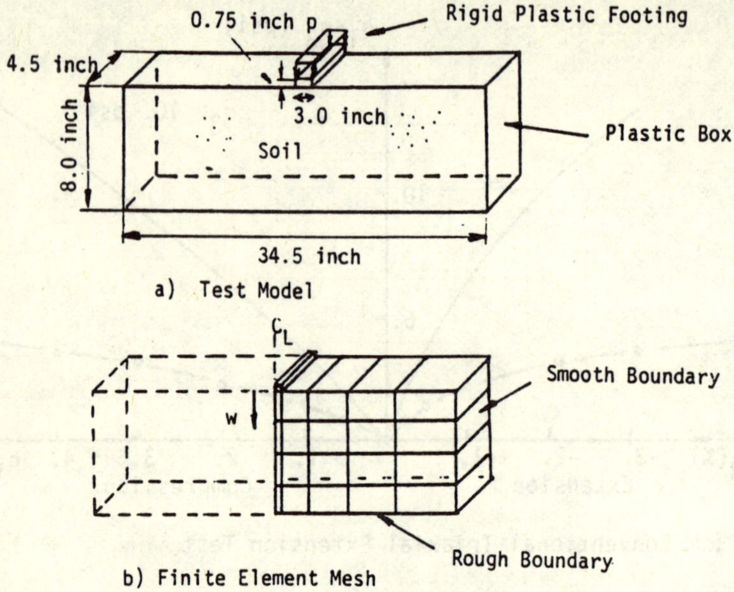

Figure 23 Footing and Finite Element Mesh

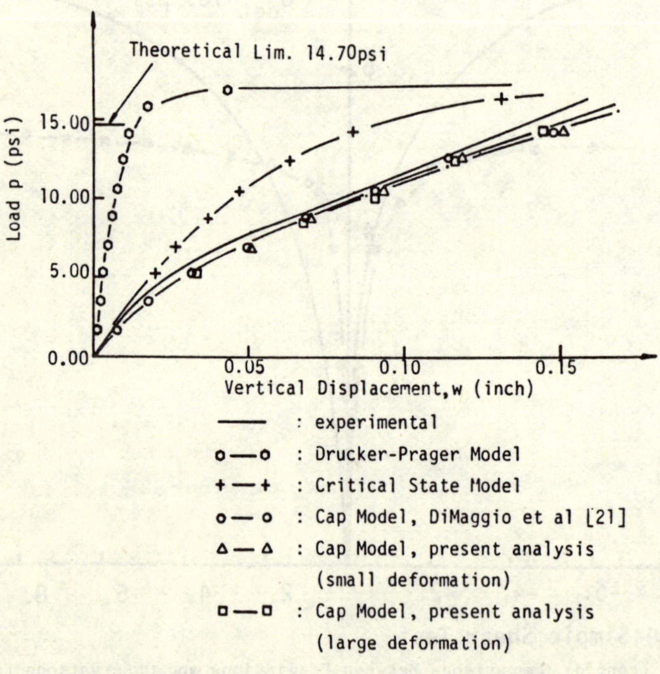

Figure 24 Comparisons Between Predictions and Test Data

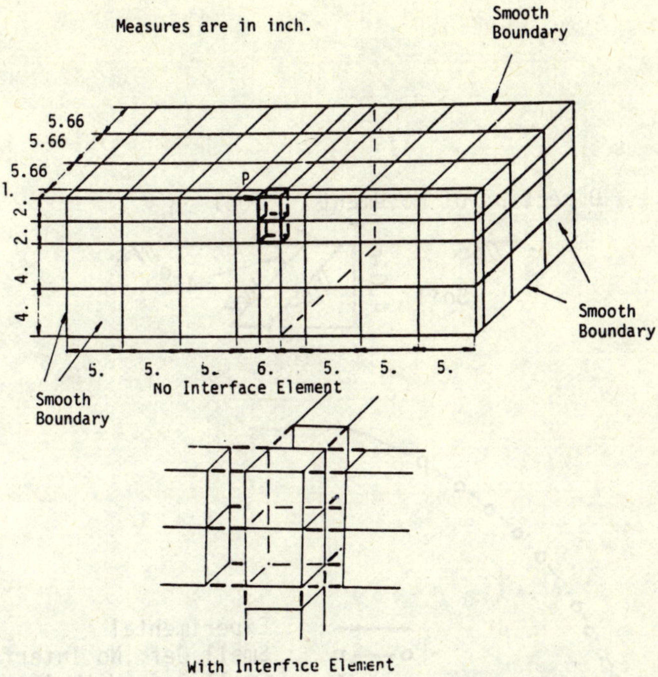

Figure 25 Mesh for Structure (Tool) - Soil Problem

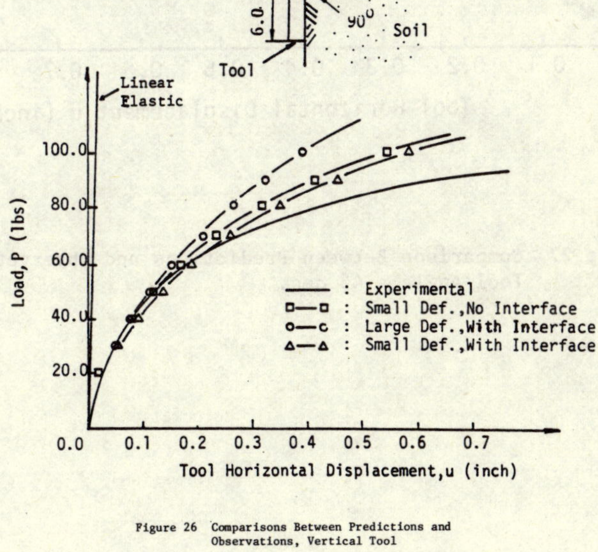

Figure 26 Comparisons Between Predictions and Observations, Vertical Tool

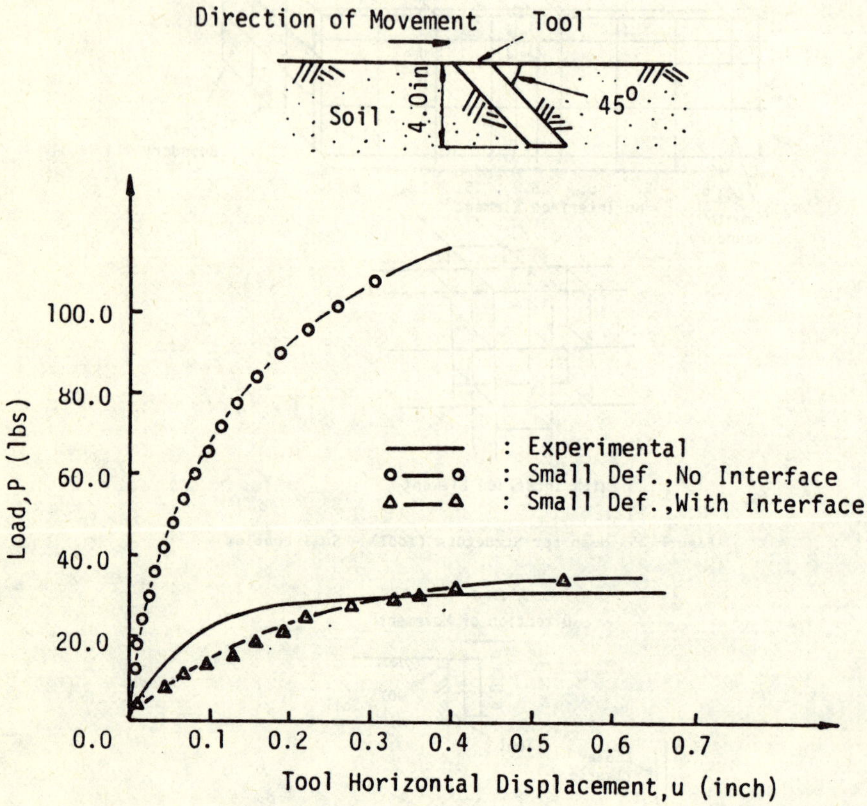

Figure 27 Comparison Between Predictions and Observations: Tool Angle = 45 deg.

volume change behavior. Also, it is necessary to allow for interaction effects and relative motion at the interface by providing the interface elements. Computations without interface elements do not provide satisfactory correlation with observations, Fig. 27.

Example 4: Three-Dimensional Analysis of Track Support Structures (20-23)

Figure 28 shows a cross-section of a (railroad) track and Fig. 29 shows a typical finite element mesh. The mesh contains solid elements for the rail, tie, ballast, subballast and subgrade, Fig. 16, and special 'thin' interfaces, Fig. 17, at the junction of the tie and ballast.

The thin interface element is found to provide satisfactory results for many soil-structure interaction problems (21-23, 31, 32, 36-38). It is found to reduce computational difficulties in comparison to the conventional interface elements base on relative displacements.

Different constitutive models are used for each component. Linear elastic, variable moduli, CAP and CAM models are used for the tie, ballast, subballast and subgrade, respectively. The parameters are determined from comprehensive series of tests by using the truly triaxial devices. A summary of the parameters is given in Table 1.

The interface shear properties are obtained by using a large cyclic multi-degree-of-freedom shear device, Fig. 18. The value of G is obtained equal to 225 psi (1553 kN/m^2). The normal properties are assumed to be E = 30,000 psi (208 kN/m^2) and ν = 0.30.

The predictions from the finite element procedures have been compared with various closed-form results and field observations. Figure 30 shows typical comparison between observed results and predictions from linear and nonlinear analysis. The results for the inner rail are considered to be nearer to the other closed-form solutions and field data (22). Overall, the numerical solutions were found to provide satisfactory correlation with a wide range of problems related to track mechanics.

Example 5: Strutted Retaining Wall and Simulation of Construction Sequences

Figure 31 shows schematic details of a supported retaining wall and Table 2 shows details of material parameters used (40, 41). The finite element mesh is shown in Fig. 32.

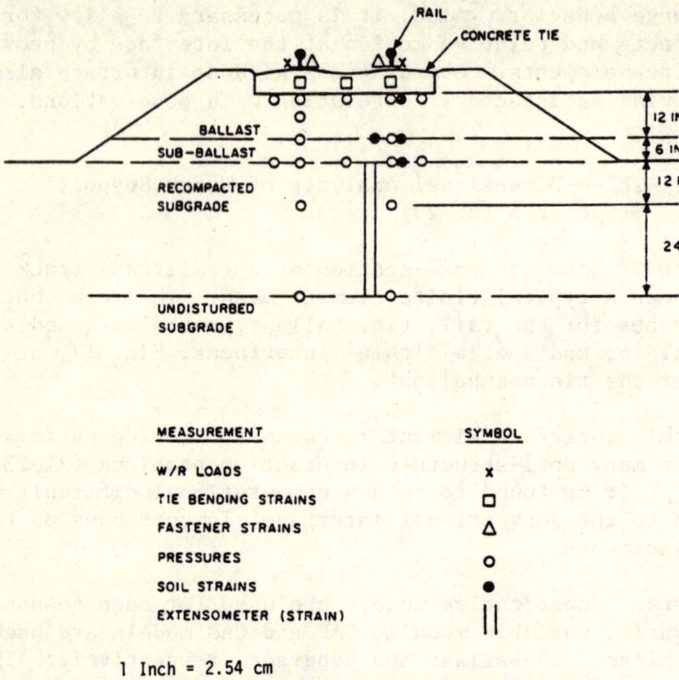

Figure 28 Field Section at UMTA Rail Track Test Site

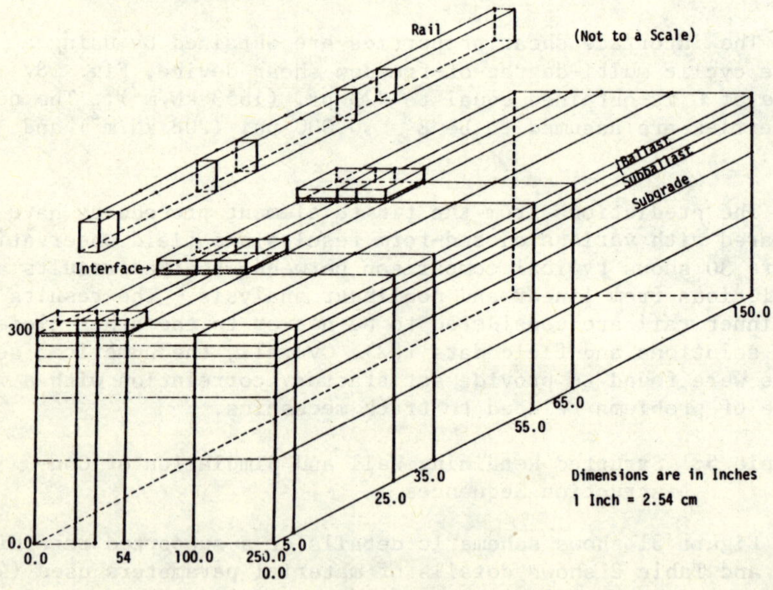

Figure 29 Mesh for Rail Track Section

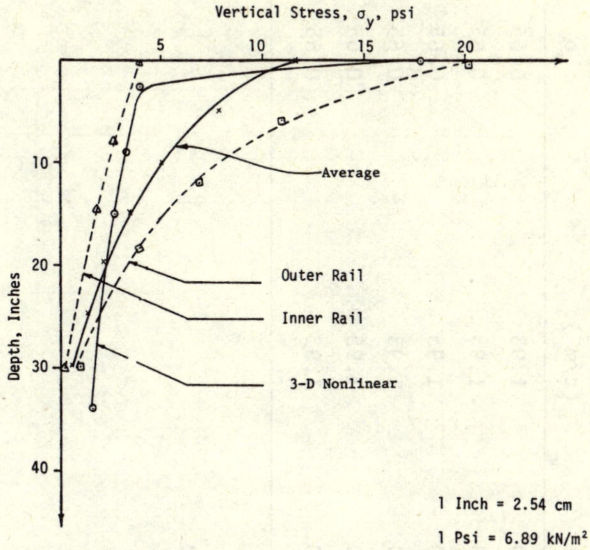

Figure 30 Comparison for Stresses Between Predictions and Field Data: UMTA Section

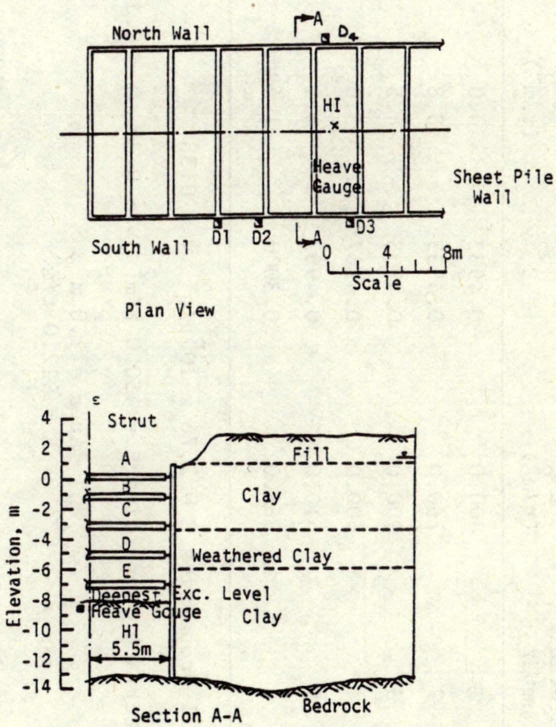

Figure 31 Details of Strutted Retaining Walls (40)

TABLE 2

Soil Properties for Vaterland 1 (40, 41)

Material Number	E (t/m^2)	ν	c (t/m^2)	ϕ	γ (t/m^3)	K_o
1	800.0	0.4933	4.0	0	1.95	0.65
2	760.0	0.4933	3.8	0	1.95	0.65
3	900.0	0.4933	4.5	0	1.95	0.65
4	500.0	0.4967	5.0	0	1.95	0.65
5	410.0	0.4967	4.1	0	1.95	0.65
6	368.0	0.3000	1.42	0	1.95	0.65

Wall Properties $E = 2.76 \times 10^6 t/m^2$ $\nu = 0.33$

Interface Properties $E = 450.0\ t/m^2$ $\nu = 0.33$ $G = 0.10\ t/m^2$

A,D Area = $1.0\ m^2$
E = $912.0\ t/m^2$

Strut Properties 1t = 1000 kg

B,C,E Area = $1.0\ m^2$
E = $4560.0\ t/m^2$

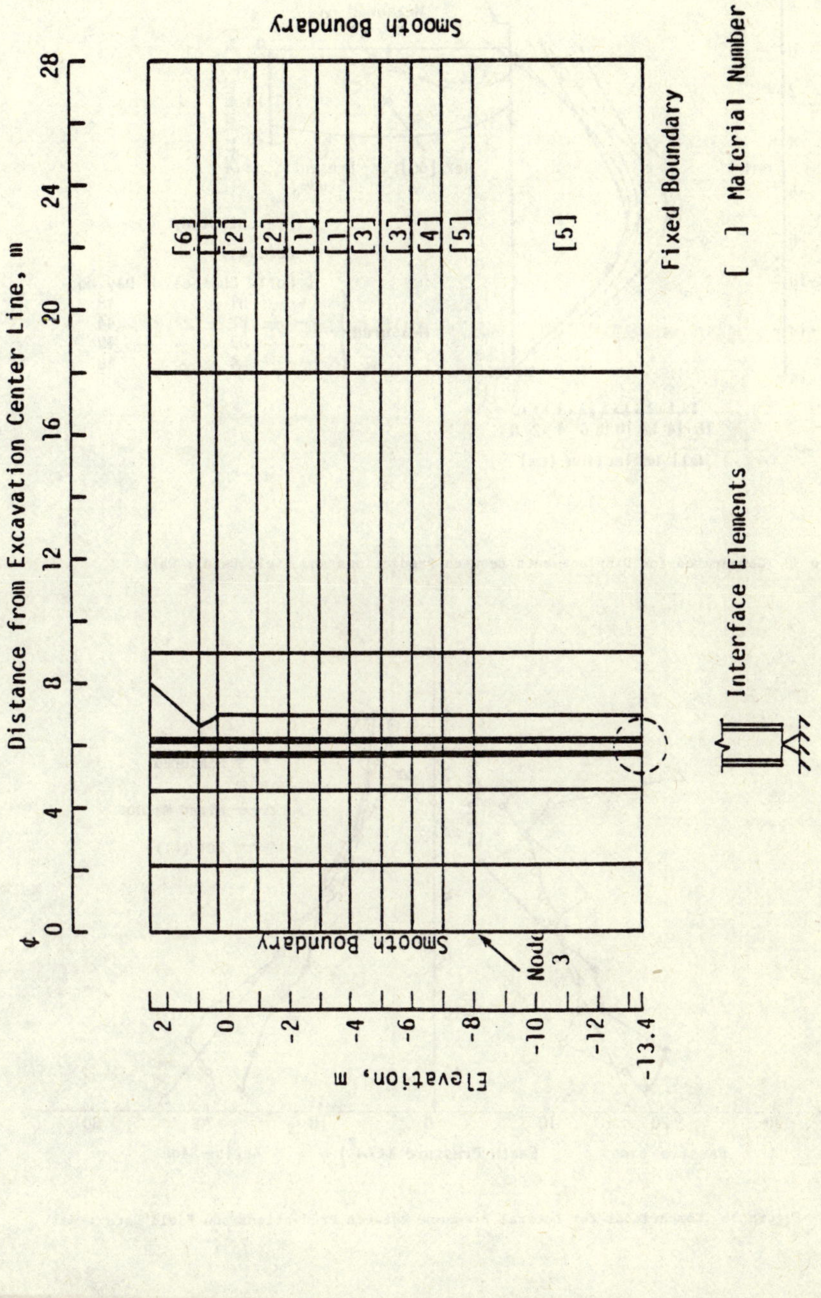

Figure 32 Mesh for Retaining Wall

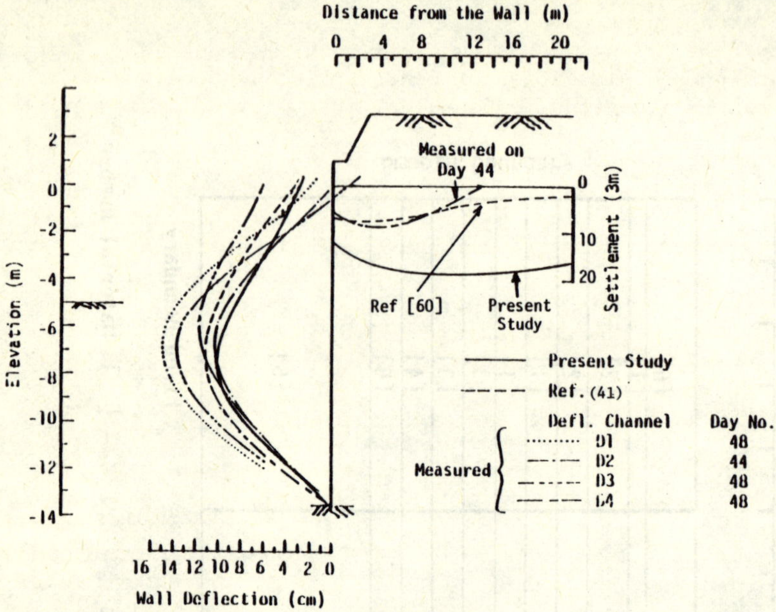

Figure 33 Comparison for Displacements Between Predictions and Field Data : Wall

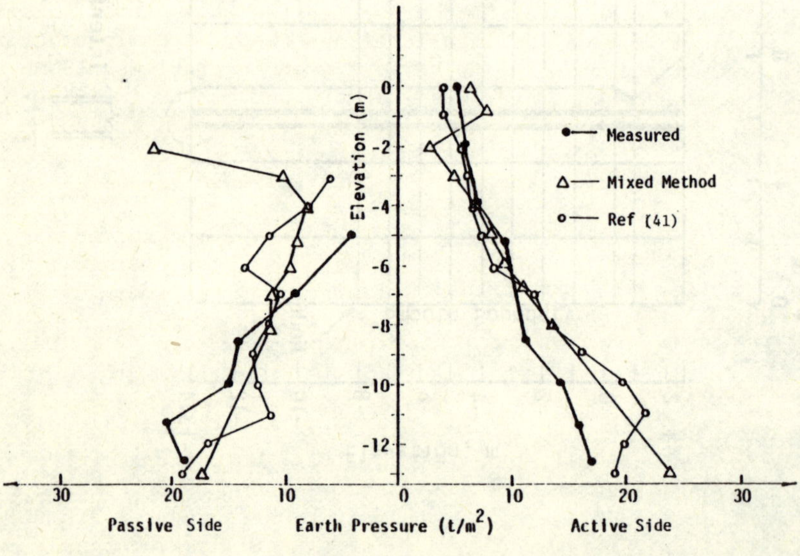

Figure 34 Comparisons for Lateral Pressure Between Predictions and Field Data : Wall

Displacement (32), mixed (35,36) and hybrid (37,38) finite element procedures have been developed and used in order to analyze this class of problems involving nonlinear soil-structure interaction and simulation of construction sequences such as dewatering, excavation, embankment, installation of supports and insitu stresses.

One of the objectives has been to investigate the usefulness of hybrid and mixed procedures for improved computation of stresses at the excavation surfaces and singularities.

Figure 33 shows a comparison for displacements between field observations (40,41) and finite element computations based on the mixed (35,36) and displacement (32,41) procedures. Similar results have been obtained by using the hybrid method (37,38).

Figure 34 shows a comparison between the results for lateral pressure in the interface. In the past, use of the conventional interface element often did not permit satisfactory and consistent evaluation of stresses in the interface. For instance, most previous results have plotted the lateral pressure as those computed in the adjoining soil elements. The proposed mixed and hybrid methods with the thin interface happen to provide satisfactory and improved correlations for pressures and stresses.

SUMMARY AND CONCLUSIONS

These lecture notes have presented the subjects of constitutive laws for solids and interfaces and their determination, and verification with respect to various example problems. Some new concepts as the thin interface elements and use of the mixed and hybrid procedures in soil-structure interaction have been included.

It is found that the constitutive laws and determination of appropriate parameters from laboratory tests play an important role in successful solution of soil-structure interaction problems. It can also be stated that the finite element method is capable of providing satisfactory solutions for a wide range of soil-structure interaction problems.

ACKNOWLEDGMENTS

The research results presented herein have been supported by grant Nos. ENG-760012, DAR-20500 and PFR-7914045 from the National Science Foundation and No. DOT-05-80013 from the Department of Transportation, Office of the University Research, Washington, D.C. H.V. Phan, H.J. Siriwardane, J.G. Lightner III, S. Sargand, R. Janardhanam, Md. Zaman and E.C. Drumm contributed to these research investigations and deserve special acknowledgement.

REFERENCES

1. Desai, C. S., Phan, H. V., and Sture, S.: 1981, Procedure, Selection and Application of Plasticity Models for a Soil, Int. J. Num. Analyt. Meth. in Geomech., Vol. 5, No. 3.

2. Desai, C. S., and Siriwardane, H. J.: Constitutive Laws of Engineering Materials; Prentice-Hall, Inc., New Jersey, under publication.

3. Nelson, I., and Baron, M. L.: 1971, Application of Variable Moduli Models to Soil Behavior, Int. J. Solids Struc., Vol. 7, pp. 399-417.

4. Drucker, D. C., and Prager, W.: 1952, Soil Mechanics and Plastic Analysis of Limit Design, Quart. Appl. Math., Vol. 10, No. 2, pp. 157-165.

5. Roscoe, K. H., and Burland, J. B.: 1968, On the Generalized Stress-Strain Behavior of 'Wet' Clays, in Engineering Plasticity, Heyman, J., and Leckie, F. A. (Eds.), Cambridge Univ. Press, England.

6. DiMaggio, F., and Sandler, I.: 1971, Material Models for Granular Soils, J. of Engg. Mech. Div., ASCE, Vol. 197, No. EM3.

7. Zienkiewicz, O. C., and Cormeau, I. C.: 1974, Visco-Plasticity, Plasticity and Creep in Elastic Solids: A Unified Numerical Solution Approach, Int. J. Numer. Methods Eng., Vol. 8, pp. 821-845.

8. Lade, P. V.: 1977, Elastic-Plastic Stress-Strain Theory for Cohesionless Soil with Curried Yield Surfaces, Int. J. Solids and Struct., Vol. 13, pp. 1019-1035.

9. Prevost, J. H.: 1977, Mathematical Modeling of Monotonic and Cyclic Undrained Clay Behavior, Int. J. Num. Analyt. Methods in Geomech., Vol. 1, No. 2, pp. 156-216.

10. Bazant, Z. P., and Krizek, R. J.: 1976, Endochronic Constitutive Law for Liquefaction of Sand, J. of Engg. Mech. Div., ASCE, Vol. 102, No. EM2, pp. 225-238.

11. Dafalias, Y. F., and Herrmann, L. R.: A Generalized Bounding Surface Constitutive Model for Clays in Applications of Plasticity and Generalized Stress-Strain Relations in Geotech. Engg. Special Publ. Series, ASCE, under publication.

12. Desai, C. S.: 1980, A General Basis for Yield, Failure and Potential Functions in Plasticity, Int. J. Num. and Analy. Methods Geomech., Vol. 4, No. 4, pp. 361-375.

13. Desai, C. S., and Siriwardane, H. J.: 1980, A Concept of Correction Functions to Account for Non-Associative Characteristics of Geological Media, Int. J. Numer. and Analyt. Methods in Geomech., Vol. 4, No.4 , pp. 377-387.

14. Baker, R., and Desai, C. S.: Consequences of Deviatoric Normality in Plasticity with Isotropic Strain Hardening, under publication in Int. J. Numer. and Analyt. Methods in Geomech.

15. Sture, S., and Desai, C. S.: 1979, Fluid Cushion Truly Triaxial or Multiaxial Testing Device, J. of Geotech. Testing, ASTM, Vol. 2, No. 1.

16. Desai, C. S., Janardhanam, R., and Sture, S.: A High Capacity Multiaxial Testing Device, Submitted to J. Geotech. Testing, ASTM.

17. Desai, C. S., Perumpral, J. V., and Sture, S.: 1980, Mechanics of Soil-Tool Interaction in Tillage, Report to NSF, Dept. of Civil Engg., Virginia Tech, Blacksburg, VA.

18. Phan, H. V.: 1979, Geometric and Material Nonlinear Analysis of Three-Dimensional Soil-Structure Interaction, Ph.D. Dissertation, Virginia Tech, Blacksburg, VA.

19. Desai, C. S., Phan, H. V., and Perumpral, J. V.: Mechanics of Three-Dimensional Soil-Structure Interaction, under publication, J. Engg. Mech. Div., ASCE.

20. Siriwardane, H. J.: 1980, Nonlinear Soil-Structure Interaction Analysis for One-, Two- and Three-Dimensional Problems Using Finite Element Method, Ph.D. Dissertation, Virginia Tech, Blacksburg, VA.

21. Janardhanam, R.: 1981, Constitutive Modeling for Media in Track Support Structures, Ph.D. Dissertation, Virginia Tech, Blacksburg, VA.

22. Desai, C. S., Siriwardane, H. J., and Janardhanam, R.: 1980, Load Transfer and Interaction in Track-Guideway Systems, Report to Office of Univ. Res., DOT, Washington, D. C.

23. Desai, C. S., and Siriwardane, H. J.: Numerical Models for Track Support Structures, accepted for publication, J. Geotech. Engg. Div., ASCE.

24. Goodman, R. E., Taylor, R. L, and Brekke, T. L.: 1968, Model for Mechanics of Jointed Rock, J. of Soil Mech. and Found. Div., ASCE, Vol. 94, No. SM1.

25. Ghaboussi, J., Wilson, E. L., and Isenberg, J.: 1973, Finite Element for Rock Joints and Interfaces, J. Soil Mech. and Found. Div., ASCE, Vol. 94, No. 3

26. Desai, C. S.: 1974, Numerical Design Analysis of Piles in Sands, J. Geotech. Engg. Div., ASCE, Vol. 100, No. GT6.

27. Katona, M. G. et al.: CANDE, 1976, A Modern Approach for the Structural Design and Analysis of Buried Culverts, Report FHWA-RD-77, Fed. Highway Admin., Washington, D. C.

28. Herrmann, L. R.: 1978, Finite Element Analysis of Contact Problems, J. Engg. Mech. Div., ASCE, Vol. 104, No. EM5.

29. Desai, C. S., and Christian, J. T. (Eds.): 1977, Numerical Methods in Geotechnical Engineering, McGraw-Hill Book Co., N.Y.

30. Desai, C. S.: 1977, Soil-Structure Interaction and Simulation Problems, Chap. 7 in Finite Elements in Geomechanics, G. Gudehus (Ed.), John Wiley, U. K.

31. Desai, C. S.: 1981, Behavior of Interfaces Between Structural and Geologic Media, Proc. Int. Conf. on Geotech. Earthquake Engg. and Soil Dynamics, Vol. II, St. Louis, MO.

32. Lightner, J. G., and Desai, C. S.: 1979, Improved Numerical Procedures for Soil-Structure Interaction Including Simulation of Construction Sequences, Report VPI-E-79. 32, Virginia Tech, Blacksburg, VA.

33. Isenberg, J., and Vaughan, D. K.: 1981, Nonlinear Effects in Soil-Structure Interaction, Proc. Symp. on Impl. of Computer Procedures and Stress-Strain Laws in Geotech. Engg., Chicago.

34. Pande, G. N., and Sharma, K. G.: 1979, On Joint Interface Elements and Associated Problems of Numerical Ill - Conditioning, Short Comm., Int. J. Numer. Analyt. Methods Geomech., Vol. 3, No. 3.

35. Lightner, J. G.: 1981, A Mixed Finite Element Procedure for Soil-Structure Interaction Including Construction Sequences, Ph.D. Dissertation, Virginia Tech, Blacksburg, VA.

36. Lightner, J. G., and Desai, C. S.: 1981, Mixed Finite Element Procedures for Soil-Structure Interaction and Construction Sequences, Report to National Science Foundation, Washington, D. C.

37. Sargand, S.: 1981, A Hybrid Finite Element Procedure for Soil-Structure Interaction Including Construction Sequences, Ph.D. Dissertation, Virginia Tech, Blacksburg, VA.

38. Sargand, S. M., and Desai, C. S.: 1981, Hybrid Finite Element Procedures for Soil-Structure Interaction and Construction Sequences, Report to National Science Foundation, Washington, D. C.

39. Desai, C. S.: A Cyclic Multi-Degree-of-Freedom Shear Device, under publication, J. Geotech. Testing, ASTM.

40. Norwegian Geotechnical Institute: 1962, Measurements at Strutted Excavation, OSLO Subway, Vaterland 1, km 1373, NGI Tech. Report No. 6, Oslo, Norway.

41. Mana, A.: 1978, Finite Element Analyses of Deep Excavation Behavior in Soft Clay, Ph.D. Dissertation, Stanford Univ., Stanford, Calif.

39. [illegible] A Contribution [illegible] Free-Field [illegible] underground [illegible], ASCE.

ROCK MASS BEHAVIOUR AND ITS MATHEMATICAL MODELLING

K. Kovári

Federal Institute of Technology, Zurich, Switzerland

ABSTRACT

The behaviour of rock masses is largely determined by the following factors: orientation and nature of its discontinuities and properties of the intact rock. Deformation and failure may take place in both parts resulting in the actual response of the rock mass to changes in stresses. Rock masses may behave both in a brittle or a ductile manner depending on the state of stress they are subjected to. It is of fundamental significance to understand the mechanism of deformation and failure as a process. For this purpose experimental evidences gained from laboratory testing will be discussed. The mathematical modelling of the observed phenomena will be done taking into consideration elasticity, internal friction, cracking and viscosity. Special attention is given to the problems of the continuum mechanics approach in tunnelling practice.

1. INTRODUCTION

It is a special characteristic of rock engineering, that its structural material - the rock mass - is not selected, but encountered in-situ, and that this material is already in a stressed state before construction even starts. Due to the construction there results in its vicinity a change in the initial state of stress. The 'loading' case may be one of unloading (e.g. tunnel excavation, rock cutting) or of loading (e.g. concrete dam). The criteria for the design of the structures results from the requirements of limiting deformations, stability and economic considerations.

J. B. Martins (ed.), Numerical Methods in Geomechanics, 145–163.

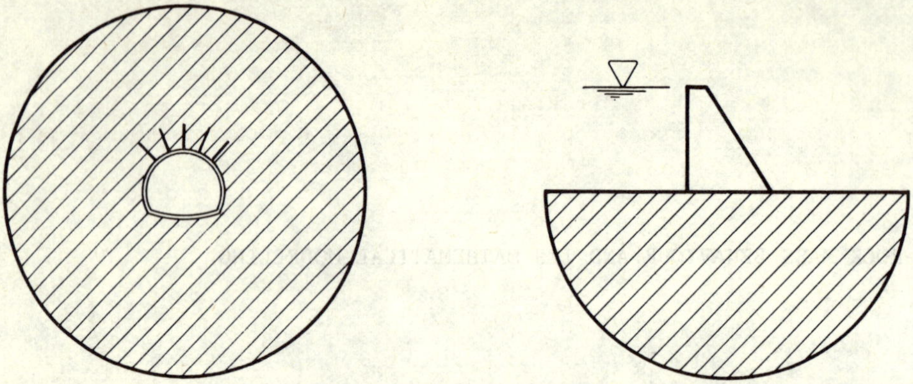

Figure 1 Structural models for tunnels and large dams

Along with engineering judgement and field observations statical computations are a valuable aid to fulfilling these requirements. The starting point is given by the structure, which in tunnelling comprises both the lining and the rock surrounding it (Fig. 1). The study of the interaction of these two structural elements helps our understanding of the structural behaviour. Similar considerations apply also to a concrete dam or a rock slope.

Although mathematical modelling is applied mostly to a plane strain problem, three dimensional studies must also be undertaken sometimes. Examples of this would be at the working face of a tunnel, the stability calculation for a rock wedge and the stress analysis of an arch dam (Fig. 2).

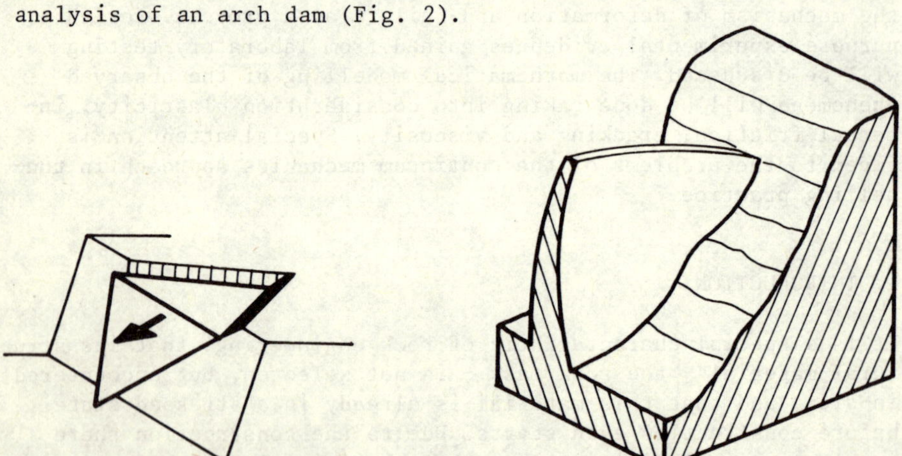

Figure 2 Three dimensional structural models, wedge failure in a rock slope and arch dam

The value of a calculation depends on the extent to which the relevant rock properties are known and may be introduced into the mathematical model. Understanding the basic mechanisms governing the behaviour of the structure, which may be regarded as an interplay between rock properties, method of construction and type of structure, is of great importance. Here,however, there are great difficulties, for, on the one hand, one has to understand thoroughly the material properties of the rock and be acquainted with methods of constructional practice and, on the other hand, one should possess a good knowledge of numerical methods in geomechanics. Statical calculations will always need to have a practical interpretation, if they are to be really useful.

The mechanical properties of rock influence considerably the design of the structure built in it. Essentially they are determined by the characteristics of the intact rock itself which may be tested in the laboratory and by the frequency and spacial orientation of the discontinuities with their properties (Fig. 3). The strength and deformational properties are of major importance for the basic material. Rocks which contain clay minerals or anhydrite can adsorb water leading either to a large volumetric strain or, if this is prevented, a pressure - the so-called swelling pressure is built up. Joints, bedding planes and schist planes in rock are as a rule plane surfaces with small or negligible tensile strength and reduced shear strength. Therefore, one has to test not only the mechanical properties of the intact rock but also the strength of the discontinuities. A special case is given by the frequently encountered anisotropy of the deformation and strength properties, which is caused by bedding planes, schistosity or by the presence of a regular joint pattern. To what extent the discontinuities or

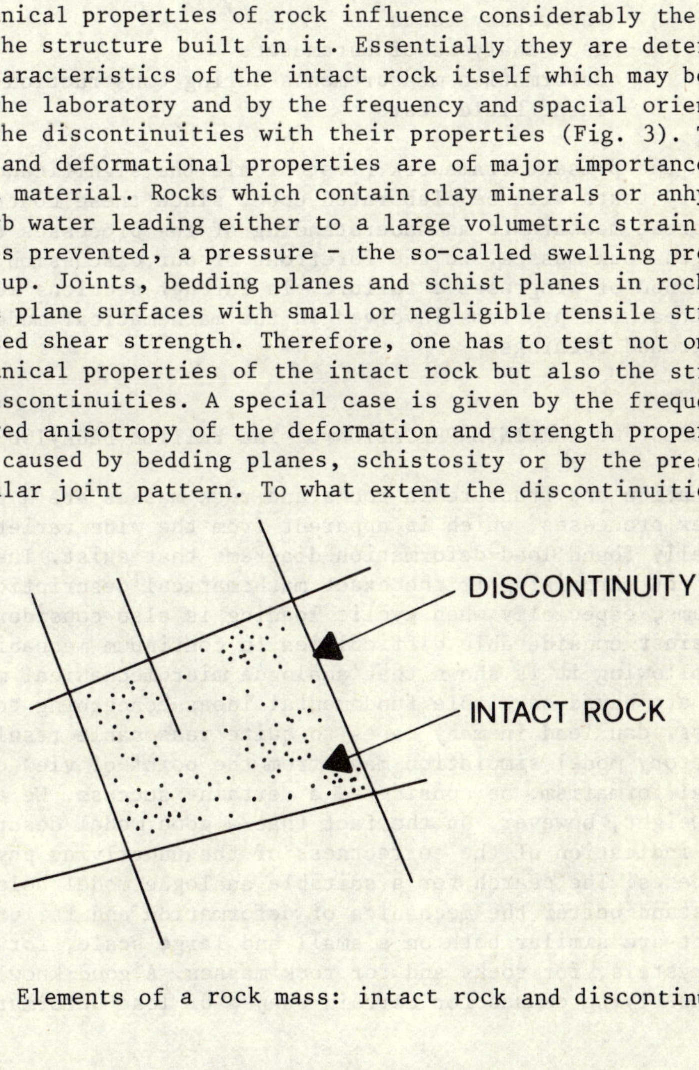

Figure 3 Elements of a rock mass: intact rock and discontinuities

the basic rock properties are responsible for the behaviour of a rock mass depends very much on the type of the planned structure. In the investigation of the stability of a rock slope attention is concentrated usually on the orientation and strength of the discontinuities. For foundation problems and tunnelling both discontinuities and intact rock properties acting together may play an important role.

The following possibilities are available for determining the relevant mechanical properties of a rock mass:

- visual inspection, experience
- geological, geophysical and rock mechanics investigations
- laboratory tests
- back analysis of parameters
- deformation measurements during construction
- large field tests

Within the present framework first of all the significance of laboratory tests will be elaborated upon, since these contribute in a great measure to an understanding of the processes taking place in rock masses. In the forefront of our discussion is the phenomenon of progressive failure. In further sections we will then treat the problems involved in the mathematical modelling of underground openings.

2. MODELS FOR THE INTERPRETATION OF THE BRITTLE BEHAVIOUR OR ROCKS

Deformation and fracture in rocks and rock masses are a result of complex processes, which is apparent from the wide variety of empirically found load-deformation diagrams that exist. Therefore, it is not surprising if the exact mathematical description of such diagrams, especially when cyclic loading is also considered,comes up against considerable difficulties in continuum mechanics. In the following it is shown that analogue micromechanical models, which are based on simple fundamental ideas concerning the failure process, can lead in many cases to quite reasonable results. A satisfactory model simulation may, from the point of view of mathematical formalism, be considered a certain success. We also lay much weight, however, on the fact that a good model description is an indication of the correctness of the underlying physical hypotheses. The search for a suitable analogue model helps us to understand better the mechanics of deformation and failure, which in fact are similar both on a small and large scale, for example for crystals, for rocks and for rock masses. A good knowledge of the underlying causes for certain shapes of load-deformation or stress-strain diagrams is especially desirable, if for practical purposes the curves have to be greatly idealized. One only has to think of the frequently used plate loading test for soil or jointed rock, for which a strongly nonlinear load-displacement curve and, in addition, for cyclic loading, hysteresis effects are obtained. In spite of this complexity only a single parameter (the so-called deformability modulus) is derived from this curve to characterize the compressibility of the ground. For the commonly used plate loading test on rock there does not even exist a generally accepted recommendation as how to evaluate this one parameter. But also, in the formulation of constitutive laws for soil and rock taking into consideration strain softening phenomena,an understanding of the nature of progressive failure would be extremely valuable.

The elementary physical phenomena, upon which the models under consideration are based, constitute elasticity, rupture (crack formation) and internal friction. In Fig. 4 simple combinations of the basic elements are represented, which result in the elastic-ideal brittle bodies. A more sophisticated combination is given by the elastic-ideal brittle body with residual strength.

Model elements representing these properties are assigned variable parameters and assembled so as to form so-called blocks. In this way the effects of the characteristic features on a micro-scale such as inhomogeneities, voids, anisotropy of individual crystals and nonuniform stress conditions can satisfactorily be considered. It has been shown, in fact, that even when ideal homogeneous external stresses are applied to the specimen the corresponding internal stresses at the scale of single grains vary considerably. An example to illustrate this is the development of tensile cracks in the uniaxial compression test, or more generally stress concentrations in the vicinity of voids and holes {1}. The observed overall failure behaviour is caused by a sequence of local material failures associated with local redistributions of the stress field. Since, in the formulation of the analogue model, we lay greatest emphasis on the simulation and quantitative description

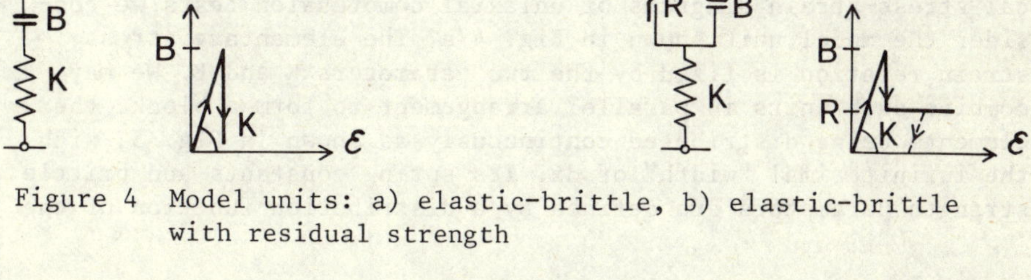

Figure 4 Model units: a) elastic-brittle, b) elastic-brittle with residual strength

of this progressive failure, one could equally well in the following speak of "process models". Alongside the mathematical description of load-deformation diagrams precise information concerning the internal state of the material at any loading stage is obtained. In this way the effect of stress paths will clearly be demonstrated.

The usefulness of analogue models to further our understanding of the mechanics of deformation is well illustrated by Prandtl's kinetic theory model {2}. He was able to establish the laws for hysteresis loops observed for certain metals under cyclic loads.His concepts were based upon processes occurring on the molecular scale and included time effects as well. Some of the basic ideas for the application of models of the kind we shall discuss here were first published by Jenkin {3}. In order to describe the stress-strain curves for metals under different test conditions he used a combination of spring and friction elements arranged in such a way that hysteresis effects could be simulated. Further developments of Jenkin's concepts are found in the work of Owen et al {4}, in which besides elasticity and friction crack elements to consider strain softening behaviour are proposed. Continuous distribution of elements with so-called "block formation" to describe empirically-found stress-strain curves for rocks under uniaxial and triaxial compression were introduced by Kovari {5}.

The following discussion will be limited to short term tests or to such tests for which time only plays a subordinate role.Further, empirically-found and not idealized load-deformation diagrams will be treated. Diagrams obtained empirically, however, are affected by various testing influences, for instance by parasitic stresses caused by imperfect loading conditions or due to different shape and size of the specimens used, etc.

2.1. The uniaxial compression test

The mathematical model is introduced here with reference to an elastic brittle material with strain softening. Based upon reliable observations we proceed upon the assumption that in this case, mainly elasticity and separation, i.e. crack formation, are important for the deformation behaviour. In order to investigate typical stress-strain diagrams of uniaxial compression tests we consider the model unit shown in Fig. 4/a. The elementary stress-strain relation is fixed by the two parameters K and B. We may combine such units in parallel arrangement to form a block, the elements being distributed continuously as shown in Fig. 5, with the infinitesimal "width" of dx. Its spring constants and brittle strength parameters are defined by a distribution function of the

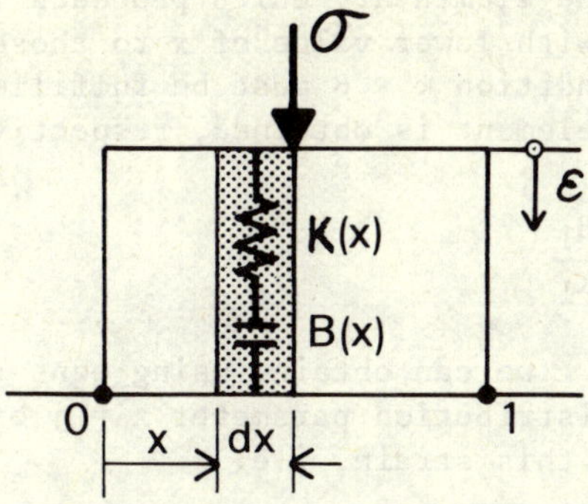

Figure 5 Continuous distribution of elastic-brittle elements forming a block

parameter x. Without restricting ourselves, x lies in the range of $o \leq x \leq 1$. The stress acting on the block as a whole, in the sense of an external load, is denoted by σ and the corresponding strain is ε. It follows from the parallel arrangement that all elementary units are subjected to the same strain ε, but, in general, the units carry different stresses. For the sake of simplicity, in the mathematical formulation we assume a distribution of element parameters, namely

$$K(x) = K_o[1+(\kappa-1)x] \ , \quad B(x) = B_o[1+(\beta-1)x] \tag{1}$$

whereby the abbreviations $K(x=1)=K_1$ and $B(x=1)=B_1$ together with

$$\kappa = \frac{K_1}{K_o} \quad \text{and} \quad \beta = \frac{B_1}{B_o} \tag{2}$$

are introduced. Under monotonic initial loading the infinitesimal element exhibits the following stress-strain relations:

$$d\sigma = \varepsilon K(x)dx \quad \text{for} \quad o \leq \varepsilon \leq \varepsilon^B(x) \ , \qquad d\sigma = o \quad \text{for} \quad \varepsilon > \varepsilon^B(x) \ . \tag{3}$$

The value $\varepsilon^B(x)$ indicates the individual failure strain of the element with the distribution parameter x and can generally be expressed by

$$\varepsilon^B(x) = \frac{B(x)}{K(x)} \tag{4}$$

To ensure that failure of the elementary units proceeds in a regular manner from elements with lower value of x to those with higher values simply the condition $\kappa < \beta$ must be fulfilled. Failure of the first and last element is obtained, respectively, at the following strains:

$$\varepsilon_o^B = \frac{B_o}{K_o} \quad , \quad \varepsilon_1^B = \frac{B_1}{K_1} \ . \tag{5}$$

For a given value of strain ε we can obtain, using eqns (1), (2) and (4), the value of the distribution parameter x = j of that element which just fails at this strain, i.e.

$$j = \frac{\varepsilon K_o - B_o}{\varepsilon(K_o - K_1) - (B_o - B_1)} \quad (o \leq j \leq 1) \ . \tag{6}$$

We call the value j a "process parameter", as it gives precise information about which part of the block has failed and which part is still intact. Alternatively we can say that the percentage of elements having reached failure is (100j)%. For steadily increasing strain in the block, the integration of eqns. (3) gives the following ranges of the stress-strain diagram

$$\begin{aligned} &\text{for } o \leq \varepsilon \leq \varepsilon_o^B \qquad \sigma = \varepsilon \int_o^1 K(x)dx \ , \\ &\text{for } \varepsilon_o^B < \varepsilon \leq \varepsilon_1^B \qquad \sigma = \varepsilon \int_j^1 K(x)dx \ . \end{aligned} \tag{7}$$

In the stage of the progressive failure $\varepsilon \geq \varepsilon_o^B$ the stress-strain diagram of the block can conveniently be represented by the two equations

$$\begin{aligned} &\varepsilon(j) = \frac{B_o - (B_o - B_1)j}{K_o - (K_o - K_1)j} \ , \\ &\sigma(j) = \frac{1}{2}[2 \ K_o(1-j) - (K_o - K_1)(1-j^2)] \ \frac{B_o - (B_o - B_1)j}{K_o - (K_o - K_1)j} \ . \end{aligned} \tag{8}$$

Parallel Coupling of Blocks:
The summation law for the effect of single elements is valid also for the effect of blocks when they too are coupled in parallel. One can for example determine the individual stress-strain diagrams by applying (14) and (15) and then by means of superposition the resultant curve is obtained. Thus the equation for N blocks in parallel arrangement is given by

$$\sigma(\varepsilon) = \sigma_1(\varepsilon) + \sigma_2(\varepsilon) + \ .... \ + \sigma_N(\varepsilon).$$

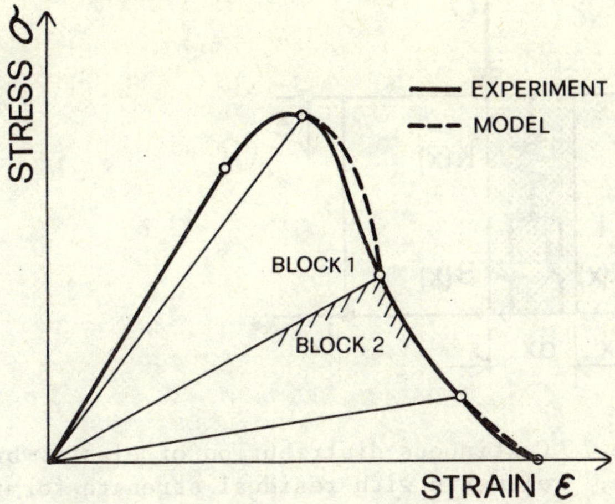

Figure 6 A two-block model approximation to a given stress strain curve

One can choose the distribution functions of the individual blocks in such a way that the block with the lowest index value fails first, the strength of this block being fully exhausted and then failure passes on to the next block. In this way the failure mechanism can be followed very easily, and what is more important, experimentally found stress-strain diagrams can be described adequately by suitable block combinations.

An example for a two-block approximation to a given experimentally find stress-strain diagram is shown in Fig. 6. Details of the determination of the individual block parameters are reported elsewhere {5}.

2.2. The triaxial compression test

In conventional triaxial tests peak strength and residual strength as well as the rupture strain of the rock specimen increase with increasing confining pressure. These properties can be taken into consideration using elastic-brittle units with residual strength, distributed continuously in a parallel arrangement (Fig. 7). For a constant confining pressure the distribution functions K(x), B(x) as well as the newly introduced function R(x) should also be linear. In addition, we assume that the strength parameters are likewise related, linearly, to the confining pressure p. Thus for a block the following relations are obtained:

$$K(x) = K_o\,[1 + (\kappa-1)\,x]\quad,$$
$$B(x) = B_o\,[1 + (\beta-1)\,x]\quad, \tag{9}$$
$$R(x) = R_o\,[1 + (\rho-1)\,x]$$

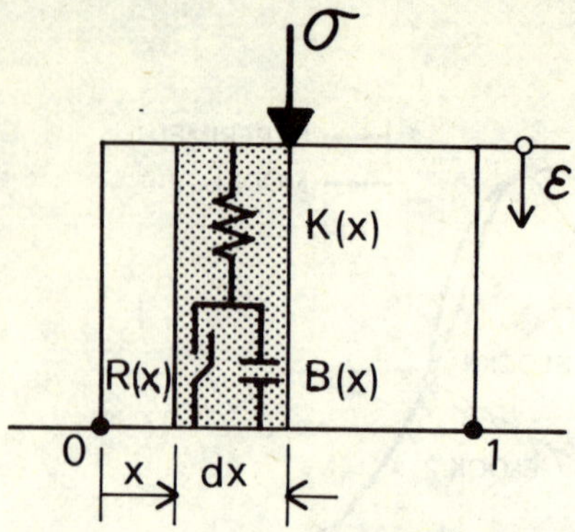

Figure 7 Continuous distribution of elastic-brittle elements with residual strength forming a block

where now the parameters B_o and R_o are given by the expressions below as a function of p

$$B_o = a + bp ,$$
$$R_o = cp .$$

The symbols κ, β and ρ are defined as

$$\kappa = \frac{K_1}{K_o} , \beta = \frac{B_1}{B_o} \text{ and } \rho = \frac{R_1}{R_o}$$

and serve as block characteristics. For continuous failure propagation inside the block the condition $\kappa < \beta$ holds again. The uniaxial compression test is seen to be a special case of the triaxial test with p = o. The process parameter j has again the form (6) and the stress-strain diagram in the stage of progressive failure ($\varepsilon \geq \varepsilon_o^B$) is now

$$\sigma = \varepsilon \int_j^1 K(x)dx + \int_o^j R(x)dx .$$

Thus the total stress is composed of the elastic resistance of the still intact units and of the residual strength of the failed units. The stress strain relationship of a block will be given by

$$\varepsilon(j) = \frac{B_o-(B_o-B_1)j}{K_o-(K_o-K_n)j} , \qquad (10)$$

$$\sigma(j) = [2K_o(1-j)-(K_o-K_1)(1-j^2)] \frac{B_o-(B_o-B_1)j}{K_o-(K_o-K_1)j} + j[2R_o-(R_o-R_1)j]\frac{1}{2} .$$

As an example for the practical application of the above equations Fig. 8 may be considered. Here the two-block model shown in Fig.6

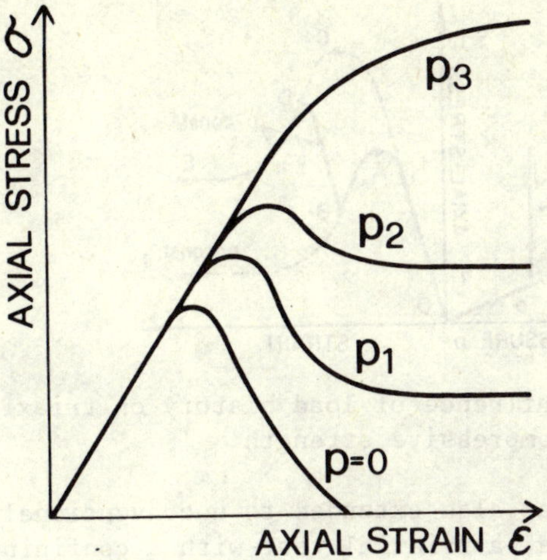

Figure 8 A two-block model simulating rock behaviour under triaxial compression

has been extended to the triaxial loading case. The curves display the major features of such tests, like increasing peak- and residual strength with increasing confining pressure and even ductility can be observed. The stress-strain curve in the progressive failure stage ($\varepsilon > \varepsilon_0^B$) can be looked upon as the result of a cohesive strength combined with frictional resistance. The distribution of these strength components as a function of strain ε is illustrated in Fig. 9 for the case of the confining pressure $p = p_1$.

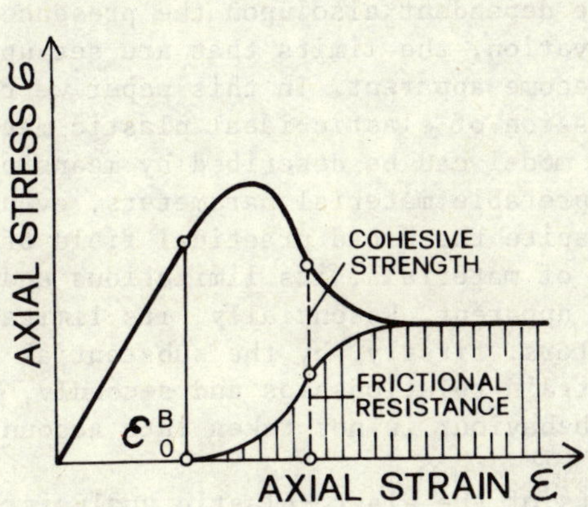

Figure 9 Development of residual strength (friction) in triaxial compression with increasing strain

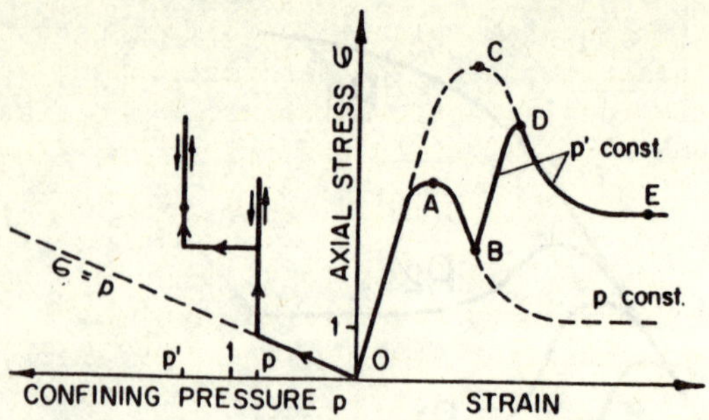

Figure 10 Influence of load history on triaxial compressive strength

The model simulation was also extended to unconventional tests. If we stop, for example, a triaxial test with a confining pressure p during the post failure stage at point B (Fig. 10) and, subsequently, increase the confining pressure to a value p', we obtain the result shown in Fig. 10. The curve BD shows that the rock, after partial loss of cohesion, exhibits a smaller peak strength (point D) and less capacity for deformation.

3. CONTINUUM MECHANICS MODELS

In the construction of underground openings and large dams all forms of occurrence of weak and hard rock can be encountered. For stress analysis it is the mechanical properties of the material on a large scale that is of consequence. When we consider that the material properties are dependent also upon the presence of water and the method of excavation, the limits that are set upon every computational method become apparent. In this paper we restrict ourselves to the discussion of elastic-ideal plastic material behaviour. This material model can be described by means of a small number of easily interpretable material parameters, even in the case of anisotropy. Despite the broad practical field of application of this concept of materials, its limitations and certain disadvantages are also apparent. Essentially, its limitations are conditioned by two factors: firstly, by the substantial simplification of the stress-strain relationships and secondly, by the fact that rheological behaviour is not taken into account.

3.1. Theoretical aspects of the elasto-plastic analysis

A material may be classified as elasto-plastic if after loading beyond a certain limit a permanent (plastic) deformation remains after unloading. A system is described as elasto-plastic if, in any part or zone of the system, plastic deformations occur. Such domains are called plastic zones. The deformation and strength behaviour of an elastic-ideal plastic material is described by the basic concepts of Hooke's law, yield condition, and flow rule.

Hooke's law:
Concerning Hooke's law, it needs only to be mentioned that an isotropic material is described by two material parameters, the transverse anisotropic material by five, and the general anisotropic material by twenty-one parameters.

The yield condition for geotechnical applications:
In general, for geotechnical applications, the Coulomb yield condition is applied, either in its usual form or in the extended Drucker-Prager {6} form. In both cases a further condition is introduced to account for the partial or complete reduction of tensile strength (tension cut-off). According to Coulomb's yield condition for an isotropic material no failure will occur if, at the point considered in the medium, the following condition is fulfilled for any arbitrary oriented element of area

$$|\tau| < \sigma \tan\phi + c \tag{11}$$

The material parameters ϕ and c are called the angles of internal friction and cohesion respectively. Shield {7} has shown that the above condition (11) leads in terms of the principal stresses to six relationships of the form:

$$\sigma_i \leq m\sigma_j + b \quad (\sigma_i > \sigma_j) \tag{12}$$

where

$$m = \frac{1+\sin\phi}{1-\sin\phi} \quad \text{and} \quad b = \frac{2c\cos\phi}{1-\sin\phi} .$$

In principal stress space $(\sigma_1, \sigma_2, \sigma_3)$ the six relationships (12) are represented by a right, irregular hexagonal pyramid whose axis is equally inclined to the principal axes.

In plane-strain problems only failure mechanisms are of interest in which the failure surface is normal to the plane. For a transverse isotropic material with tension cut-off, three conditions in the form of inequalities

$$4\tau_{xy}^2 + (\sigma_x - \sigma_y)^2 - [(\sigma_x + \sigma_y) + 2\cot\phi]^2 \sin^2\phi \leq o ,$$
$$4\tau_{xy}^2 + (\sigma_x - \sigma_y)^2 - (\sigma_x + \sigma_y + 2\beta_z)^2 \leq o ,$$
$$|\tau_{xy}| - \sigma_y \tan\bar{\phi} - \bar{c} \leq o$$

must be formulated, namely for shear failure across the stratification in any direction, for the inability to sustain tensile stresses greater than $\sigma_i \geq \beta_z$ and for failure in the planes with the predetermined direction and with the friction angle $\bar{\phi}$ and cohesion $\bar{c}$. It was shown {8} that these conditions may be illustrated effectively in the 3-dimensional stress space $(\sigma_x, \sigma_y, \sqrt{2}\tau_{xy})$ by means of three surfaces (Fig. 11). The Coulomb condition for an isotropic material is represented by a circular cone, while the anisotropic condition is identified with two planes. The condition of no-tension (β_z = o) fixes the vertex of a second cone at the origin o.

The well-known Drucker-Prager yield condition

$$F = -\alpha I_\sigma + \sqrt{II_\sigma} - k = 0 \qquad (13)$$

is an extended form of the Coulomb condition. The quantities I_σ and II_σ represent the basic stress invariants of the stress tensor and the stress deviator respectively; α and κ are material constants. In the principal stress space, Equation (13) represents a cone with the same axis as the pyramid resulting from Shield's representation. Since the Coulomb condition is more deeply anchored in engineering practice and in addition, it allows

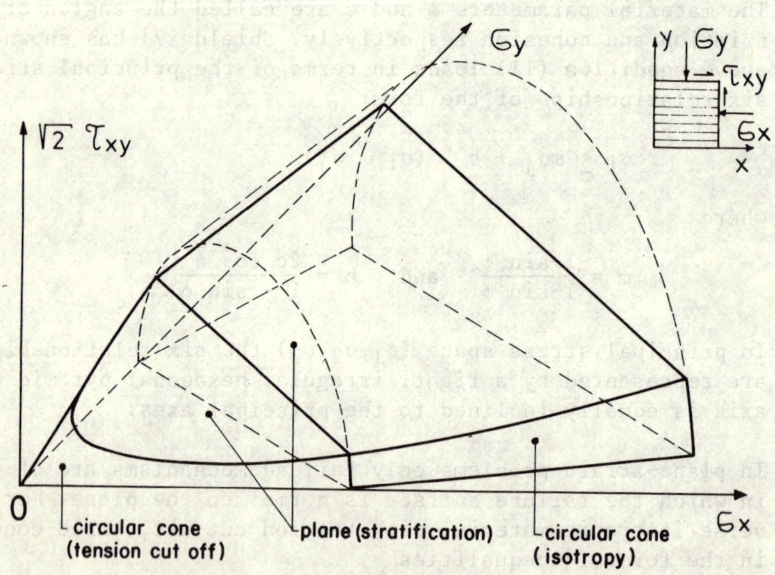

Figure 11 The Coulomb yield criterion for plane strain conditions in the stress space for a material with "tension cut-off" and with planes of weakness parallel to the σ_x-direction

for a unique description of the anisotropic material, it is given preference over the Drucker-Prager condition. The Coulomb condition also permits a simple clear and rapid graphical check to be made on the state of stress with the aid of the Mohr stress circle.

The flow rule
The flow rule gives information about the plastic strain rate $\{\dot{\varepsilon}\}^p$ in the state of flow. For the elastic-ideal plastic material (such as different metals) it appears that the so called normality condition

$$\{\dot{\varepsilon}\}^p = \lambda \frac{\partial F}{\partial\{\sigma\}} \qquad (14)$$

is appropriate both from the theoretical point of view and from the results of experimental investigations. In Equation (14) λ is a constant of proportionality not yet determined. The normality condition implies that the plastic strain rate $\{\dot{\varepsilon}\}^p$ lies in the stress space in a direction normal to the yield surface. A direct consequence of this result is the fact that, theoretically, a volume increase in the state of flow results for Coulomb materials. The validity of the normality condition is a matter of much dispute for materials of the earth's crust but there is no doubt about the occurrence of a volume dilatancy when the rock is in the state of failure.

4. COMPUTATIONAL MODELS IN TUNNELLING

Depending on the type of rock pressure phenomena anticipated the computational models for a tunnel must be adopted section by section. The differentiation between rock pressures in terms of the following three phenomena, namely

- loosening pressure
- genuine rock pressure
- swelling pressure

corresponds to traditional engineering practice and also gives the first important indications of the kind of statical problems to be expected. The variations in model concepts are also brought about by the different ways in simulating the interaction between tunnel and surrounding rock {9}.

4.1. Computational models for rock masses with loosening pressure

Such a rock mass is basically self-supporting around the excavated opening. The rock samples have a uniaxial compressive strength which is many times higher than the overburden pressure. Thus the excavation might not induce new fractures in the rock. Its behaviour is characterized mainly by the original joint pattern and by the shear strength along those joints or other surfaces of weakness. The overall deformations of the rock take place essentially before the introduction of the supports. The corresponding reduction in the cross-section is negligible, and generally hardly noticeable. The danger of loosening and rock fall due to the presence of local defects, however, call for permanent and, in many cases, for temporary support as well. Unfavourable combinations of layers and joints, shattering effects of blasting and percolation of ground water can all contribute to the settling of a certain volume of material on to the lining, which exerts a pressure according to its deadweight. Because the process of loosening occurs in parts of the tunnel which cannot be determined in advance and is influenced by uncontrollable factors, its prediction computationally appears to be out of question. In this type of rock a reasonable approach is to assume a loading in the roof-zone of the tunnel which corresponds to the estimated volume of the loosened material. The interaction of the lining with the rock mass is taken into account by the partial embedment of the tunnel lining in an elastic continuum (Fig. 12). The modulus E, of the continuum corresponds to the deformability of the rock mass. In the computation the condition is applied that no tensile stress is transmitted between the rock and the lining. Generally it necessitates an iterative procedure. A criticism of this computational model is that all computed results are dependent upon the estimated loading and are thus questionable. One must consider, however, that by means of such computations useful relationships between important factors, like the shape and size of the cross-section, the stiffness of the lining and the deformability of the rock, may be established. In this way a reasonable and rewarding parametric study can be carried out. Optimistic and pessimistic

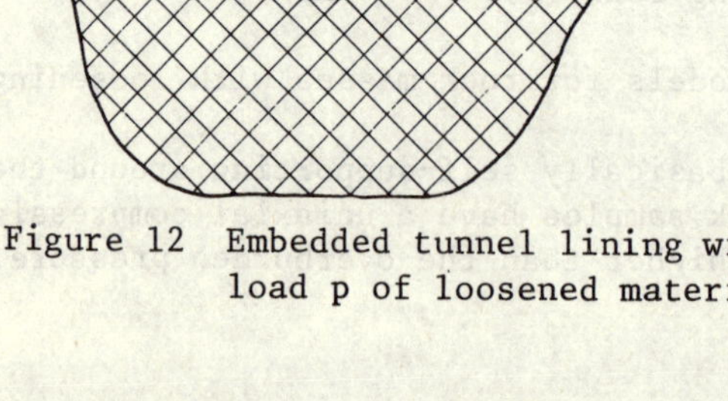

Figure 12 Embedded tunnel lining with assumed dead load p of loosened material

estimates of the loading can be made by considering the constructional method (protective or destructive excavation), the rock structure (jointing etc.) and the size of the opening. Similar considerations are valid for the deformability modulus, E, of the rock mass.

4.2. Computational models for rock masses with genuine rock pressure

In contrast to the rock masses characterized by loosening pressure, in this kind of rock it is not the occurrence of accidental details such as jointing that is important, but the properties of the basic material surrounding the tunnel. Its low strength, its susceptibility to creep, the presence of water and high overburden stresses favour the development of genuine rock pressure.The deformations occur in all directions and make their presence felt in the form of a reduction in cross-section. The ground exerts pressure on the tunnel support from all sides even from the bottom. The movements are strongly time-dependent but reach a standstill as a consequence of the support measures. In order to follow, by means of computations, the occurrence of failure in the rock,i.e. the spread of the plastic zones, the redistribution of stresses and associated deformations, a continuum model is generally used.

For the sake of simplicity we restrict in the following our attention to the case of axy-symmetric conditions. For tunnels with approximately circular cross section and under an approximately hydrostatic state of initial stress the so-called "characteristic-line method" has rendered worthwhile service {10}. The associated computational model assumes - in the strict sense - a homogeneous, isotropic, elasto-plastic continuum in which the initial stresses are in a hydrostatic state of magnitude $\bar{p}$. The model (Fig. 13) further assumes a circular hole of diameter corresponding to the dimensions of the tunnel. In this way only the simple case of an axi-symmetric stress analysis has to be dealt with. In order now to be able to calculate the interaction between the rock and the tunnel lining a uniformly distributed internal pressure p is assumed to act on the boundary of the opening. It is quickly recognized that for a value of $p = \bar{p}$ the undisturbed state of initial stress is maintained and thus no deformations result.A successive reduction of the internal pressure leads to radial rock deformations. The relationship between the rock displacement δ at the excavation boundary and the internal pressure p yields a curve, which represents the characteristic line for the rock. In the case of genuine rock pressure, after falling below a certain threshold value of p, failure occurs in the rock, so that the characteristic line is no longer linear. The tunnel lining (in the form

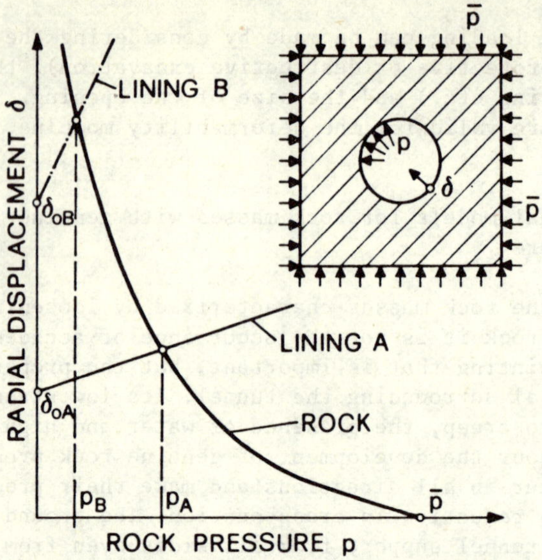

Figure 13 Rock pressure assessment by means of the "characteristic-line method"

of a closed ring) also exhibits a certain deformation response to an all-round pressure. Assuming elastic behaviour the "characteristic line" of the concrete lining is linear in form, as shown in Fig. 13. If the rock has already undergone a radial deformation before placing the permanent lining, then the corresponding straight line will begin at that value on the δ-axis. The intersection point of the straight line with the curve yields the theoretical value of the rock pressure and the corresponding value of the total rock deformation. In Fig. 13 we recognize further, that the characteristic line of the tunnel lining B leads to a lower value of rock pressure (p_B) than that of A. The reason for this is,firstly, the higher value of permited deformation of the rock δ_{oB} and, secondly, the greater deformability of the lining (B) compared to A. Even with this greatly simplified representation the basic aspects of the interaction of lining and rock under the condition of genuine rock pressure are brought out. A better approach to problems in practice may be achieved by the introduction of visco-elastic and visco-plastic material behaviour into the mathematical model.

5. CONCLUDING REMARKS

The aim of statical computations in rock engineering practice is not to obtain results for one particular set of input data, but to carry out a parametric study. Many of the available computer programs have today the possibility, at reasonable additional cost, to vary the most important parameters involved in order to check the sensitivity of the design to these variations. In this way a quantitative assessment of the performance of an underground opening a rock slope or foundation can be obtained.However, prior to any computations a good knowledge of the general behaviour pattern of the structure under the particular rock conditions is important. As an example, in tunnelling the classification of the rock mechanics problems in loosening type of rock pressure,genuine rock pressure and swelling pressure provide a useful guide to choose adequate computational models.

REFERENCES

{1} Griffith, A.A. 1921, The Phenomena of Rupture and Flow in Solids, Phil. Trans.

{2} Prandtl, L. 1928, Ein Gedankenmodell zur kinetischen Theorie der festen Körper, Zeitschrift für angewandte Mathematik und Mechanik, H.2

{3} Jenkin, C.F. 1922, A Mechanical Model Illustrating the Behaviour of Metals under Static and Alternating Loads, Engineering, p.603

{4} Owen, D.R.J., Prakash, A., Zienkiewicz, O.C. 1974, Finite Element Analysis of Nonlinear, Composite Materials by Use of Overlay Systems, Computer & Structures Vol. 4

{5} Kovári, K. 1977, Micromechanics Model of Progressive Failure in Rock and Rock-Like Materials, Proc. Int. Symp. on Geotechnique of Structurally Complex Formations, Capri, Italy

{6} Drucker, D.C., Prager, W. 1952, Soil Mechanics and Plastic Analysis or Limit Design, Quart. Appl. Math. 10

{7} Shield, R.T. 1955, On Coulomb's Law of Failure in Soils, Journal of Mechanics and Physics of Solids, 4

{8} Kovári, K. 1977, The Elasto-Plastic Analysis in the Design Practice of Underground Openings, in Finite Elements in Geomechanics, Gudehus, G. editor, John Wiley & Sons, London

{9} Kovári, K. 1979, Basic Consideration on the Design of Underground Openings, Int. Assoc. for Bridge and Structural Engineering, IABSE Periodica 3

{10} Lombardi, G.1971, Zur Bemessung der Tunnelauskleidung mit Berücksichtigung des Bauvorganges, Schweiz. Bauzeitung 32

CHAPTER II

FINITE ELEMENTS TECHNIQUES

NUMERICAL ANALYSIS OF SHALLOW FOUNDATIONS

I.M. Smith

University of Manchester, England, U.K.

INTRODUCTION

The numerical method which will be considered in these notes is the finite element method. It is at present by far the most widely used and versatile aid to obtaining solutions of linear and nonlinear problems in foundation engineering. Alternative methods, in particular boundary integral techniques, are considered at length in other lectures to the Institute.

The first question to be answered must clearly be - why use numerical methods in foundation engineering at all? In order to answer it, some of the standard, classical solutions of soil mechanics are reproduced below by numerical means. It is then possible to see where modest extensions to current practice can be achieved, for example in the areas of inhomogeneous soils, complicated quasi-static loading states and cyclic and dynamic loading.

However, what will also emerge is that <u>in detail</u> the solutions produced by numerical means, using rather simple constitutive assumptions, can be quite erroneous. For example skeleton volume changes can be quite inaccurate and this can be far more important in some problems than in others. The engineer may have to ask if he is primarily interested in collapse loads, or in deformations at working load.

The solutions presented below all assume that soil is a continuum in which compatibility of strains always exists. Of course bifurcations are known to occur in practice with subsequent discontinuous deformation in localised zones. Since

J. B. Martins (ed.), Numerical Methods in Geomechanics, 167–185.

computations relating to this kind of material behaviour have not yet been carried out, our interest will lie in how closely the continuum solutions, perhaps involving strain-softening, can reproduce observed behaviour.

Load-deflection and bearing capacity of shallow footings for vertical loading will be considered in some detail. Since other lectures to the Institute deal with structure/soil interaction, only perfectly rigid structures will be assumed.

Combined loadings, leading to interaction diagrams, will then be discussed.

Finally some problems of cyclic loading and shakedown, with reference to offshore foundations, will be considered.

STANDARD SOIL MECHANICS SOLUTIONS

The solutions described in this lecture were obtained using meshes of 8-node isoparametric quadrilateral elements. Reduced, 2x2, integration was used to evaluate the stiffness integrals. No attempt is made here to justify this choice theoretically, the interested reader being referred elsewhere (1). What is of immediate concern is the nature of solutions obtained for cohesive and cohesionless soils. The former are approximated as elastic-plastic, yielding according to the von Mises criterion. The latter are also approximated as simple elastic-plastic although yielding according to the Mohr-Coulomb criterion and in this case, both non-associated (zero volume change) and associated (excessive plastic dilation) flow rules are considered. Before shallow foundations in bearing are dealt with in detail, retaining walls and slopes are considered.

RETAINING WALL SOLUTIONS

A comprehensive set of analyses has been conducted (2) for both rough and smooth walls retaining cohesive and cohesionless soils. For example, Figure 1 shows that the Rankine active and passive failure states in cohesionless soil are very accurately captured when the retaining wall is perfectly smooth.

For rough walls there are no "exact" solutions but Figure 2 shows a comparison between finite element solutions and other approximate solutions due to Caquot and Kerisel (3). The finite element solutions for this c-ϕ material also show that the influence of dilatancy is negligible for a smooth wall but has a considerable influence on failure loads for rough walls.

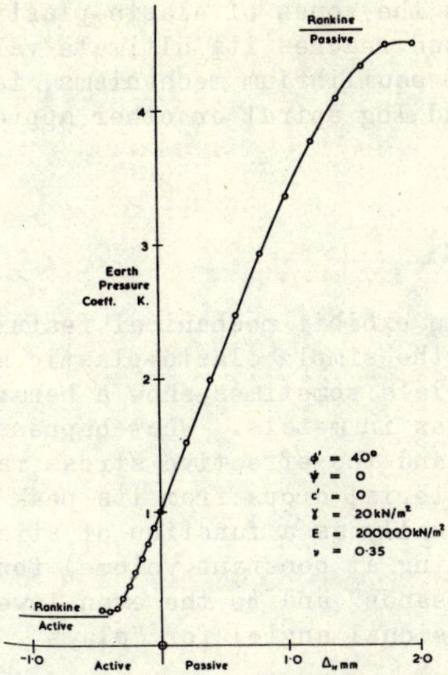

Figure 1. Mobilisation of active and passive pressures.

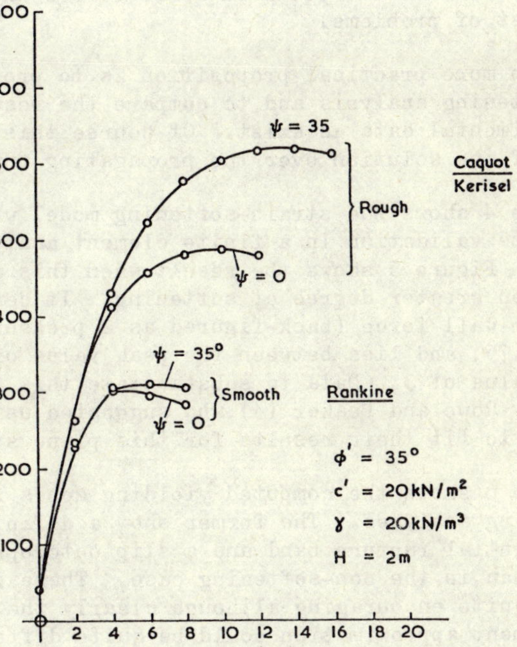

Figure 2. Influence of wall roughness and flow rule

Figure 3 shows the zones of elasto-plastic soil which have yielded when the load reaches its ultimate value. These can be compared with limit equilibrium mechanisms, ie. the $45^{o} - \phi/2$ line for smooth walls and log spiral or other approximations for rough walls.

PROGRESSIVE FAILURE

Dense sands can exhibit mechanical features which differ significantly from the simple elasto-plastic models described above. These materials sometimes show a behaviour which resembles brittleness in metals. What happens is that locally a rupture band forms and the effective stress ratio which can be sustained by the material drops from its peak value "progressively" (usually as a function of strain) to an ultimate value of ϕ_{cv} (shearing at constant volume) for rotund particulate materials such as "sands" and to the even lower value of ϕ_r (shearing at the residual angle) for "clays".

Mathematically this is a bifurcation, detectable by studying the eigenvalues of the numerical approximation to the soil mass. However, any post-bifurcation analysis is really quite complicated and has not yet been achieved for anything other than the simplest of problems.

A much more practical proposition is to proceed with a strain-softening analysis and to compare the results with such good experimental data as exist. Of course this implies a smearing of the solution over the propagating rupture zone.

Figure 4 shows the strain-softening model which was used for sand and its validation in a finite element analysis of a passive wall test. Figure 5 shows the result when this model is used with an even greater degree of softening. It can be seen that the maximum wall force (back-figured as a pressure coefficient K) is about 3.75, and lies between the peak value of 4.6 and the residual value of 3. Data to substantiate this result were produced by Rowe and Peaker (4) who suggested using the triaxial value of ϕ to fit their results for this plane strain problem.

Figure 6 shows the computed yielding zones for softening and non-softening analyses. The former show a definite narrowing into a potential rupture band and a slip outcropping further from the wall than in the non-softening case. These results are therefore quite encouraging although clearly the adequacy of the finite element approximation could be quite different in different boundary value problems. The difficulty in foundation analysis, as we shall see below, is the lack of experimental data of the quality of Rowe and Peaker's for walls.

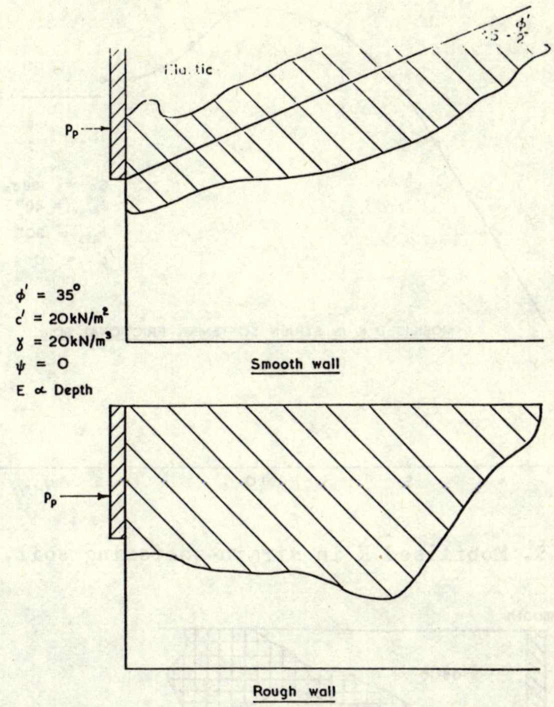

Figure 3. Yield zones in c-ϕ soil.

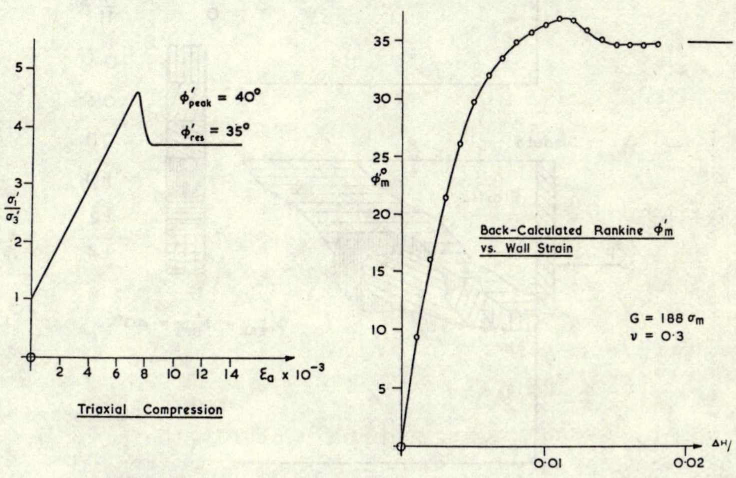

Figure 4. Strain-softening frictional soil.

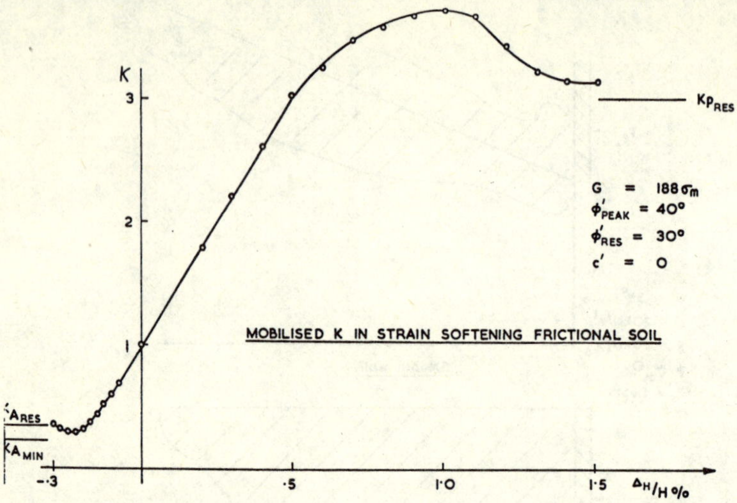

Figure 5. Mobilised K in strain-softening soil.

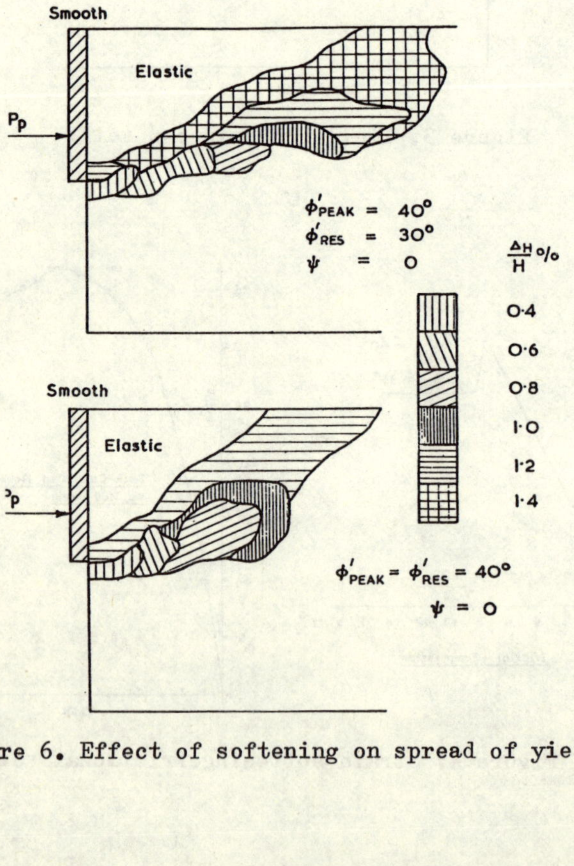

Figure 6. Effect of softening on spread of yield.

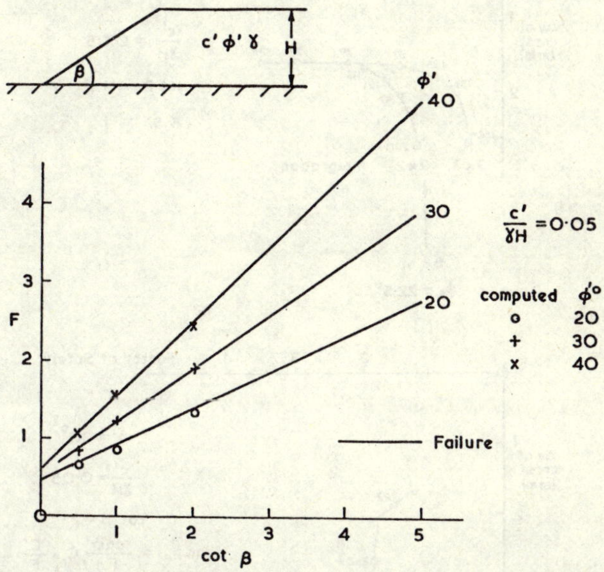

Figure 7. Comparison of FE results with Ref.(6).

SLOPE STABILITY SOLUTIONS

The ability of the finite element solutions to recapture Bishop and Morgenstern's (6) data is demonstrated briefly in this section. Figure 7 shows the geometry of the elasto-plastic slope resting on a firm base. The method of load application was "gravity switch-on" and the factors of safety were assessed by keeping c' constant and reducing ϕ' until collapse occurred. Figure 7 shows that the results compare very well with Bishop and Morgenstern's which were obtained assuming cylindrical rupture surfaces.

Figure 8 shows the progress of the slope towards "collapse" as defined by vertical movement of the crest. The superiority of reduced, 2x2, integration over exact, 3x3, integration should be noted.

Figure 9 shows the displacement vectors at collapse for frictional and frictionless slopes. The zones of yielding (not shown) are broad and smear over any supposed cylindrical discontinuity, but the agreement with classical solutions is again most encouraging.

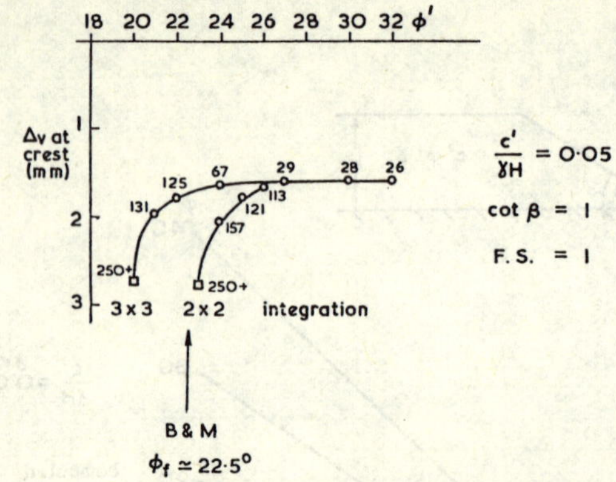

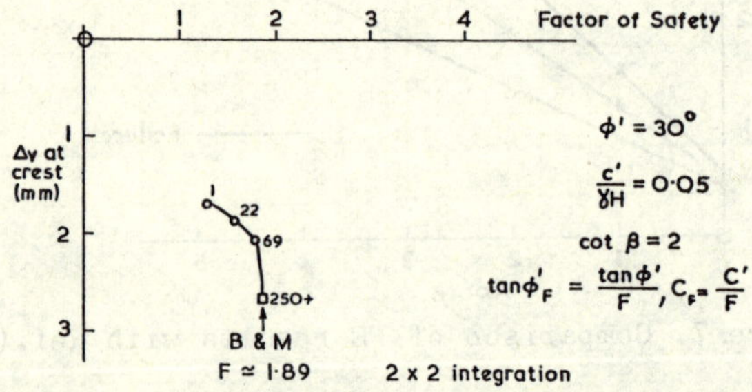

Figure 8. Movement of slope crest with decreasing strength.

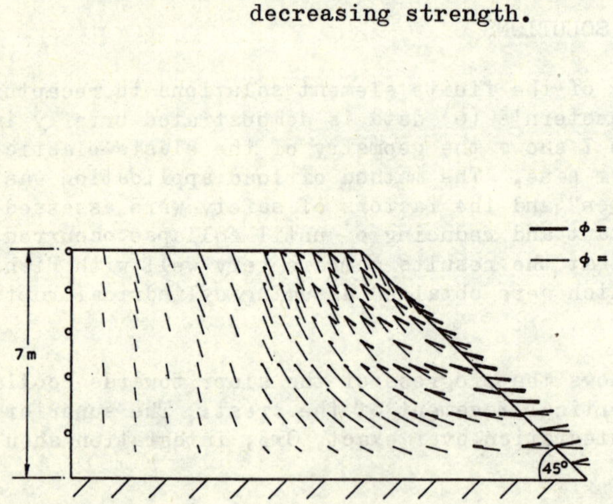

Figure 9. Displacements at collapse.

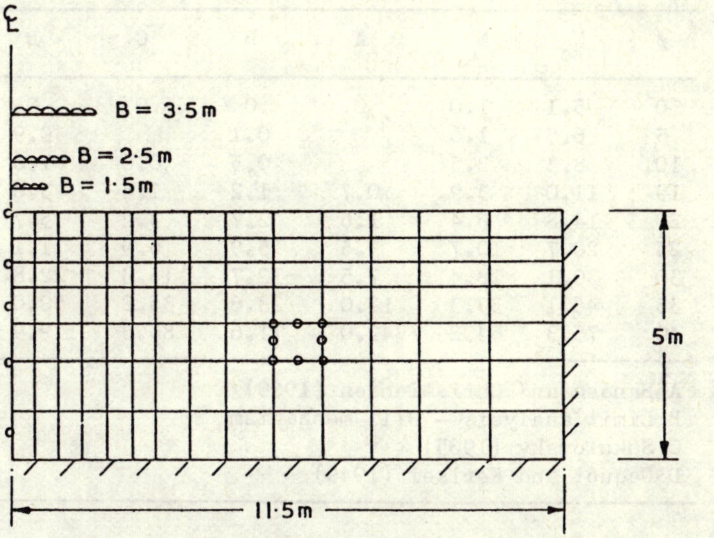

Figure 10. Mesh for bearing capacity.

SHALLOW FOUNDATIONS SOLUTIONS: VERTICAL LOADING

A typical finite element mesh is shown in Figure 10, for various footing breadths, B. The classical solutions to be recaptured in this case are listed as Table 1 (for smooth footings) and Table 2 (for rough footings) respectively. Note the very wide scatter in the bearing capacity factor N_γ. Therefore, one would expect solutions to be most easily reproduced for weightless soils.

Figure 11 shows the finite element solutions for N_c for smooth footings in comparison with theory (8)

$$N_c = \cot\phi \left[\tan^2(45 + \phi/2)\, e^{\pi\tan\phi} - 1 \right]$$

while Figure 12 shows the solutions for N_q for comparison with the theoretical

$$N_q = \tan^2(45 + \phi/2)\, e^{\pi\tan\phi}.$$

The agreement can be seen to be very close in both cases.

The nature of the load-displacement curves is illustrated by Figures 13 and 14 for weightless soil without and with surcharge respectively.

SMOOTH FOOTINGS

N_c, N_q from Sokolovsky. Several N_γ values given

ϕ	N_c	N_q	A	B	C	D
0	5.1	1.0	0	0	0	0
5	6.5	1.6		0.1		0.9
10	8.3	2.5		0.5	0.6	1.6
15	11.0	3.9	0.7	1.2	1.4	3.0
20	14.8	6.4	1.6	2.7	3.2	5.7
25	20.7	10.7	3.5	5.9	6.9	11.1
30	30.1	18.4	7.5	12.7	15.3	22.8
35	46.1	33.3	18.0	28.6	35.2	50.0
40	75.3	64.2	42.0	71.6	86.5	119.9

A Hansen and Christiensen (1969)
B Limit analysis - Hill mechanism
C Sokolovsky (1965)
D Caquot and Kerisel (1949)

Table 1

ROUGH FOOTINGS

N_c, N_q from Terzaghi. Several N_γ values given

ϕ	N_c	N_q	A	B	C
0	5.7	1.0	0	0	0
5	7.0	1.6		0.1	0.5
10	9.5	2.7		0.7	1.3
15	13.0	4.5	1.2	2.0	2.9
20	17.0	7.5	2.9	4.8	6.2
25	24.0	13.0	7.0	9.8	13.0
30	37.0	23.0	15.0	20.0	27.7
35	58.0	42.0	35.0	43.0	61.5
40	98.0	77.0	85.0	98.0	145.3

A Hansen and Christiensen (1969)
B Terzaghi
C Prandtl mechanism

Table 2

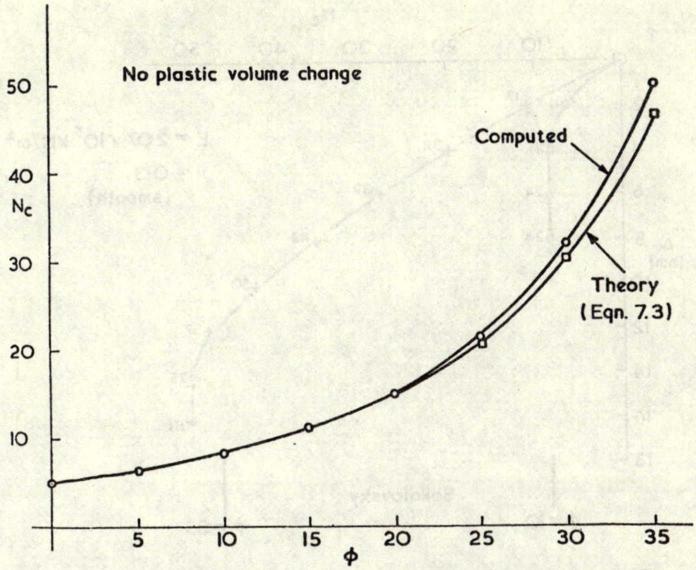

Figure 11. N_c for smooth strip footings.

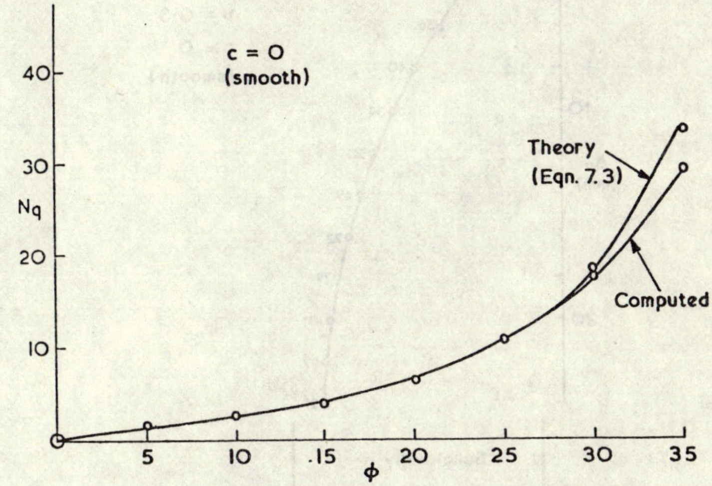

Figure 12. N_q for smooth strip footings.

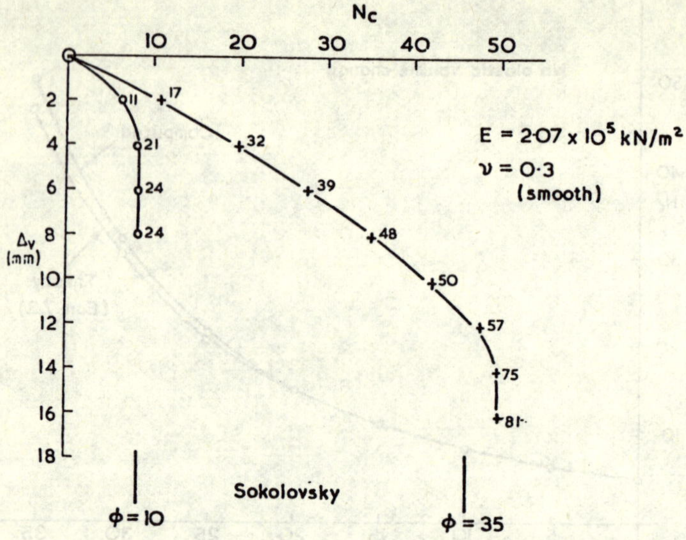

Figure 13. Convergence for weightless c-ϕ soil.

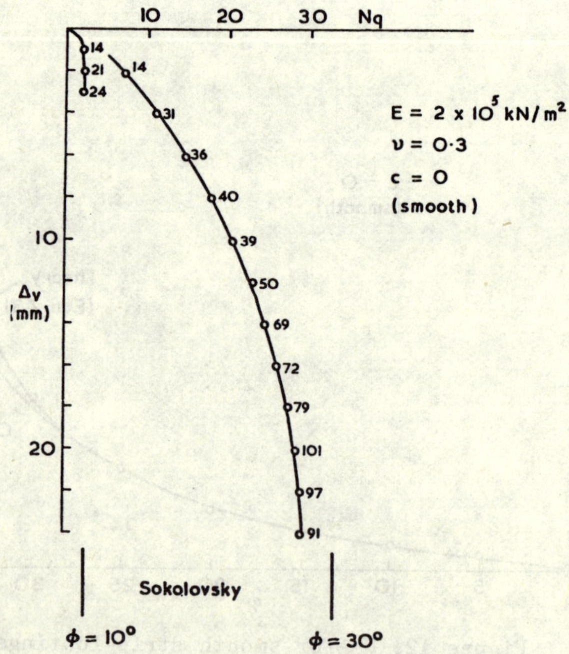

Figure 14. Convergence for weightless soil with surcharge.

It can be seen that a peak load is computed although as ϕ increases for the surcharged soil there is a tendency for the load to creep up even at very large displacements. Of course the same phenomenon would be measured in any experiment to determine "failure" loads in laboratory or field.

Turning to soil with weight, Figure 15 compares computed N_γ values with some empirical values from the literature (7) for smooth footings. The finite element values show some dependence of N_γ on B, which is not unreasonable given the arbitrary non-dimensionalisation inherent in the selection of N_γ as the bearing capacity factor.

For rough footings Figure 16 compares the computed and empirical factors. For both smooth and rough footings the computed values lie within the range of the other results, and this deduction is reinforced by Figure 17 which shows the total bearing capacity of a smooth, rigid footing as computed, and as obtained from classical theories. Two computed solutions are shown - one assuming zero plastic volume change and the other assuming associated Mohr-Coulomb plastic flow. The range of the computed results falls neatly in the middle of the range of all results.

The rather unrealistic displacement field produced from associated flow calculations is shown in Figure 18.

LOAD-DEFLECTION BEHAVIOUR

Until now, we have considered only the ability of computations to model ultimate load behaviour of footings. In the writer's opinion, the simple elasto-plastic model assumed for soil is quite adequate in the present examples (but see below). A modulus proportional to the square root of mean stress reproduces the distribution of stiffness quite well, and any further refinement is curve fitting which might as well be done on the basis of whatever field and/or laboratory data exist, using as simple a constitutive model as possible.

POREPRESSURES

The major omission from the previous discussion is any mention of the role of porewater pressures in the analyses. It is implied that loads are put on so slowly that no substantial excess water pressures arise. Fortunately this is likely to be true in the materials for which analysis is most difficult - ie. highly frictional soils. Referring to Figure 17, it is quite straightforward to compute the <u>undrained</u> capacity of such a

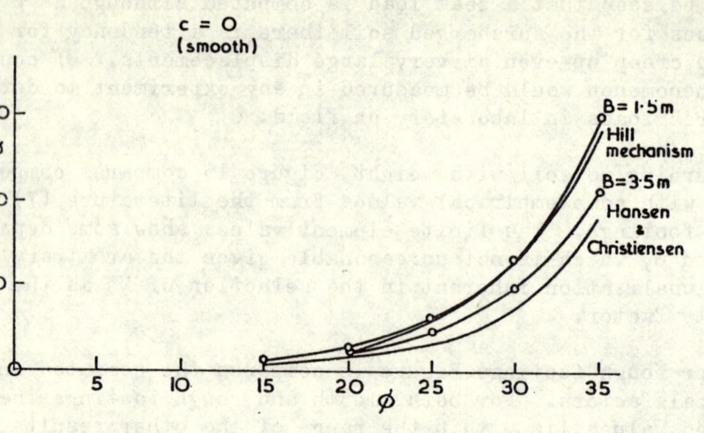

Figure 15. N_γ for smooth strip footings.

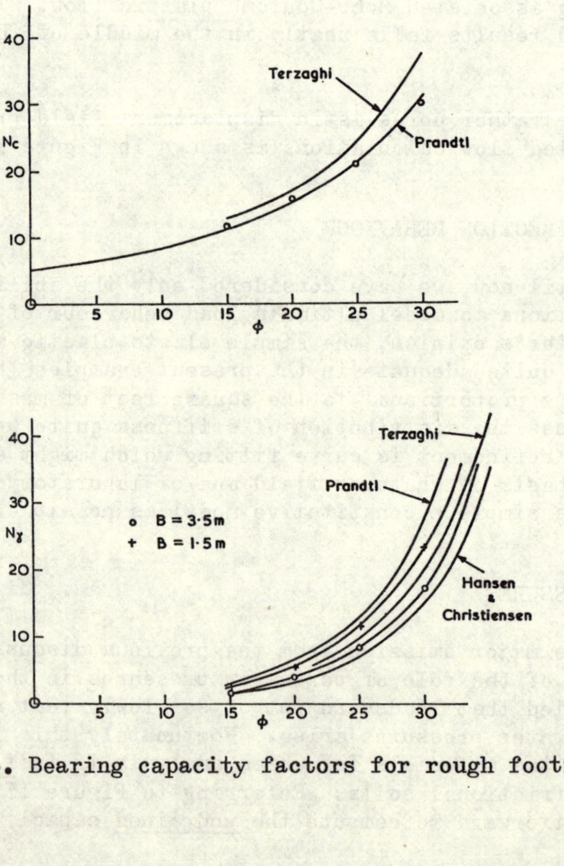

Figure 16. Bearing capacity factors for rough footings.

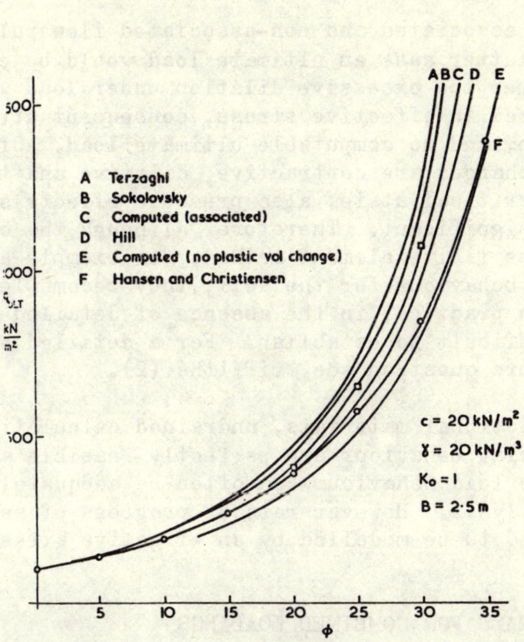

Figure 17. Bearing capacity of smooth rigid footing.

Figure 18. Displacements at collapse.

footing assuming associated and non-associated flow rules. However, in the latter case an ultimate load would be computed while in the former the excessive dilation under load would lead to continuing rises in effective stress, consequent strengthening of the foundation, and no computable ultimate load. Of course in practice volume changes are contractive, dilative and then zero approaching the critical state; also pressure effects such as crushing become significant. Therefore, although the calculations can be done by the finite element method, for example assuming stress-dilatancy behaviour for the soil, they become lengthy and their relation to practice, in the absence of detailed measurements, difficult to establish. For a detailed examination of the porepressure question see Griffiths (2).

For less frictional materials, undrained calculations using for example Cam-clay behaviour are perfectly feasible although then the ultimate load behaviour can often be adequately inferred from a $\phi_u = 0$ analysis. However rate of progress of settlement with time may need to be modelled by an effective stress analysis.

INTERACTION DIAGRAMS FOR COMBINED LOADINGS

Developments in the offshore industry have given impetus recently to the study of foundations under combined loadings. For the "gravity" type of offshore structure, lateral forces of the order of $\frac{1}{3}$ of the structure weight can be encountered. The classical work of Brinch Hansen (9) can be referred to in the case of homogeneous foundations, but a common occurrence is to find clay foundations of greatly varying strength with depth, or layered foundations of sands interspersed with clays.

An example of an interaction diagram for failure is given in Figure 19. Depending on the degree of inhomogeneity (due to layering, weakening after cyclic loading etc) the diagram becomes distorted, reflecting the greater contribution from the upper layers to stability in the horizontal loading mode (10). In this situation, approximations of the inhomogeneous foundation by an equivalent homogeneous counterpart are unlikely to succeed.

CYCLICALLY LOADED OFFSHORE FOUNDATIONS: SHAKEDOWN

Observation of structures in the field and in the laboratory indicates that gravity (shallow foundation) structures settle vertically into the sea bed under predominantly laterally applied cyclic loading. The reasons are examined in detail elsewhere (11) but can briefly be stated thus: cyclic loading into the plastic region causes a redistribution of (initially antisymmetric) shear stress beneath the structure base; ultimately a symmetric plastic

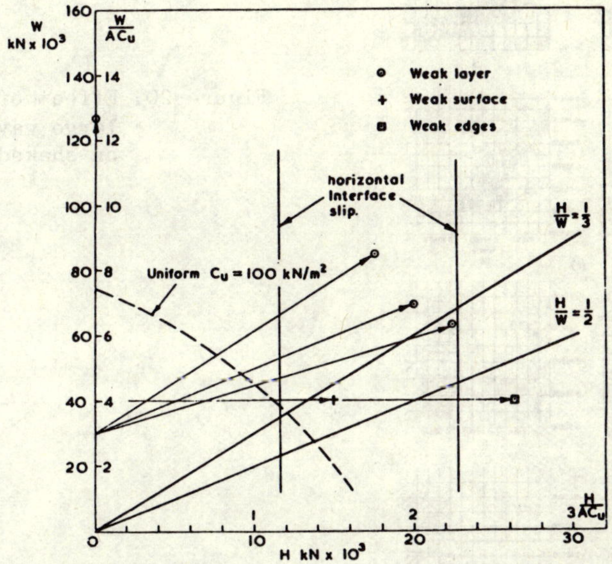

Figure 19. Interaction diagram (with weakening).

shear stress results and if this can be resisted by the soil, the structure will remain stable. Of course the critical point in design is what the cyclically weakened shear strength of the ground below the structure really is.

To illustrate the computed behaviour for an elastic-plastic soil, Figure 20 shows the vertical settlement of a sinusoidally loaded structure, both as a function of time and of horizontal cyclic force H. At various points in the loading process, one, two and four large waves are inserted into the programme. It can be seen that large waves have a stabilising effect after the settlement they cause has occurred. It should be emphasised that these results could not be predicted by any non-numerical calculation method, and indeed any calculation ignoring plastic effects could not reproduce the permanent settlements.

CONCLUSIONS

Applications of the finite element method to the solution of shallow foundation problems have been described. Classical solutions have been recovered using 8-node quadrilateral elements

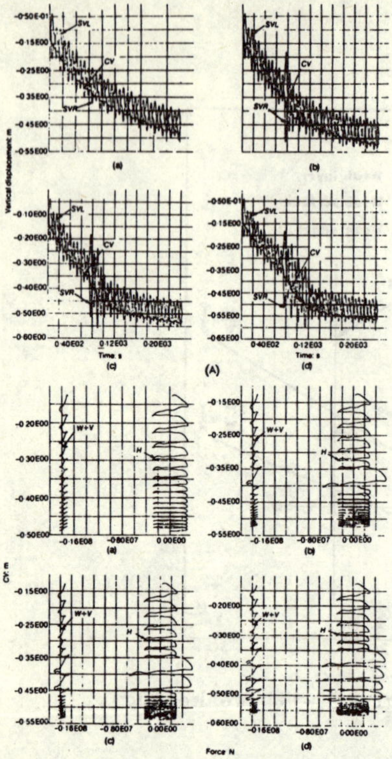

Figure 20. Effect of large waves on shakedown.

with 2x2 integration. Therefore the refinements advocated by Sloan (12) are unnecessary in practice. Despite reservations about progressive failure, eg. (13), it seems quite likely that the continuum calculations are adequate in the large at least for problems not dominated by porepressure considerations. The numerical method enables modest extensions of current practice to be made, for example to inhomogeneous soils, combined loading and cyclic loading.

REFERENCES

1. Zienkiewicz, O.C.: 1971, "The finite element method in engineering science," McGraw-Hill, London/New York.

2. Griffiths, D.V.: 1980, "Finite element analyses of walls, footings and slopes," Ph.D Thesis, University of Manchester.

3. Caquot, A. and Kerisel, J.: 1949, "Traite de mechanique des sols," Gauthiers-Villars, Paris.

4. Rowe, P.W. and Peaker, K.: 1965, "Passive earth pressure measurements," Geotechnique 15, No.1, pp.57-78

5. Griffiths, D.V.: 1980, "Finite element analyses of walls, footings and slopes," Symp. Comp. Appl. to Geotech. Probs. in Highway Eng., Cambridge University.

6. Bishop, A.W. and Morgenstern, N.: 1960, "Stability coefficients for earth slopes," Geotechnique 10, pp.129-150.

7. Hansen, B. and Christiensen, N.H.: 1969, "Discussion of theoretical bearing capacity of very shallow footings," J. Soil Mech. Found. Eng. Div., A.S.C.E., Vol.95, No.SM6, pp.1568-1572.

8. Sokolovsky, V.V.: 1965, "Statics of granular material," Pergamon Press, New York.

9. Hansen, B.: 1970, "A revised and extended formula for bearing capacity," Danish Geotechnical Institute Publication No.28, Copenhagen.

10. Smith, I.M.: 1976, "Aspects of the analysis of gravity offshore structures," Proc. 2nd Int. Conf. Num. Meth. Geomechanics, Blacksburg, Vol.2.

11. Smith, I.M. and Molenkamp, F.: 1980, "Dynamic displacements of offshore structures due to low frequency sinusoidal loading," Geotechnique 30, No.2, pp.179-205.

12. Sloan, S.W. and Randolph, M.F.: 1981, "Numerical prediction of collapse loads using finite elements," Int. J. Num. Anal. Meth. Eng., Vol.5.

13. Cleary, M.P. and Bathe, K.J.: 1979, "On tractable constitutive relations and numerical procedures for structural analysis in masses of geological materials," Proc. 3rd Int. Conf. Num. Meth. Geomechanics, Aachen, Vol.1.

NUMERICAL ANALYSIS OF DEEP FOUNDATIONS

I.M. Smith

University of Manchester, England, U.K.

INTRODUCTION

Deep foundations - piles, piers, caissons etc. - derive their support from end bearing and from resistance mobilised along their sides, often called skin "friction" whatever its physical basis. Traditionally these are assessed independently and then added to give the total resistance. Using numerical methods, it becomes possible to establish the rates at which the various components of resistance become mobilised with increasing load. This is a function of the relative stiffness of the ground and the foundation.

Before proceeding with such an interaction analysis, however, the ability of numerical methods to recapture classical end bearing formulae for deep foundations must be established. This involves the problems of the effects of strain-softening and crushing on mobilised friction angles which we saw in the previous lecture.

Mobilisation of skin "friction" with load will then be considered using discrete spring and continuum (finite element) idealisations of the soil; the ability of these models to cope with combined, cyclic loading, as met in offshore conditions, will then be mentioned.

A major feature of deep foundations not shared by shallow ones is the difficulty of installing them. In particular, attention will be paid to the role of numerical methods in assessing the drivability of piles, including possible effects of plugging. Finally some remarks on dynamic effects will be made.

J. B. Martins (ed.), Numerical Methods in Geomechanics, 187–201.

Most of the lecture will deal with single piles, but extensions to the behaviour of groups will be mentioned.

END BEARING CAPACITY OF DEEP FOUNDATIONS

The numerical models are the same as were discussed for shallow foundations. A typical finite element mesh (1) is shown in Figure 1. Since we are not concerned at present with any resistance mobilised along the "shaft" AB the simple alternatives are to leave this boundary free (called herein "open hole" analysis) or to place it on frictionless rollers (called herein "closed hole" analysis). Figure 1 may represent either a plane cross-section through a long strip footing or a radial section through an axisymmetric pile or pier. For the plane strain case and a cohesive soil the computed values of N_c are shown in Figure 2 where they are compared with classical values. The open hole results agree closely with Skempton's while the closed hole ones show no ultimate load, indicating the difficulty the failure mechanism has in forming. Clearly compressibility of the ground near the base of the hole will play a big role in load-displacement behaviour of such a foundation. This point is reinforced by Figures 3 and 4 which show the displacement vectors at collapse for the closed and open hole analyses respectively.

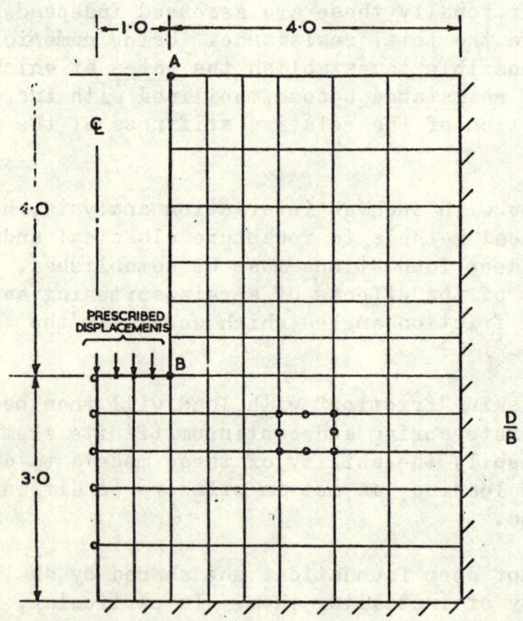

Figure 1. F.E. mesh for deep foundations.

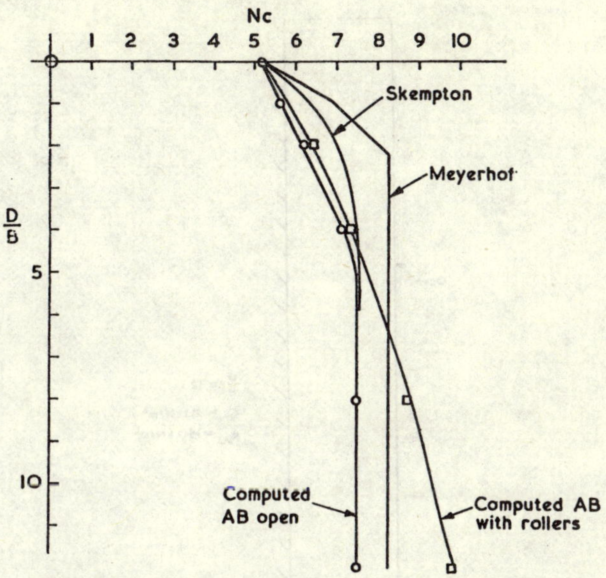

Figure 2. Bearing capacity of deep strip in cohesive soil.

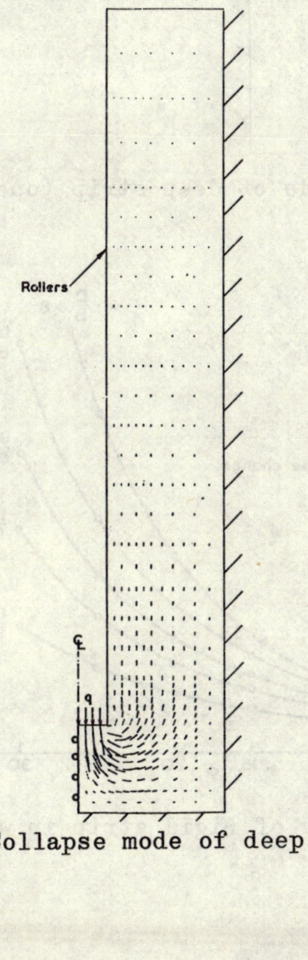

Figure 3. Collapse mode of deep strip (closed hole).

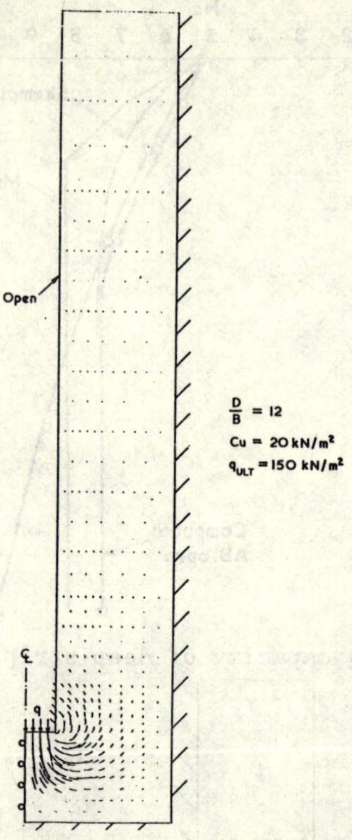

Figure 4. Collapse mode of deep strip (open hole).

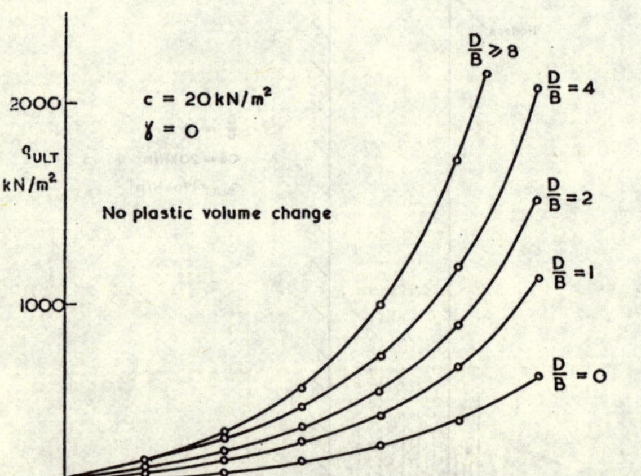

Figure 5. Bearing capacity of rigid strip in weightless soil.

In the former the displacements have to dissipate vertically while in the latter they rotate into the hole (2).

Continuing the discussion of open hole strip footings the influence of friction on N_c is illustrated in Figure 5 for weightless soil. There is no depth effect beyond D/B of about 8. When self-weight is included, the cN_c component of resistance can be subtracted to give a result of the form

$$q_{ult} = cN_c + \gamma \frac{B}{2} N_{\gamma q}$$

and is plotted as Figure 6. These results are found to be much higher than those published by Hu (3) but somewhat lower than Meyerhof's (4).

For piles where B is small, bearing capacity can be given closely by the simple expression

$$q_{ult} = q\ N_q$$

where q is the effective vertical stress at the pile tip. Values of N_q computed for a strip foundation are compared with those of Meyerhof and Berezantsev (5) for circular piles in Figure 7. In general the computations show less well-

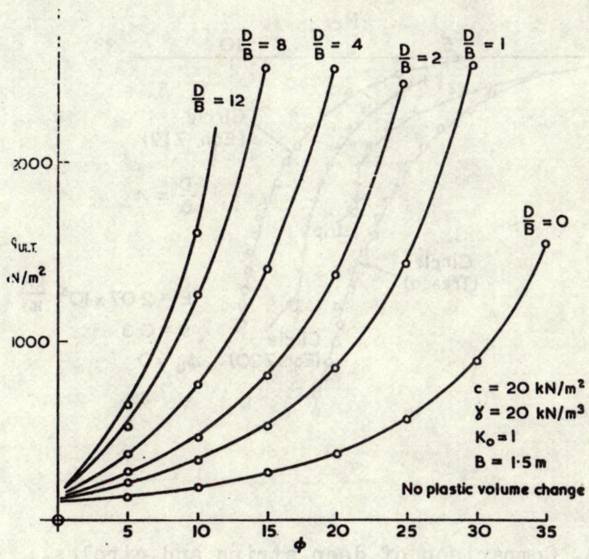

Figure 6. Bearing capacity of rigid strip in heavy soil.

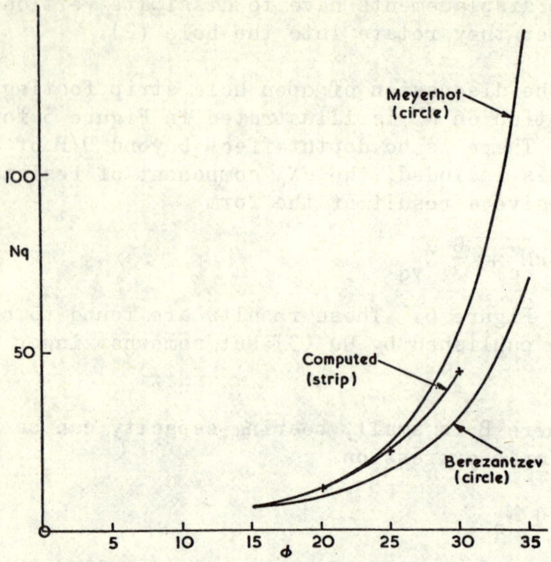

Figure 7. N_q for cohesionless soil D/B = 4.

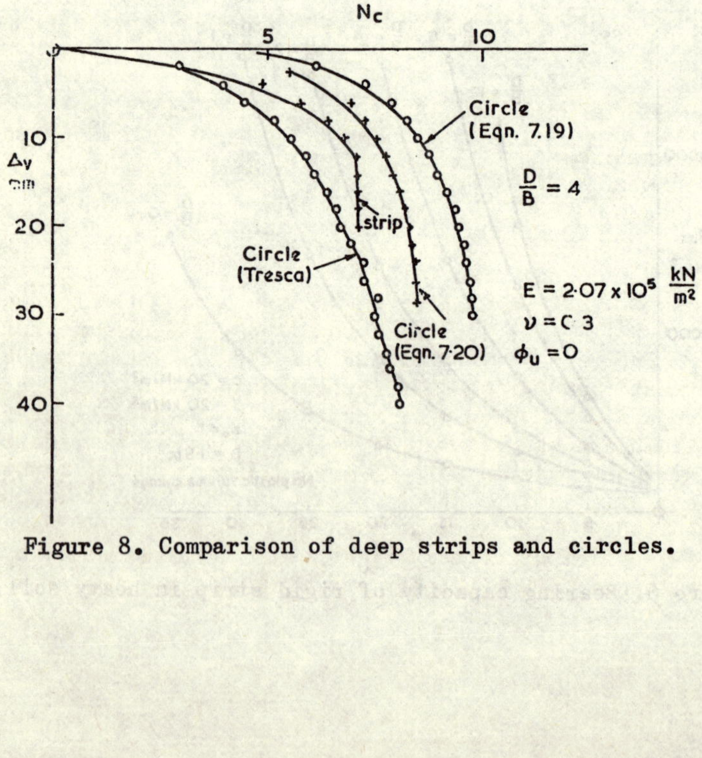

Figure 8. Comparison of deep strips and circles.

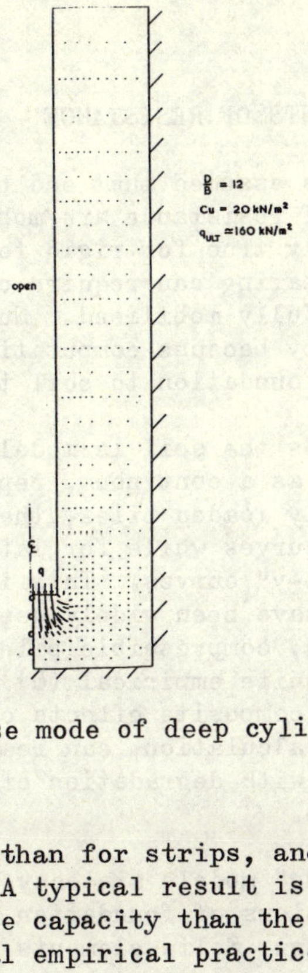

Figure 9. Collapse mode of deep cylinder (cohesive).

defined failure loads than for strips, and these failure pressures are somewhat higher. A typical result is shown in Figure 8 where the circle has 20% more capacity than the equivalent strip, confirming conventional empirical practice.

The displacement vectors at collapse for a circular footing at depth are shown in Figure 9 where the much more localised nature of the deformations is apparent.

PRESSURE AND STRAIN EFFECTS

In all of the above analyses, ϕ and c have been assumed to be constant. In fact, in deep foundations, very large tip loads can occur and progressive/crushing effects are likely to be much more dominant than for shallow foundations. Clearly computations can be conducted in which ϕ is varied as a function of pressure and of strain and the effects of dilatancy can be considered (all of the foregoing results were for non-dilatant soil). Further, the effects of elastic compressibility on the results can be determined. However, experimental laboratory or field data is lacking in sufficient detail for meaningful comparisons to be made with the different effects isolated. This is an area in which detailed experimentation, perhaps using a centrifuge, would be fruitful.

MOBILISATION OF COMPONENTS OF RESISTANCE

Traditionally it is assumed that end bearing and skin "friction" components of resistance are mobilised simultaneously, which will be essentially true for rigid foundations. However, even in that case end bearing can require considerably greater movements before being fully mobilised. Numerical methods help to remove this difficulty because computations enable the process of load diffusion from foundation to soil to be monitored.

In such calculations the soil is modelled either as discrete "spring" resistances or as a continuum, represented by finite elements. For vertically loaded piles, the spring resistances are specified by "t-z" curves while for laterally loaded piles they are specified by "p-y" curves. This information and the methods of computation have been widely used in the offshore industry where very long, compressible piles are common. Of course the approach is quite empirical (6) but the exercise of attempting to model the composite effects of pile and soil can be most instructive. The calculations can readily be extended to consider cyclic loading with degradation of the soil's stiffness and resistance with time.

Although the "spring" models are easy to use and cheap, true finite element idealisations of foundation and ground are ultimately more appealing. Solid elements can be used for foundation and surrounding ground with the interface between them being represented by special line elements. The stiffness and strength of the ground can then be described by parameters well understood in soil mechanics: c, ϕ, bulk and shear moduli and so on. An example of computed load transfer during loading and comparison with field results is shown in Figure 10.

COMBINED LOADINGS WITH CYCLING

An increasingly important analysis which has to be conducted with regard to offshore foundations involves combined axial and lateral loading with cycling. The foundation may be subjected to sustained compression or, in the case of "tension leg" platforms, to sustained tension. There is then the chance of the foundation gradually being pulled out of the ground. Clearly the spring idealisations are of very limited use for this combined loading problem. However the finite element continuum approximation can be used and indicates reduced axial capacity in the presence of cyclic lateral load, even without soil degradation as shown in

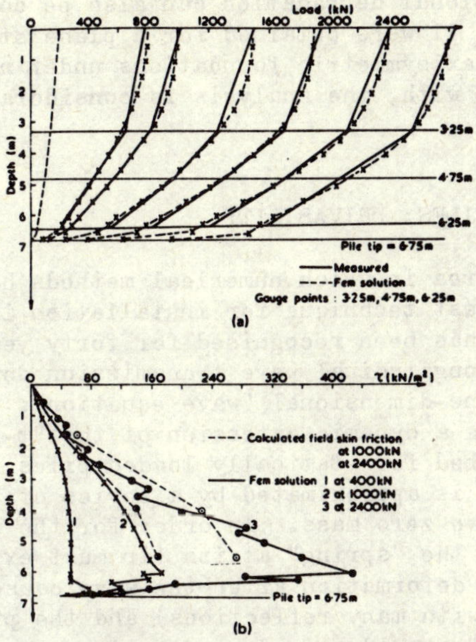

Figure 10. Back-analysis of field test by FEM.

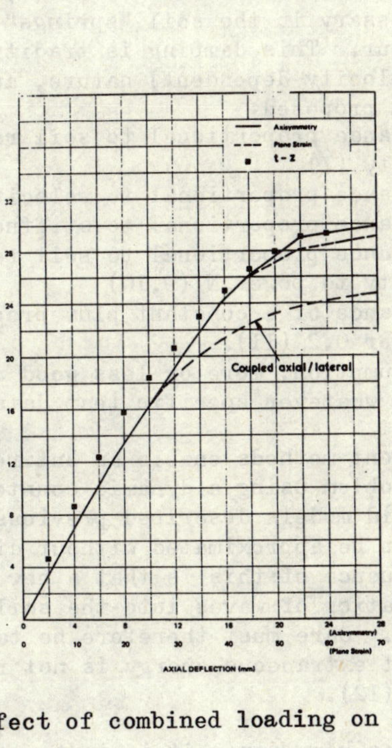

Figure 11. Effect of combined loading on axial failure.

Figure 11. Additional degradation can also be considered. The results in Figure 11 were obtained for a plane strain strip model, because although axisymmetric foundations under non-axisymmetric load can be dealt with, the analysis is considerably more cumbersome.

INSTALLATION OF PILES: DRIVABILITY

This is an area in which numerical methods have long been used. The commonest technique for installation is by impact driving and this has been recognised for forty years or more to be a process of longitudinal wave transmission down the pile, governed by the one-dimensional "wave equation". The commonest solution method is a dynamic extension of the "t-z" approach previously described for statically loaded piles. That is, the ground resistance is approximated by a series of "springs" which are assumed to have zero mass. In order for the pile to be installed at all, the "spring" at its tip must experience a permanent plastic deformation after the wave caused by the hammer blow has passed (with many reflections) and the pile come to rest. Thus the "springs" are idealised as elastic-perfectly plastic. Even then, the early work in this field (7) showed that extra damping was necessary in the soil "springs" to account for observed behaviour. This damping is traditionally assumed to be of a viscous (velocity dependent) nature, and many different models have been proposed:

(a) Resistance proportional to soil resistance times velocity (7)
(b) Resistance proportional to velocity (8)
(c) Resistance proportional to stiffness times velocity
(d) Resistance proportional to soil resistance times velocity to power N (9,10)
(e) Resistance of a constant plus proportional to velocity to power 0.2 (11).

Whatever the assumption, more or less good agreement can usually be obtained with whatever specific test data is being analysed.

Finite element methods enable an entirely new look to be taken at this problem using a dynamic counterpart of the axisymmetric solid models described previously. Then the mass of pile and soil can be approximated without difficulty. An immediate consequence of this is that a new source of damping, namely from radiation of waves into the soil mass surrounding the pile, is present. Care must therefore be taken in the computations that extraneous energy is not reflected back from mesh boundaries (12).

When this has been done, it is instructive to compare the responses of the one-dimensional and three-dimensional piles during driving. In the three-dimensional model, viscous damping is assumed along the pile shaft only. Figures 12, 13 and 14 show the results for three different ground resistances, namely 25 kN/m^2, 100 kN/m^2 and 500 kN/m^2. The ground was assumed to be uniform clay in all cases.

It can be seen that the one-dimensional and three-dimensional analyses agree quite closely in the early part of the penetration curve. However, the extra radiation damping in the 3-d case is very apparent at later times, and in fact the 1-d model never comes to rest at all. In practice, the "set" of a pile is traditionally assessed by stopping the computer program once the tip displacement first shows signs of decreasing. The Figures show that this may or may not be a good approximation to the true set of the model. For very weak soil, the 3-d model indicates that the 1-d one may underestimate the set while for strong soil it may overestimate it. The latter would lead to poor estimates of drivability in the field and there is evidence that predictions have been worse in very stiff clays.

The 3-d model also allows an independent estimate of soil "quake", ie. the elastic recoverable compression of the ground at the pile tip. By experience this is taken to be 0.1 inch (2.5mm) with a range of 1mm to 4mm. The 3-d calculations fall exactly in

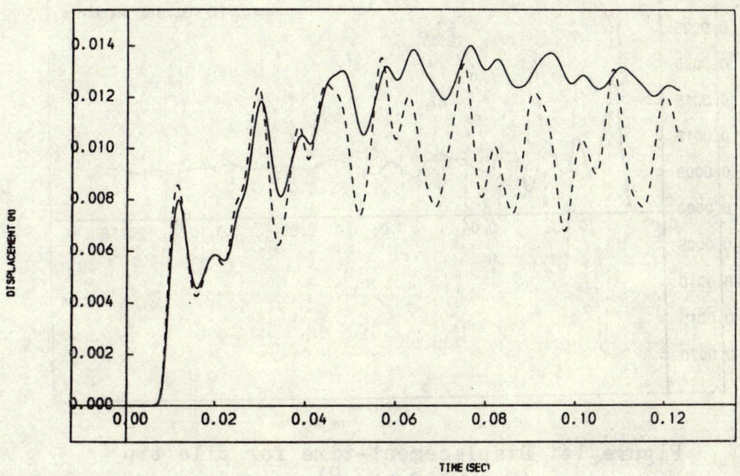

Figure 12. Displacement-time for pile tip (C_u = 25 kN/m^2).

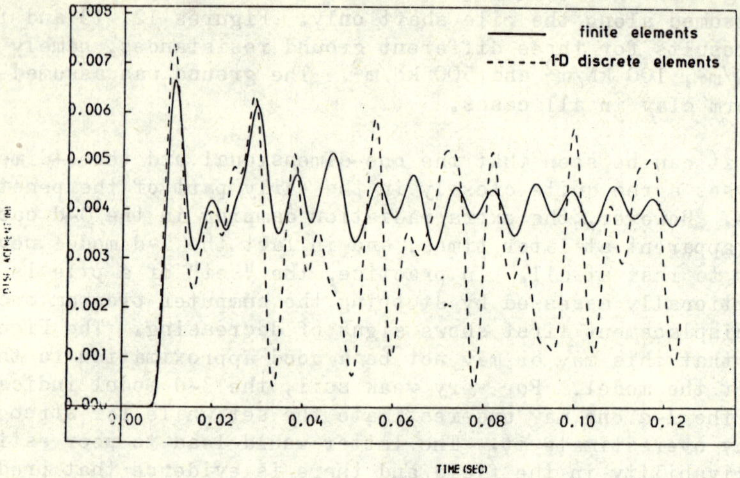

Figure 13. Displacement-time for pile tip ($C_u = 100$ kN/m^2).

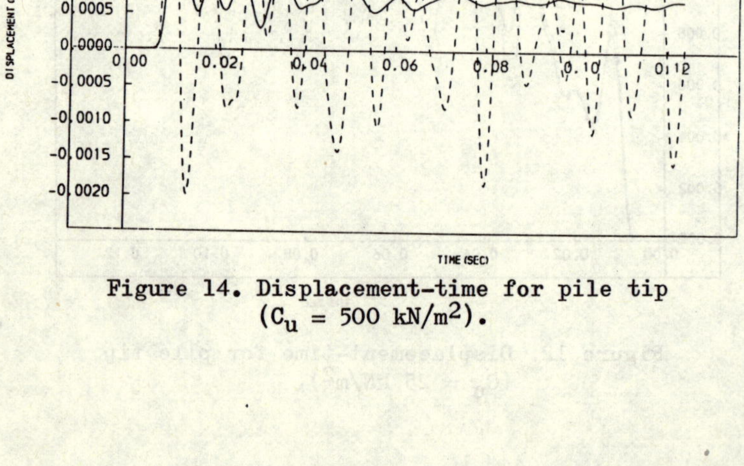

Figure 14. Displacement-time for pile tip ($C_u = 500$ kN/m^2).

this range.

PLUGGING

Large piles driven offshore are usually open ended steel pipes and the question arises of whether the soil advances up the pipe during static loading or driving (unplugged pile) or whether it does not (plugged pile). The 3-d finite element model enables this to be investigated. The mesh is shown in Figure 15 (13). Under static loading, at failure it is found that outer shaft "friction" is fully mobilised while inner shaft "friction" is small. Thus the pile behaves as plugged during static loading.

During driving, however, it is found that the inner skin "friction" does become mobilised and the pile behaves as unplugged. This has been recognised in the field and somewhat cumbersome modifications to 1-d models are sometimes made to account for it.

DYNAMIC EFFECTS

For single piles subjected to low frequency wave loading, dynamic amplifications seem to be small (less than 10 percent)

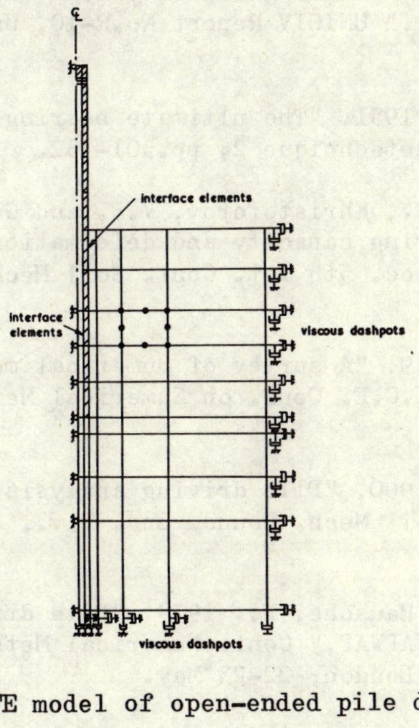

Figure 15. FE model of open-ended pile drivability.

unless the pile is close to failure in which case amplifications of up to 40% have been calculated. Viscosity can also account for some 15% additional resistance during long wave excitation in plastic clays.

PILE GROUPS

In the offshore situation, with which the writer has largely been concerned, group action is not very significant and group efficiencies are close to 100%. In closer spacing of course group action has to be considered. Finite element models are expensive in this area and interaction factors are usually worked out on the basis of simple elastic theory for displacements and equivalent pier approximations for collapse.

REFERENCES

1. Griffiths, D.V.: 1980, "Finite element analyses of walls, footings and slopes," Ph.D Thesis, University of Manchester.

2. Griffiths, D.V.: 1981, "Elasto-plastic analyses of deep foundations in cohesive soil," Int. J. Num. Anal. Meth. Geomechanics, Vol.5.

3. Hu, G.C.Y.: 1965, "Bearing capacity of foundations with overburden shear," UNICIV Report No.R-10, University of New South Wales.

4. Meyerhof, G.G.: 1951, "The ultimate bearing capacity of foundations," Geotechnique 2, pp.301-332.

5. Berezantzev, V.G., Khristoforov, V.S. and Golubkov, V.M.: 1961, "Load bearing capacity and deformation of piled foundations," Proc. 5th Int. Conf. Soil Mech. Found. Eng., 2, Paris.

6. Smith, I.M.: 1979, "A survey of numerical methods in offshore piling," Proc. I.C.E. Conf. on Numerical Methods in Offshore Piling, pp.1-8.

7. Smith, E.A.L.: 1960, "Pile driving analysis by the wave equation," J. Soil Mech. Found. Eng. Div., A.S.C.E., Vol.86, No.SM4, Aug.

8. Goble, C.G. and Rausche, F.: 1979, "Pile driveability predictions by CAPWAP," Conf. Numerical Methods in Offshore Piling, I.C.E., London, 22-23 May.

9. Coyle, H.M. and Gibson, G.C.: 1970, "Empirical damping constants for sands and clays," J. Soil Mech. Found. Div., A.S.C.E., Vol.96, No.SM3, May, pp.949-965.

10. Litkouhi, S. and Poskitt, T.J.: 1980, "Damping constants for pile driveability calculations," Geotechnique 30, No.1, pp.77-86.

11. Heerema, E.P.: 1979, "Relationships between wall friction, displacement, velocity and horizontal stress in clay and in sand for pile driveability analysis," Ground Engineering, January.

12. Chow, Y.K. and Smith, I.M.: 1981, "Static and periodic infinite solid elements," Int. J. Num. Meth. Eng., Vol.17, No.4, pp.503-526.

13. Chow, Y.K.: 1981, "Dynamic behaviour of piles," Ph.D Thesis, University of Manchester.

FINITE ELEMENT ANALYSIS OF COUPLED DEFORMATION AND FLUID FLOW IN POROUS MEDIA

Ranbir S. Sandhu

The Ohio State University, Columbus, Ohio, U.S.A.

Application of the finite element method to problems of coupled deformation and fluid flow in porous media is briefly reviewed. Equations governing the physical process, variational formulations appropriate for use with finite element methods, and finite element discretization as well as solution strategies are discussed. To illustrate various methods and procedures, darcy flow of an incompressible fluid in a linear elastic solid is considered in some detail. Other cases are occasionally referred to.

1. INTRODUCTION

Since the first application of the finite element method to analysis of seepage in elastic media (1,2), considerable progress has been made in the theoretical formulation as well as the computational procedures. Surveys of literature and descriptions of the state of the art have been attempted from time to time (e.g. 3,4). These expositions, however, need periodic updates due to the fact that significant developments continue in the field. Recent advances include variational formulations admitting limited smoothness of finite element bases (5,6), experimentation with several different spatial interpolation schemes (7-11) and investigation of various temporal approximation methods (8,11-17). The finite element method has been applied to soils exhibiting secondary compression (10,13,15), nonlinear soil behavior (15,18-22), and to finite deformation (18,20,22). Developments in solution procedures include use of Laplace transforms (12,13,16), automatic selection of the time-step size (15) and the use of single variable formulations (23,24).

J. B. Martins (ed.), Numerical Methods in Geomechanics, 203–227.

Other developments cover use of boundary element methods, allowance for infinitely distant boundaries, material interfaces etc.

We note the fact that there have been several efforts at developing a consistent theory of flow through deformable porous media (e.g. 18,20,22,25,26). However, we do not intend to discuss these aspects of the problem in detail and shall confine our attention to the application of finite element methods assuming darcy flow of an incompressible fluid through a saturated porous medium following Sandhu's formulation (1,27).

2. GOVERNING EQUATIONS

2.1 Preliminaries

Let R be the open connected region occupied by the saturated soil. Let ∂R be the boundary of R and $\overline{R}$ the closure of R. The domain of definition of all functions of interest is the cartesian product of the spatial domain with the positive interval of time, $[0, \infty)$.

Let u_i, e_{ij}, σ_{ij}, f_i, q_i and θ_i denote the cartesian components, respectively, of the soil displacement vector, the soil strain tensor, the soil effective stress tensor, the body force vector per unit mass, the relative fluid flux vector and the vector conjugate to the relative fluid flux. Also, let π denote the isotropic fluid stress above a reference state (say the atmospheric pressure). Let ρ denote the bulk density of the saturated soil. ρ equals the sum of ρ_2, the intrinsic density of the fluid and ρ_1 the 'buoyant' bulk density of the soil. We assume π and σ_{ij} to act over the entire area of any surface element. This is in line with Terzaghi's effective stress concept as well as the theory of superposed continua. Assumption of ρ_2 as the intrinsic density of the fluid corresponds to the notion of the soil skeleton being buoyant in the fluid (1).

The indicial notation is used herein. Summation on repeated indices is implied except where otherwise indicated. Subscripts following a subscripted comma denote differentiation with respect to the reference cartesian coordinate indicated by the subscript. Parentheses around two subscripts denote symmetric part with respect to the subscripts. A superposed dot denotes differentiation will respect to either the time variable or a scalar parameter. δ_{ij} is Kronecker's delta. As a rectangular cartesian frame is used for description of tensor components, no destinction is made between covariant and contravariant components.

2.2 The Field Equations

The strain displacement relationship for the soil is

$$e_{ij} = \tfrac{1}{2}(u_{i,j} + u_{j,i}) + \tfrac{1}{2} u_{m,i} u_{m,j} \qquad (2.1)$$

and, for small deformations i.e. linear theory,

$$e_{ij} = \tfrac{1}{2}(u_{i,j} + u_{j,i}) = u_{(i,j)} \qquad (2.2)$$

Equilibrium of forces on infinitesimal volumes, assuming small deformation, neglecting inertia effects and for isotropic fluid stress is expressed as

$$\sigma_{ji,j} + \pi_{,j}\,\delta_{ij} + \rho f_i = 0 \qquad (2.3)$$

Also, for no body couples

$$\sigma_{ji} = \sigma_{ij} \qquad (2.4)$$

Referring to material frame associated with the soil skeleton, for incompressible pore-fluid and no chemical action, continuity implies continued saturation of the soil i.e. the rate of outflow from a material volume of the soil equals the rate of decrease of the volume:

$$q_{i,i} = -\dot{e}_{ii} \qquad (2.5)$$

Constitutive equations have to be written for the stresses as well as for the diffusive resistance.

Stress-strain laws. The components of effective stress in the soil, for linear isothermal elasticity can be expressed in terms of strain as

$$\sigma_{ij} = E_{ijkl}\, e_{kl} \text{ on } \bar{R} \times [0, \infty) \qquad (2.6)$$

where E_{ijkl}, the components of the isothermal elasticity tensor satisfy the symmetry relationships

$$E_{ijkl} = E_{jikl} = E_{klij} = E_{ijlk} \qquad (2.7)$$

Conversely, the constitutive equations may be written as

$$e_{ij} = C_{ijkl}\, \sigma_{kl} \qquad (2.8)$$

where C_{ijkl} are components of the isothermal elastic compliance tensor such that

$$C_{ijkl} = C_{jikl} = C_{klij} = C_{ijlk} \qquad (2.9)$$

For non-isothermal situations, the Neumann-Duhamel postulate implies

$$e_{ij} = C_{ijkl}\, \sigma_{kl} + e_{ij}^{T} \qquad (2.10)$$

when e_{ij}^{T} are components of strain associated with the temperature change.

For a linear hereditary material, the constitutive equations for components of effective stress are

$$\sigma_{ij} = J_{ijkl} * de_{kl} \qquad (2.11)$$

where * denotes Stieltjes' integral i.e.

$$u{*}dv = \int_{\tau=0}^{\tau=t} u(t-\tau)\, dv(\tau) \qquad (2.12)$$

and J_{ijkl} is a set of relaxation functions for the material. Conversely, the constitutive equations may be written as

$$e_{ij} = G_{ijkl} * d\sigma_{kl} \qquad (2.13)$$

where G_{ijkl} is a set of creep compliance functions for the soil.

For a nonlinear material, if it is hyperelastic, an energy function $U(e_{ij})$ exists such that

$$\sigma_{ij} = \frac{\partial U}{\partial e_{ij}} \qquad (2.14)$$

Conversely, if a complementary energy function $V(\sigma_{ij})$ exists

$$e_{ij} = \frac{\partial V}{\partial \sigma_{ij}} \qquad (2.15)$$

In (2.14) and (2.15) the notation ∂/∂ denotes gradients in the appropriate space. If the material is nonlinear and U,V do not exist, its mechanical behavior is generally represented in incremental form as

$$\dot{\sigma}_{ij} = E_{ijkl}\, \dot{e}_{kl} \qquad (2.17)$$

or, conversely,

$$\dot{e}_{ij} = C_{ijkl}\, \dot{\sigma}_{kl} \qquad (2.18)$$

where E_{ijkl}, C_{ijkl} may be functions of stress (strain) history.

Diffusive Resistance. Diffusive resistance in a binary

mixture of soil and water, for no inertia effects, is defined by the components(1)

$$\pi_i \equiv -\pi_{,i} - \rho_2 f_i \qquad (2.19)$$

For the linear theory, assuming irrotational flow, the diffusive resistance is linearly proportional to the relative velocity of flow (1,27) i.e.

$$\pi_i = -C_{ij}\, q_i \qquad (2.20)$$

If K_{ij} is the inverse of C_{ij}, we may write (2.20) as

$$q_i = K_{ij}\, \theta_j \qquad (2.21)$$

where we define θ_j as

$$\theta_j \equiv -\pi_j \qquad (2.22)$$

(2.21) is the generalized form of Darcy's law. K_{ij} are components of the permeability tensor.

Combining (2.5), (2.19), (2.21) and (2.22), the continuity equation is

$$\left[K_{ij}(\pi_{,j} + \rho_2 f_j)\right]_{,i} = -\dot{e}_{kk} \qquad (2.23)$$

An alternative form of the continuity equation is obtained by integrating (2.5) from the initial state to the current state. This is (1, 27)

$$\int_o^t q_{i,i}(\tau)d\tau = -\int_o^t \dot{u}_{k,k}(\tau)\, d\tau$$

$$= u_{k,k}(o) - u_{k,k}(t) \qquad (2.24)$$

This form does not contain time derivatives and was found to be suitable (2,5,6,27) for setting up variational formulations for the problem. Of course, another alternative would be to write the equilibrium equations and the strain-displacement equations in 'rate' or incremental form to get a consistent formulation suitable for construction of variational principles.

2.3 The Boundary Conditions

Associated with the system of field equations is a set of boundary conditions. Here, it is not our intent to discuss the necessity and sufficiency of these conditions for the problem to be well-posed. We merely state these conditions and refer the reader to literature on the subject (e.g. 9,15,27,28) for details.

The displacement boundary condition is

$$u_i (\underset{\sim}{x}, t) = \hat{u}_i(\underset{\sim}{x}, t) \text{ on } S_{1i} \times [o,\infty) \qquad (2.25)$$

where S_{1i} is a subset of ∂R and $\underset{\sim}{x}$ denotes the description of an arbitrary point in ∂R

The traction boundary condition is

$$t_i (\underset{\sim}{x},t) = \hat{t}_i (\underset{\sim}{x},t) \text{ on } S_{2i} \times [o, \infty) \qquad (2.26)$$

where t_i is the total traction on surface elements defined as

$$t_i \equiv (\sigma_{ij} + \pi \, \delta_{ij})\, n_j \qquad (2.27)$$

S_{2i} is a subset of ∂R complementary to S_{1i} and n_j are components of the unit outward normal to S_{2i}. We note that it is not necessary to separately prescribe $\sigma_{ij}\, n_j$ and π on S_{2i} i.e. it is only necessary to prescribe total traction on the portion of the boundary on which the displacements are not prescribed.

The fluid pressure boundary condition is

$$\pi (\underset{\sim}{x},t) = \hat{\pi} (\underset{\sim}{x},t) \text{ on } S_3 \times [o,\infty) \qquad (2.28)$$

where S_3 is a subset of ∂R. S_3 is independent of S_{1i}, S_{2i}.

The flux boundary condition is

$$Q(\underset{\sim}{x},t) = \hat{Q} (\underset{\sim}{x},t) \text{ on } S_4 \times [o,\infty) \qquad (2.29)$$

where

$$Q \equiv q_i\, n_i \qquad (2.30)$$

and S_4 is a subset of ∂R complementary to S_3.

2.4 The Initial Conditions

The initial conditions on the problem are

$$u_i (\underset{\sim}{x},o) = d_i (\underset{\sim}{x}) \text{ on } R \qquad (2.31)$$

For an initially undeformed state d_i (x) vanishes. Assuming that the initial state is a steady state in which no fluid flow is taking place relative to the soil skeleton, q_i vanishes and as a consequence of (2.21), for positive definite K_{ij}, θ_j vanishes. This implies, in view of (2.19), that

$$\pi_{,i}\ (\underset{\sim}{x},o) = -\ \rho_2\ f_i\ (\underset{\sim}{x},o) \qquad (2.32)$$

It has been shown (9,24,27,28) that the fluid pressure field existing immediately after application of a load cannot be independently prescribed and in fact is directly related to the load and the initial condition on displacements, being the solution for fluid pressures for the undrained case. As we shall see in Section 4, some finite element schemes may fail to provide a solution for the undrained problem.

2.5 The Internal Discontinuity Conditions

In seeking approximation to the correct solution by the finite element method, it is customary to use functions having limited smoothness. In particular, the space of admissible approximants may have discontinuities across inter-element boundaries. To ensure that the approximation be constrained to satisfy continuity requirements or prescribed jump discontinuities, (5,6,29) the following internal boundary conditions must be adjoined to the problem.

$$u'_i = g_i \text{ on } S^*_{1i} \times [o,\infty) \qquad (2.33)$$

$$t'_i = h_i \text{ on } S^*_{2i} \times [o,\infty) \qquad (2.34)$$

$$\pi' = g \text{ on } S^*_3 \times [o,\infty) \qquad (2.35)$$

$$Q' = h \text{ on } S^*_4 \times [o,\infty) \qquad (2.36)$$

Here the superscript prime denotes the magnitude of the jump discontinuity in the quantity superscripted. While introducing these discontinuity conditions, Sandhu (5,6,29) stipulated that S^*_{1i}, S^*_{2i} be disjoint and S^*_3, S^*_4 be disjoint. Bufler (30) pointed out that this constraint was unnecessary. However, in a finite element solution scheme, the surfaces S^*_{1i}, S^*_{2i}, S^*_3, S^*_4 must be imbedded in the interior of R and must be contained in the union of the intersections of element boundaries.

2.6 A Comment on the Theoretical Formulation

Many of the modern theories of soil consolidation have been derived (e.g. 1,22,25,26,31) from recently developed theories of mixtures (e.g. 32-39). Sandhu (10) expressed concern regarding some of the postulates on which these theories are based. For example, the notion of the mixture of several superposed constituents as a continuum in motion is inadmissible. Consequently the introduction of thermodynamic quantities like an energy function for the mixture is questionable except as the sum of corresponding quantities for individual constituents. Thus, the constitutive equations involving coupling between various constituents cannot be derived on the basis of an energy function existing for the mixture as a continuum but must be based on other considerations.

3. VARIATIONAL FORMULATION

For isothermal consolidation of linear elastic soils, Sandhu (1,2) proposed a general variational principle. Yokoo (40) proposed variational formulations relaxing the need for satisfying specified boundary conditions on the fluid pressure. Sandhu presented (5,6) a comprehensive discussion of variational principles for soil consolidation including extensions to visco-elastic and nonlinear soils. Although complementary and mixed formulations have been developed, the variational form commonly used for finite element approximation has been the direct form. For linear elastic soils undergoing small deformations, for m finite elements, the general governing function for the variational formulation is (5,6)

$$
\begin{aligned}
\Omega = \sum_{e=1}^{m}\Big[\int_{R_e} \Big\{ & -u_i * (\sigma_{ij,j} + \pi_{,i} + 2\rho f_i) \\
& + e_{ij} * (E_{ijkl}\, e_{kl} - \sigma_{ij} + 2\bar{\sigma}_{ij}) \\
& + \sigma_{ij} * (u_{(i,j)} - e_{ij}) \\
& + q_i * g' * (\theta_i - \pi_{,i} - 2\rho_2 f_i) \\
& + \theta_i * g' * (q_i - K_{ij}\,\theta_j) \\
& + \pi * (u_{i,i} + g' * q_{i,i}) \Big\} dR_e \Big] \\
& - \int_{S_{1i}} \sigma_{ij}\, n_j * (u_i - 2\hat{u}_i)\, ds \\
& + \int_{S_{2i}} u_i * (\sigma_{ij}\, n_j - 2\hat{t}_i)\, ds \\
& + \int_{S_3} q_i n_i * g' * (\pi - 2\hat{\pi})\, ds \\
& - \int_{S_4} \pi * g' * (q_i n_i - 2\hat{Q})\, ds \\
& - \int_{S_{1i}^*} \sigma_{ij}\, n_j * (u'_i - 2g_i)\, ds
\end{aligned}
$$

$$+\int_{S^*_{2i}} u_i^* \, (t'_i - 2h_i) \, ds$$

$$+\int_{S^*_3} q_i n_i^* g'^* \, (\pi' - 2g) \, ds$$

$$-\int_{S^*_4} \pi * g'^* \, (Q' - h) \, ds \qquad (3.1)$$

Here R_e is the part of the domain R represented by the eth finite element. The function $g'(t) = 1$ and consequently the notation

$$g'^*u = \int_0^t u(\tau) \, d\tau \qquad (3.2)$$

The quantity $\bar{\sigma}_{ij}$ represents 'initial stress in the soil corresponding to the undeformed state. The Gateaux differential of Ω with respect to variables u_i, e_{ij}, σ_{ij}, π, θ_i, q_i vanishes if and only if all the field equations, the boundary conditions and the internal constraints are satisfied. Various extensions of the variational formulation can be developed (5,6,27) using the following equalities to eliminate certain quantities from the governing function.

$$\int_{R_e} u_i \, \sigma_{ij,j} \, dR_e = -\int_{R_e} u_{i,j} \, \sigma_{ij} \, dR_e + \int_{\partial R_e} u_i \sigma_{ij} n_j \, ds \qquad (3.3)$$

$$\int_{R_e} \pi q_{i,i} \, dR_e = -\int_{R_e} \pi_{,i} \, q_i dR_e + \int_{\partial R_e} \pi q_i n_i \, ds \qquad (3.4)$$

$$\int_{R_e} \pi u_{i,i} \, dR_e = -\int_{R_e} \pi_{,i} \, u_i dR_e + \int_{\partial R_e} \pi u_i n_i \, ds \qquad (3.5)$$

The extension which has been in general use is based on elimination of terms containing $u_i\sigma_{ij,j}$, $\pi q_{i,i}$ and $\pi_{,i}u_i$. Substitution of (3.3) through (3.5) in (3.1) gives, for vanishing initial stress,

$$\Omega_1 = \sum_{e=1}^{m} \left[\int_{R_e} \left\{ -2 \, u_i^* \rho f_i + e_{ij}^* \, (E_{ijkl} \, e_{kl} - \sigma_{ij}) \right. \right.$$

$$+ \sigma_{ij}^* (2u_{i,j} - e_{ij}) + q_i^* g'^* (\theta_i - 2\pi_{,i} - 2\rho_2 f_i)$$

$$+ \theta_i^* g'^* (q_i - K_{ij}\theta_j) + 2\pi^* u_{i,i} \Big\} dR_e \Big]$$

$$- 2\int_{S_{1i}} \sigma_{ij} n_j * (u_i - \hat{u}_i)\, ds$$

$$-2\int_{S_{2i}} u_i^* \hat{t}_i ds + 2\int_{S_3} q_i n_i^* g'^* (\pi - \hat{\pi})\, ds$$

$$+2\int_{S_4} \pi^* g'^* \hat{Q}\, ds - 2\int_{S_{1i}^*} \sigma_{ij} n_i^* (u'_i - g_i)\, ds$$

$$-2\int_{S_{2i}^*} u_i^* h_i ds + 2\int_{S_3^*} q_i n_i^* g'^* (\pi' - g) ds$$

$$+2\int_{S_4^*} \pi^* g'^* h\, ds \qquad (3.6)$$

Many specializations of this extended form (similar specializations are possible with other forms) are obtained by requiring the set of admissible solutions in the approximation process to identically satisfy one or more of the field equations, the boundary conditions and the internal continuity constraints. For instance, specialization to the case where there are no internal discontinuities i.e., g_i, h_i, g, h all vanish, and q_i, θ_i are derived from π by identically satisfying (2.19) thru (2.22) and σ_{ij}, e_{ij} are derived from u_i by identically satisfying (2.2) and (2.6), we get the two-field formulation governed by

$$\Omega_2 (u_i, \pi) = \sum_{e=1}^{m} \Big[\int_{R_e} \Big\{ -u_i^* \rho f_i + \tfrac{1}{2} u_{i,j}^* E_{ijkl} u_{k,l}$$

$$+ \pi^* u_{i,i} - \tfrac{1}{2} \theta_i^* g'^* K_{ij} \theta_j \Big\} dR_e \Big]$$

$$- \int_{S_{1i}} \sigma_{ij} n_j^* (u_i - u_i) ds$$

$$- \int_{S_{2i}} u_i^* \hat{t}_i\, ds + \int_{S_3} q_i n_i^* g'^* (\pi - \hat{\pi})\, ds$$

$$+\int_{S_4} \pi^* g'^* \hat{Q} ds - \int_{S_{1i}^*} \sigma_{ij} n_j^* u_i' ds$$

$$+\int_{S_3^*} q_i n_i^* g'^* \pi' ds \qquad (3.7)$$

Further restricting π, u_i to be continuous over the interior of R and u_i to satisfy the boundary conditions on S_{1i}, we get Yokoo's formulation (40)

$$\Omega_3 (u_i, \pi) = \sum_{e=1}^{m} \Big[\iint_{R_e} \{ -u_i^* \rho f_i + \tfrac{1}{2} u_{i,j}^* E_{ijkl} u_{k,l}$$

$$+ \pi^* u_{i,i} - \tfrac{1}{2}\theta_i^* g'^* K_{ij} \theta_j \} dR_e \Big]$$

$$-\int_{S_{2i}} u_i^* \hat{t}_i ds + \int_{S_3} q_i n_i^* g'^* (\pi - \hat{\pi}) ds$$

$$+\int_{S_4} \pi^* g'^* \hat{Q} ds \qquad (3.8)$$

This formulation, not requiring the fluid pressure to satisfy the prescribed condition on S_3 is expected to avoid the errors in the approximate solution, immediately after application of load, caused by forcing the excess fluid pressure at a free draining boundary to vanish (9). If we do enforce the condition on S_3, the variational formulation implemented by Sandhu (1,2,27), and several others, in finite element procedures is realized.

Ghaboussi (7,41) generalized Sandhu's two-field formulation (2) to admit compressibility of the fluid as well as inertia effects. For compressible fluid, following Biot's assumption (42) of the existence of an energy function for the mixture, a symmetric constitutive coupling was postulated i.e.

$$\sigma_{ij} = E_{ijkl}\, e_{kl} + M_{ij} \zeta \qquad (3.9)$$

$$\pi = M_{ij}\, e_{ij} + M \zeta$$

where ζ is the isotropic fluid strain and M_{ij}, M are material constants. Sandhu (5,6,10) extended the original formulation (1,2) to admit viscoelastic soil behavior as well as incremental formulation of the governing equations. Recently, Moussa (22,43) has studied variational formulations for the case of finite deformation allowing for saturated as well as unsaturated conditions.

In setting up variational formulations for the problem of fluid flow in deformable media, Sandhu (1,5,6,27) and Ghaboussi (7) had followed Sandhu and Pister's (44) extension of Gurtin's method (45-47) based on a generalization of Mikhlin's basic variational theorem (48). Herrera (49,50) has pointed out that simpler formulations, involving fewer convolutions, arise if the differential (in the time domain) form of the field equations is used. This is in line with Tonti (51,52) and Magri (53).

Alternatives to construction of functions such that vanishing of their Gateaux differentials is equivalent to the governing equations being satisfied, include use of the principle of virtual work, or method of weighted residuals. Krause (23,24) proposed virtual work formulations for soil consolidation. Zienkiewicz (54), Hwang (55) and Prevost (11), among others, used Galerkin's method.

Booker (16) set up the variational formulation after taking Laplace tranforms. Corresponding to a specialization of (3.8) to the case $\pi \equiv \hat{\pi}$ on S_3, $\hat{Q} \equiv o$ on S_4, $f_i \equiv o$ on R, Booker's functional, for finite element approximation, would be

$$\Omega_4 (u_i, \pi) = \sum_{e=1}^{m} \Big[\int_{R_e} \Big\{ \tfrac{1}{2} E_{ijkl} \bar{u}_{i,j} \bar{u}_{k,l} + \bar{\pi} \bar{u}_{i,i} - \tfrac{1}{2} \pi_{,i} K_{ij} \pi_{,j} \Big\} dR_e \Big] - \int_{S_{2i}} \bar{u}_i \hat{t}_i \, ds \qquad (3.10)$$

where a superscript bar denotes Laplace transform of the relevant quantity. Generalization of (3.10) to admit more complicated boundary conditions etc. is straightforward.

4. FINITE ELEMENT APPROXIMATION

4.1 Spatial Interpolation

In developing the variational formulation (3.8), from the general formulation (3.1), the following assumptions were made:

a. $u_i \sigma_{ij}$, πq_i, πu_i are continuously differentiable over each finite element.

b. θ_i, q_i are derived from π by identically satisfying (2.19) thru (2.22) and σ_{ij}, e_{ij} are derived from u_i by identically satisfying (2.2) and (2.6).
c. π, u_i are continuous over each finite element and across inter-element boundaries.
d. u_i satisfy the prescribed boundary condition on S_{1i}.

In the specialization used by Sandhu (1,2,27) and others, an additional assumption is that π satisfy the prescribed boundary condition on S_3. The finite element bases used for setting up approximate solutions are subject to all these constraints. Further, in evaluation of integrals appearing in the governing function, Sandhu (1) proposed that the order of terms appearing in a product or convolution be the same e.g. π and $u_{i,i}$ have to be of the same order. This produced the 'composite' element in which the order of polynomial interpolation for displacements was higher than that for fluid pressures. Evidently, the same consideration in the case of other variational formulations might require other types of composite elements. The composite elements first proposed by Sandhu (1,2), and used by Hwang (55) and others, was the '63' element with quadratic interpolation for displacements and linear interpolation for fluid pressures over triangular elements. Later, Sandhu (9,27) introduced the '84' quadrilateral element which had eight point biquadratic interpolation for displacements and was a four point isoparametric quadrilateral for fluid pressures. This element was used by Runesson (15). Buchmaier (60) experimented with five, six and seven point quadrilaterals for fluid pressure as transition elements near loaded surfaces.

Several spatial interpolation schemes, besides the composite elements, have been tried by various investigators. Yokoo (56), Booker (16), and Vermeer (17) used triangular elements with linear interpolation for both the displacement and the fluid pressure fields. Ghaboussi (41) used four point isoparametric quadrilaterals for both the fields. However, an additional incompatiable mode was included in the displacement approximation. Smith's (57) formulation was similar to Ghaboussi's except that no incompatible modes were used. Prevost (11) proposed cautious use of 'reduced integration' in conjunction with Smith's '44' element. Other interpolation schemes, based on use of quadrilaterals built up from linear and four or five point triangles were tried by Sandhu (10).

All investigators have generally reported success with whatever scheme they used. Comparative studies of different elements are rare. Some comparisons of numerical performance were attempted by Sandhu (9,10). In evaluating various candidate schemes, we require that an acceptable method should meet the following requirements.

i. The interpolation scheme must conform with the assumptions regarding continuity and differentiability used in setting up the governing variational formulation.
ii. It should be possible to generate the 'undrained' solution i.e. the state of fluid pressures and displacements at time t = 0+
iii. For sufficiently small time-steps, the scheme should be insensitive to the choice of the time-step size.

Figure 1 shows some elements that satisfy i.

After discretization in space, a time-stepping scheme has the form (1,2,9,27).

$$\begin{bmatrix} K_1 & C \\ C^T & -\alpha\Delta t\, K_2 \end{bmatrix} \begin{Bmatrix} u_1 \\ \pi_1 \end{Bmatrix} = \begin{Bmatrix} P \\ Q + C^T u_o + (1-\alpha)\Delta t\, K_2\, \pi_o \end{Bmatrix} \quad (4.1)$$

where u_1, π_1 are values of the displacement and the fluid pressure at the end of the time-step and u_o, π_o are the corresponding values at the beginning of the time step. P,Q are the load vector and the flow vector respectively. K_1, K_2 are the stiffness matrix and the flow matrix, and C is the coupling matrix. Δt is the time step and α is a factor reflecting the time-domain approximation. For $\Delta t = o$, the solution for the displacements is

$$u_1 = K_1^{-1} (P - C\pi_1) \quad (4.2)$$

Substitution of (4.2) in the second of (4.1) yields

$$C^T K_1^{-1} C\, \pi_1 = C^T K_1^{-1} P - C^T u_o - Q \quad (4.3)$$

This equation will give a solution for the fluid pressure π_1 immediately after loading if the matrix $C^T K^{-1} C$ is invertible. As K_1^{-1} exists, the invertibility of $C^T K_1^{-1} C$ is dependent upon the character of the coupling matrix C. This is defined by the choice of the spatial interpolation for the two field variables. It was found (9) that the elements using higher order interpolation for displacements as well as fluid pressures (elements 66 and 88) did not give a solution to the undrained case. Four point isoparametric interpolation for both the field variables was found to result in an undrained solution in which there was a spatially oscillatory error in fluid pressures. These errors dissipated with progress in the time domain. Elements designated as 1,7,8 in Figure 1 were found not to satisfy iii. There was a 'numerical' consolidation involved at each time step i.e. a large number of very small time steps could give large dissipation of fluid pressure (10).

#	TYPE	REFERENCE	DISPLACEMENTS	FLUID PRESSURE
1	33	16,17		
2	63	1,2,9,27,55		
3	66	9,27		
4	44	7,8,10,41		
5	84	9,15,27,60		
6	88	9,27,58		
7		10		
8		10		

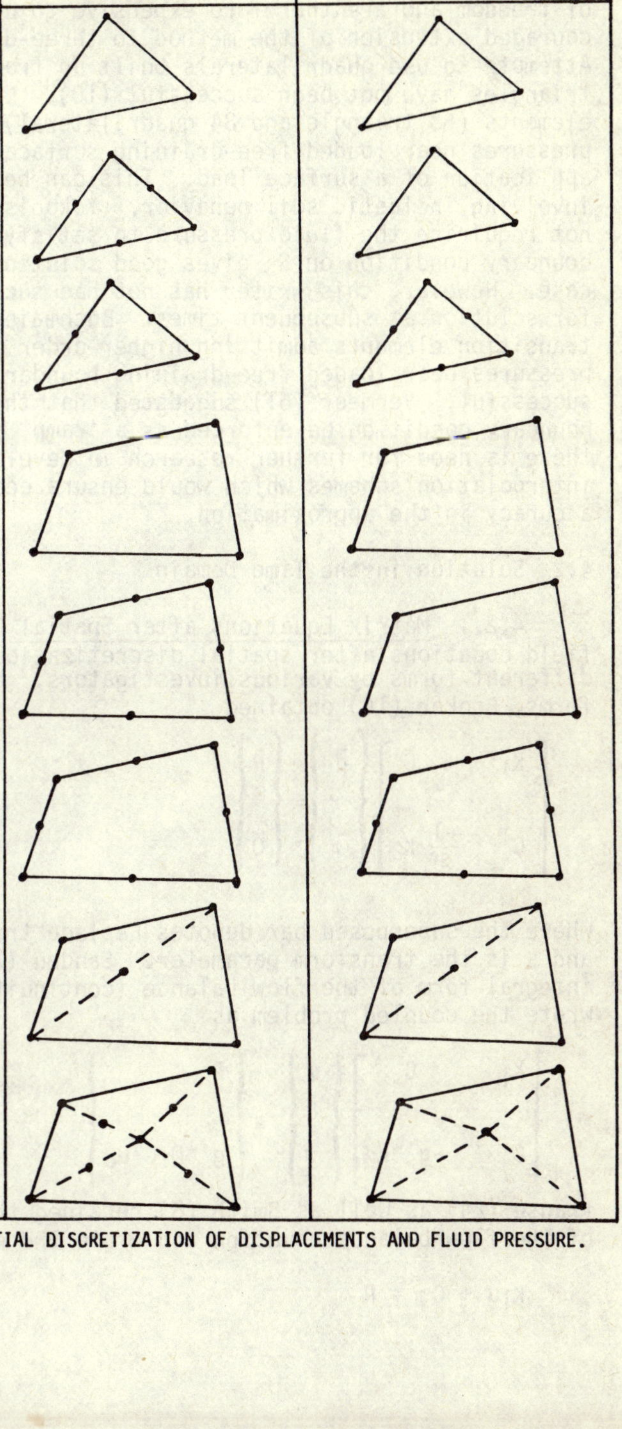

Figure 1. SPATIAL DISCRETIZATION OF DISPLACEMENTS AND FLUID PRESSURE.

The composite elements have, apparently, been the most successful so far. However, these have a large number of degrees of freedom and are therefore expensive to use. This has discouraged extension of the method to three-dimensional consolidation. Attempts to use quadrilaterals built up from four and five point triangles have not been successful (10). Even the composite elements (63 triangle and 84 quadrilateral)give inaccurate pressures near loaded free-draining surfaces immediately after application of a surface load. This can be troublesome in problems involving inelastic soil behavior. Yokoo's (56) formulation, not requiring the fluid pressure to satisfy the prescribed boundary condition on S_3 gives good solution for the undrained case. However, this writer has not had success with this procedure for solution at subsequent times. Buchmaier's (60) studies with transition elements admitting higher order interpolation of fluid pressures near loaded free-draining boundaries also were not successful. Vermeer (61) suggested that the fluid pressure boundary condition be enforced as a 'ramp' condition. Apparently, there is need for further research in development of suitable interpolation schemes which would ensure economy as well as accuracy in the approximation.

4.2 Solution in the Time Domain

4.2.1 Matrix Equations after Spatial Discretization. The field equations after spatial discretization have been stated in different forms by various investigators. Using Laplace transforms, Booker (16) obtained

$$\begin{bmatrix} K_1 & C \\ C^T & \frac{-1}{s} K_2 \end{bmatrix} \begin{Bmatrix} \bar{u} \\ \bar{\pi} \end{Bmatrix} = \begin{Bmatrix} \bar{P} \\ \bar{Q} \end{Bmatrix} \tag{4.4}$$

where the superposed bar denotes Laplace transform of a quantity and s is the transform parameter. Sandhu (1,2,27) used an integral form of the flow balance (continuity) equation and wrote the coupled problem as

$$\begin{bmatrix} K_1 & C \\ C^T & -g'*K_2 \end{bmatrix} \begin{Bmatrix} u \\ \pi \end{Bmatrix} = \begin{Bmatrix} P \\ g'*Q+C^T u_0 \end{Bmatrix} \tag{4.5}$$

Krause (24) as well as Smith (8) retained the differential form of the flow balance equation and wrote the equations as

$$K_1 u + C\pi = P \tag{4.6}$$

$$C^T\dot{u} - K_2 \pi = Q \tag{4.6}$$

For solution of nonlinear problems, Runesson (15), Prevost (11) and Siriwardane (21), have used the incremental form

$$\left.\begin{aligned} K_1\dot{u} + C\dot{\pi} &= P \\ C^T\dot{u} - K_2 \pi &= \dot{Q} \end{aligned}\right\} \tag{4.7}$$

4.2.2 Temporal Discretization. Sandhu (1,2,27) used an interpolation scheme in the time domain to evaluate the quantity $g'*\pi$. This leads to

$$g'*\pi = \alpha\ \pi_1 + (1-\alpha)\ \pi_0 \tag{4.8}$$

where the subscripts one and zero denote the values at the end and the beginning of a time-step. The coefficient α depends upon the type of interpolation. For example, $\alpha = \frac{1}{2}$ corresponds to linear interpolation. Sandhu (1,27) suggested use of $\alpha = \frac{1}{2}$ or

$$\alpha = 1 + \frac{1}{\Delta t} - \frac{1}{\ln\ (1 + \Delta t)} \tag{4.9}$$

where Δt is the time-step size. This corresponds to logarithmic interpolation. Hwang (55) proposed

$$\alpha = 1 + \frac{t_n}{\Delta t} - \frac{1}{\ln(1 + \frac{\Delta t}{t_n})} \tag{4.10}$$

where t_n is the total elapsed time up to the beginning of the time step. Sandhu (9) reported a comparison of numerical performance of some choices of values of α. Booker (16) showed that this time-stepping scheme is unconditionally stable for $\alpha \geqslant \frac{1}{2}$. Using this discretization scheme, Sandhu (1,2,27) set up the following time-stepping procedure:

$$\begin{bmatrix} K_1 & C \\ C^T & -\alpha\,\Delta t K_2 \end{bmatrix} \begin{Bmatrix} u_1 \\ \pi_1 \end{Bmatrix} = \begin{bmatrix} 0 & 0 \\ C^T & (1-\alpha)\,\Delta t K_2 \end{bmatrix} \begin{Bmatrix} u_0 \\ \pi_0 \end{Bmatrix} + \begin{Bmatrix} P_1 \\ (\alpha Q_1 + (1-\alpha)\ Q_0)\ \Delta t \end{Bmatrix} \tag{4.11}$$

These equations ensure force equilibrium at the end of and flow balance in the sense of an integral over the time interval.

Smith (8) and Krause (24) assumed u and π in (4.6) to be the mean values for the interval. These were written in terms of values at the beginning and the end of an interval as

$$\pi = \alpha \pi_1 + (1-\alpha)\,\pi_o$$
$$u = \alpha u_1 + (1-\alpha)\,u_o \tag{4.12}$$

This led Smith (8) to the formulation

$$\begin{bmatrix} \alpha K & \alpha C \\ \alpha C^T & -\alpha^2 \Delta t K_2 \end{bmatrix} \begin{Bmatrix} u_1 \\ \pi_1 \end{Bmatrix} = \begin{bmatrix} (\alpha-1)K_1 & (\alpha-1)C \\ \alpha C^T & -\alpha(\alpha-1)\Delta t K_2 \end{bmatrix} \begin{Bmatrix} u_o \\ \pi_o \end{Bmatrix} + \begin{Bmatrix} \alpha P_1 + (1-\alpha)\,P_o \\ (\alpha^2 Q_1 + \alpha(1-\alpha)\,Q_o)\Delta t \end{Bmatrix} \tag{4.13}$$

The second equation here is identical with that in (4.11). However, the first equation is different, ensuring force equilibrium in tne sense of a mean value or, equivalently, in the sense of an integral over the time step. For $\alpha = 1$ the scheme is identical that proposed by Sandhu (1,2,27).

Prevost (11) satisfied the incremental equilibrium equation at the end of the interval i.e.

$$K_1 \dot{u}_1 + C\,\dot{\pi}_1 = \dot{P}_1 \tag{4.14}$$

For the second of equations (4.7), Prevost set

$$\pi = \pi_1 = \pi_o + \Delta t\;\dot{\pi} \tag{4.15}$$

where $\dot{\pi}$ is the mean rate i.e.

$$\dot{\pi} = \alpha\,\dot{\pi}_1 + (1-\alpha)\,\dot{\pi}_o \tag{4.16}$$

Combining (4.15) and (4.16)

$$\pi_1 = \pi_o + \alpha \Delta t\;\dot{\pi}_1 + (1-\alpha)\Delta t\;\dot{\pi}_o \tag{4.17}$$

Then,

$$C^T \dot{u}_1 - \alpha \Delta t K_2\;\dot{\pi}_1 = \dot{Q} + K_2\,(\pi_o + (1-\alpha)\Delta t\;\dot{\pi}_o) \tag{4.18}$$

Together, equations (4.14) and (4.18) are

$$\begin{bmatrix} K_1 & C \\ C^T & -\alpha\Delta t K_2 \end{bmatrix} \begin{Bmatrix} \dot{u}_1 \\ \dot{\pi}_1 \end{Bmatrix} = \begin{Bmatrix} \dot{P}_1 \\ \dot{Q}+K_2(\pi_o +(1-\alpha)\Delta t\, \dot{\pi}_o) \end{Bmatrix} \quad (4.19)$$

Of course, in the nonlinear problem, K_1,C,K_2 are functions of the history of deformation.

4.2.3 <u>Size of the time-step</u>. In unconditionally stable algorithms, size of the time-step can apparently be selected arbitrarily. However, very large time steps or large changes in the size of the time-step have been found (27) to give erroneous solutions. Of course the error rapidly decays. For constant applied loads, gradually increasing time-steps can be used. However, any change in the time-step size involves reduction of a new set of equations. In actual analysis, therefore, a few changes in the time-step size along with use of several steps of equal size at each stage is the most economical. Obviously where the applied load is changing as a function of time, small time-steps may be necessary throughout.

It might be reasonable to expect that use of a smaller time-step should result in improved accuracy of the solution. For the case of linear interpolation over triangular elements for both the field variables, Vermeer (17) determined that for stability a minimum size of time-step is necessary.

In nonlinear problems, with use of implicit formulations, it is possible to relate the time-step size to the need for convergence of the iterative procedure (62). Runesson's (15) method for time-integration of nonlinear problems has a built-in time-step selection procedure.

4.2.4 <u>Solution Process</u>. Christian and Boehmer (59) solved the coupled equations by first obtaining the undrained solution (solution for zero volumetric strain in the soil). This would give the solution for fluid pressures. These fluid pressures define $C^T\dot{u}$, the rate of volume change. For a selected time step this gives the volume change over the interval. Next, the equilibrium equation is solved for the known volume change. An iterative procedure might be employed to ensure that the volume change assumed equals that computed for the time step. The cycle of operations is repeated each time step.

Sandhu (1,2,27), Hwang (55), and others used the algorithm defined by (4.11) for appropriate choice of Δt and α to simultaneously generate a history of displacements and fluid pressures. Smith (8) used (4.13). Krause (23,24) and Smith (8) proposed reducing the number of degrees of freedom by elimination of either the fluid pressures or the soil displacements. The second equation in (4.6) can be solved to give

$$\pi = K_2^{-1} (C^T \dot{u} - Q) \quad (4.20)$$

Substitution in the first equation of the same set gives

$$CK_2^{-1} C^T \dot{u} = -K_1 u + P + CK_2^{-1} Q \quad (4.21)$$

As α - algorithm for this equation was given by Smith (8) as

$$(CK_2^{-1} C^T + \alpha \Delta t K_1) u_1 = (CK_2^{-1} C^T - (1-\alpha) \Delta t K_1) u_o + \Delta t (P + CK_2^{-1} Q) \quad (4.22)$$

This is the same as Krause's 'displacement' formulation if we set $\Delta t = 1/2$. Alternatively, solving the first of equations (4.6) for the beginning as well as the end of the time-step,

$$u_1 = K_1^{-1} (P_1 - C \pi_1) \quad (4.23)$$

$$u_o = K_1^{-1} (P_o - C \pi_o) \quad (4.24)$$

Setting $\dot{u} = (u_1 - u_o)/\Delta t$, for $P_o = P_1$, and substituting (4.23) and (4.24) in the second equation of the set (4.6), we get Krause's 'pressure' formulation

$$(C^T K_1^{-1} C + \frac{\Delta t K_2}{2}) \pi_1 = (C^T K_1^{-1} C - \frac{\Delta t}{2} K_2) \pi_o - \Delta t Q \quad (4.25)$$

The principal merit of the 'displacement' and the 'pressure' formulation was expected to be the reduction in the number of degrees of freedom. However, the matrices to be solved are no longer narrow-band as they were in the coupled formulation. This makes the 'displacement' formulation rather inefficient.

Booker (16) eliminated the displacements from equations (4.4), in terms of Laplace transforms, to get

$$(K_2 - s\, C^T K_1^{-1} C)\, \bar{\pi} = s\, (C^T K_1^{-1} \bar{P} - \bar{Q}) \quad (4.26)$$

This directly leads to the solution

$$\bar{\pi} = \sum_{n=1}^{N} \frac{s}{s - sn} \bar{a}_n \pi_n$$

where s_n are eigenvalues from (4.26), π_n the corresponding eigenvectors, and

$$\bar{a}_n = - \frac{\pi_n^T (C^T K_1^{-1} \bar{P} - \bar{Q})}{\pi_n^T C^T K_1^{-1} C \pi_n} \quad (4.27)$$

The solution for Laplace transform of the displacements is,

$$\bar{u} = \bar{u}_0 - \sum_{n=1}^{N} \frac{s}{s - s_n} \bar{a}_n K_1^{-1} C \pi_n \qquad (4.28)$$

where N is the total number of degrees of freedom in (4.4). The pressure and displacement solutions can now be generated by inverting the transforms.

Smith (8) proposed direct integration of (4.21) in the time domain using rational polynomials to approximate the exponential term. Norsett type algorithms were proposed as more efficient then the α-algorithms. Of the α- algorithms, Smith (8) found the fully implicit type (α =1) to be superior to the one with α=1/2. The distinction between the role of α in different formulations should be carefully noted. For α=1 Smith's and Sandhu's formulations coincide. However for $\alpha \neq 1$ the two algorithms are different.

For nonlinear problems, Siriwardane (21) used an algorithm similar to Sandhu's (1,2,27) time-stepping technique. Two alternative schemes, one based on initial values of K_1, K_2 being assumed to hold throughout a time step and the other based on an implicit procedure using average values for the time-step were considered. Prevost (11) used α=1 i.e. a backward difference scheme with a predictor - corrector type algorithm. It was shown, that $\alpha \geq 1/2$ implies unconditional stability of the algorithm.

Time-domain integration has been discussed in some detail by Runesson (16). Herein we shall not dwell on the details of this procedure which is essentially a generalization of Runge-Kutta methods.

4.3 Additional Comments on Finite Element Approximation.

In using the finite element method, it is useful to note that plane strain problems can be regarded as limiting cases of axisymmetric problems. Thus, a computer program such as RC63 written for analysis of consolidation of axisymmetric problems can be used for plane-strain consolidation using an artificial modification of data viz. adding a large number to all the radial coordinates.

Often, in finite element discretization, we find it convenient to draw a rectangular grid and then subdivide it into right angled triangles. Use of one set of diagonals to generate the triangles may be expected to lead to a 'bias' in the solution. This is easily taken care of (27,21,63) by using two sets of triangles using both diagonals in each rectangle. This device is inexpensive for linear consolidation. However for nonlinear consolidation, where the element properties have to be repeatedly computed this could be expensive.

For the undrained solution, the sequence of numbering of nodes can be important. This is to avoid a zero term on the diagonal of the governing equation. An alternative is to use an extremely small value of Δt instead of setting it exactly equal to zero. Another device is to use pivoting. However this would, in most cases, essentially correspond to using the 'pressure' formulation and could be quite expensive.

5. CURRENT AND FUTURE RESEARCH

Although considerable progress continuesto be made in developing finite element procedures for analysis of fluid flow through deformable media, the techniques available to date are still too expensive and not sufficiently reliable. There is apparent need for development of procedures that would avoid the local error in fluid pressures associated with application of loads discontinuous in the time domain. Even though sophisticated time-domain solution schemes have been developed and theoretically the extension of available procedure to nonlinear and three-dimensional problems is quite straightforward, the solution process needs to be made much more efficient to bring down the cost of the analysis. Role of reduced integration (64) in improving solutions to problems of elasticity has been pointed out. These ideas have been implemented (11) to some extent in analysis of soil consolidation. Further investigation is necessary. Other items needing and receiving attention of investigators are treatment of infinite domains, use of boundary element methods etc.

No matter how sophisticated a solution process, we cannot hope to get correct solutions to a physical problem if the physical processes are not correctly modeled. In the context of flow through porous media, there is still need for development of consistent theories. Also, the finite element or any other method will give solutions depending upon the data provided. This emphasizes the need for development of proper consitutive models for deformation and diffusion.

6. REFERENCES

1. Sandhu, R.S.: 1968, 'Fluid Flow in Saturated Porous Elastic Media; Ph.D. Thesis, Univ. of California at Berkeley, Berkeley, California.
2. Sandhu, R.S., and Wilson, E.L.: 1969, J.Engrg. Mech. Div., Am. Soc. Civ. Engrs., 95, pp. 641-652.
3. Sandhu, R.S.: 1972, 'Proc. Symp. Appl. Finite Element Methods in Geotech, Engrg.; Ed. C.S. Desai, U.S. Army Waterways Experiment Station, Vicksburg, Mississippi, pp. 697-738.
4. Christian, J.T.: 1977; Numerical Methods in Geotechnical Engineering, Eds. C.S. Desai and J.T. Christian, McGraw-Hill, pp. 399-426.
5. Sandhu, R.S.: 1975, 'Variational Principles for Soil Consol-

idation; Report OSURF-3570-75-2 to National Science Foundation, Dept. of Civil Engrg., The Ohio State Univ. Columbus, Ohio.
6. Sandhu, R.S.; 1976, 'Numerical Methods in Geomechanics; Proc., Second Intl. Conf. Numer. Methods in Geomech., Ed. C.S. Desai, Am. Soc. Civ. Engrs., 1, pp. 20-40.
7. Ghaboussi, J., and Wilson, E.L.: 1973, Int. J. Num. Methods in Engrg., 5, pp. 419-442.
8. Smith, I.M.: 1978, 'Numerical Methods in Offshore Engineering; Eds. O.C. Zienkiewicz, R.W. Lewis, and K.G. Stagg, J. Wiley, pp. 483-513.
9. Sandhu, R.S., Liu, H., and Singh, K.J.: 1977, Int. J. Num. Anal. Methods in Geomech., 1, pp. 177-194.
10. Sandhu, R.S. and Liu, H.: 1979, 'Numerical Methods in Geomechanics', Proc. Third Intl. Conf. Numer. Methods in Geomech., Ed. W. Wittke, A.A. Balkema, pp. 1255-1263.
11. Prevost, J.H.: 1981, J. Engrg. Mech. Div., Am. Soc. Civ. Engrs., 107, pp. 169-186.
12. Booker, J.R.: 1973, Quar. J. Mech. Appl. Math., XXVI, 4, pp. 445-470.
13. Booker, J.R., and Small, J.C.: 1977, Int. J. Solids Struct., 13, pp. 137-149.
14. Smith, I.M.: Siemieniuch, J.L., and Gladwell, I., 1977, Int. J. Num. Anal. Methods Geomech, 1, pp. 57-74.
15. Runesson, K.: 1978, 'On Nonlinear Consolidation of Soft Clay: Publ. 78:1, Dept. Struct. Mech., Chalmers Institute of Technology, Goteborg, Sweden.
16. Booker, J.R., and Small, J.C.: 1975, Int. J. Solids Struct., 11, pp. 907-917.
17. Vermeer, P.A., and Verruijt, A.: 1981, Int. J. Num. Anal. Methods Geomech., 5, pp. 1-14.
18. Carter, J.P., Booker, J.R., and Small, J.C.: 1979, Int. J. Num. Anal. Methods Geomech., 3, pp. 107-129.
19. Suklje, L.: 1978, Int. J. Num. Anal. Methods Geomech., 2, pp. 129-158.
20. Carter, J.P., Small, J.C., and Booker, J.R.: 1977, Int. J. Solids Struct., 13, pp. 467-478.
21. Siriwardane, J.H., and Desai, C.S.: 1981, Int. J. Num. Methods in Engrg., 17, pp. 405-426.
22. Moussa, A.B., 1980: 'A Coupled Problem of Finite Deformation and Flow in Porous Media', Ph.D. Thesis, The Ohio State Univ., Columbus, Ohio.
23. Krause, G.: 1975, 'Untersuchungen uber numerische Verfahren fur elastisch, porose Medien', Dr. Ing. Thesis, Technische Universitat, Berlin, West Germany.
24. Krause, G.: 1978, J. Engrg. Mech. Div., Am. Soc. Civ. Engrs., 104, pp. 605-620.
25. Morland, L.W.: 1978, Geophys. J.R. Astro. Soc., 55, pp. 393-410.
26. Prevost, J.H.: 1978, Int. J. Engrg. Sci., 18, pp. 787-800.
27. Sandhu, R.S.: 1976, 'Finite Element Analysis of Soil Consol-

idation', Report OSURF-3570-76-3 to National Science Foundation, Dept. of Civ. Engrg., The Ohio State Univ., Columbus, Ohio.
28. Sandhu, R.S.: 1979, J.Engrg. Mech. Div., Am.Soc. Civ. Engrs., 105, pp. 355-356.
29. Sandhu, R.S., and Salaam, U.: 1975, Computer Methods in Appl. Mech. and Engrg., 7, pp. 75-91.
30. Bufler, H.: 1979, Computer Methods in Appl. Mech. and Engrg., 19, pp. 235-255.
31. Tabaddor, F.: 1968, 'Constitutive Equations and the Solution of Some Problems of Interacting Continua', Ph.D. Thesis, Michigan State Univ., East Lansing, Michigan.
32. Truesdell, D., and Toupin, R.A.: 1960, 'The Classical Field Theories', Handbuch der Physik, III/1, Ed. S. Flugge, Springer-Verlag.
33. Truesdell, C.: 1962, J. Chem. Phys., 37, pp. 2336-2344.
34. Green, A.E., and Naghdi, P.M.: 1965, Int. J. Engrg. Sci., 3, pp. 231-240.
35. Green, A.E., and Naghdi, P.M.: 1967, Arch. Rat. Mech. Anal., 24, pp. 243-263.
36. Green, A.E., and Naghdi, P.M.: 1968, Int. J. Engrg. Sci., 6, pp. 631-635.
37. Green, A.E., and Naghdi, P.M.: 1969, Quar. J. Mech. Appl. Math., XXII, pp. 427-438.
38. Crochet, M.J., and Naghdi, P.M.: 1966, Int. J. Engrg. Sci., 4, pp. 383-401.
39. Atkin, R. J.: 1967, Z.A.M.P. 18, pp. 803-825.
40. Yokoo, Y., Yamagata, K., and Nagaonka, H.: 1971, Soils and Foundations, 11, pp. 25-35.
41. Ghaboussi, J. and Wilson, E.L.: 1972, J. Engrg. Mech. Div., Am. Soc. Civ. Engrs., 98, pp. 947-963.
42. Biot, M.A.: 1972, Indiana University Mathematics Journal, 21, pp. 597-620.
43. Moussa, A.B.: 1981, 'Nonlinear Formulation for the Coupled Problem of Flow and Large Deformation in Unsaturated Porous Media, Int. Conf. on Numerical Methods for Coupled Problems, University College, Swansea, Wales.
44. Sandhu, R.S., and Pister, K.S.: 1971, Int. J. Solids Struct., 7, pp. 639-654.
45. Gurtin, M.E.: 1963, Arch. Rat. Mech. Anal., 13, pp. 179-191.
46. Gurtin, M.E.: 1964, Arch. Rat. Mech. Anal., 16, pp. 234-254.
47. Gurtin, M.E.: 1964, Quar. Appl. Math., 12, pp. 252-256.
48. Mikhlin, S.G.: 1965, 'The Problem of the Minimum of a Quadratic Functional', Holden-Day, San Francisco, California.
49. Herrera, I., and Bielak, J.: 1973, J. Engrg. Mech. Div., Am. Soc. Civ. Engrs., 99, pp. 1097-1098.
50. Herrera, I. and Bielak, J.: 1977, Int. J. Solids Struct., 13, pp. 377-378.
51. Tonti, E.: 1972, 'On the Variational Formulation for Linear Initial Value Problems', Report, Instituto di Matematica del Politechnico di Milano, Milan, Italy.

52. Tonti, E.: 1972, 'Variational Methods in Engineering', Eds. C.A. Brebbia and H. Tottenham, Southampton Univ. Press, pp. 1/1 - 1/12.
53. Magri, F.: 1974, Int. J. Engrg. Sci., 12, pp. 537-549.
54. Zienkiewicz, O.C.: 1977, 'The Finite Element Method; McGraw-Hill, New York.
55. Hwang, C.T., Morgenstern, N.R., and Murray, D.W.; 1971, Canadian Geotech. J., 8, pp. 109-118.
56. Yokoo, Y., Yamagata, K., and Nagaonka, H.: 1971, Soils and Foundations, 11, pp. 29-46.
57. Smith, I. M., and Hobbs, R.: 1976, Geotechnique, 26, pp. 149-171.
58. Valliappan, S., Lee, I.K., and Boonlaulohr, P: 1974, 'Finite Element Methods in Flow Problems, Ed. J.T. Oden, O.C. Zienkiewicz, R.H. Gallagher, and C. Taylor, Univer. of Alabama at Huntsville Press. pp. 741-755.
59. Christian, J.T., and Boehmer, J.W.: 1970, J.Soil Mech. Found. Div., Am. Soc. Civ. Engrs., 96, pp. 1435-1457.
60. Buchmaier, R., 1980, Personal communication.
61. Vermeer, P.A., 1980, Personal communication.
62. Sandhu, R.S., Rai, I.S., and Desai, C.S.: 1974, 'Finite Element Methods in Flow Problems', Ed. J.T. Oden, O.C. Zienkiewicz, R.H. Gallagher, and C. Taylor, Univ. of Alabama at Huntsville Press, pp. 573-579.
63. Sandhu, R.S., and Schiffman, R.L.: 1977, Implementation of Computer Programs RC 63 and FECONB for Plane Strain and Axisymmetric Consolidation of Elastic Soils, Report 51405-10, Norwegian Geotechnical Institute, Oslo, Norway.
64. Sandhu, R.S., and Singh, K.J.: 1978, Computer Methods in Appl. Mech. Engrg., 14, pp. 23-37.

FINITE ELEMENTS AND SLOPE STABILITY

NAYLOR, D.J.

Department of Civil Engineering, Univ. Col. of SWANSEA

ABSTRACT

Limit equilibrium methods can be misleading when used to compute factors of safety. Also they do not provide information about the development of failure. The finite element method is shown to be particularly useful for showing how failure initiates, and the likely collapse mechanism. It can also be used to determine the safety factor. There are two methods - the "direct" and the "enhanced limit" method. These have a research role for checking safety factors computed using conventional limit analyses. The direct method appears to have the most potential.

INTRODUCTION

As the consequences of a collapse are usually more serious than with other forms of malfunction greater care is justified to ensure stability. Failures are relatively common in embankments and excavated slopes. Safety may be assured by attending to qualitative aspects of the design (choice of materials, drainage, geometry, timing of construction), and by the use of appropriate methods of analysis. Limit methods are well established for the latter, but can in certain situations be misleading. Also, they may not provide sufficient information.

In the context of slopes and embankments the "safety factor" is invariably the factor by which the *strength* must be reduced to cause collapse. When the material has both cohesion (c) and friction (ϕ) it is usual to apply the same factor to c and to Tanϕ. Denoting the reduced parameters by an asterisk, and

J. B. Martins (ed.), Numerical Methods in Geomechanics, 229–244.

the factor by λ_s:

$$\left.\begin{aligned} c^* &= c/\lambda_s \\ \mathrm{Tan}\,\phi^* &= \mathrm{Tan}\,\phi/\lambda_s \end{aligned}\right\} \qquad (1)$$

$\lambda_s = F_s$ when c^* and ϕ^* are associated with incipient failure.

This definition of safety factor is quite arbitrary. There are a variety of different (and equally arbitrary) ways by which it could be defined. Thus for footings and piles it is the factor by which the *load* must be increased to cause collapse. Alternatively the factor may be defined in terms of the *geometry*: eg the ratio of the collapse height to actual height of an embankment or slope. In general quite different factors will apply to a particular situation according to the definition. The differences may be marked with frictional materials, but the safety factor will usually be the same irrespective of definition for purely cohesive materials (8). This has bearing when the f.e.m. is used to compute safety factors as will be seen below.

Apart from a safety factor it will often be desirable to have some information about the development of failure (should it occur), ie about the *mechanism*. Instrumentation (slope indicators, surface measuring points) can then be suitably placed to give prior warning of the onset of a dangerous situation. Although common sense and experience will often suffice here, some situations spring surprises. Burland (3) describes an excavation in stiff clay where unexpectedly deep seated movement occurred. This could not have been predicted using limit techniques but could by means of analysis which revealed the development of failure. It will be argued that the f.e.m. has a role in situations such as this.

We first examine briefly the validity of limit methods for assessing collapse and consider what the f.e.m. has to offer here. The use of the f.e.m. to predict mechanisms is next considered, followed by a description of techniques for using the method to compute factors of safety.

LIMITATIONS OF LIMIT METHODS

Let it first be made clear that limit equilibrium methods (ie the rigorous or approximate upper bound solutions obtained from the conventional sliding wedge or curved slip surface methods of analysis) are very powerful techniques which will not be superseded by the more sophisticated methods referred to here. Their strength is their simplicity. They are so widely used, however, that there is a danger that the qualifications associated with them when first proposed will be forgotten.

First, since the limit analyses are usually "upper bounds", the analytically exact safety factor will be less than that computed. The analyses are unconservative. It seems that as knowledge increases so also does awareness of the possible extent of this source of error. Thus Cooling and Golder (4) thought that the Swedish slip circle method was associated with an error of the order of 4% when applied to earth dams. Later, Bishop (1) demonstrated that the Swedish method could significantly underestimate the safety factor compared with his simplified method which was purportedly in close agreement with the rigorous solution. More recently Davis and Booker (5) have shown that the slip circle method when applied to a footing on level ground with cohesion that increases with depth is in error by a factor of up to four times, ie 400%. The slip circle analysis here is an upper bound and for soils with cohesion which increases significantly with depth, a rather poor one. Figure 1 taken from their paper, illustrates this. Safety factors computed using conventional limit methods must be treated with caution.

Another and more obvious limitation of limit methods is that they provide no information about the development of failure. They do not indicate where yield initiates. Such information can be valuable as has been indicated.

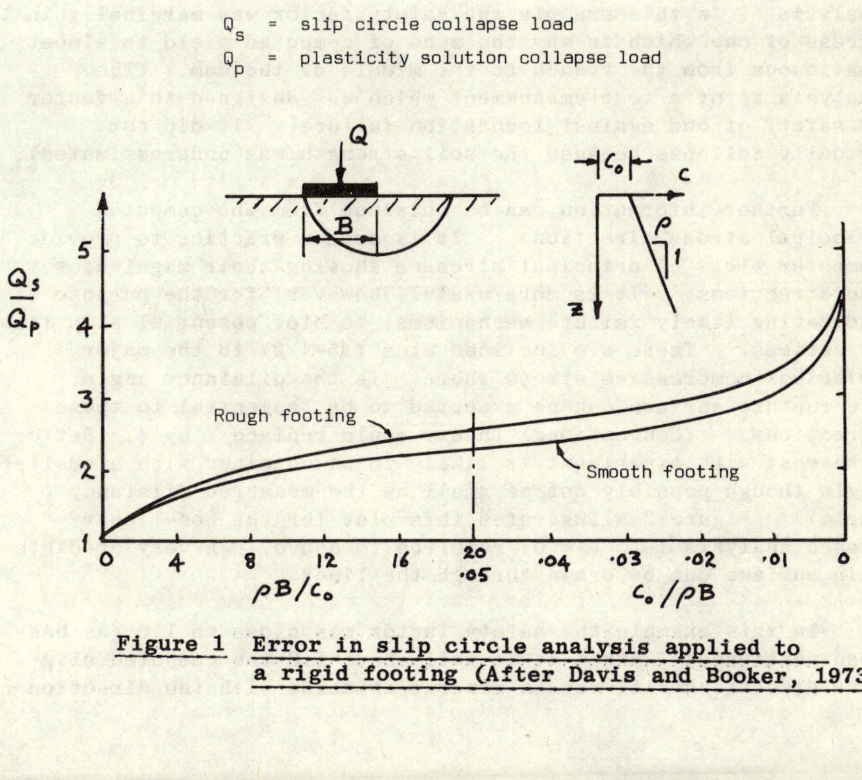

Figure 1 Error in slip circle analysis applied to a rigid footing (After Davis and Booker, 1973)

DEVELOPMENT OF FAILURE

Important information about the development of failure can be obtained from a linear elastic finite element analysis. This may seem surprising since such an analysis cannot model yield. It can, however, indicate *potential* yield. This can conveniently be done by defining an *over-stress ratio* (O.S.R.) as:

$$\text{O.S.R.} = \frac{\sigma_1 - \sigma_3}{2c\text{Cos}\phi + (\sigma_1 + \sigma_3)\ \text{Sin}\phi} \qquad (2)$$

in which σ_1 and σ_3 are the major and minor principal stresses. (Compression positive, $\sigma_1 > \sigma_3$).

An O.S.R. value of less than one indicates that yielding will not occur at the point to which it applies. Regions of potential yield in an elastic analysis are identified by O.S.R. values greater than one. A properly constituted non-linear program will indicate a yielded region as a "plateau" with the O.S.R. limited to one. Figure 2 illustrates this for an embankment on a stiff clay. Comparison of the upper and lower diagrams shows how the potential yielded zones of the elastic analysis become modified and extended in the non-linear (Tresca) analysis. In this example the safety factor was marginally in excess of one which is why the zone of computed yield is almost continuous from the trench to the middle of the dam. (The analysis is of a test embankment which was designed to a factor of safety of one against foundation failure. It did not actually collapse because the soil strength was underestimated).

Further information can be obtained from the computed principal stress directions. It is common practice to provide computer plots of principal stresses showing their magnitudes and directions. It is more useful, however, for the purpose of indicating likely failure mechanisms, to plot potential slip line directions. These are inclined at $\pm$ $(45-\psi/2)$ to the major principal compressive stress where ψ is the dilatancy angle. The rupture surface can be expected to be tangential to these directions. (Conventional theory would replace ψ by ϕ. Better agreement with experiment is likely to be obtained with a smaller angle though possibly not as small as the measured dilatancy angle). Figure 3 illustrates this plot for the non-linear Tresca analysis ($\psi = \phi = 0$) referred to above. A very credible slip surface can be drawn through the lines.

In this example the safety factor was close to 1.0 (as has been mentioned). When it is well above 1.0 the computed slip line directions will not in general coincide with the direction

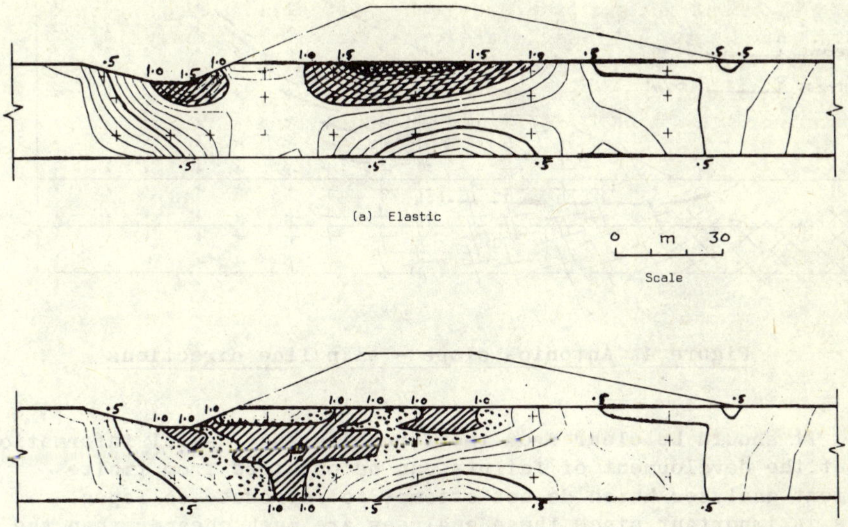

Figure 2 Embankment on stiff clay - overstress ratios

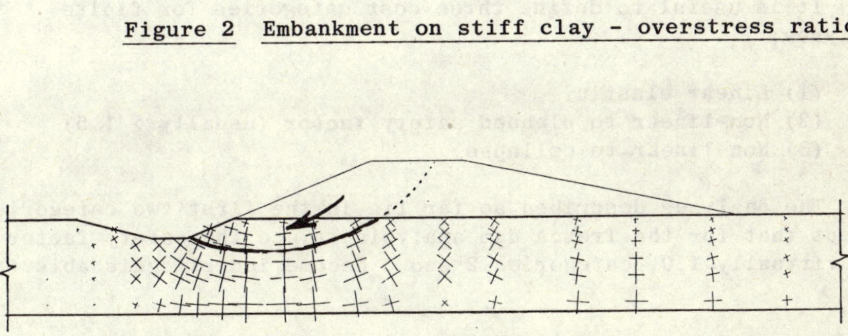

Figure 3 Embankment on stiff clay - slip line directions

of the eventual slip surface since principal stress rotation would take place if the soil strength was reduced to cause failure. This is illustrated in figure 4. In this example yield is indicated in the relatively weak clay layer but has not spread through the overlying material. It is clear that an anti-clockwise rotation of the principal stress directions would need to occur in the overlying material before a continuous slip surface involved the complete slope. Care must be taken, therefore, in deriving failure surfaces from the slip line directions when the safety factor is large, or when the analysis is linear elastic.

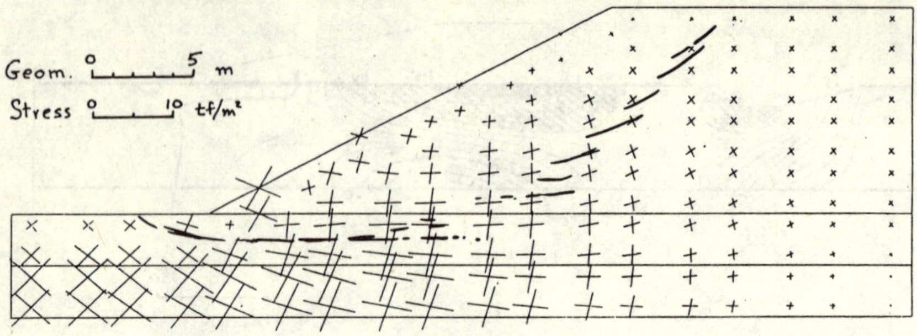

Figure 4 Antonios slope - slip line directions

It should be clear from the foregoing that useful information about the development of failure can be obtained from finite element analyses which do not attempt to reproduce collapse. This is important since these analyses are much cheaper than the non-linear analyses described below which model collapse.

It is useful to define three cost categories for finite analyses:

(1) Linear elastic.
(2) Non-linear to planned safety factor (usually > 1.5)
(3) Non-linear to collapse.

The analyses described so far lie in the first two categories except that for the Tresca dam analysis, since the safety factor was virtually 1.0, categories 2 and 3 become indistinguishable.

COLLAPSE SIMULATION

The f.e.m. can be used to obtain a direct measure of the safety factor by either progressively reducing the strength or progressively increasing the load until failure is indicated. This involves analysis in cost category 3. Failure is indicated by the development of very large displacements of salient points on the body. This may also be associated with non-convergence of iterative processes, depending on the formulation. Not every non-linear formulation can achieve this. The writer has found that "initial stress" based methods (10) applied to frictional materials often require a very large number of iterations as failure is approached, so that it becomes impractical to determine collapse. His colleagues have had better success using the visco-plastic method, albeit at high computing cost (11).

Using conventional formulations it will generally be necessary to carry out a number of separate analyses each with the strength reduced by a different value of the factor λ_s of equation (1). A graph such as that shown in figure 5(a) may then be used to estimate the safety factor. It is less trouble to increment loads (by a factor λ_Q) than reduce the strengths since this can be done in a single analysis. For purely cohesive soils the safety factor can be obtained by this means since in this case the load and safety factors are the same. At failure, then, $\lambda_Q = \lambda_s = F_s$ (= F_Q). Figure 5(b) illustrates the procedure.

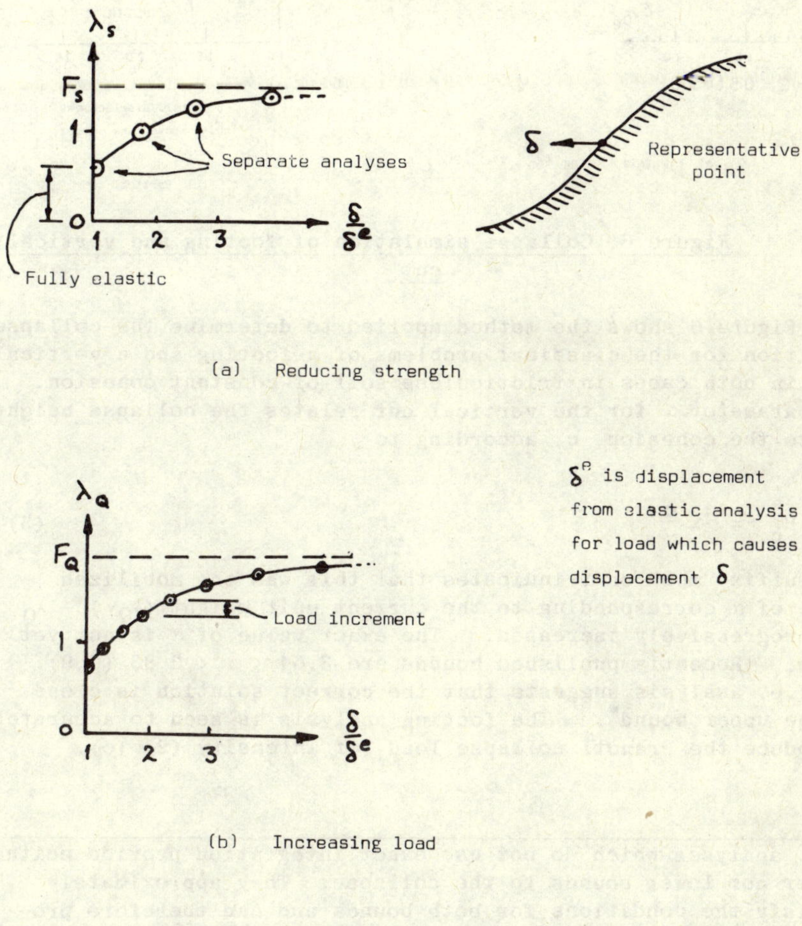

Figure 5 Collapse by the direct method

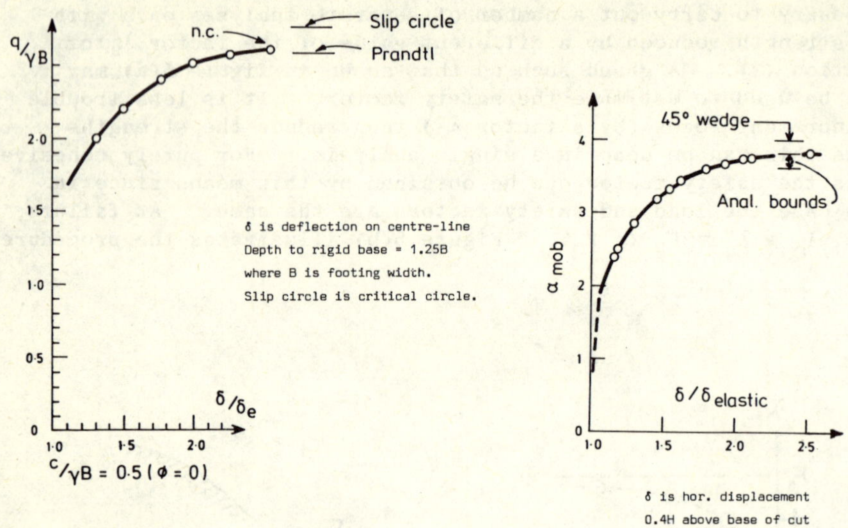

(a) Flexible footing

(b) Vertical cut

Figure 6 Collapse simulation of footing and vertical cut

Figure 6 shows the method applied to determine the collapse condition for the classical problems of a footing and a vertical cut, in both cases in frictionless soil of constant cohesion. The parameter α for the vertical cut relates the collapse height, H_c, to the cohesion, c, according to

$$H_c = \alpha \frac{c}{\gamma} \qquad (3)$$

The suffix "mob" on α indicates that this was the mobilized value of α corresponding to the current unit weight ($\lambda_Q \gamma$). λ_Q was progressively increased. The exact value of α is not yet known. Recently published bounds are $3.64 < \alpha < 3.83$ (6,9). The f.e. analysis suggests that the correct solution is close to the upper bound*. The footing analysis is seen to accurately reproduce the Prandtl collapse load (of intensity $(2+\pi)c$).

*F.e. analyses which do not use exact integration provide neither upper nor lower bounds to the collapse. They approximately satisfy the conditions for both bounds and can therefore produce an approximation of the theoretical collapse condition.

Figure 7 shows the method applied to the embankment on clay results for which have already been presented in figures 2 and 3. Here again advantage could be taken of the equality of load and safety factors for cohesive materials. The indicated safety factor of about 1.0 was in good agreement with conventional analysis.

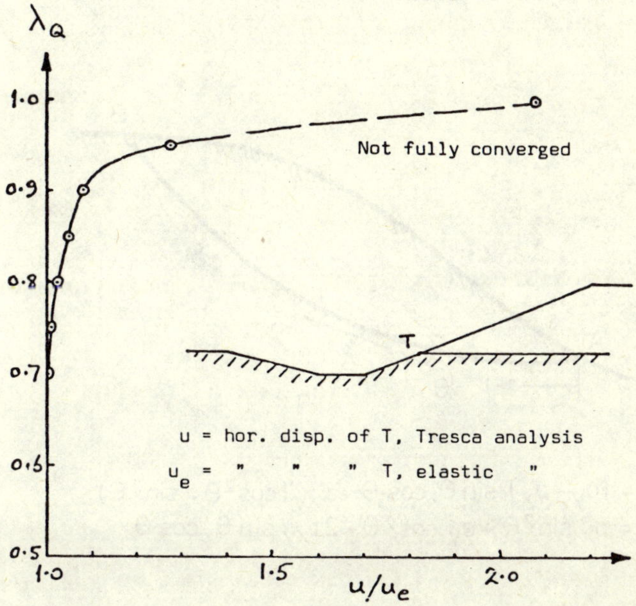

Figure 7 Load-displacement curve for embankment on stiff clay

ENHANCED LIMIT METHODS

A major shortcoming of the conventional limit equilibrium methods is that for frictional materials, in which the strength depends on the stress level, some assumption must be made to determine the stresses. The finite element method has the reputation of producing relatively reliable stresses*, so it

*This may be illusory. Comparisons have tended to be made on vertical stresses which in the present context are strongly governed by the weight of overburden and are insensitive to the method of calculation. This is not the case with horizontal stresses which are hard to predict whatever the method.

appears an attractive proposition to use these in conjunction with a conventional limit method such as Bishop's (simplified) method of slices. This is the enhanced method. It appears to have first been used by Brown and King (2). It involves f.e. analyses in cost categories 1 or 2 so is not unduly expensive. Its advantages are offset by the additional work involved in determining the computed normal and shear stresses on a potential slip surface, as will be seen.

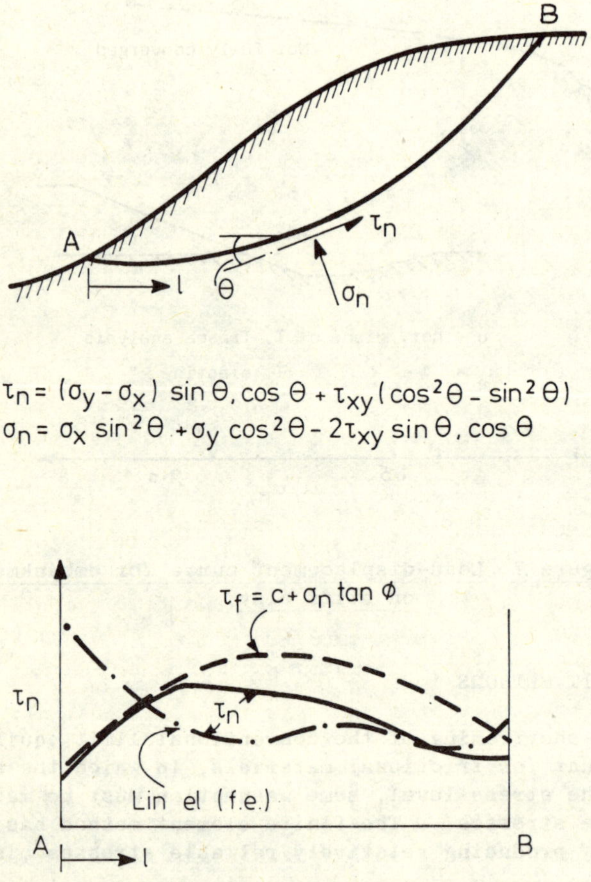

$$\tau_n = (\sigma_y - \sigma_x)\sin\theta.\cos\theta + \tau_{xy}(\cos^2\theta - \sin^2\theta)$$
$$\sigma_n = \sigma_x \sin^2\theta + \sigma_y \cos^2\theta - 2\tau_{xy}\sin\theta.\cos\theta$$

Figure 8 The enhanced limit method

Figure 8 illustrates the method. AB in the upper diagram is a trial slip surface. The dashed line in the lower diagram shows the variation in shear strength, τ_f, along the surface AB calculated as $c + \sigma_n \tan\phi$ where σ_n is the computed normal stress.

The actual shear stress distribution is indicated by a full line which must lie on or under the strength line. Where they coincide local yielding occurs. A properly constituted non-linear analysis will produce such a line. The chain dotted line typifies the shear stress distribution obtained from a linear elastic f.e. analysis. It lies above the strength line in regions of potential yield. For both linear and non-linear analyses the safety factor is determined as the ratio of the area under the strength curve to the area under the computed shear stress curve, ie:

$$F_s = \frac{\int \tau_f \, d\ell}{\int \tau_n \, d\ell} \tag{4}$$

Clearly a more accurate estimate of the safety factor can be expected from a non-linear analysis (in cost category 2) than from a linear. When the yielded zone is not extensive (implying a fairly high safety factor) the total area under the τ_n curve will not differ much between linear and non-linear analyses, and both will produce much the same value of F_s.

Figure 9 illustrates the method applied to the vertical cut in clay to which we had previously applied the direct method. The final surface taken was a 45° wedge. (This provides an upper bound solution with α = 4 compared with the more accurate value of 3.83 referred to earlier.) $c/\gamma H$ was chosen as 0.4, so that expressing the safety factor as

$$F_s = \frac{H_c}{H} = \frac{\alpha c}{\gamma H} \tag{5}$$

gives $F_s = 1.6$ using $\alpha = 4$. This is the value which should result from equation (4) using the enhanced method. The actual results obtained were 1.65 and 1.64 from a linear elastic and Tresca non-linear analysis respectively. The shear stress distributions are shown by the two full lines in the lower figure.

The dotted lines in the figure 9 graphs illustrate a pitfall which can lead to poor results with this method. They were obtained using the mesh with node at A fixed. There was therefore a boundary reaction at A which affected the stresses in the triangular region above AB. This source of error was eliminated by freeing node A as shown in the inset diagram.

This example (which was used to test the method) overcame the difficulty of determining the stresses on the trial surface by arranging for this to pass through the middle of elements. In more general situations it is necessary to obtain contour plots of the three stress components and add the trial slip

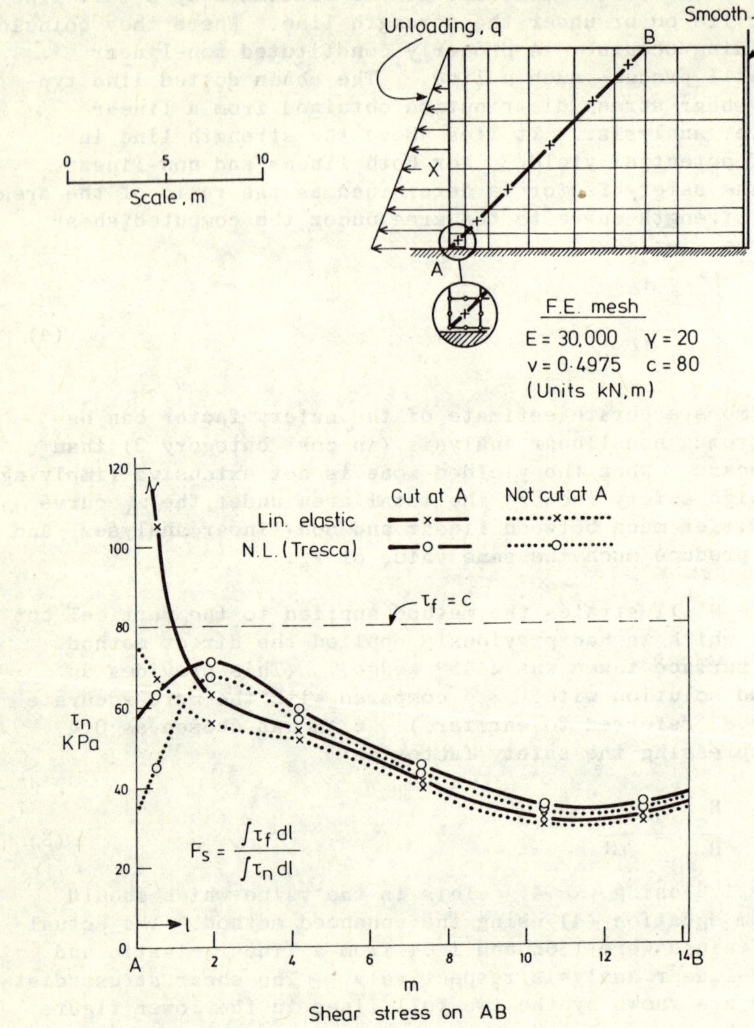

Figure 9 Enhanced limit method applied to vertical cut

surface onto these. The triplet ($\sigma_x, \sigma_y, \tau_{xy}$) must then be obtained by interpolation from the contours at a number of points along the trial surface. These must be transformed to give σ_n, τ_n and then integrated to provide the numerator and denominator of equation (4). The procedure is illustrated in figure 10. Here the method is applied to "Antonios slope"* for which a slip line direction has already been presented as figure 4.

* So named since the problem was obtained from Antonio Compos e Matos of the University of O'Porto.

Figure 11 shows the τ_f and τ_n stress distributions obtained. The resulting factor of safety is 2.24 which may be compared with 1.92 obtained independently using a method of slices based on linear programming (obtained by B. Reis of University of O'Porto (7)) and a direct method analysis which, due to limitations on computing time, did not reach collapse but indicated that the factor was in the region of two.

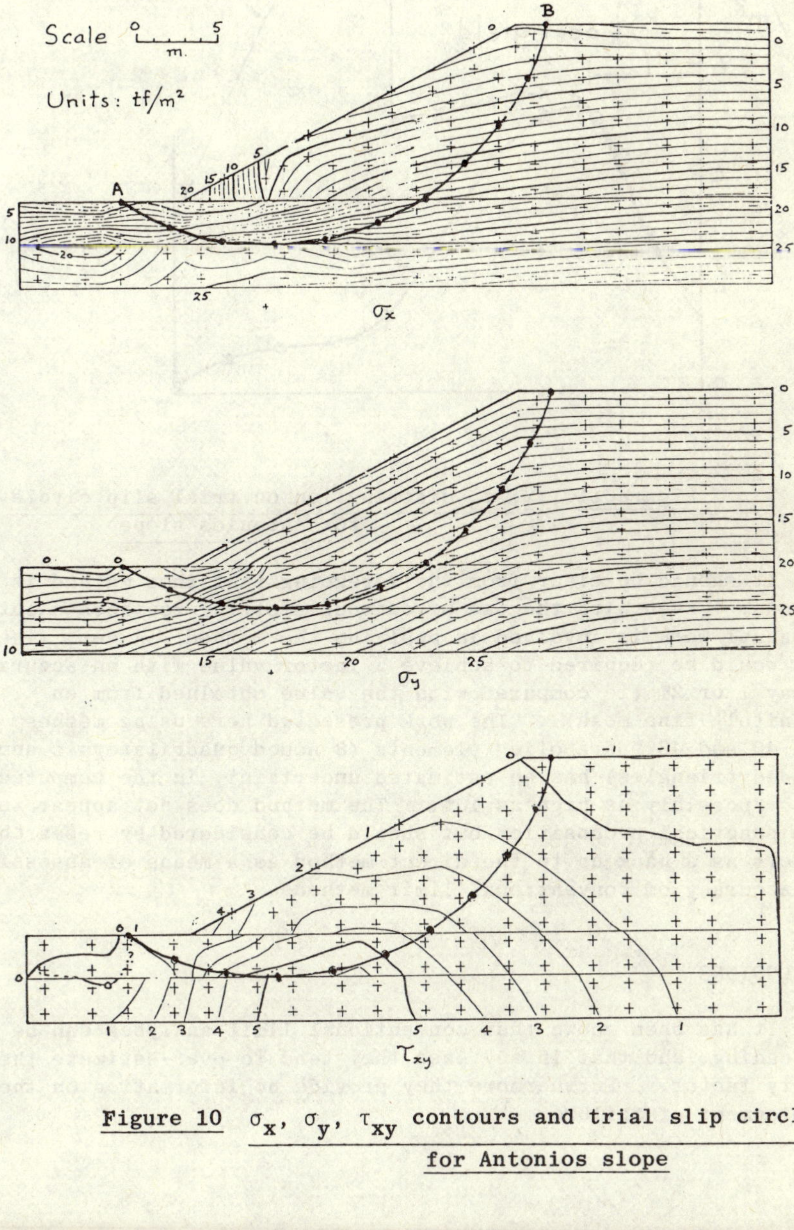

Figure 10 σ_x, σ_y, τ_{xy} **contours and trial slip circle for Antonios slope**

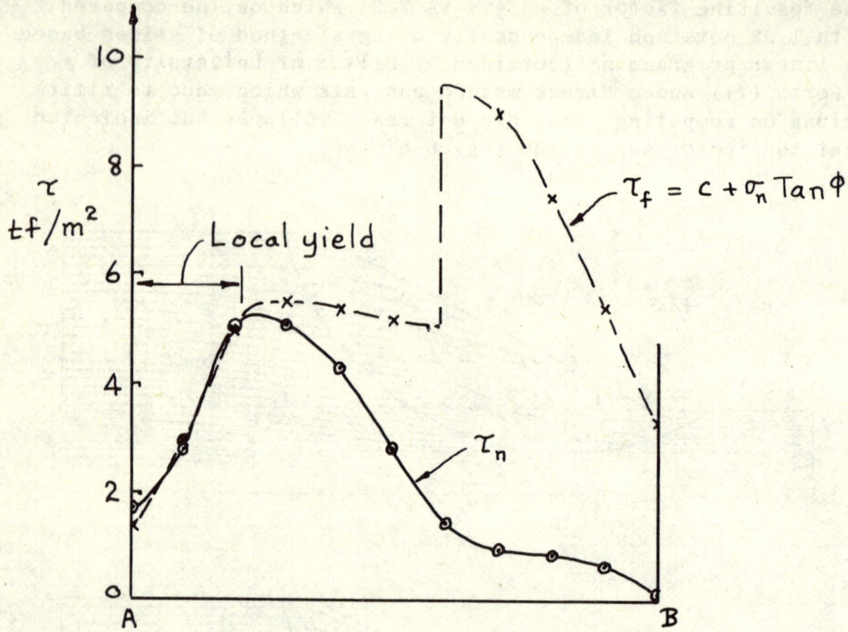

Figure 11 τ_f, τ_n distribution on trial slip circle for Antonios slope

It should be clear from the foregoing that this method is tedious. Even with the use of contour plotting routines a lot of manual work is involved in applying the method. A very fine mesh would be required to achieve a factor value with an accuracy of say 1 or 2% (ie compared with the value obtained from an infinitely fine mesh). The work presented here using meshes with 40 and 48 "parabolic" elements (8 noded quadrilaterals and 6 noded triangles) has an estimated uncertainty in the computed factor possibly as high as 10%. The method does not appear to be a practical proposition but should be considered by research workers as a back up to the direct method as a means of assessing the accuracy of conventional limit methods.

CONCLUSIONS

It has been shown that conventional limit analyses can be misleading, and that in any case they tend to over-estimate the safety factor. Furthermore they provide no information on the development of failure.

The finite element method is particularly useful for assessing the development of failure. Graphical output in the form of O.S.R. and slip line direction plots should be used. Much information can be obtained from linear analyses, and this can be enhanced by relatively cheap (cost category 2) non-linear analyses.

The f.e.m. can also be used to compute safety factors. The "direct method" is a powerful research tool, given good computing facilities and a properly formulated non-linear program, which can be used to check conventional limit equilibrium methods. The alternative "enhanced limit" method is less attractive but has a limited research role.

REFERENCES

1. Bishop,A.W. (1977): *The use of the slip circle in the stability analysis of slopes*, Geotechnique, 5, No.1, pp.1-17.

2. Brown,C.B. and King,I.P. (1966): *Automatic embankment analysis: Equilibrium and instability conditions*, Geotechnique 16, No.3, pp. 209-219.

3. Burland,J.B., Longworth,T.I. and Moore,J.F.A. (1977): *A study of ground movement and progressive failure caused by a deep excavation in Oxford clay*, Geotechnique, 27, No.4, pp. 557-591.

4. Cooling,L.F. and Golder,H.Q. (1942): *The analysis of the failure of an earth dam during construction*, J. Inst. Civ. Engrs. (London), 19, pp. 38-55.

5. Davis,E.H. and Booker,J.R. (1973): *The effect of increasing strength with depth on the bearing capacity of clay*, Geotechnique 23, No.4, pp. 551-563.

6. Josselin de Jong,G. (1980): *Application of the calculus of variations to the vertical cut off in cohesive frictionless soil*, Geotechnique, 30, No.1, pp. 1-16.

7. Martins,J.B.: Private communication, March 1981.

8. Naylor,D.J. (1981): *Quantify safety*, article to be published in "Talking Point" in Ground Engineering in late 1981.

9. Pastor,J. (1978): *Analyse limite: Determination de solutions statiques completes. Application au talus vertical*, J. de Mechanique Appliquee 2, No.2, pp. 167-197.

10. Zienkiewicz,O.C. (1977): *The Finite Element Method*, McGraw Hill (3rd edition)

11. Zienkiewicz,O.C., Humpheson,C. and Lewis,R.W. (1975): *Associated and non-associated visco-plasticity and plasticity in soil mechanics*, Geotechnique, 25, No.4, pp. 671-689.

CHAPTER III

MATHEMATICAL PROGRAMMING METHODS

PLASTIC ANALYSIS IN GEOMECHANICS BY MATHEMATICAL PROGRAMMING

John Munro

Imperial College of Science and Technology
University of London, England

The theory of plasticity is being applied increasingly to geomechanical problems. The fundamental relations of statics, kinematics and plastic properties may be organised into a set of Karush-Kuhn-Tucker conditions and a pair of primal-dual mathematical programs may be deduced when certain convexity conditions are satisfied. Thus the theory of mathematical programming may be used to generate consistently the relevant extremal principles for the considered class of geomechanical problems. The formalism is illustrated with respect to a problem of statistical mechanism of soils and the limit analysis of slope stability.

1. INTRODUCTION

Since the early 1950s, Mathematical Programming (MP) |1-4| has been applied successfully to a widening range of problems in the analysis and design of structural systems |5-7|. The theory of plasticity is being applied increasingly to geomechanics |8-10| and it seems likely that MP could also be of value in this field but, so far, relatively few applications |11-15| have been reported.

One particularly promising approach to structural problems has exploited the Karush-Kuhn-Tucker (KKT) theory of MP |16|. The fundamental structural relations (statics, kinematics and constitutive relations) for any problem class have been organised in the form of KKT conditions and (where the appropriate convexity restrictions apply) the corresponding primal-dual MPs have been derived directly from the KKT theory. It has been

J. B. Martins (ed.), Numerical Methods in Geomechanics, 247–272.

found that the purely static and kinematic statements must be stated in a dual way and attention must also be devoted to the dual aspects of the constituent relations if a consistent theory is to be developed. Thus, MP has become a formalism for the generation of consistent primal-dual extremal principles as well as a vehicle for numerical computation.

The object of the present discussion is to give an introduction to the KKT theory, to make some comments on common computational devices and to present some brief illustrative examples with relevance to geomechanics.

Consequently the second section contains a brief summary of some fundamental aspects of KKT theory and this is used to demonstrate the concept of duality with particular reference to linear programs (LPs) - the simplest form of MP. The third section contains a discussion of some elementary concepts of computation, again with particular emphasis on LPs. The fourth section contains the derivation of a classical optimisation problem which is fundamental to the statistical mechanics of soils. The fifth section shows how the KKT theory can be applied to the limit analysis of a rather idealised problem in slope stability. It is hoped that this discussion will be suggestive of a wider range of potential MP applications to geomechanics.

2. KARUSH-KUHN-TUCKER THEORY AND DUALITY

For the present purposes the problems of engineering optimisation will be discussed under three headings. First the unconstrained optimisation (minimisation) of a function will be considered

$$\text{Min } z(\underset{\sim}{x}) \tag{2.1}$$

The necessary condition for minimality is

$$\nabla_{\underset{\sim}{x}}(z) = 0 \tag{2.2}$$

However, this condition is satisfied by all local minima and maxima. It becomes a sufficient condition for global minimality when z is a convex function and it becomes sufficient for unique global minimality when z is strictly convex.
The second class of problem is the optimisation (minimisation) of a function when subjected to equality constraints

$$\begin{aligned} &\text{Min } z(\underset{\sim}{x}) \\ &\underset{\sim}{g}(\underset{\sim}{x}) = \underset{\sim}{0} \end{aligned} \tag{2.3}$$

and where the variables $(\underset{\sim}{x})$ are unrestricted in sense.

This problem is frequently referred to as classical (or lagrangean) optimisation. A new set of variables $(\underset{\sim}{\lambda})$ is introduced and there is one such variable (termed a lagrange multiplier) for each constraint. A function $L(\underset{\sim}{x},\underset{\sim}{\lambda})$ is then introduced such that

$$L(\underset{\sim}{x},\underset{\sim}{\lambda}) \equiv z + \underset{\sim}{\lambda}^T\underset{\sim}{g} \tag{2.4}$$

and L is termed the lagrangean.

It can be shown that the stationarity conditions

$$\begin{bmatrix} \nabla_{\underset{\sim}{x}} \\ \hline \nabla_{\underset{\sim}{\lambda}} \end{bmatrix} (L) = \underset{\sim}{0} \tag{2.5}$$

are the necessary conditions for this problem of equality-constrained minimality. These conditions become sufficient for global minimality when the objective function (z) and the constraint functions are convex and a unique minimal solution is obtained when z is strictly convex.

The first of (2.5) becomes

$$\nabla_{\underset{\sim}{x}}(z) + \nabla_{\underset{\sim}{x}}(\underset{\sim}{g}^T)\underset{\sim}{\lambda} = \underset{\sim}{0} \tag{2.6}$$

whilst the second of (2.5) recovers the equality constraints of (2.3).

It will be seen that this form of constrained optimisation becomes equivalent to a problem of unconstrained optimisation after introducing the lagrange multipliers.

The third class of problem is the optimisation (minimisation) of a function subjected to inequality constraints, for example,

$$\begin{aligned} &\text{Min } z(\underset{\sim}{x}) \\ &\underset{\sim}{g}(\underset{\sim}{x}) \leqslant \underset{\sim}{0} \\ &\underset{\sim}{x} \geqslant \underset{\sim}{0} \end{aligned} \tag{2.7}$$

Such a problem is called a "mathematical program" (MP). The function z is termed the "objective function" and the functions g are the constraint functions. The necessary conditions for a solution of this MP are now sought. First the inequality constraints are converted to equations by introducing slack variables $(\underset{\sim}{s})$

$$\underset{\sim}{g} + \underset{\sim}{s} = \underset{\sim}{0} \qquad \underset{\sim}{s} \geqslant \underset{\sim}{0} \tag{2.8}$$

The langrangean (L) is defined in terms of the generalised lagrange multipliers $(\underset{\sim}{\lambda})$ as

$$L \equiv z + \underset{\sim}{g}^T \underset{\sim}{\lambda} \tag{2.9}$$

Since this is not a classical optimisation problem, the partial derivatives of the lagrangean with respect to the variables $(\underset{\sim}{x})$ are not necessarily zero

$$\underset{\sim}{t} \equiv \nabla_{\underset{\sim}{x}}(L) \tag{2.10}$$

It can be shown that the necessary conditions for solution of MP (2.7) are

$$\begin{aligned} &\nabla_{\underset{\sim}{x}}(z) + \nabla_{\underset{\sim}{x}}(\underset{\sim}{g}^T)\underset{\sim}{\lambda} - \underset{\sim}{t} = \underset{\sim}{0} \\ &\underset{\sim}{g} + \underset{\sim}{s} = \underset{\sim}{0} \\ &\underset{\sim}{x}^T \underset{\sim}{t} + \underset{\sim}{s}^T \underset{\sim}{\lambda} = 0 \\ &\underset{\sim}{x} \geqslant \underset{\sim}{0} \quad \underset{\sim}{t} \geqslant \underset{\sim}{0} \quad \underset{\sim}{s} \geqslant \underset{\sim}{0} \quad \underset{\sim}{\lambda} \geqslant \underset{\sim}{0} \end{aligned} \tag{2.11}$$

These are known as the "Karush-Kuhn-Tucker conditions" or more briefly as the KKT conditions. When the constraints define a convex set and when the objective function is convex, the KKT conditions are sufficient for global minimality. Thus in general any solution of the mathematical program (2.7) satisfies the set of conditions (2.11) but the argument can only be reversed under the above convexity conditions. In addition, the KKT conditions are sufficient for a unique global minimum when the constraint set is convex and the objective function is strictly convex.

The simplest class of MP is the linear program (LP) in which the constraint and objective functions are all linear. A particular form of LP is:

$$\begin{aligned} &\text{Min } z = \underset{\sim}{c}^T \underset{\sim}{x} \\ &\underset{\sim}{A}\,\underset{\sim}{x} \leqslant \underset{\sim}{b} \end{aligned} \tag{2.12}$$

$$\underset{\sim}{x} \geq \underset{\sim}{0}$$

The coefficients (c_j) of the linear objective function are generally termed "cost coefficients" whilst the coefficients (a_{ij}) of the constraints are termed "structural coefficients" and their matrix is the "structural matrix". The quantities on the right hand side (b_i) of the constraints are generally termed "stipulations".

The KKT conditions for this LP can be obtained as a special case of (2.11) in the following form

$$\begin{aligned} &\underset{\sim}{A}^T\underset{\sim}{\lambda} + \underset{\sim}{c} - \underset{\sim}{t} = \underset{\sim}{0} \\ &\underset{\sim}{A}\,\underset{\sim}{x} + \underset{\sim}{s} - \underset{\sim}{b} = \underset{\sim}{0} \\ &\underset{\sim}{x}^T\underset{\sim}{t} + \underset{\sim}{s}^T\underset{\sim}{\lambda} = 0 \\ &\underset{\sim}{x} \geq \underset{\sim}{0}, \quad \underset{\sim}{t} \geq \underset{\sim}{0}, \quad \underset{\sim}{s} \geq \underset{\sim}{0}, \quad \underset{\sim}{\lambda} \geq \underset{\sim}{0} \end{aligned} \qquad (2.13)$$

Since the linear constraint set is convex and since a linear function is convex (but not strictly so), it follows that any solution of (2.13) is also a solution of (2.12) and multiple optimal solutions may exist. The KKT conditions (2.13) are all linear except for the third which is in complementary form. For this reason (2.13) is said to define a "linear complementarity problem" (LCP). It should be noted that the complementary condition ensures that if x_i is positive then t_i must be zero and vice versa. Similarly if λ_i is positive then s_i must be zero and vice versa.

Consider now another LP of the following form

$$\begin{aligned} &\text{Max } w = \underset{\sim}{b}^T\underset{\sim}{y} \\ &\underset{\sim}{A}^T\underset{\sim}{y} \leq \underset{\sim}{c} \\ &\underset{\sim}{y} \leq \underset{\sim}{0} \end{aligned} \qquad (2.14)$$

It is convenient to transform w and $\underset{\sim}{y}$ so that

$$\bar{w} \equiv -w$$

$$\underset{\sim}{\pi} \equiv -\underset{\sim}{y}$$

and LP (2.14) becomes

$$\text{Min } \bar{w} = \underset{\sim}{b}^T\underset{\sim}{\pi}$$

$$-\tilde{A}^T\tilde{\pi} \leq \tilde{c} \tag{2.15}$$

$$\tilde{\pi} \geq \tilde{0}$$

and from (2.11) the KKT conditions for LP (2.15) are

$$\begin{aligned}
&-\tilde{A}\,\tilde{\mu} + \tilde{b} - \tilde{\tau} = \tilde{0} \\
&-\tilde{A}^T\tilde{\pi} + \tilde{\sigma} - \tilde{c} = \tilde{0} \\
&\tilde{\pi}^T\tilde{\tau} + \tilde{\sigma}^T\tilde{\mu} = \tilde{0} \\
&\tilde{\pi} \geq \tilde{0}, \quad \tilde{\tau} \geq \tilde{0}, \quad \tilde{\sigma} \geq \tilde{0}, \quad \tilde{\mu} \geq \tilde{0}
\end{aligned} \tag{2.16}$$

In the above KKT conditions, the symbols $\tilde{\mu}$, $\tilde{\tau}$ and $\tilde{\sigma}$ replace $\tilde{\lambda}$, $\tilde{t}$ and $\tilde{s}$ respectively in (2.11). It will be seen that the two sets of KKT conditions (2.13) and (2.16) have identical form if the following equivalence is imposed.

$$\tilde{\mu} \to \tilde{x} \quad \tilde{\tau} \to \tilde{s} \quad \tilde{\sigma} \to \tilde{t} \quad \tilde{\pi} \to \tilde{\lambda} \tag{2.17}$$

Thus the two LPs (2.12) and (2.14) have identical KKT conditions (2.13) and (2.16) and any solution of the LCP representing these conditions is a solution to both LPs. The LPs are said to be a primal-dual pair. For convenience the minimising LP (2.12) will be dubbed the primal LP and the maximising LP (2.14) will be dubbed the dual. The KKT theory can be extended to cover a constraint set which contains both types of inequality (≥, ≤) and also equalities. Similarly the variables can be a mixture of non-negative, non-positive and unrestricted. Table 1 indicates the general primal-dual relations for LPs.

PRIMAL		DUAL
Min z	Optimality Criterion	Max w
$\tilde{c}^T$	Cost Coefficients	$\tilde{b}^T$
$\tilde{b}$	Stipulations	$\tilde{c}$
$\tilde{A}$	Structural Matrix	$\tilde{A}^T$
≥	i^{th} Primal Constraint and i^{th} Dual Variable	$y_i \geq 0$
=		y_i Unrestricted
≤		$y_i \leq 0$
$x_j \geq 0$	j^{th} Primal Variable and j^{th} Dual Constraint	≤
x_j Unrestricted		=
$x_j \leq 0$		≥

TABLE 1.

If any set of primal variables satisfies the primal constraints then the corresponding value of the primal objective function is said to be "feasible". Similarly any set of dual variables which satisfies the dual constraints leads to a feasible value of the dual objective function.

The von Neumann duality theorem |1| of linear programming states that if both primal and dual LPs possess feasible solutions then the optimal value of the primal objective function (z_{min}) is equal to the optimal value of the dual objective function (w_{max}).

It will be seen that feasible solutions (z_f, w_f) bound the optimal value from above and below respectively.

$$z_f \geqslant z_{min} = w_{max} \geqslant w_f \qquad (2.18)$$

3. COMPUTATION

The most important computational tool in linear programming is the simplex algorithm devised by Dantzig |1|. The method is generally presented in two phases. The first phase is concerned with generating a feasible solution to the constraints or identifying that no such feasible solution exists. The second phase is concerned with iterating from one feasible solution corresponding to an extreme point of the feasible polytope to another in such a way that the minimisation (maximisation) objective function is monotonically non-increasing (non-decreasing). The iterations proceed until optimality is reached. More compact forms of the algorithm have been discussed by Lasdon |4|. Many important problems of limit analysis can be encoded in LP format and hence can be solved by the simplex algorithm.

MPs with quadratic objective functions and linear constraint functions are termed quadratic programs (QPs) |17|. The set of KKT conditions for a QP is also an LCP. The extension of the simplex algorithm to QPs has been developed by Markowitz |18| and Wolfe |19|. Many problems in elastoplastic analysis |5| may be formulated as LCPs or equivalently as QPs. Smith |20| has adapted the Wolfe-Markowitz algorithm to problems of non-holonomic elastoplasticity which may be expressed as parametric linear. Complementarity problems (PCLP) or equivalent parametric quadratic programs (PQP). Alternative procedures have been developed by Maier et al |21|.

4. THE ENTROPY OF GRANULAR MATERIAL

Granular material is most frequently modelled as a continuum but some theoretical studies have been made of the influence of its particulate micro-structure |22|. These have usually been made as generalisations of the Hertzian contact theory applied to regular packings (such as face-centred-cubic and simple-cubic) of equal-sized particles. The analysis of such an assembly subjected to hydrostatic pressure is complicated and that complication would greatly increase if the theory were extended to random packings. The theory does not consider the change of packing during the loading.

Presumably a naturally occurring cohesionless soil has no regular micro-structure and it would appear that the phenomenological behaviour of a macrosample requires an appropriate random modelling of the micro-structure. A study |23,24| has been made of 28 distinct micro-states spanning across the co-ordination number range 4 through 12. The ith microstate has an associated porosity (α_i) and co-ordination number (N_i). The macrosample is assumed to be made up of a complex assembly of cells each of which will correspond to one of the 28 microstates. Let p_i be the fraction of the macrosample in the ith microstate. Equivalently p_i is the probability that a particle selected at random is in the ith microstate.

The Shannon measure of uncertainty (H) is given by

$$H = -k \sum p_i \ln p_i \tag{4.1}$$

where k is a scaling constant and H is the entropy.

If the only prior knowledge is the value of the porosity of the macrosample then this can be equated to the mean porosity of the microstates ($\bar{\alpha}$).

The Jaynes-Shannon |25,26| method of obtaining the least-biased estimate of p_i for the given prior knowledge leads to the following non-linear program

$$\begin{aligned} &\text{Max } \frac{H}{k} = -\sum p_i \ln p_i \\ &\sum p_i = 1 \\ &\sum p_i \alpha_i = \bar{\alpha} \\ &p_i \geqslant 0 \quad i = 1, 2, \ldots, 28 \end{aligned} \tag{4.2}$$

It can be shown that the non-negativity of the probabilities need not be stated explicitly and, since the remaining constraints are equalities, the problem reduces to a classical optimisation problem.

$$\text{Min } z = \Sigma\, p_i \ln p_i$$

$$\Sigma\, p_i = 1 \tag{4.3}$$

$$\Sigma\, p_i\, \alpha_i = \bar{\alpha}$$

Using the theory of section 2, the lagrangean (L) is defined as

$$L = z + \lambda_1 \{\Sigma p_i - 1\} + \lambda_2 \{\Sigma p_i \alpha_i - \bar{\alpha}\} \tag{4.4}$$

and differentiating,

$$\frac{\partial L}{\partial p_i} = \frac{\partial z}{\partial p_i} + \lambda_1 + \lambda_2\, \alpha \qquad \frac{\partial z}{\partial p_i} = \ln p_i + 1$$

Thus the stationarity condition $\left[\frac{\partial L}{\partial p_i} = 0\right]$ leads to

$$p_i = \exp\{-\mu_o - \mu_i\, \alpha_i\} \tag{4.5}$$

where

$$\mu_o \equiv \lambda_1 + 1 \qquad \mu_1 \equiv \lambda_2$$

The stationarity conditions $\left[\frac{\partial L}{\partial \lambda_1} = 0 = \frac{\partial L}{\partial \lambda_2}\right]$ recover the two constraints. Substitution from (4.5) into the first constraint gives a relation which allows μ_o (and hence p_i) to be calculated for any value of μ_1. The corresponding value of $\bar{\alpha}$ can then be calculated from the second constraint. Thus the least-biased probability distribution (p_i) corresponding to a particular value of mean (or macrosample) porosity has been calculated. By repeating this calculation for varying values of μ_1, the probability distribution for any such porosity can be evaluated. The mean co-ordination numbet ($\bar{N}$) can now be calculated for any porosity and this has been found to be in good agreement with experiemntal evidence. The maximum attainable entropy (and its corresponding porosity and void ratio) can also be calculated and it has been found to be close to the experimentally derived critical void ratio.

Thus the classical optimisation theory has given a theoretical basis for important experimental results and the maximum entropy formalism has given a starting point for a new theory |27| of the statistical mechanics of soils.

5. LIMIT ANALYSIS AND SLOPE STABILITY

The problem of slope stability |28-31| is a particularly important example of limit analysis. In the following discussion a vertical section will be considered. The plane will be divided into triangular-elements and each element will be considered to be rigid and with rough edges. The elements will be connected through thin layers of rigid-plastic material possessing a Mohr-Coulomb yield locus.

First, consider the kinematics of the rigid elements (Figure 1). The displacements $\{\Delta u_1, \Delta u_2, \Delta u_3, \Delta w_1, \Delta w_2, \Delta w_3\}$ of the element sides can be evaluated in terms of the three degree of freedom $\{\Delta q_1, \Delta q_2, \Delta q_3\}$ in the following form:

$$\begin{bmatrix} A_1^e \\ \hline \underset{\sim}{A_2^e} \end{bmatrix} \underset{\sim}{q} = \begin{bmatrix} \underset{\sim}{\Delta u} \\ \hline \underset{\sim}{\Delta w} \end{bmatrix} \tag{5.1}$$

Now consider the statics of a triangular element. The body forces ($\underset{\sim}{p}$) are applied at the centroid of the element. The equipollent tangential ($\underset{\sim}{t}$) and normal ($\underset{\sim}{n}$) forces applied by the element to the Mohr-Coulomb layers at its edges are related to the body forces in the following way:

$$\begin{bmatrix} (\underset{\sim}{A_1^e})^T & \vdots & (\underset{\sim}{A_2^e})^T \end{bmatrix} \begin{bmatrix} \underset{\sim}{t} \\ \hline \underset{\sim}{n} \end{bmatrix} = \underset{\sim}{p} \tag{5.2}$$

The transpose relation linking equations (5.1) and (5.2) is a confirmation of the consistency of these transformations with respect to a static-kinematic duality (SKD) |32|. The Mohr-Coulomb layers are given arbitrary numbering and orientation as indicated on Figure 4 and the sign conventions of the normal and tangential forces (t_i and n_i) on the ith layer are related to the mesh orientation as indicated in Figure 5. The application of this convention to layer 6 is shown on Figure 6. The rigid elements displace (Figure 7) and the layers deform to produce a mode of collapse. For example the deformations of layer 6 (Figure 8) are related to the displacements of the sides of elements 3 and 4 in the following way

$$u_6 = \Delta u_2 - \Delta u_3$$

$$w_6 = -\Delta w_2 - \Delta w_3$$

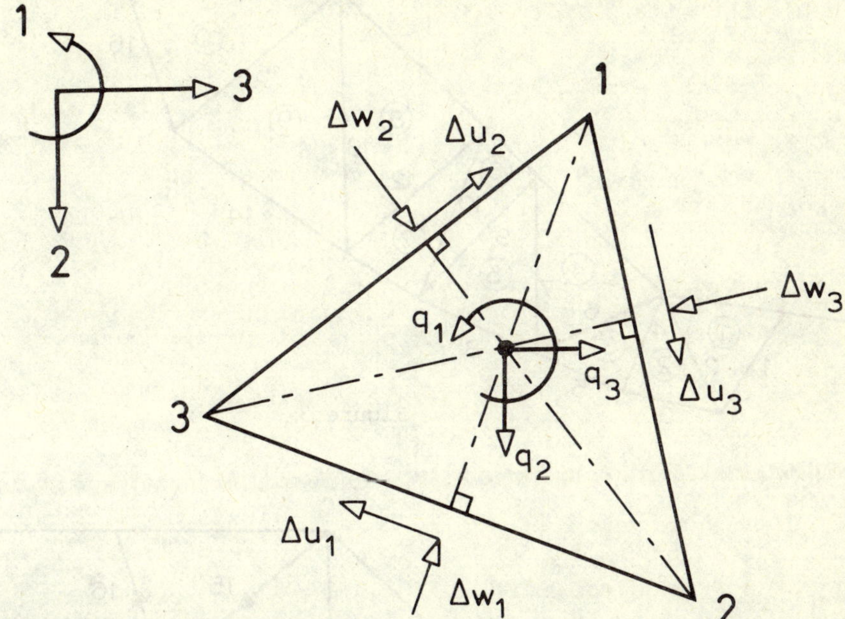

Figure 1

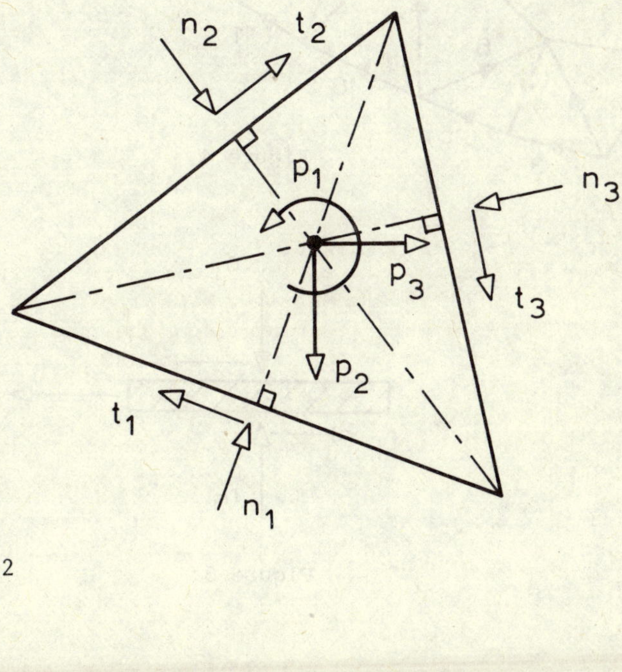

Figure 2

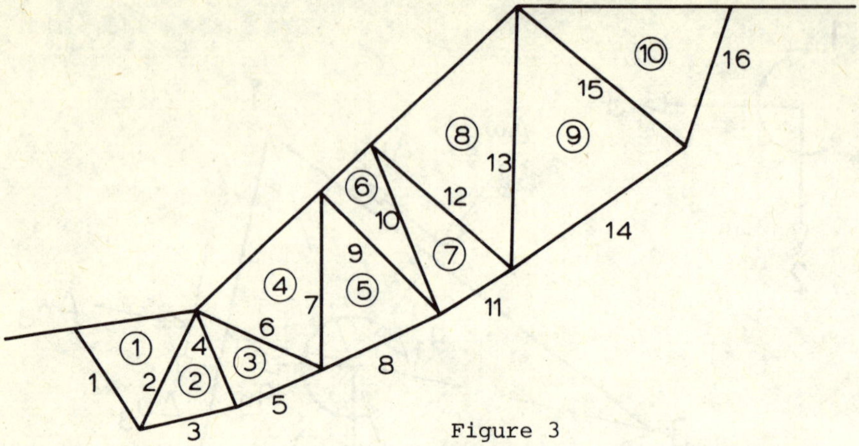

Figure 3

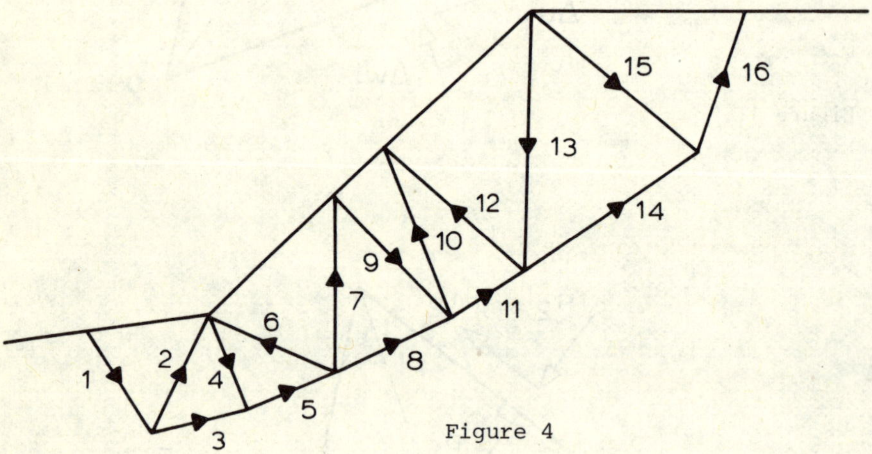

Figure 4

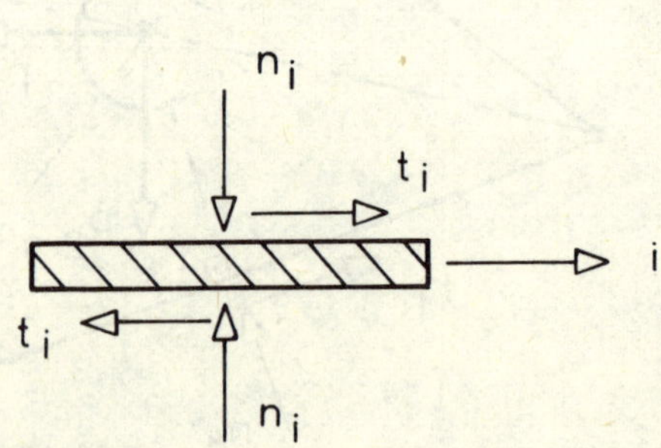

Figure 5

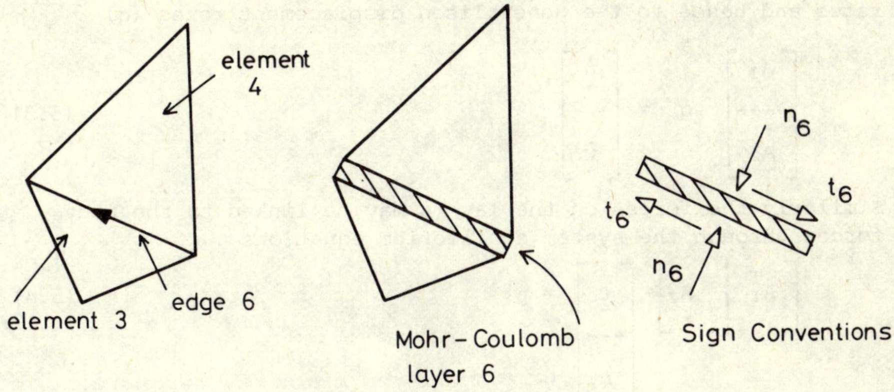

Figure 6

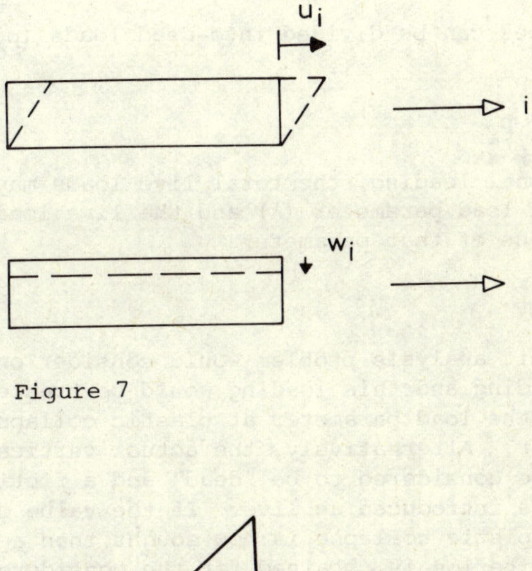

Figure 7

Figure 8

From the totality of such relations, the deformation rates ($\dot{\tilde{u}}$, $\dot{\tilde{w}}$) for the 16 layers may be linked to the element side displacement rates and hence to the generalised displacement rates ($\dot{\tilde{q}}$)

$$\begin{bmatrix} \tilde{A}_1 \\ \hline \tilde{A}_2 \end{bmatrix} \dot{\tilde{q}} = \begin{bmatrix} \dot{\tilde{u}} \\ \hline \dot{\tilde{w}} \end{bmatrix} \tag{5.3}$$

Similarly the forces on the layers may be linked to the body forces through the system equilibrium equations

$$\begin{bmatrix} \tilde{A}_1^T & \vdots & \tilde{A}_2^T \end{bmatrix} \begin{bmatrix} \tilde{t} \\ \hline \tilde{n} \end{bmatrix} = \tilde{p} \tag{5.4}$$

The transpose relationship linking equations (5.3) and (5.4) is a further manifestation of SKD.

The body forces can be divided into dead loads ($\tilde{p}_d$) and live loads ($\tilde{p}_\ell$)

$$\tilde{p} = \tilde{p}_d + \tilde{p}_\ell \tag{5.5}$$

For proportional loading, the total live loads may be expressed in terms of a load parameter (λ) and the live loads ($\tilde{p}_o$) associated with unit value of that parameter

$$\tilde{p}_\ell = \lambda \tilde{p}_o \tag{5.6}$$

A typical limit analysis problem would consider only vertical (gravity) loading and this loading would be treated as "live". The value of the load parameter at plastic collapse ($\lambda = \lambda_c$) will then be sought. Alternatively, the actual vertical (gravity) loading can be considered to be "dead" and a fictitious horizontal loading can be introduced as live. If the value of the load parameter at plstic collapse is now sought then a form of the Sarma |33| criterion is obtained for the considered problem. Conventionally the safety factor is linked to the reserve strength along the potential failure surface when the actual (gravity) loads are applied. Sarma |33| has described the relations between this conventional safety factor and his own criterion.

The yield conditions (Figure 9) for a typical Mohr-Coulomb layer are

$$t_i \leqslant c_i + \mu_i n_i$$

$$-t_i \leqslant c_i + \mu_i n_i$$

Hence for all such layers,

$$\underset{\sim}{t} - \underset{\sim}{M}\,\underset{\sim}{n} \leqslant \underset{\sim}{c} \qquad (5.7)$$
$$-\underset{\sim}{t} - \underset{\sim}{M}\,\underset{\sim}{n} \leqslant \underset{\sim}{c}$$

where $M \equiv \text{diag}\,\{\mu_1, \ldots\ldots, \mu_i, \ldots\ldots, \mu_{16}\}$

The matrix inequations (5.7) may be combined in a single matrix inequality

$$\left[\begin{array}{c|c} \underset{\sim}{I} & -\underset{\sim}{M} \\ \hline -\underset{\sim}{I} & -\underset{\sim}{M} \end{array}\right] \left[\begin{array}{c} \underset{\sim}{t} \\ \hline \underset{\sim}{n} \end{array}\right] \leqslant \left[\begin{array}{c} \underset{\sim}{c} \\ \hline \underset{\sim}{c} \end{array}\right] \qquad (5.8)$$

where $\underset{\sim}{I}$ is the identity matrix. Inequations (5.8) may now be combined with equations (5.4), (5.5) and (5.6) into the static admissibility conditions

$$\left[\begin{array}{c|c|c} \cdot & \underset{\sim}{I} & -\underset{\sim}{M} \\ \hline \cdot & -\underset{\sim}{I} & -\underset{\sim}{M} \\ \hline \underset{\sim}{p}_o & -\underset{\sim}{A}_1^T & -\underset{\sim}{A}_2^T \end{array}\right] \left[\begin{array}{c} \lambda \\ \hline \underset{\sim}{t} \\ \hline \underset{\sim}{n} \end{array}\right] \begin{array}{c} \leqslant \\ \leqslant \\ = \end{array} \left[\begin{array}{c} \underset{\sim}{c} \\ \hline \underset{\sim}{c} \\ \hline -\underset{\sim}{p}_d \end{array}\right] \qquad (5.9)$$

It will be convenient to eliminate the inequations by introducing slack variables. In addition, the material properties render negative (i.e. tensile) values of n inadmissible. The static admissibility conditions now become

$$\left[\begin{array}{c|c|c} \cdot & \underset{\sim}{I} & -\underset{\sim}{M} \\ \hline \cdot & -\underset{\sim}{I} & -\underset{\sim}{M} \\ \hline \underset{\sim}{p}_o & -\underset{\sim}{A}_1^T & -\underset{\sim}{A}_2^T \end{array}\right] \left[\begin{array}{c} \underset{\sim}{\lambda} \\ \hline \underset{\sim}{t} \\ \hline \underset{\sim}{n} \end{array}\right] + \left[\begin{array}{c} \underset{\sim}{s}^+ \\ \hline \underset{\sim}{s}^- \\ \hline \underset{\sim}{0} \end{array}\right] = \left[\begin{array}{c} \underset{\sim}{c} \\ \hline \underset{\sim}{c} \\ \hline \underset{\sim}{0} \end{array}\right] \qquad (5.10)$$

$$\underset{\sim}{n} \geqslant \underset{\sim}{0} \qquad \underset{\sim}{s}^+ \geqslant \underset{\sim}{0} \qquad \underset{\sim}{s}^- \geqslant \underset{\sim}{0}$$

The mechanism deformation and displacement rates have an arbitrary magnitude and hence a scaling must be introduced so that the variables are restricted to finite magnitudes. This can be achieved through the following normalisation

$$\underset{\sim}{p}_o^T\,\dot{\underset{\sim}{q}} = 1 \qquad (5.11)$$

It is necessary to introduce constraints which eliminate inadmissible combinations of static and kinematic solutions. Thus plastic tangential mechanism deformations ($\dot{u}_i > 0$) can only take place at a Mohr-Coulomb layer which is fully plasticised in a positive sense ($s_i^+ = 0$) and similarly negative deformations ($\dot{u}_i < 0$) can only take place at a layer which is fully plasticised negatively ($s_i^- = 0$). First the tangential deformation rates ($\dot{u}_i$) are expressed in terms of non-negative components

$$\dot{u}_i = \dot{u}_i^+ - \dot{u}_i^- \qquad \dot{u}_i^+ \geqslant 0 \qquad \dot{u}_i^- \geqslant 0$$

and then the required conditions can be expressed in the following complementary way

$$s_i^+ \dot{u}_i^+ = 0 \qquad s_i^- \dot{u}_i^- = 0$$

This condition will be termed the "parity rule". For all layers, the systems parity rule becomes

$$\begin{aligned} &(\underset{\sim}{s}^+)^T \dot{\underset{\sim}{u}}^+ + (\underset{\sim}{s}^-)^T (\dot{\underset{\sim}{u}}^-) = 0 \\ &\underset{\sim}{s}^+ \geqslant \underset{\sim}{0} \quad \underset{\sim}{s}^- \geqslant \underset{\sim}{0} \quad \dot{\underset{\sim}{u}}^+ \geqslant \underset{\sim}{0} \quad \dot{\underset{\sim}{u}}^- \geqslant \underset{\sim}{0} \end{aligned} \qquad (5.12)$$

It will be noted that since the slack variables (s_i^+ and s_i^-) of the yield conditions cannot both be zero, it follows that the non-negative components ($\dot{u}_i^+$ and $\dot{u}_i^-$) of the tangential deformation rates cannot both be non-zero.

Now consider the normality condition with respect to the plastic deformation rates (Figure 10)

For $n_i > 0$, $s_i^+ = 0$

$$\frac{-\dot{w}_i}{\dot{u}_i} = \tan \phi_i \equiv \mu_i$$

or $-\dot{w}_i - \mu_i(\dot{u}_i^+ - \dot{u}_i^-) = 0$ (5.13)

Similarly for $n_i > 0$, $s_i^- = 0$

$$-\dot{w}_i + \mu_i(\dot{u}_i - \dot{u}_i^-) = 0 \qquad (5.14)$$

However

$$\left[s_i^+ = 0\right] \rightarrow \left[s_i^- > 0\right] \rightarrow \left[\dot{u}_i^- = 0\right]$$

$$\left[s_i^- = 0\right] \rightarrow \left[s_i^+ > 0\right] \rightarrow \left[\dot{u}_i^+ = 0\right]$$

and it follows that equations (5.13) and (5.14) can be written as a single equation

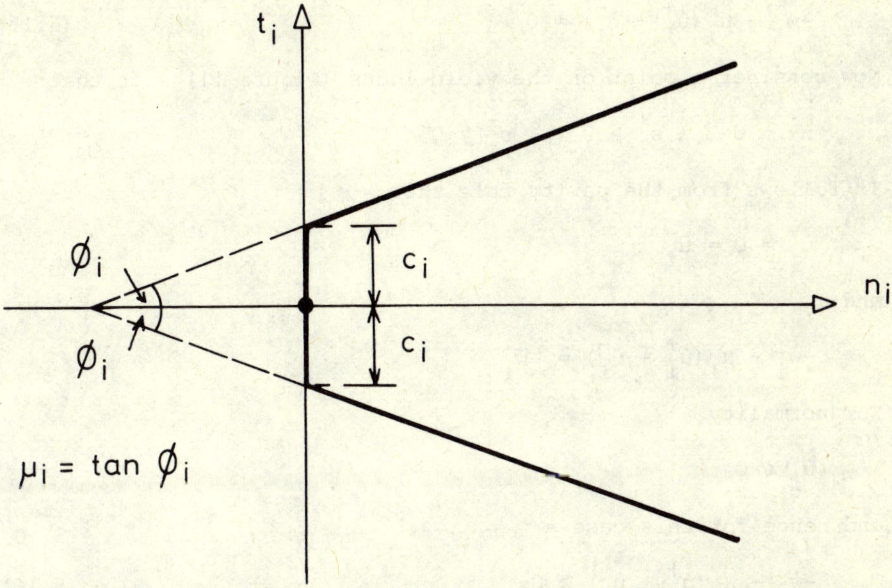

Figure 9

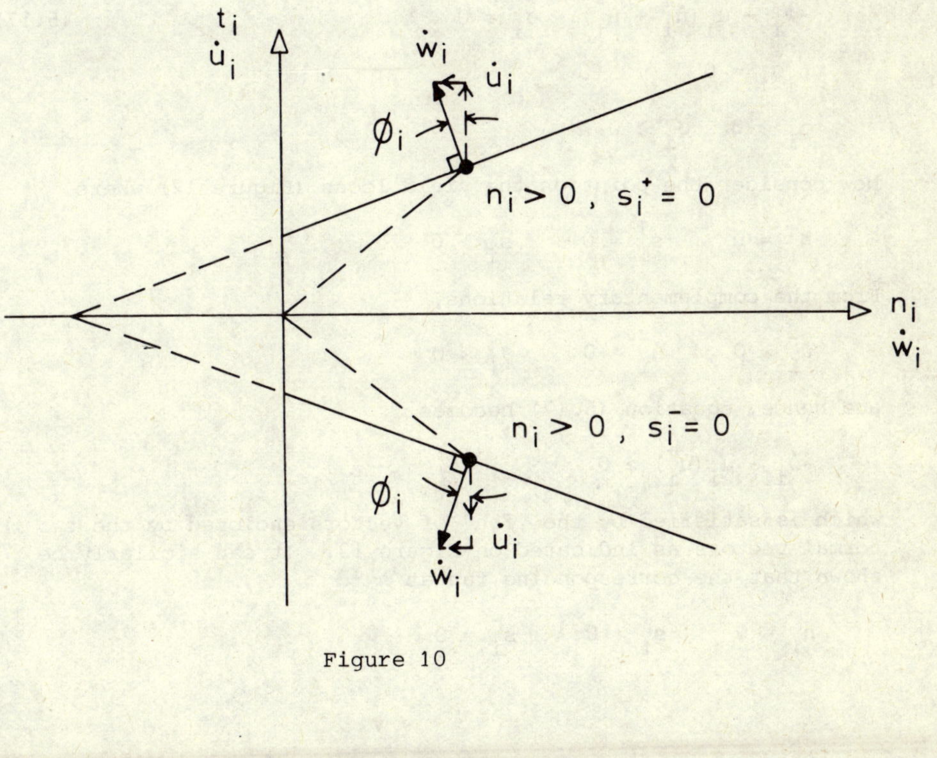

Figure 10

$$-\dot{w}_i - \mu_i(\dot{u}_i^+ + \dot{u}_i^-) = 0 \qquad (5.15)$$

Now consider a point on the yield locus (Figure 11) such that

$$n_i = 0 \qquad s_i^+ > 0 \qquad s_i^- > 0$$

It follows from the parity rule that

$$\dot{u}_i^+ = 0 = \dot{u}_i^-$$

and

$$-\dot{w}_i - \mu_i(\dot{u}_i^+ + \dot{u}_i^-) = -\dot{w}_i$$

For normality

$$\dot{w}_i \leqslant 0$$

and hence for this case

$$-\dot{w}_i - \mu_i(\dot{u}_i^+ + \dot{u}_i^-) \geqslant 0 \qquad (5.16)$$

It will be seen that both (5.15) and (5.16) are covered by the following additional constraints

$$-\dot{w}_i - \mu_i(\dot{u}_i^+ + \dot{u}_i^-) - \sigma_i = 0 \qquad (5.17)$$

$$n_i\sigma_i = 0$$

$$n_i \geqslant 0 \quad \sigma_i \geqslant 0$$

Now consider the point on the yield locus (Figure 12) where

$$n_i = 0 \qquad s_i^+ = 0 \qquad s_i^- > 0$$

From the complementary relations,

$$\dot{u}_i^- \geqslant 0 \qquad \dot{u}_i^- = 0 \qquad \sigma_i \geqslant 0$$

and hence, equation (5.17) becomes

$$-\dot{w}_i - \mu_i(\dot{u}_i^+) \geqslant 0$$

which is satisfied by the "fan" of vectors enclosed by the two normal vectors as indicated on Figure 12. It can similarly be shown that the corresponding fan at

$$n_i = 0 \qquad s_i^+ > 0 \qquad s_i^- = 0$$

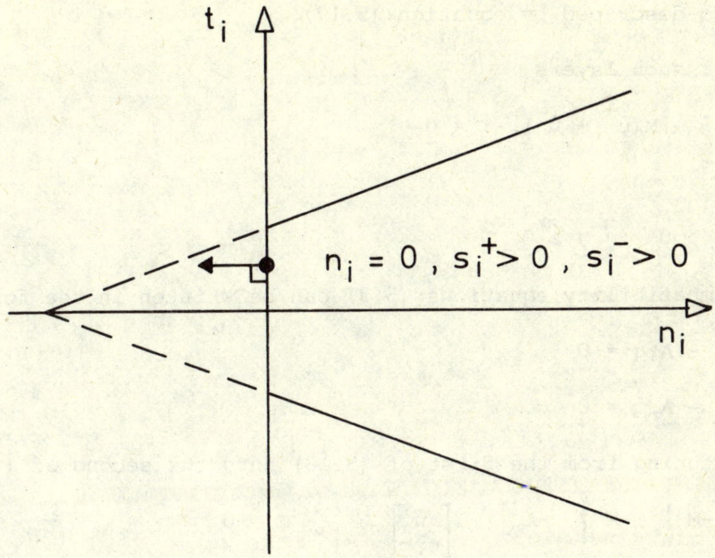

Figure 11

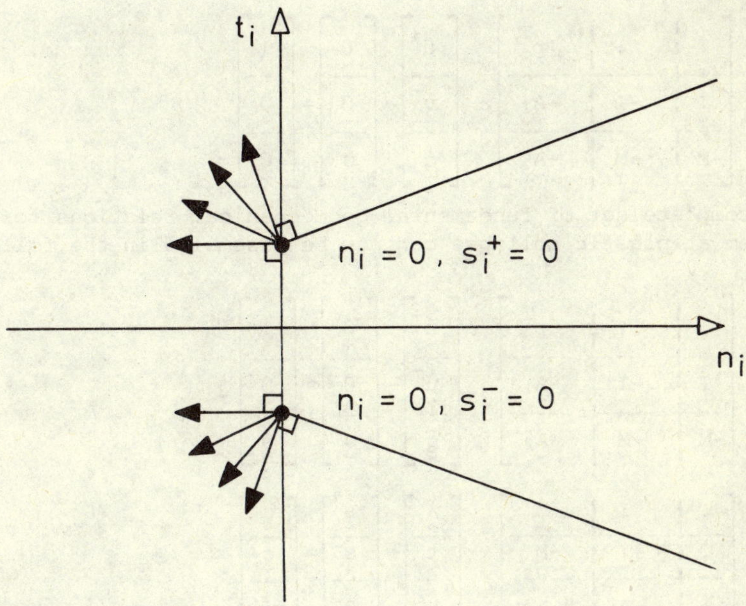

Figure 12

is also described by equation (5.17).

For all such layers,

$$-\dot{w} - M(\dot{u}^{+} + \dot{u}^{-}) - \sigma = 0 \qquad (5.18)$$

$$n^{T}\sigma = 0$$

$$n \geqslant 0 \qquad \sigma \geqslant 0$$

The compatibility equations (5.3) can be written in the form,

$$\dot{u} - A_1\dot{q} = 0 \qquad (5.19)$$

$$\dot{w} - A_2\dot{q} = 0$$

Substituting from the first of (5.18) into the second of (5.19)

$$\left[\begin{array}{c|c|c} -M & -M & -A_2 \end{array}\right] \left[\begin{array}{c} \dot{u}^{+} \\ \hline \dot{u}^{-} \\ \hline \dot{q} \end{array}\right] - \sigma = 0 \qquad (5.20)$$

Equation (5.11), the first of equations (5.19) and equation (5.20) are now assembled as the systems kinematic admissibility conditions

$$\left[\begin{array}{c|c|c} \cdot & \cdot & p_o^{T} \\ \hline I & -I & -A_1 \\ \hline -M & -M & -A_2 \end{array}\right] \left[\begin{array}{c} \dot{u}^{+} \\ \hline \dot{u}^{-} \\ \hline \dot{q} \end{array}\right] - \left[\begin{array}{c} 0 \\ \hline 0 \\ \hline \sigma \end{array}\right] = \left[\begin{array}{c} 1 \\ \hline 0 \\ \hline 0 \end{array}\right] \qquad (5.21)$$

The complete set of fundamental geomechanical relations for the system at plastic collapse can now be assembled in the following form

$$\left[\begin{array}{c|c|c} \cdot & \cdot & p_o^{T} \\ \hline I & -I & -A_1 \\ \hline -M & -M & -A_2 \end{array}\right] \left[\begin{array}{c} \dot{u}^{+} \\ \hline \dot{u}^{-} \\ \hline \dot{q} \end{array}\right] - \left[\begin{array}{c} 0 \\ \hline 0 \\ \hline \sigma \end{array}\right] = \left[\begin{array}{c} 1 \\ \hline 0 \\ \hline 0 \end{array}\right]$$

$$\left[\begin{array}{c|c|c} \cdot & I & -M \\ \hline \cdot & -I & -M \\ \hline p_o & -A_1^{T} & -A_2^{T} \end{array}\right] \left[\begin{array}{c} \lambda \\ \hline t \\ \hline n \end{array}\right] + \left[\begin{array}{c} s^{+} \\ \hline s^{-} \\ \hline 0 \end{array}\right] = \left[\begin{array}{c} c \\ \hline c \\ \hline -p_d \end{array}\right] \qquad (5.22)$$

$$(\underset{\sim}{s}^+)^T \dot{\underset{\sim}{u}}^+ + (\underset{\sim}{s}^-)^T \dot{\underset{\sim}{u}}^- + \underset{\sim}{n}^T \underset{\sim}{\sigma} = 0$$

$$\underset{\sim}{s}^+ \geqslant \underset{\sim}{0} \qquad \underset{\sim}{s}^- \geqslant \underset{\sim}{0} \qquad \dot{\underset{\sim}{u}}^+ \geqslant \underset{\sim}{0} \qquad \dot{\underset{\sim}{u}}^- \geqslant \underset{\sim}{0} \qquad \underset{\sim}{n} \geqslant \underset{\sim}{0} \qquad \underset{\sim}{\sigma} \geqslant \underset{\sim}{0}$$

It will be seen from the earlier discussion of KKT theory that the solution of this LCP can be obtained through either of the following primal-dual LPs

$$\text{Min } z = \left[\underset{\sim}{c}^T \;\middle|\; \underset{\sim}{c}^T \;\middle|\; -\underset{\sim}{p}_d^T\right] \begin{bmatrix} \dot{\underset{\sim}{u}}^+ \\ \hline \dot{\underset{\sim}{u}}^- \\ \hline \dot{\underset{\sim}{q}} \end{bmatrix}$$

subject to

$$\left[\begin{array}{c|c|c} \cdot & \cdot & \underset{\sim}{p}_o^T \\ \hline \underset{\sim}{I} & -\underset{\sim}{I} & -\underset{\sim}{A}_1 \\ \hline -\underset{\sim}{M} & -\underset{\sim}{M} & -\underset{\sim}{A}_2 \end{array}\right] \begin{bmatrix} \dot{\underset{\sim}{u}}^+ \\ \hline \dot{\underset{\sim}{u}}^- \\ \hline \dot{\underset{\sim}{q}} \end{bmatrix} \begin{array}{c} = \\ = \\ \geqslant \end{array} \begin{bmatrix} 1 \\ \hline \underset{\sim}{0} \\ \hline \underset{\sim}{0} \end{bmatrix} \tag{5.23}$$

$$\dot{\underset{\sim}{u}}^+ \geqslant \underset{\sim}{0} \qquad \dot{\underset{\sim}{u}}^- \geqslant \underset{\sim}{0}$$

Primal Linear Program

$$\text{Max } w = \left[1 \;\middle|\; \underset{\sim}{0}^T \;\middle|\; \underset{\sim}{0}^T\right] \begin{bmatrix} \lambda \\ \hline \underset{\sim}{t} \\ \hline \underset{\sim}{n} \end{bmatrix}$$

subject to

$$\left[\begin{array}{c|c|c} \cdot & \underset{\sim}{I} & -\underset{\sim}{M} \\ \hline \cdot & -\underset{\sim}{I} & -\underset{\sim}{M} \\ \hline \underset{\sim}{p}_o & -\underset{\sim}{A}_1^T & -\underset{\sim}{A}_2^T \end{array}\right] \begin{bmatrix} \lambda \\ \hline \underset{\sim}{t} \\ \hline \underset{\sim}{n} \end{bmatrix} \begin{array}{c} \leqslant \\ \\ = \end{array} \begin{bmatrix} \underset{\sim}{c} \\ \hline \underset{\sim}{c} \\ \hline -\underset{\sim}{p}_d^T \end{bmatrix} \tag{5.24}$$

$$\underset{\sim}{n} \geqslant \underset{\sim}{0}$$

Dual Linear Program

The LP duality theory ensures that the two programs have the same optimal values of objective function. The optimal value of the dual objective function is the load parameter at plastic collapse. The optimal primal objective function must therefore be equal to that collapse load parameter. The LP requiring less computational effort will be solved but both sets of variables will be obtained through the simplex multipliers. The primal-dual LPs also encode the two limit theorems for this class of problem. The consistency of the fundamental geomechanical relations as KKT constraints was obtained from two forms of duality. The first duality (SKD) followed directly from the statics and kinematics of the rigid elements. The second duality was imposed by selecting a flow rule which was dual to the yield condition.

$$\begin{bmatrix} \underset{\sim}{I} & -\underset{\sim}{M} \\ -\underset{\sim}{I} & -\underset{\sim}{M} \end{bmatrix} \begin{bmatrix} \underset{\sim}{t} \\ \underset{\sim}{n} \end{bmatrix} \leqslant \begin{bmatrix} \underset{\sim}{c} \\ \underset{\sim}{c} \end{bmatrix} \qquad \text{Yield} \qquad \underset{\sim}{n} \geqslant \underset{\sim}{0} \tag{5.25}$$

$$\begin{bmatrix} \underset{\sim}{I} & -\underset{\sim}{I} \\ -\underset{\sim}{M} & -\underset{\sim}{M} \end{bmatrix} \begin{bmatrix} \underset{\sim}{u}^{+} \\ \underset{\sim}{u}^{-} \end{bmatrix} \begin{matrix} = \\ \geqslant \end{matrix} \begin{bmatrix} \dot{\underset{\sim}{u}} \\ \dot{\underset{\sim}{w}} \end{bmatrix} \qquad \text{Flow} \qquad \dot{\underset{\sim}{u}}^{+} \geqslant \underset{\sim}{0} \quad \dot{\underset{\sim}{u}}^{-} \geqslant \underset{\sim}{0} \tag{5.26}$$

If flexible elements are introduced then SKD will not follow automatically and must be imposed when selecting the static and kinematic finite element models.

The yield locus must be piece-wise linearised if the limit analysis programs are to remain linear and the consistent dual flow rule must be introduced by imposing normality with respect to the piece-wise linearised yield locus.

If both these dualities are imposed then the KKT conditions and primal-dual LPs will be of the same general form as those which have been obtained for the simpler model discussed earlier.

The kinematic aspects of the constitutive relations corresponding to a perfectly plastic Mohr-Coulomb material do not apply to materials in which energy is dissipated by friction. Palmer has shown how a new yield surface (the so-called "g-surface") can be generated for materials with non-associated flow rules. Palmer |34| has suggested a form of safe theorem for such materials.

The same formalism can be applied to other classes of problems. For example if the constitutive relations of holonomic plasticity are employed then the fundamental relations can be organised in the form of the KKT conditions corresponding to a pair of primal-dual QPs.

6. CLOSURE

The consistent formulation of plastic problems requires the enforcement of two forms of duality. The first is static-kinematic duality and the second is material duality linking the static and kinematic aspects of the plastic constitutive relations. If the fundamental geomechanical relations are organised in the form of a set of Karush-Kuhn-Tucker conditions, then a pair of primal-dual mathematical programs can be inferred. These programs encode the appropriate extremal principles for the considered class of problems.

The application of this formalism to the plastic limit analysis of slope stability for materials with a piece-wise linearised yield locus has led to a pair of primal-dual linear programs which encode the corresponding limit theorems. The solution to either program provides the optimal value of static and kinematic variables. Thus the value of the load parameter is evaluated and the failure line is determined.

Materials which dissipate energy by friction do not conform with this material duality and consequently these limit theorems do not apply to this class of materials.

7. REFERENCES

1. Dantzig, G.B., *Linear programming and extensions.* Princeton, 1963.

2. Kunzi, H.P., and Krelle, W., *Nonlinear programming.* Blaisdell, 1966.

3. Mangasarian, O.L., *Nonlinear programming.* McGraw-Hill, 1969.

4. Lasdon, L.S., *Optimisation theory for large systems.* McMillan, 1970.

5. Cohn, M.Z., Maier, G., and Grierson, D.E., (Editors) *Engineering plasticity by mathematical programming.* Pergamon, 1979.

6. Munro, J., "The application of mathematical programming to structural problems". Procs. World Congress IASS. Madrid, 1979, 5, pp. 89-207.

7. Maier, G., and Munro, J., "Mathematical programming application to engineering plastic analysis". (to appear in Appl. Mechs. Revs).

8. Chen, W.F., *Limit analysis and soil plasticity.* Elsevier, 1975.

9. Palmer, A.C., (Editor) Procs. Symposium on role of plasticity in Soil Mechanics. Cambridge, 1973.

10. Schofield, A.N., and Wroth, C.P., *Critical state Soil Mechanics.* McGraw-Hill, 1968.

11. Lysmer, J., "Limit analysis of plane problems in Soil Mechanics". *J. Soil Mech. Found. Div.*, ASCE, 96, 1970, pp. 1311-1334.

12. Fremond, M., and Salencon, J., "Limit analysis by finite-element methods". Proc. Symp. on role of plasticity in Soil Mechanics. Cambridge, 1973, pp. 297-309.

13. Zavelani-Rossi, A., Gatti, G., and Gioda, G., "On finite-element stability analysis in Soil Mechanics". Int. Symp. on discrete methods in Engineering. CISE Segrate., Milano, 1974.

14. Gioda, G., and de Donato, O., "Elastic-plastic analysis of geotechnical problems by mathematical programming". *Int. J. Num. An. Meth. Geomechanics.*, 3, 1979, pp. 381-401.

15. Martins, J.B., Reis, E.B., and Matos, A.C., "New methods of analysis for stability of slopes". Procs. Tenth Intl. Conf. on Soil Mecs. and Foundation Engineering, Stockholm. 3, 1981, pp. 463-467.

16. Kuhn, H.W., and Tucker, A.W., "Nonlinear programming". Procs. 2nd Berkeley Symp. Math. Statistics and Probl. 1951, pp. 481-

17. Cottle, R.W., "Fundamentals of quadratic programming and linear complentarity". Chapter 12., *Engineering Plasticity by Mathematical Programming*. Pergamon, 1979.

18. Markowitz, H.M., "The optimisation of a quadratic function subject to linear constraints". *Naval Res. Log. Quart.* 3, 1956, pp. 111-133.

19. Wolfe, P. "The simplex method for quadratic programming". *Econometrica.*, 27, 1959, pp. 382-298.

20. Smith, D.L., "The Wolfe-Markowitz algorithm for non-holonomic elastoplastic analysis". *Eng. Structs.*, 1, October 1978, pp. 8-16.

21. Maier, G., Giacomini, S., and Paterlini, F., "Combined elastoplastic and limit analysis via restricted basis linear programming". Procs. 3rd Cong. Italian Assoc. for Theo. Appl. Mech. Cagliari, 1976.

22. Deresiewicz, H., "Mechanics of granular media". *Advances in Appl. Mech.*, 5, Academic Press, 1958, pp. 236-306.

23. Munro, J., and Jowitt, P.W., "Information theory and granular media". Res. Rep. SAM 74/1. Civ. Eng. Dept., Imperial College, London, 1974.

24. Jowitt, P.W. and Munro, J., "The influence of void distribution and entropy on the engineering properties of granular media". Proc. ICASP 2. Aachen, 1975, pp. 365-385.

25. Shannon, C.E., "The mathematical theory of communication". *Bell System Tech. Journ.*, 27, 1948, pp. 379-428, pp. 623-656.

26. Jaynes, E.T., "Information theory and statistical mechanics". *Phys. Rev.* 106, 1957, pp. 620-630.

108, 1957, pp. 171-190.

27. Jowitt, P.W., and Munro, J., "The principle of maximum

entropy and some applications to granular media". Procs. Conf. Prob. Theo. and Reliability An. Geotech. Eng. Rensselaer Poly. Inst. New York, 1977.

28. Bishop, A.W., "The use of the slip circle in the stability analysis of slopes". *Geotechnique,* 5, 1, 1955, pp. 7-17.

29. Janbu, N., Earth pressures and bearing capacity calculations by generalised procedure of slices. Proc. 4th Intl. Conf. Soil Mechs. and Found. Eng. 2, 1957, pp. 207-212.

30. Morgenstern, N.R., and Price, V.E., "The analysis of the stability of general slip surfaces". *Geotechnique,* 15, 1, 1965, pp. 79-93.

31. Skempton, A.W., and Hutchinson, J., "Stability of natural slopes and embankment foundations". Proc. 7th Intl. Conf. Soil Mechs. Found. Eng. State-of-the-art Vol. pp. 291-340.

32. Munro, J., and Smith, D.L., "Linear programming duality in plastic analysis and synthesis". Proc. Intl. Conf. Computer-aided Design. University of Warwick, 1972, pp. A1.22-54.

33. Sarma, S.K., "Stability analysis of embankments and slopes". *Journ. Geotech. Eng. Div.*, ASCE, 105, GT12, December 1979, pp. 1511-1524.

34. Palmer, A.C., "A limit theorem for materials with non-associated flow laws". *J. Mecanique,* 5, (2), 1966, pp. 217-222.

OPTIMIZATION METHODS FOR PARAMETRIC IDENTIFICATION OF GEOTECHNICAL SYSTEMS

Giulio Maier and Giancarlo Gioda

Department of Structural Engineering
Technical University (Politecnico), Milan, Italy

Abstract

Parameters which characterize geotechnical systems and are representative of the behaviour of soil or rock masses are often known with a high degree of uncertainty. A way of reducing uncertainties and thus improving the mathematical models for analysis and design purposes is provided by a systematic adjustment of parameters so that theoretical predictions through the model match observational data. This identification or "inverse" problem frequently implies recourse to techniques of mathematical optimization and particularly of mathematical programming. This paper concerns the role played in this context by direct search methods, least squares procedures and quadratic and nonlinear programming for identifying parameters in purely numerical models, linear models and linear complementarity models, respectively. A procedure of statistical identification is briefly mentioned at the end. The scope is not to survey the field but merely to elucidate some aspects and potentialities of identification methods in geomechanics on the basis of a variety of recent results on particular problems.

1. INTRODUCTION

The numerical analysis of a geotechnical (or structural) system rests on a "mathematical model" where suitably idealized description of the local geometrical and physical properties are assumed as completely known. The main purpose is a quantitative prediction of the response of the system to given external actions (such as loads), i.e. the determination of the output supplied by the model when a simulation of those actions is fed in the model as input data.

J. B. Martins (ed.), Numerical Methods in Geomechanics, 273–304.

In design problems geometrical and physical aspects inside the system are to be determined, in such a way that its responses to given external actions fulfill prescribed behavioural constraints and some optimality criterion (e.g. minimum cost) be complied with.

The main purpose in the identification or "inverse" problems to consider here, is the improvement of mathematical models by means of experimental data concerning the response of the system to given external actions. Because of their "regressive" nature (exterior → interior, as distinct from the cause → effect sequence of analysis, "progressive" problems), identification as understood here bears some similarity to design (or "synthesis"). More specifically, the variables in an identification problem are represented by both the usual variables, such as displacements, which define the response of the system to given loads, and parameters which characterize local features of the system, such as constitutive laws. In order to compensate for these additional unknowns (additional with respect to analysis) further relationships, at least in principle, are generated by minimizing a suitable measure of the discrepancy between experimental data and predictions (via the model), all concerning the response to assigned external actions. Thus, quite naturally, some sort of optimization becomes central to the theory and solution methodology of identification problems, as it is in the traditional areas of optimum design, limit analysis and energy approaches to stress analysis.

These brief lecture notes are primarily concerned with the role of mathematical optimization, in particular of mathematical programming, (i.e. optimization under inequality constraints in finite-dimensional spaces) in some statical inverse problems related to soil mechanics.

No attempt will be made here to survey the available notions and techniques for solving inverse problems in geotechnical engineering. It is worth stressing that the identification of engineering systems represents nowadays a broad, multifaced and interdisciplinary area of great potentialities. Its main aspects and implications are concisely mentioned in the following remarks, in order to avoid that the drastically limited coverage of these notes give rise to a distorted image.

(a) The experimental data must be such that the sought parameters are identifiable, i.e. susceptible, at least in principle, to be uniquely identified. The "design of the experiments" to this and other aims (optimal selection of quantities to measure, suitable choice of external actions, etc.) represents a crucial preliminary problem, for which, however, theoretical or methodological frameworks[1] are scarce even in relatively simple solutions of civil engineering relevance. These aspects are totally out of the scopes of this paper.

(b) An identification process is basically intended to remove or reduce uncertainties in a mathematical models. But uncertainties affecting quantities different from parameters to identify, often play an essential role and need to be described as random disturbances or "noises". This may hold for experimental data, and/or for inaccuracies embodied in the model. Then the context in which inverse geomechanical problems must be imbedded is the stochastic identification, a growing area of active research in the last few years at the intersection of system engineering, statistics and probability theory [2] . Most of what follows is confined to the deterministic identification, where the minimization of the discrepancy is reduced to the optimization of a "loss function" by means of techniques basically similar to those customary in optimum design and limit analysis, although endowed with peculiar and partly novel features. In stochastic identification problems the minimization of the discrepancy between experimental data and model predictions acquires the more subtle meaning of optimization of the estimators in a statistical sense (a particular method in this category will be dealt with in Sec.5).

(c) A mathematical model involves, besides parameters to identify, several ingredients assumed by engineering judgement, first of all the mathematical structure of the model. The parameter identification process may lead to an assessment of the adequacy of the model as for its "a priori" assumptions and can be supplemented by a model identification or "bias removal" process. This is another subarea of system identification not considered here and not yet explored, to the writers' knowledge, in the context of soil or structural mechanics.

After decades of dramatic developments in computational mechanics and computing tools, mathematical modelling of engineering systems, generally speaking, is likely to be enhanced more by better understanding and descriptions of local aspects (such as material behaviour) than by further refinement of numerical solution procedures.This appears to hold particularly in soil and rock engineering; in fact,"in situ" measurements are common practice and laboratory tests on small scale samples are not always meaningful or conclusive for soil or rock masses (because of the presence of macro cracks, faults, etc.); uncertainties of various nature are frequently very significant. Therefore, identification methods (as natural developments from "parametric studies" and "backanalyses") seem to exhibit remarkable potentialities in geotechnical engineering, at least like they already do in structural dynamics (e.g. for damping or damage assessment). After all, parametric identification methods in geomechanics can be looked at as a combination of some modern developments in applied mathematics and system engineering with the practice advocated by Terzaghi of "flexible", observation-guided design methods [3]

2. IDENTIFICATION OF "NUMERICAL" MODELS BY DIRECT SEARCH TECHNIQUES

2.1. Problem formulation.

We will consider first the frequent situations where the main analysis tool is a computer program used as a "black-box": this will be referred to by the expression "numerical" model, as distinct from "analytical" models to be considered in the subsequent Sections. In order to predict the response of a geotechnical system, the data to be supplied will concern the loading process, the geometry of the discrete (e.g. finite element) model implemented in the code and values of some parameters which e.g. are contained in the constitutive laws adopted. These parameters, collected in vetor $\underline{P}$, will represent the objectives of the identification. Let $\underline{u}^m$ be the column-vector of the quantities, e.g. displacements, measured on the real system under the load condition generated and simulated in the model. These values are taken as deterministic, i.e. free from random errors or disturbances. Let $\underline{u}^c$ denote the vector of corresponding quantities calculated by means of the model and, hence, generally depending on $\underline{P}$. A quite natural "loss function", i.e. a measure of the discrepancy between experimental and calculated values, is the Euclidean norm of the differences or its square. Thus the following problem arises:

$$\min \{\omega(\underline{P}) \equiv [\tilde{\underline{u}}^m - \tilde{\underline{u}}^c(\underline{P})] [\underline{u}^m - \underline{u}^c(\underline{P})]\} \tag{1}$$

where a tilde denotes transpose. Clearly, alternative definitions of loss functions are possible, e.g. as the maximum absolute value of the differences.

2.2. Solution methods.

Since $\omega(\underline{P})$ is a complicate function which can be evaluated, for any $\underline{P}$, through the computer code only, optimization methods which do not require derivatives of the objective function lend themselves naturally to the numerical solution of problem (1), as proposed in [4]. Among these methods, called "direct search" techniques, the following ones are the best known (see e.g. [5] for a comprehensive treatment).

(a) Simplex (or "flexible polyhedron") method (Nelder-Mead, 1964). The procedure is initialized by evaluating ω in the $p + 1$ vertices of a simplex in the p-dimensional space of the parameters. By comparing these values and those calculated in further suitably chosen points, a new simplex is generated, etc. The process terminates when both the maximum difference between the values at the vertices and the maximum distance between vertices become smaller than assigned tolerances.

(b) Alternating variable method. The function ω is minimized

(by "linear search") with respect to each variable in turn. This easy procedure is efficient only when, with terminology appropriate for $p = 2$, the geometric representation of $\omega(\underline{P})$ would exhibit a long narrow "valley" in the direction of one axis.
(c) Rosenbrock method. The basic idea is to perform moves in the $\underline{P}$ space according to a variable orthogonal axis set, and to place one of this axis in the direction of a possible "valley", as suggested by the behaviour of the relaxation procedure (b). Some variants have been proposed in the implementation of this basic strategy.
(d) Powell method. The strategy of this more recent and sophisticated technique tends to introduce in the direct search context the philosophy of the conjugate gradient methods.

Inequality constraints are easily dealt with, in general, by penalization: when a point is reached where some constraint is violated, the loss function is assigned a large addend so that the usual comparison-based strategy automatically drives back the optimization itinerary into the feasible domain. Constraints may arise e.g. from the limitation of the search to a parameter range of physical meaning or engineering interest; or from the need to avoid parameter domains where the loss function is constant, as the response of the system is not affected by the parameters (e.g. is elastic while yield limits are sought); in this case the penalization may depend on a nonlinearity test performed in solving the current analysis problem [6] .

2.3. Applications.

Direct search solutions have been studied for the identification of cohesion c , friction angle ϕ and "in situ" stresses σ_I through a pressure tunnel test [6] and the permeability coefficients in an anisotropic clay deposit through displacement and pore pressure measurements during consolidation [7] . In both cases finite element models were adopted and the experiments simulated (generating "experimental" data by a preliminary analysis with the "real" values of the parameters, to be recovered by the identification procedure).

In the former case [6] , basic assumptions were elastic perfectly-plastic behaviour with Mohr-Coulomb yield criterion, axisymmetric, plane strain, homogeneity and isotropy (also of the "in situ" stress state σ_I). The "experimental" information consist of average increases of the tunnel diameter at various pressures of the water pumped in the tunnel, well beyond the elastic limit in the nonlinear range. Fig.1 shows the contour lines of $\omega(c, \phi)$ for the "in situ" stress $\sigma_I = 50$ kg/cm^2, the first 3 simplexes and the trajectory of the minimum-loss vertices of the subsequent simplexes up to the optimum in a numerical test. It is worth noting that the loss function exhibits nonconvexity and a valley almost orthogonal to the bisectrix of axes. Fig.2a provides an isometric view of the optimization

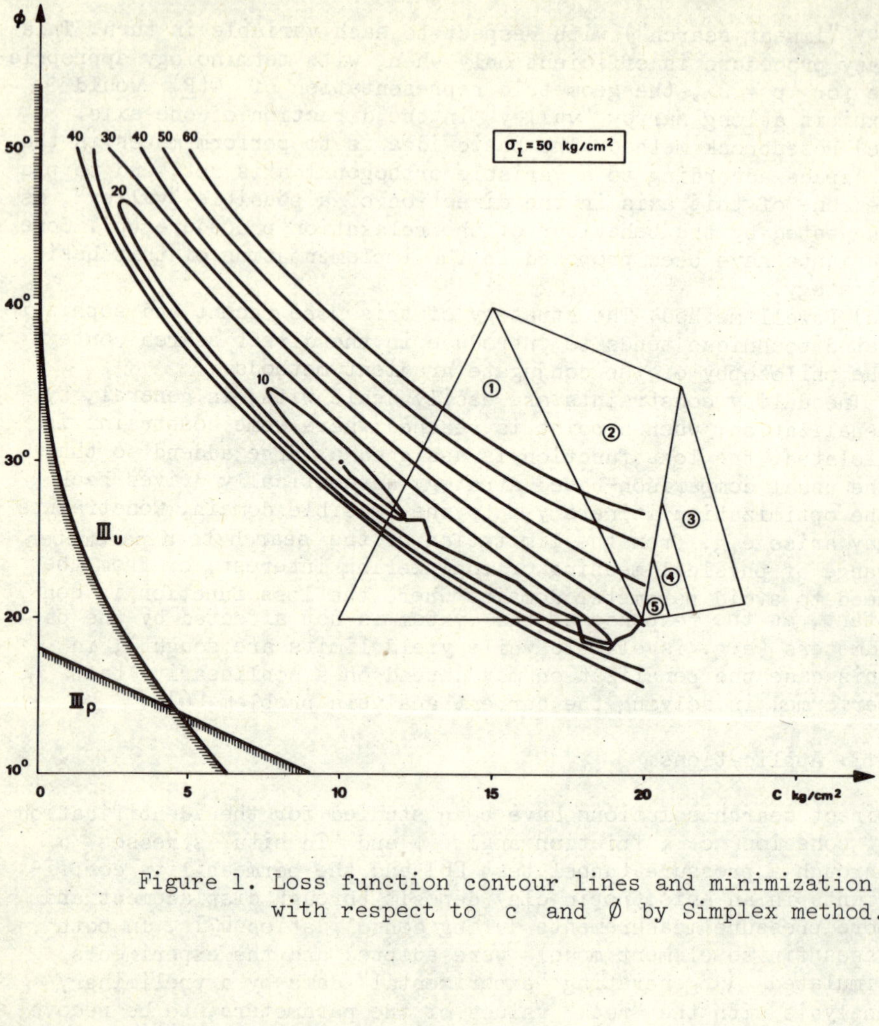

Figure 1. Loss function contour lines and minimization with respect to c and Ø by Simplex method.

trajectory for all three parameters. Fig.2b shows the discrepancy reduction versus the iteration number for various algorithms comparatively.

The following remarks are justified or confirmed by the limited computational experience achieved so far. The pros of direct search techniques are: full generality (no regularity requirements on the objective function), ease of accomodating inequality constraints and simplicity of programming. The cons are due to various circumstances: computing times are generally long, as each function evaluation requires to solve an analysis problem, and increase more than proportionally to the number of parameters; termination at a local but non-global minimum can be avoided only by

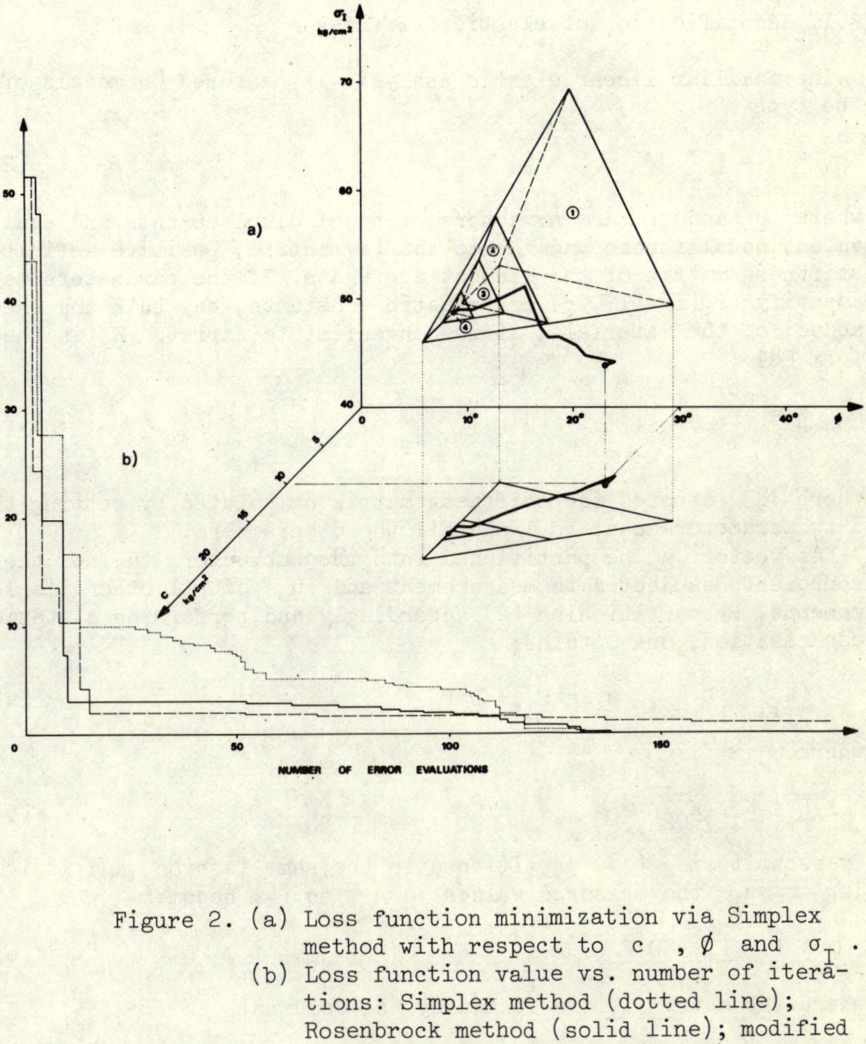

Figure 2. (a) Loss function minimization via Simplex method with respect to c , $\emptyset$ and σ_I .
(b) Loss function value vs. number of iterations: Simplex method (dotted line); Rosenbrock method (solid line); modified Rosenbrock method (dashed line).

changing the initialization (but it hardly occurs in practice); purely numerical methods like direct searches are of no use for any "theory" of inverse problems, which would be desirable in order to establish e.g. criteria of identifiability and for the design of experiments.

3. INVERSE PROBLEMS IN LINEAR MODELS BY LEAST SQUARES METHODS

3.1. Identification of elastic stiffness .

The familiar linear elastic analysis is centered on models of the type

$$\underline{K}\ \underline{u} = \underline{F} \tag{2}$$

where $\underline{u}$ and $\underline{F}$ are n-vectors of nodal displacements and equivalent nodal forces and $\underline{K}$ is the (symmetric, positive definite) stiffness matrix of the element aggregate. If the parameters to identify P_i $(i = 1...p)$ are elastic constants, say bulk and shear moduli of the materials, it is convenient to express $\underline{K}$ in the form [8] :

$$\underline{K} = \sum_{1}^{p}{}_i\ \underline{K}^i\ P_i \tag{3}$$

where $\underline{K}^i$ denotes the stiffness matrix calculated by setting the i-th parameter equal to 1 and all the others zero.

Let vector $\underline{u}$ be partitioned into the subvectors $\underline{u}_1$ of the components subjected to measurement and $\underline{u}_2$ of all other displacements. By partitioning (2) accordingly and performing a static condensation, one obtains:

$$(\underline{K}_{11} - \underline{D}\ \underline{K}_{21})\ \underline{u}_i = \underline{F}_1 - \underline{D}\ \underline{F}_2 \tag{4}$$

where

$$\underline{D} \equiv \underline{K}_{12}\ \underline{K}_{22}^{-1} \tag{5}$$

By substituting (3), partitioned in the same fashion, and assigning to $\underline{u}_1$ the measured values $\underline{u}^m$, eq.(4) becomes

$$\underline{V}\ \underline{P} = \underline{F}_1 - \underline{D}\ \underline{F}_2 \tag{6}$$

where the matrix $\underline{V}$ is formed by the columns:

$$\underline{V}_i \equiv (\underline{K}_{11}^i - \underline{D}\ \underline{K}_{21}^i)\ \underline{u}^m \qquad (i = 1...p) \tag{7}$$

Eq.(6) can be solved with respect to $\underline{P}$ in the least squares sense, i.e. minimizing

$$\min \{ \omega \equiv \| \underline{V}\,\underline{P} - \underline{F}_1 + \underline{D}\,\underline{F}_2 \|^2 \} \tag{8}$$

provided the number m of the measures is not less than the number p of the parameters. Matrix $\underline{D}$ defined by eq.(5) depends on $\underline{P}$ so that the objective function in (8) is not quadratic. This suggests to solve problem (8) iteratively as follows [8] . The best guess $\underline{P}^o$ of $\underline{P}$ is used to evaluate $\underline{D}$ and to solve (8) taking $\underline{D}$ constant; this means to solve the linear equation:

$$\tilde{\underline{V}}\,\underline{V}\,\underline{P} = \tilde{\underline{V}}(\underline{F}_1 - \underline{D}\,\underline{F}_2) \tag{9}$$

Eq.(9) supplies a new vector $\underline{P}^1$, which replaces $\underline{P}^o$ in a repeated operation cycle. The step sequence terminates when some measure of the improvement of $\underline{P}$ goes below an assigned tolerance.

3.2. Applications.

The above procedure was adopted in [9] for the identification of the shear and bulk moduli of the rock mass beneath a hypothetical earth dam (fig.3a) on the basis of vertical and horizontal displacements measured in the underlaying rock (squares in fig. 3b) at the completion of a cofferdam, before the construction of the main dam. The same problem was solved by the simplex method. The results are shown in fig.4 . The least square iterative method (curves L) turns out to exhibit faster and smoother convergence than the more general simplex procedure (curves S).

3.3. Identification of interaction pressures.

Mathematical models ameneable to eq.(2) are linear from the geomechanics standpoint; they become "linear" also in the identification terminology, when the parameters $\underline{P}$ to identify are linearly contained in the load vector $\underline{F}$, so that the dependence of the measurable quantities $\underline{u}_1$ on $\underline{P}$ becomes linear as well. This twofold linearity, rare in structural identification, occurs in geotechnical engineering when the interaction distribution between a structure (assumed as elastic) and the surrounding soil is sought on the basis of displacements measured on the former. Note that to this purpose the soil behaviour is immaterial. Various specific problems of this category, concerning tunnel liners and earth retaining walls have been solved in [10] by means of (direct, not iterative) least squares procedure.

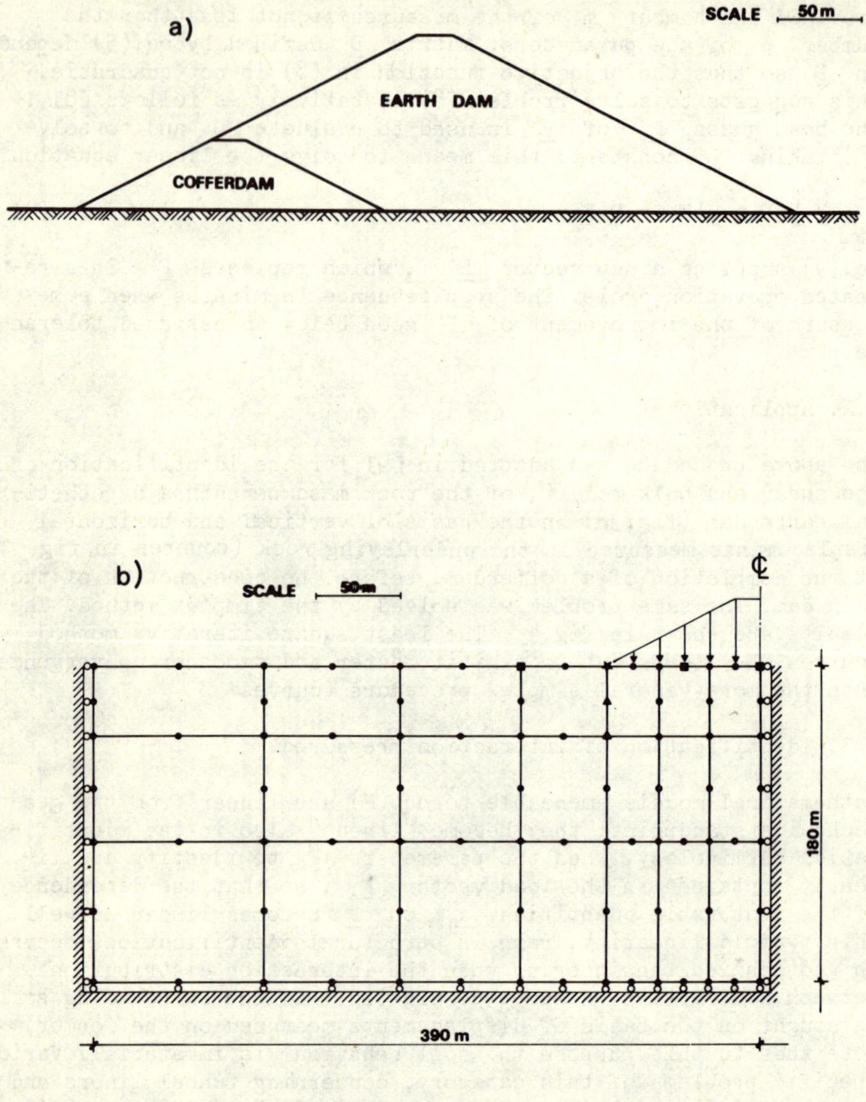

Figure 3. (a) Schematic representation of the example geometry.
(b) Finite element mesh and load distribution adopted for the identification of the elastic moduli.

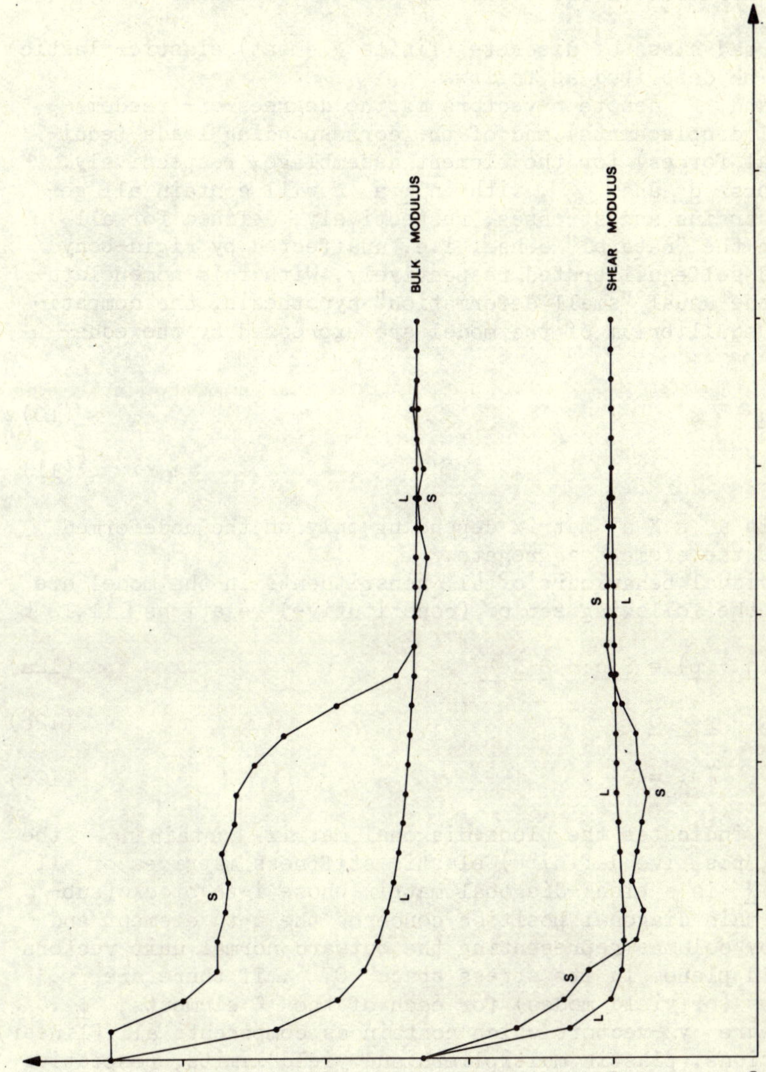

Figure 4. Variation of the bulk and shear moduli during the error minimization process (L refers to least square minimization; S refers to Simplex method).

4. IDENTIFICATION OF PIECEWISELINEAR ELASTOPLASTIC MODELS BY QUADRATIC AND NONLINEAR PROGRAMMING

4.1. Model formulation for perfectly plastic behaviour.

A fairly broad class of discrete (finite element) elastic-plastic models can be described as follows.

Let $\underline{u}$ and $\underline{F}$ denote n-vectors of the degrees-of-freedom (free nodal displacemnts) and of the corresponding loads (equivalent nodal forces) for the element assemblage, respectively. The $\bar{n}$-vectors $\underline{q}$ and $\underline{Q}$, with $\bar{n} \geq n$, will contain all generalized strains and stresses, respectively, defined for all ℓ elements in the "natural" sense, i.e. unaffected by rigid-body motions and selfequilibrated, respectively. With this nomenclature, under the usual "small deformation" hypothesis, the compatibility and equilibrium of the model are expressed by the equations:

$$\underline{q} = \underline{C}\,\underline{u} \tag{10}$$

$$\tilde{\underline{C}}\,\underline{Q} = \underline{F} \tag{11}$$

where $\underline{C}$ is a $\bar{n} \times n$ matrix depending only on the undeformed geometry of the element aggregate.

The individual behaviours of all constituents in the model are defined by the following set of (constitutive) relations [11,12]:

$$\underline{Q} = \underline{S}\,(\underline{q} - \underline{p}) = \underline{S}\,\underline{q} - \underline{S}\,\underline{N}\,\underline{\lambda} \tag{12a}$$

$$\underline{\phi} = \tilde{\underline{N}}\,\underline{Q} - \underline{r} \leq \underline{0} \tag{12b}$$

$$\underline{\lambda} \geq \underline{0}\quad,\quad \tilde{\underline{\phi}}\,\underline{\lambda} = 0 \tag{12c}$$

In (12) $\underline{S}$ indicates the block-diagonal matrix containing the (symmetric, positive definite) elastic stiffness matrices of all elements; $\underline{N}$ is a block-diagonal matrix whose i-th block (submatrix) in main diagonal position concerns the i-th element and is formed by columns representing the outward normal unit vectors to the yield planes in its stress space $\underline{Q}^i$. If there are y yield planes (or yield modes) for each of the ℓ elements, $\underline{\phi}$, $\underline{\lambda}$ and $\underline{r}$ are $y\ell$-vectors which contain as components all (linear) yield functions, plastic multipliers and yield limits, respectively. The $\bar{n}$-vector of (generalized) plastic strains for all elements is $\underline{p} = \underline{q} - \underline{e} = \underline{N}\,\underline{\lambda}$, so that eq.(12) implicitly enforces the normality rule of associative plasticity theory ("stability" in Drucker's sense). The relation set (12) provides a compact analytical description of <u>piecewiselinear, holonomic</u>, elastic-perfectly plastic behaviour of all elements in the model. The meaning of this statement is clarified in the subsequent Sec. 4.2.

The determination of the stress, strain and/or displacement

response of the above modelled mechanical system to given loads $\underline{F}$, is the solution of the analysis problem. Substitute vectors $\underline{q}$, $\underline{Q}$ from the governing relations (10)-(12a) and set:

$$\underline{K} \equiv \underline{\tilde{C}}\,\underline{S}\,\underline{C}\,, \quad \underline{u}^e \equiv \underline{K}^{-1}\,\underline{F}\,, \quad \underline{G} \equiv \underline{K}^{-1}\,\underline{C}\,\underline{S}\,\underline{N} \tag{13}$$

The displacement vector is thus expressed as:

$$\underline{u} = \underline{u}^e + \underline{G}\,\underline{\lambda} \tag{14}$$

Substitute now $\underline{Q}$ in eq.(12b) through (12a), (10) and (14) and set:

$$\underline{Q}^e \equiv \underline{S}\,\underline{C}\,\underline{K}^{-1}\,\underline{F}\;;\quad \underline{A} \equiv -\underline{\tilde{N}}\,\underline{Z}\,\underline{N} = \underline{\tilde{N}}\,\underline{S}\,\underline{N} - \underline{\tilde{G}}\,\underline{K}\,\underline{G} \tag{15}$$

Thus the analysis problem is reduced to the determination of the plastic multiplier vectors $\underline{\lambda}$ which satisfy the relations [11,12]:

$$\underline{\phi} = \underline{\tilde{N}}\,\underline{Q}^e - \underline{A}\,\underline{\lambda} - \underline{r} \tag{16a}$$

$$\underline{\phi} \leq \underline{0}\,,\quad \underline{\lambda} \geq \underline{0}\,,\quad \underline{\tilde{\phi}}\,\underline{\lambda} = 0 \tag{16b}$$

Relations (16) form, in the mathematical jargon, a "linear complementarity" problem (LCP). Since it is associated to the positive semidefinite matrix $\underline{A}$, the LCP (16) is fully equivalent, by virtue of the theorem of Kuhn and Tucker, to the quadratic programming (QP), as it has been pointed out in [11]:

$$\min \{\tfrac{1}{2}\,\underline{\tilde{\lambda}}\,A\,\underline{\lambda} - \underline{\tilde{\lambda}}(\underline{N}\,\underline{Q}^e - \underline{r})\} \tag{17a}$$

$$\text{subject to: } \underline{\lambda} \geq \underline{0} \tag{17b}$$

The physical meaning, coverage and limitations of the preceding class of mathematical models are easily understood with reference to particular cases as those which follow.

(a) The foundation beam of fig.5a is idealized by assuming Winkler soil and elastic perfectly-plastic behaviour for both the soil and the beam in bending. The unilateral nature of the support, i.e. the possibility of lifting-off, is allowed for. The law assumed to govern the i-th equivalent "spring" is depicted in fig.5b and is readily seen to be described by a relation set of type (12) marked by index i, provided that $\underline{Q}^i$, $\underline{S}^i$, $\underline{q}^i$, $\underline{p}^i$, $\underline{e}^i$, become scalars, and $\underline{\phi}^i$, $\underline{\lambda}^i$, $\underline{r}^i$ are vectors of two components, one for each of the two "yielding modes" (for the positive one corresponding to lifting-off, clearly $r_1^i = 0$); $\underline{N}^i = [1, -1]$. Similar remarks hold for the flexural behaviour of the beam represented in fig.5c. If this is interpreted by a lumped-compliance model, i.e. as an articulated, piecewise-rigid system with rotational elastoplastic springs at the hinges or nodes, to where the contact with the supporting springs is restricted, then eqs.(12)

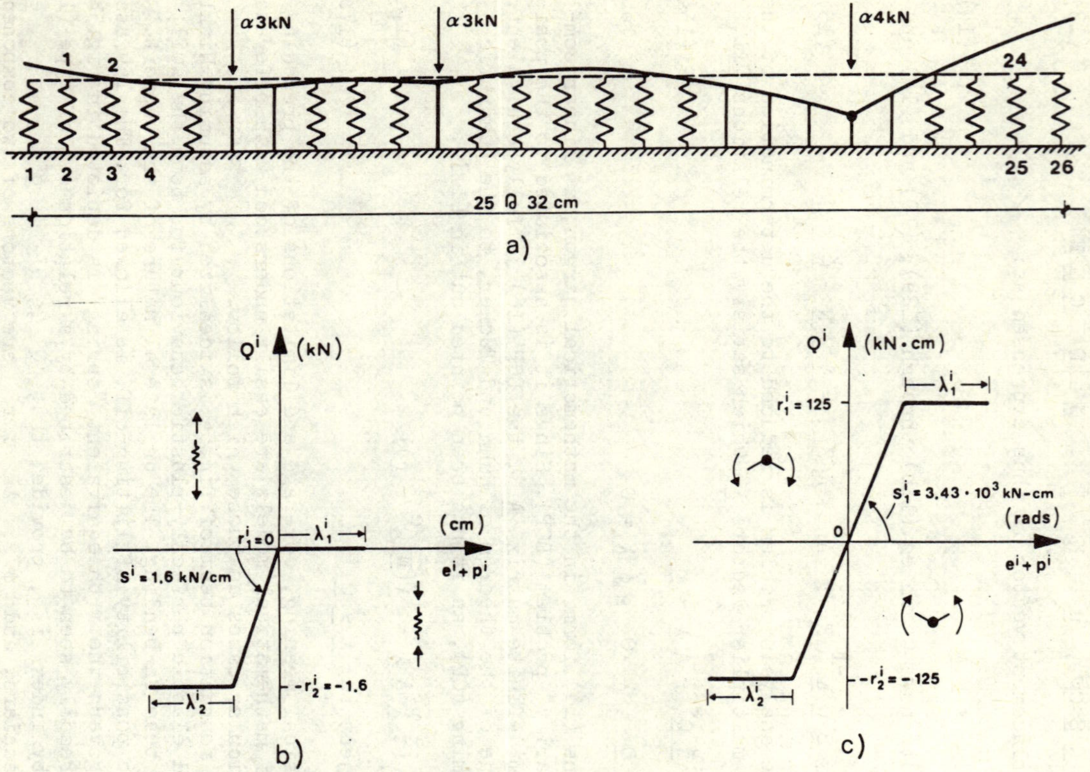

Figure 5. (a) Foundation beam on Winkler soil.
(b) Force-displacement law for the springs.
(c) Bending moment-rotation law for the beam.

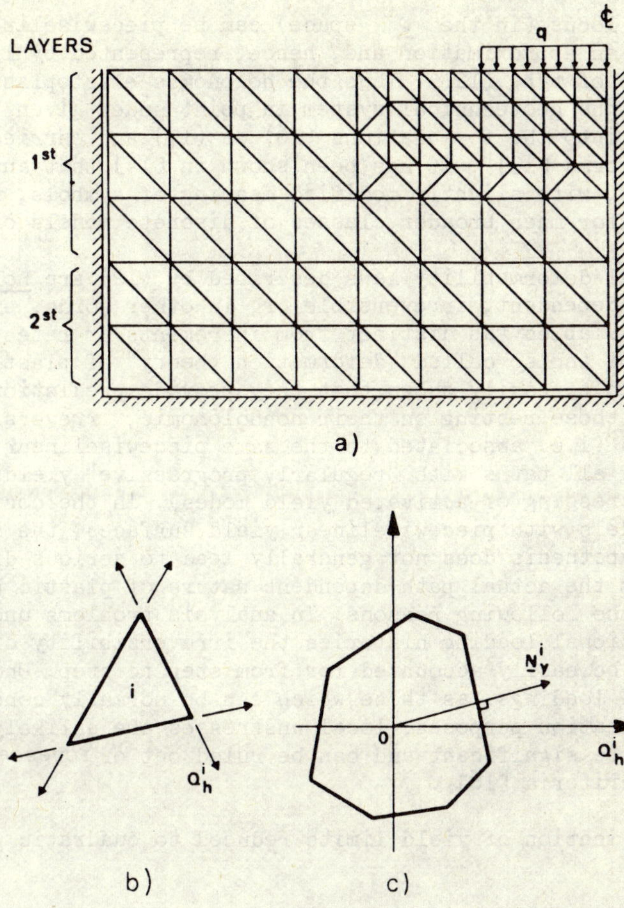

Figure 6. (a) Discretization by C.S.T. elements.
(b) i-th element and side forces as generalized natural stresses.
(c) Yield locus of the i-th element.

can be conceived as a condensed matrix formulation of the constitutive laws of all deformable elements, whereas Eqs.(10)-(11) can be written immediately.

(b) Let a two layer soil deposit subjected to a strip load be interpreted as a plane strain system modelled as the aggregate of constant strain triangle (CST) finite elements shown in fig.6a . For these elements the components of strain and stress in the plane are related by one-to-one contragradient correspondence to the three natural generalized strains (side elongations) and stresses (side forces shown in fig.6b), respectively; therefore, the material constitutive laws can be translated into an "element law", whose yield locus (in the $\underline{Q}^i$ space) can be piecewiselinearized (fig.6c) as an approximation and, hence, represented by linear inequalities of type (12b). Thus the holonomic elastoplastic analysis of the geotechnical system in point under given loads can be cast into the formulations (16) or (17) and carried out by QP algorithms [13] . It has been shown in [14] that analogous formulations, with slightly modified meaning of symbols, may be constructed for much broader classes of discrete models of continua.

(c) The local deformability laws described by (12) are holonomic, i.e. path-independent, irreversible or, in other words, expressed in finite variables (as distinct from increments or rates) in the spirit of the so called "deformation theory" of plasticity [15] . It can be easily shown that they provide predictions coincident with those resting on their nonholonomic, irreversible counterparts (i.e. associated to the same piecewiselinear yield locus) along all paths with "regularly progressive" yielding (i.e. without unstressing of activated yield modes). In the context of discrete models with piecewiselinear yield surfaces, the above holonomic hypothesis does not generally lead to serious discrepancies from the actual path-dependent nature of plastic behaviour because of the following reasons. In analysis problems under stepwise proportional loading histories the irreversibility of plastic strains can be easily accounted for from step to step. Under proportional loadings, as those which can be normally considered for identification purposes, local unstresses are unlikely to occur or to be significant and can be ruled out or foreseen by "a priori" criteria [16] .

4.2. Identification of yield limits reduced to quadratic programming.

The elastic and geometric properties of the system are assumed here to be known, whereas the local resistance properties (yield limits) have to be determined by an identification process. Measurements on the real system subjected to given loads within the nonlinear range are assumed to provide some (say m) displacements collected in vector $\underline{u}^m$. Let $\underline{u}^c = \underline{B}\,\underline{u}$ denote the vector of the corresponding (nodal) displacements calculated by means of the mathematical model formulated in Sec.4.1 , $\underline{B}$ being a binary matrix which selects the measurable components in $\underline{u}$
The loss function to minimize can be defined as:

$$\omega \equiv (\underline{\tilde{u}}^c - \underline{\tilde{u}}^m)(\underline{u}^c - \underline{u}^m) \tag{18}$$

By substituting (14) and setting:

$$\underline{M} \equiv \underline{\tilde{G}}\,\underline{\tilde{B}}\,\underline{B}\,\underline{G} \;, \quad \underline{b} \equiv 2\,\underline{\tilde{G}}\,\underline{\tilde{B}}\,(\underline{B}\,\underline{u}^e - \underline{u}^m), \qquad c \equiv (\underline{\tilde{u}}^m - \underline{\tilde{u}}^e\,\underline{\tilde{B}})(\underline{u}^m - \underline{B}\,\underline{u}^e) \tag{19}$$

ω becomes a convex quadratic function of $\underline{\lambda}$:

$$\min \ \{ \omega(\underline{\lambda}) \equiv \tilde{\underline{\lambda}} \ \underline{M} \ \underline{\lambda} + \tilde{\underline{b}} \ \underline{\lambda} + c \} \qquad (20a)$$

If the parameters $\underline{P}$ to identify govern the yield limits through a linear relation $\underline{r} = \underline{R} \ \underline{P}$ (sometimes $\underline{R}$ will be a Boolean identification matrix) and are bracketed by known or conjectured bounds $\underline{P}^L$, $\underline{P}^U$, the optimization constraints are, account taken of (4.7a):

$$\underline{\phi} = \tilde{\underline{N}} \ \underline{Q}^e - \underline{A} \ \underline{\lambda} - \underline{R} \ \underline{P} \leq \underline{0} \qquad (20b)$$

$$\underline{\lambda} \geq \underline{0} \ , \ \underline{P}^L \leq \underline{P} \leq \underline{P}^U \ , \tilde{\underline{\phi}} \ \underline{\lambda} = 0 \qquad (20c)$$

It was demonstrated in [17] that this nonlinear programming problem (NLP), made difficult to solve by the nonconvex "complementarity" constraint $\tilde{\underline{\phi}} \ \underline{\lambda} = 0$, is equivalent to the following <u>nonconvex quadratic program</u>:

$$\min \ \{\psi(\underline{\lambda},\underline{\phi}) \equiv \omega(\underline{\lambda}) - \rho \ \tilde{\underline{\phi}} \ \underline{\lambda}\} \quad \text{subject to:} \qquad (21a)$$

$$\underline{\phi} = \tilde{\underline{N}} \ \underline{Q}^e - \underline{A} \ \underline{\lambda} - \underline{R} \ \underline{P} \leq \underline{0} \ , \underline{\lambda} \geq \underline{0} \ , \ \underline{P}^L \leq \underline{P} \leq \underline{P}^U \qquad (21b)$$

where ρ represents a sufficiently large positive number. The equivalence of problems (20) and (21) can be rigorously proved by a rather long path of reasoning (here omitted for brevity) starting from recent mathematical results [18]. It looks intuitive, as the bilinear form $\tilde{\underline{\phi}} \ \underline{\lambda}$ is nonnegative over the feasible domain (21b) and, hence, the complementarity relation is enforced by the minimization process.

4.3. A solution method.

The QP problem (21) turns out to be more tractable because its constraints are all linear. However the nonconvexity, removed from the constraints, shows up in the objective function and makes the numerical solution still non trivial. The two-phase "ad hoc" method developed in [17] is briefly outlined below.

<u>Phase 1</u>. Anyone of the familiar algorithms for QP or NLP with linear constraints applied to (21) generates a local minimum $\bar{\underline{P}}$, $\bar{\underline{\lambda}}$. If the corresponding value of the objective function is judged to approach zero in a manner consistent with the experimental and modelling inaccuracies, then the solution process is terminated as the local minimum achieved is global. Otherwise another phase follows.

<u>Phase 2</u>. <u>Convex</u> QP problems depending on binary variables (vector $\underline{\beta}$) as parameters are constructed as follows:

$$\min \ \{\tilde{\underline{\lambda}} \ \underline{M} \ \underline{\lambda} + \tilde{\underline{c}}'(\underline{\beta})\underline{\lambda} + \tilde{\underline{c}}''(\underline{\beta})\underline{\phi} \ \} \ \text{subject to (21b)} \qquad (22)$$

by setting:

$$\underline{c}' \equiv \underline{b} + \rho' \underline{\beta} \quad , \quad \underline{c}'' \equiv \rho'' (\underline{U} - \underline{\beta}) \tag{23}$$

where ρ' and ρ'' are large positive numbers and $\underline{U}$ a vector with all elements equal to 1. This problem can be conceived as generated from (20) by enforcing complementarity through binary complementarity penalization of the variables ϕ and λ in the objective (instead of imposing it explicitly in the constraints). It can be proved [17], as it is intuitive, that the minimum over all binary vectors $\underline{\beta}$, of the minima of problems (22), equals the minimum of (20). The procedure is initialized by deriving a vector $\overline{\underline{\beta}}$ from the solution ($\overline{\underline{P}}$, $\overline{\underline{\lambda}}$, $\overline{\underline{\phi}}$) of phase 1 through the rule of setting $\overline{\beta} = 1$ for any zero component of $\overline{\underline{\lambda}}$, $\overline{\underline{\beta}} = 0$ otherwise. The solution of the dual to QP (22) for $\underline{\beta} = \overline{\underline{\beta}}$ permits to single out through linear inequalities, on the basis of duality theory of QP, a set $\overline{A}$ of binary vectors such that the minimum of the QP (22) corresponding to any $\underline{\beta} \in \overline{A}$ is not less than the minimum of the QP corresponding to $\overline{\underline{\beta}}$. Hence, the set $\overline{A}$ is ruled out from further consideration. A vector $\underline{\beta}'$ not belonging to $\overline{A}$ is chosen and a new cycle centered on the corresponding convex QP is performed, and so on. This decomposition procedure is easily seen to terminate at the solution of NLP (20) in a finite number of steps. Various termination criteria (no eligible $\underline{\beta}$ left out of the union of successive sets $\overline{A}$; close upper and lower bounds on the global minimum) and various improvements have been envisaged in [17], where also alternative, substantially different approaches were proposed.

4.4. Computational aspects and experience.

Several numerical tests using the above method have been performed and partly expounded in [17] with reference to the foundation beam model described in Sec. 4.1, fig.5a, assuming: 26 degrees of freedom; 50 deformable elements (26 springs, 24 "critical sections" or elastoplastic hinges) and, hence, 100 yield modes; 3 parameters to identify, i.e. sagging and hogging yield limit in the beam, limit compression of the springs. Error-free experimental data were simulated by solving preliminarily the elastoplastic analysis problem (17) under the given loads by means of the same mathematical model, on the basis of assumed parameters to be recovered by the identification process.

On the basis of this limited experience, but also of the mathematical and mechanical nature of the problem and solution procedures, the following conclusions can be drawn.

(a) The two-phase method (quadratic programming and decomposition procedure) normally requires a computational effort of the same order of magnitude as a single analysis by QP. In other terms, the local minimum generated by the first phase is most likely to represent the global minimum, particularly if the parameters are sought within ranges preliminarily guessed by engineering judgement as it is always possible in practice.

(b) If the number and locations of displacement measurements are affected by the local yielding processes and if these activate the yield limits to identify, the identification conducted by the proposed method leads to results which are unique and stable with respect to disturbances of the experimental data (as shown by sensitivity studies of the type mentioned in Sec.5).

(c) The elastoplastic inverse problem in the context of piecewiselinear holonomic discrete elastoplastic models exhibit a typical, new mathematical structure (quadratic minimization under complementarity constraints) susceptible to a variety of solution approaches not fully investigated yet.

4.5. Model formulation for linear hardening behaviour.

The perfectly plastic idealization is often inadequate and grossly unrealistic in describing the local behaviour of geotechnical systems. In the context of piecewiselinear constitutive laws, strainhardening (or softening) can be allowed for by introducing in the expression (12b) of the yield functions $\underline{\phi}$ a term which depends linearly on the plastic multipliers $\underline{\lambda}$ and alters the yield limits from the original values $\underline{r}$:

$$\underline{\phi} = \tilde{\underline{N}}\,\underline{Q} - \underline{r} - \underline{H}\,\underline{\lambda} \qquad (24)$$

where the matrix of hardening moduli $\underline{H}$ defines the (linear) hardening rule. Considering a single, say the i-th, deformable element with one natural generalized strain component and two yielding modes such as the Winkler spring in fig.5a, we have $\underline{H}^i = \mathrm{diag}\,[H^i_1 , H^i_2]$, $H^i_{1,2}$ being the slope in the $Q^i \div p^i$ plane of the hardening branch 1 or 2. For multicomponent constituents such as the CST finite elements of fig.6a: $\underline{H}^i = \mathrm{diag}\,[H^i_j]$ corresponds to non interacting, independently translating yield planes in the Q^i space (Koiter's rule), $H^i_j < 0$ implying local shrinking of the yield surface at the activation (softening); $\underline{H}^i = h\,\tilde{\underline{N}}^i\,\underline{N}^i$, h being a scalar, represents Prager's kinematic hardening rule (yield surfaces translating in the direction of $\dot{\underline{p}}$). These and other representations of linear or piecewiselinear hardening were developed in [11] .

Clearly, the analysis problem can still be formulated as the LCP (16), provided the definition (15) of matrix $\underline{A}$ is replaced by:

$$\underline{A} = \underline{H} - \tilde{\underline{N}}\,\underline{Z}\,\underline{N} \qquad (25)$$

4.6. Identification of yield limits and hardening moduli.

If the parameters $\underline{P}$ to identify in the model (16) govern the hardening moduli contained in matrix $\underline{H}$ of (25), alone or together with the yield limits $\underline{r}$, the minimization of the loss function, problem (20), becomes substantially different, inasmuch the inequality constraints (20b) are no longer linear, nor convex. The modified formulation (21) now involves a nonquadratic nonconvex objective function, as readily seen by substituting vector $\underline{\phi}$ in (21a). Starting from this formulation, in [19] the following two solution procedures were proposed and tested with reference to the foundation beam model of fig.5a (two yield limits and two hardening moduli to identify) and on the basis of a set of simulated experiments similar to those mentioned in Sec.4.4.
(a) A feasible-conjugate-direction algorithm for general (nonconvex) NLP problems [20] . Computing times turn out to be much longer than the corresponding analysis problem, but satisfactory convergence towards stable results is generally found.
(b) An iterative procedure called "least-squares, linear-Taylor differential-correction" technique [21] . This is easily programmable and requires rather limited storage, but its computing times and convergence are satisfactorily compared to (a) only if "a priori" knowledge permits to initialize the solution procedure not far from the actual values of the sought parameters.

More extensive numerical experience in the domain of large size problems, and possibly recourse to alternative methods, would be needed for any conclusive assessment of merit.

5. ON THE ROLE OF AND ALLOWANCE FOR UNCERTAINTIES

5.1. Influence of uncertainties of the experimental data.

Perhaps to a larger extent than in any other branch of engineering, uncertainties affect most quantities the geotechnical engineer has to deal with, not only those which are or can be selected as unknowns in an identification problem. In the face of this circumstance, many formal tools for the analysis of randomness have been recently developed from probability theory and statistics, as aids to, not substitute for, geotechnical design experience and sound engineering judgement. As parts of these developments, stochastic identification methods have been successfully applied to some soil and rock systems [22] , [23] .

This Section is intended to provide nothing more than some introductory remarks to this important topic, as a critical supplement to the preceeding discussion of optimization techniques in inverse problems. Reference is made to the following simple situation [24]. A weathered rock mass contains an inclusion of softer material. A plane-strain linear elastic finite element model is constructed (fig.7) and used first to simulate a plate load test for generating

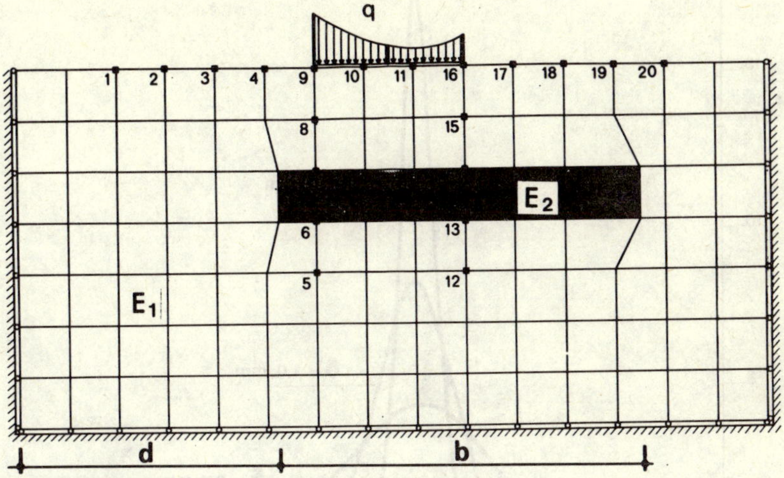

Figure 7. Finite element model for the Plate Load Test.

"experimental" data: these are m displacements calculated at the nodes numbered in fig.7 (or at some of them), when the elastic moduli and the geometry parameters are assigned the "real" values. The variability of the experimental data is simulated by adding to the m-vector $\underline{u}^m$ of the above displacements a vector $\Delta\,\underline{u}$ of "errors", produced by m independent generators of random numbers with zero mean value and Gaussian probability distribution. A first series of numerical tests was performed considering $m = 10$ measurements and three samples, each of 1000 error vectors, with standard deviations of: 1/6, 1/3, 2/3 mm ; these are equivalent, in engineering terms, to inaccuracies or instrument "resolutions" δ of $\pm 1/2$, 1, 2 mm, respectively. The maximum and minimum component in the vector $\underline{u}^m$ of mean values of the measured displacement were 12,0 and 0,35 mm, respectively.

The inverse problem considered on the basis of these "noisy" experimental data is the identification of the elastic moduli of the rock mass E_1 and of the lens E_2. The solutions obtained by (a simplified version of) the optimization procedure of Sec.3 (least squares, iterative) for all thousand measure vectors $\underline{u}^m + \Delta\underline{u}$, obviously lead to the estimates of the moduli as mean values (which always practically coincided with the values assumed in the analyses simulating experiments). The same samples of solutions permit to construct the histograms of the identified parameters. Fig.8 shows the log-normal distributions obtained from the histograms by means of a best-fitting procedure (a slight lack of symmetry with respect to the mean value suggests to assume log-normal instead of normal distribution). These frequency diagrams visualize the well expected circumstance that the modulus E_i exhibits less scattering (i.e. its identification is less

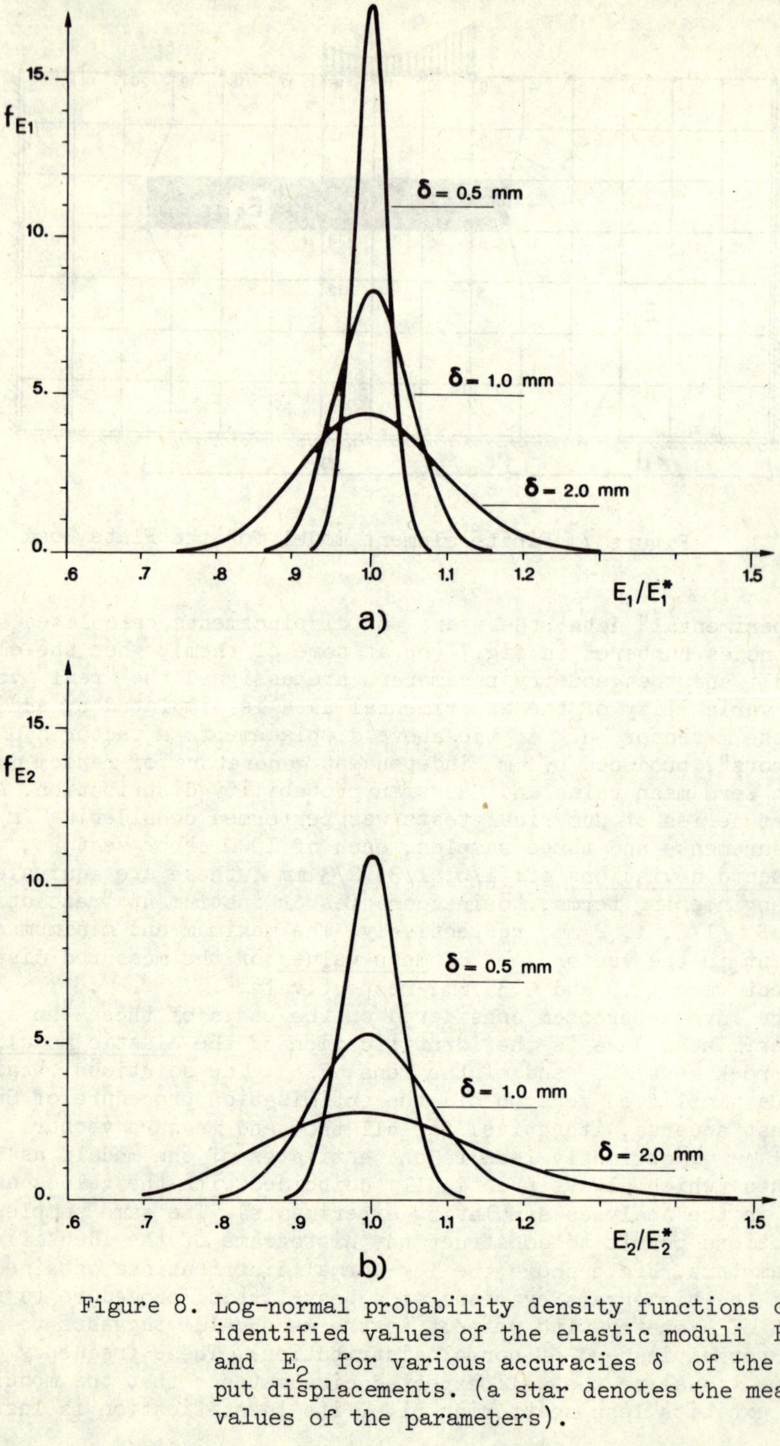

Figure 8. Log-normal probability density functions of the identified values of the elastic moduli E_1 and E_2 for various accuracies δ of the input displacements. (a star denotes the mean values of the parameters).

uncertain, or has a higher confidence level) than the modulus E_2 of the inclusion, since the former has a larger influence on the measured displacements. The standard deviations evaluated from the above mentioned best fitting log-normal distributions are indicated in fig.9 , for three different numbers of measurement (m = 6, 10, 20). These plots visualize how an increase of the experimental data basis improves the accuracy of the identification. It is also apparent that the ratios of the standards deviations to the mean values for the identified parameters increase more than linearly with the measure δ of the experimental data dispersion, as expected.

Results of this type are useful both for a critical assessment of the identification and for the design of the experiments. E.g., one can read in fig.9 that, in the case in point, either 6 "in situ" displacement measurements with resolution ±1 mm or 20 with ±2 mm, are needed in order to identify the Young modulus of the inclusion with a nondimensional standard deviation of 0,1 [24] .

Clearly, the above elementary, empirical approach resting on repeated deterministic identification becomes prohibitive for most practical problems. The preceding considerations, however, focus the interest of formal methods of parameter identification capable to take into account random variability.

5.2. An iterative procedure for stochastic identification of static systems.

Methods of stochastic identification have been applied to geotechnical uncertain systems which vary in time [22] , [23] . The models considered in the preceding Sections 3 and 4 are statical (as distinct from dynamical), in the sense that they correlate output and input vectors which are constant in time. For this kind of simpler models, a stochastic version of the inverse problems dealt with in Sections 3 and 4 by optimization techniques can be tackled by the approach briefly outlined below and extensively discussed in [25] , [26] .

The stochastic experimental information to be exploited for the identification will now be assumed to be a vector $\underline{u}^m$ of "measured" values (or mean values derived from samples of measurements) and the relevant covariance matrix $\underline{C}_u$. More specifically, the measured displacements are conceived as affected by random errors $\Delta\underline{u}$, i.e. $\underline{u}' = \underline{u}^m + \Delta\underline{u}$, with zero mean $E[\Delta\underline{u}] = \underline{0}$ and specified covariance $\underline{C}_u = E[\Delta\underline{u}\,\Delta\tilde{\underline{u}}]$. Usually $\underline{C}_u$ is diagonal, i.e. the experimental data are regarded as statistically independent; the main diagonal entries represent the variances assessed by the experimentalist for the measurements.

In the absence of, or prior to, any identification process, let the parameters be assigned some values $\underline{P}_o$ on the basis of engineering judgement, along with some measures of their uncertainties in form of variances. These estimates will be collected

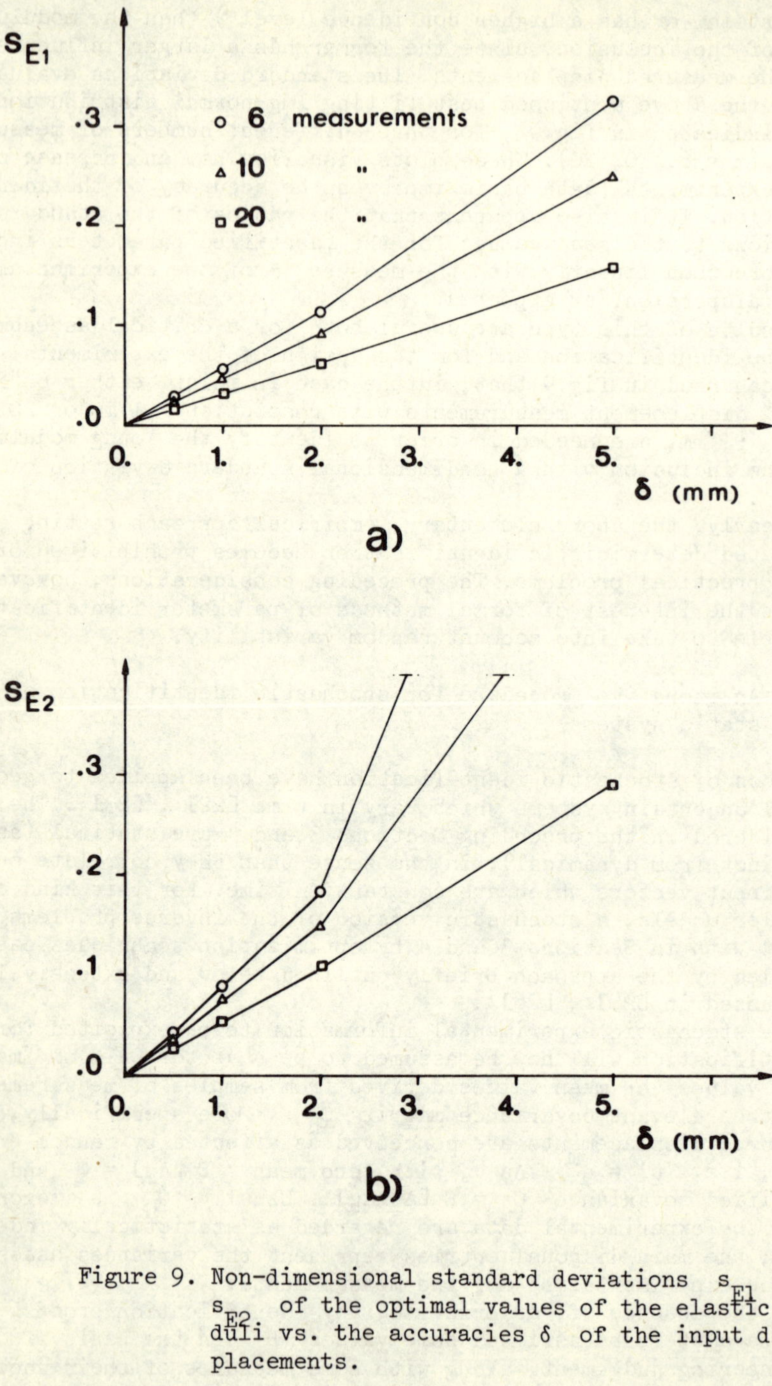

Figure 9. Non-dimensional standard deviations s_{E1} and s_{E2} of the optimal values of the elastic moduli vs. the accuracies δ of the input displacements.

in a diagonal "a priori" covariance matrix $\underline{C}_p^o$ and the parameters initially regarded as normally distributed random quantities with mean $\underline{P}_o$. On the basis of this information the estimation theory would provide a classical procedure of stochastic identification <u>if</u> the model were linear in the parameters, i.e. if it were possible to write(the prime reminds that only the displacements susceptible to measurements are collected here in the displacement vector $\underline{u}'$):

$$\underline{u}' = \underline{L}\,\underline{P} \tag{26}$$

$\underline{L}$ being a suitable matrix.
Precisely, see e.g. [27] , the "best" estimated values $\underline{P}'$ of $\underline{P}$, i.e. those with minimum variances, can be proved to be expressed in the form:

$$\underline{P}' = \underline{P}_o + \underline{W}_o\,(\underline{u}^m - \underline{L}\,\underline{P}_o) \tag{27}$$

where the "mapping matrix" $\underline{W}_o$ reads:

$$\underline{W}_o = \underline{C}_p^o\,\tilde{\underline{L}}\,(\underline{L}\,\underline{C}_p^o\,\tilde{\underline{L}} + \underline{C}_u)^{-1} \tag{28}$$

The covariance matrix associated to the best estimator (27) is defined by the expression [26] :

$$\underline{C}'_p = \underline{C}_p^o - \underline{W}_o\,\underline{L}\,\underline{C}_p^o$$

Since the mathematical models to calibrate in the present context are not linear in the parameters, it is quite natural to apply the above notions iteratively, along a sequence of linearized steps. This linearization is carried out by truncating after the linear terms a Taylor's series expansion (around the current best estimates of the parameters) of the dependence, through the model, of the measured quantities on the parameters. Thus, at the h-th step one writes:

$$\underline{u}' = \underline{u}'\,(\underline{P}_h) + \underline{L}\,(\underline{P}_h)\,(\underline{P} - \underline{P}_h) \tag{30}$$

where $\underline{L}$ is no longer constant as in (26), but represents a "sensitivity matrix" to be calculated by means of analytical or numerical (secant) derivatives.

This procedure, already successfully used in structural dynamics, [28] , [29] , is briefly outlined below.

1. Calculate, through the model, the displacements $\underline{u}_o$ due to the given loads if the parameters were equal to $\underline{P}_o$. Calculate, through numerical derivatives, the current sensitivity matrix $\underline{L}\,(\underline{P}_o)$.
2. By assuming (30) for $h = 0$ as a linear model, set $\underline{u} = \underline{u}^m$ in it and apply the above optimal estimation theory; namely, calculate $\underline{W}_o$ by means of (28) and $\underline{P}_1$ and $\underline{C}'_p$ by (27) and (29), respectively.

3. Perform a convergence test. If this is not passed, go to 1 replacing $\underline{P}_o$ by $\underline{P}_1$ and subsequently to 2 using again C_P^o (not $\underline{C}'_P$); etc.

Note that the convergence criterion may concern the rate of change of the unknowns, not only the difference $\underline{u}^m - \underline{u}^c$, whose norm may not go below a preassigned tolerance because of modelling inaccuracies. The last estimated vector $\underline{P}^h = \underline{P}'$ represents the identified parameters; the covariance matrix $\underline{C}'_P$ gives an idea (not a quantitative measure) of the "quality" of the information achieved.

If no indications are available or considered about the uncertainties of the parameters, i.e. if the only random quantities are the measures, the mapping matrix $\underline{W}_o$ supplied by estimation theory becomes [26]:

$$\underline{W}_o = (\tilde{\underline{L}}\, \underline{C}_u\, \underline{L})^{-1}\, \tilde{\underline{L}}\, \underline{C}_u^{-1} \tag{31}$$

Then the method could be easily seen to coincide with a weighted least square procedure. This can be considered as the minimization of a "loss function" defined as in (1), (8) and (18), weighted by the inverse of the variances:

$$\omega \equiv (\tilde{\underline{u}} - \tilde{\underline{u}}^m)\, \underline{C}^{-1}\, (\underline{u} - \underline{u}^m) \tag{32}$$

5.3. Applications.

Consider first piecewiselinear models of the category dealt with in Sec 4 in a strictly deterministic context. The sensitivity matrix $\underline{L}$ relates the variations of the measurable displacements $\underline{u}$ to perturbations of the parameters $\underline{P}$ governing the yield limits $\underline{r}$; it can be evaluated as follows for the current vector $\underline{P}_h$ (at the h-th iteration step) [26]:

$$\underline{L}(\underline{P}_h) \equiv \left[\frac{\partial\, \underline{u}}{\partial\, \underline{P}}\right]_{\underline{P}_h} = \left[\frac{\partial\, \underline{u}}{\partial\, \underline{\lambda}}\right]_{\underline{P}_h} \cdot \left[\frac{\partial\, \underline{\lambda}}{\partial\, \underline{P}}\right]_{\underline{P}_h} \tag{33}$$

For given parameters $\underline{P}_h$ and, hence, given yield limits $\underline{r}_h$, the LCP (16) or the QP problem (17) can be solved by usual algorithms. In the solution of either of these analysis problems, let $\underline{\lambda}_h$ denote the subvector of the λ components which correspond to components $\phi = 0$. Account taken of (13c) and (14), if $\underline{G}_h$ indicates the submatrix of $\underline{G}$ corresponding to $\underline{u}'$ and $\underline{\lambda}_h$, we may simplify (33) into:

$$\underline{L}(\underline{P}_h) = \underline{G}_h \left[\frac{\partial\, h}{\partial\, \underline{P}}\right]_{\underline{P}_h} \tag{34}$$

since the yield modes which are inactive ($\phi < 0$) in $\underline{P}_h$ will be so for $\underline{P}_h + d\,\underline{P}$ and, hence, can be neglected in calculating the sensitivity matrix $\underline{L}$. Let a subscript h still denote subvectors and submatrices which correspond to the above defined $\underline{\lambda}_h$. Suppose for the moment that the principal submatrix $\underline{A}_h$ thus defined is nonsingular. Then one obtains from (20b) for $\underline{\phi}_h = \underline{0}$:

$$\underline{\lambda}_h = \underline{A}_h^{-1}(\underline{\tilde{N}}\,\underline{Q}^e)_h - \underline{A}_h^{-1}\underline{R}_h\,\underline{P}_h \tag{35}$$

This provides the derivative matrix in the espression (34) of $\underline{L}$, which thus acquires the explicit closed form [26] :

$$\underline{L}(\underline{P}_h) = -\underline{G}_h\,\underline{A}_h^{-1}\underline{R}_h \tag{36}$$

For perfectly-plastic holonomic models, matrix $\underline{A}$ according to eq.(15b) is symmetric positive semidefinite (not definite) and, hence, generally singular. This gives rise to two computationally important consequences:

(a) the principal submatrix $\underline{A}_h$ may be singular (it is so in rare but possible situations of lack of uniqueness or "bounded mechanisms" below collapse), cfr. [12] ;

(b) the iteration sequence may lead to a parametric vector $\underline{P}_h$ for which the LCP (16) is not solvable (mechanically speaking this means that the local strength of the system model is so low that the test loads exceed its carrying capacity).

In both these occurrences the provision of adding to $\underline{A}$ a diagonal matrix $\underline{H}$ with very small positive diagonal entries turned out to be an effective and easy remedy [26] . Clearly this computational device can be interpreted as a temporary assumption of hardening ("numerical hardening") in the model (see Sec. 4.5).

The theory summarized in Sec. 5.2 supplemented by the above "ad hoc" developments, has been applied in [26] to the local strength identification for the foundation beam model depicted in fig.5 and already referred to in the deterministic context (Sec.4). The assumed statistical "a priori" knowledge (prior to any measurement) consists of values for parameters $\underline{P}_o$ (beam limit moment, spring limit compression) assumed as mean values, and of their covariance matrix $\underline{C}_p$, which quantifies their uncertainty or variability and is diagonal, since the parameters are regarded as statistically idependent; this information are generated by engineering judgement. The measures $\underline{u}^m$ are provided by the experimentalist as mean values together with their variances which define the measurement accuracy and form the diagonal matrix $\underline{C}_u$. Some numerical data and results appear in fig.10; details can be found in [26] .

If compared with the deterministic identification of the same parameters on the basis of the same model and simulated experimental data, the present stochastic identification turns out to be more substantially laborious and costly, as it implies the solution of an analysis problem at each iteration step. However, it supplies richer, more meaningful results and permits, in the spirit

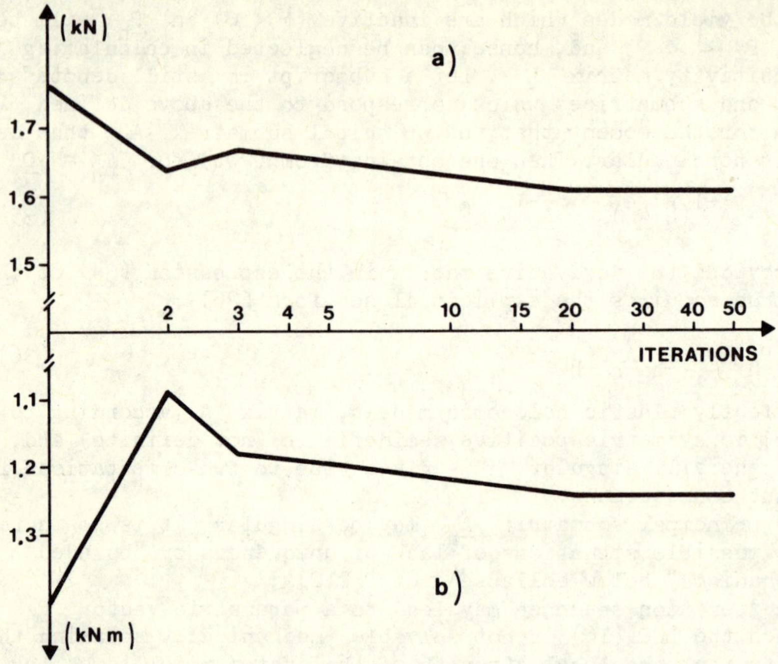

Figure 10. Parameter values vs. number of iterations for the stochastic identification of yield limits for the foundation beam problem depicted in fig.5 . (a) foundation compressive strength (b) beam yield limit

of Bayesian statistics, to exploit "a priori" engineering guesses and subjective evaluations and to combine them with experimental data.

The same stochastic Bayesian approach has been applied in [25] to the linear elastic model of fig.7 , by considering as parameters to identify first the elastic moduli of the inclusion E_2 and of the surrounding rock E_1 as in Sec. 5.1 ; subsequently, the moduli and simultaneously the two lenghts b and d of fig.7, which are geometric characteristics of the lens. The sensitivity matrices $\underline{L}$ for these problems have to be evaluated numerically by approximating derivatives through finite differences. Results concerning the former case are shown in fig.11. The latter (4 parameters) case led to results some of which appear in fig.12 .

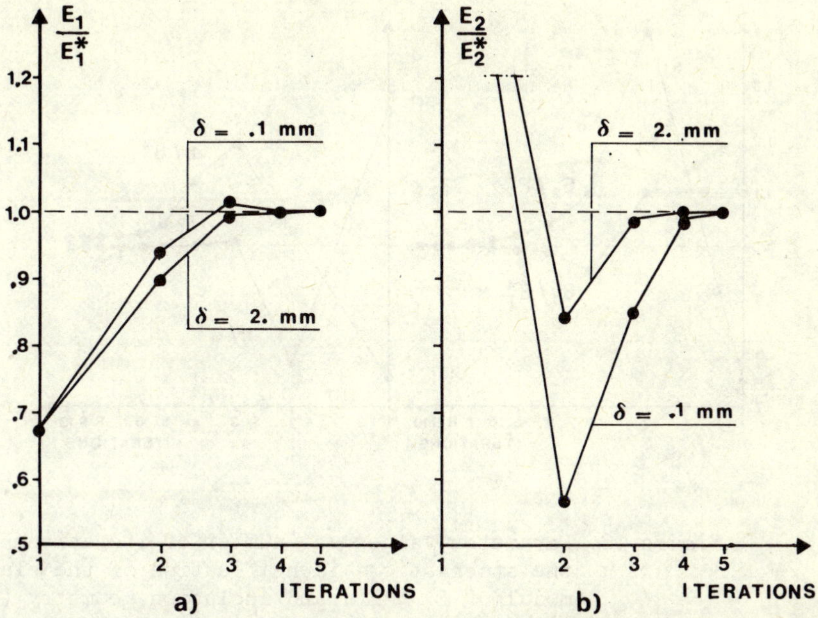

Figure 11. Stochastic identification of the elastic parameters for the problem shown in fig.7 ; elastic moduli E_1 (a) and E_2 (b) vs. number of iterations for different values of the measurement accuracy δ (a star denotes the "true" values of the parameters).

6. CONCLUSIONS

The diverse identification problems briefly surveyed in this paper concern mathematical models which are characterized by the following features: nonlinearity with respect to the parameters to identify and mostly to the output variables as well; the number of the latter quantities by far exceedes that of the former; the processes considered are not dynamic, also in the sense that the experimental data are used in "batch" rather than in sequential ("recursive") manner (besides the quasistatic, time-independent nature of the mechanical phenomena concerned). Parameter identification or calibration of mathematical models turns out to be generally more laborious and more demanding than analysis and design in terms of optimization techniques; direct search, quadratic programming and nonlinear-programming (both convex and nonconvex) are required, depending on the peculiarities of the problem. Piecewiselinear models have been found to form a special but noteworthy category where deterministic identification appears to be closer to analysis as for solution procedures and computational effort. However, mathematical programming techniques represent only one of several

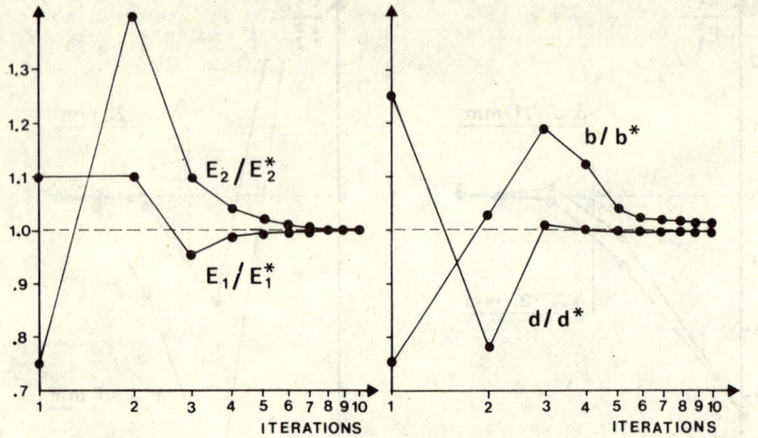

Figure 12. Parameter values vs. number of iterations for the stochastic identification of the elastic moduli (E_1 , E_2) and inclusion geometry (b, d) of the problem shown in fig.7 (measurement accuracy $\delta = \pm 1.$ mm; a star denotes the "true" values of the parameters).

conceptual ingredients and tools of potential use in system parameter estimation (in particular for geotechnical systems), a complex field where, at the present state of development, "each application exhibits elements of both art and sciences" (Eykhoff [2] p.28).

ACKNOWLEDGEMENT

This paper has been written in the framework of a research project sponsored by M.P.I.

REFERENCES

[1] Goodwin, G.C., and Payne, R.T., 1977, "Dynamic System Identification: Experiment Design and Data Analysis", Academic Press, New York.

[2] Eykhoff, P., 1974, "System Identification: Parameter and State Estimation", J.Wiley, New York.

[3] Bjerrum, L.: 1960, "Some Notes on Terzaghi's Method of Working", in 'From Theory to Practice in Soil Mechanics', J.Wiley, New York.

[4] Jurina, L., Maier, G. and Podolak, K.: 1977, "On Model Identification Problems in Rock Mechanics", Proc. Int. Symp. Geotechnics of Structurally Complex Formations, Capri, Vol.1, A.G.I.

[5] Himmelblau, D.M.: 1972, "Applied Nonlinear Programming", McGraw-Hill, New York.

[6] Gioda, G. and Maier, G.: 1980, "Direct Search Solution of an Inverse Problem in Elastoplasticity: Identification of Cohesion, Friction Angle and In Situ Stress by Pressure Tunnel Tests", Int. J. Num. Meth. Engrg., Vol.15, pp.1823-1848.

[7] Gioda, G.: 1979, "A Numerical Procedure for Defining the Values of Soil Parameters Aggecting Consolidation", Proc. 7th Europ. Conf. Soil Mech. Found. Engrg., Brighton.

[8] Kavanagh, K.T.:1973, "Experiment Versus Analysis: Computational Techniques for the Description of Static Material Response", Int. J. Num. Meth. Engrg., Vol.5, pp.503-515.

[9] Gioda, G.:1980, "Indirect Identification of the Average Elastic Characteristics of Rock Masses", Int. Conf. on Structural Foundations on Rock, Sydney, Australia.

[10] Gioda, G. and Jurina L.:1981, "Numerical Identification of Soil Structure Interaction Pressures", Int. J. Num. Anal. Meth. Geomechanics, Vol.5, pp.33-56.

[11] Maier, G.:1970, "A Matrix Structural Theory of Piecewise-Linear Plasticity with Interacting Yield Planes", Meccanica, Vol.5, No.2, pp.54-66.

[12] Cohn, M.Z. and Maier, G.:1979, "Engineering Plasticity by Mathematical Programming", Proc. NATO/ASI, Univ. of Waterloo, Pergamon Press, New York.

[13] Gioda, G. and De Donato, O.:1979, "Elastic Plastic Analysis of Geotechnical Problems by Mathematical Programming", Int. J. Num. Anal. Methods Geomech., Vol.3, pp.381-401.

[14] Corradi, L.: 1978, "On Compatible Finite Element Models for Elastoplastic Analysis", Meccanica, Vol.13, pp.133-150.

[15] Martin, J.B.:1975, "Plasticity: Fundamentals and General Results", MIT Press, Boston.

[16] De Donato, O. and Maier, G.:1976, "Local Unloading in Piecewiselinear Elasto-Plasticity", J. Engrg. Mech., Div.ASCE, EM3, pp.383-394.

[17] Maier, G., Giannessi, F. and Nappi, A.:1981, "Indirect Identification of Yield Limits by Mathematical Programming", Engineering Structures, to appear.

[18] Giannessi, F. and Nicolucci, F.:1976, "Connection Between Nonlinear and Integer Programming Problems", Ist. Naz. Alta Matematica, Symp. Mathem., Academic Press, Vol.19, pp.161-176.

[19] Nappi, A.:1981, "System Identification for Yield Limits and Hardening Moduli in Discrete Elastic-Plastic Structures by Nonlinear Programming", Appl. Math. Modelling, submitted to.

[20] Best, M.S.:1975, "A Feasible Conjugate Direction Method to Solve Linearly Constrained Minimization Problems", J. Opt. Theor. appl., Vol.16, pp.25-38.

[21] McCalla, T.R.:1967, "Introduction to Numerical Analysis and Fortran Programming", J.Wiley, New York.

[22] Asoaka, A. and Matsuo, M.:1979, "Bayesian Approach to Inverse Problem in Consolidation and its Application to Settlement Prediction", Proc. 3rd Int. Conf. Num. Meth. Geomechanics, Aachen.

[23] Tomizawa, M.:1980, "Identification of a One-Dimensional Model for a Soil-Layer-Bedrock System During an Earthquake", Earthquake Engineering and Structural Dynamics, Vol.8.

[24] Cividini, A., Jurina, L. and Gioda, G.:1981, "Some Aspects of 'Characterization' Problems in Geomechanics", Int. J. Rock Mech. Min. Sciences, to appear.

[25] Cividini, A. and Maier, G., "Stochastic Parameter Identification of Some Structural and Geotechnical Static Models", to appear.

[26] Maier, G. and Nappi, A., "Statistical Identification of Parameters in Piecewiselinear Structural Models", to appear.

[27] Lewis, T.O. and Odell, P.L.:1971, "Estimation in Linear Models", Prentice-Hall, Englewood Cliff.

[28] Collins, J.D., Hart, G.C., Hasselmann, T.K. and Kennedy, B.: 1974, "Statistical Identification of Structures", AIAA Journal, Vol.12, pp.185-190.

[29] Hart, G.C. and Torkamani, M.A.M.:1974, "Structural System Identification", in Stochastic Problems in Mechanics, Ariaratuam, S.T. and Leipholz, M.M.E. Eds., University of Waterloo, Canada. pp. 207-228.

EMBANKMENTS AND SLOPES BY MATHEMATICAL PROGRAMMING

J.B. Martins

Civil Engineering, University of Minho,
Braga, Portugal

1. INTRODUCTION

Since Coulomb (1776) and Rankine (1857) that the stability of slopes has been calculated resorting to limit equilibrium methods, using friction and cohesion as strenght characteristics for the soils. However the major advance has been obtained by Fellenius (1927) with his method of slices.The further important step was done by Bishop (1956) who made a better static approach and introduced pore pressures. Both these methods assume circular surfaces of sliding. Janbu (1957) and Morgenstern and Price (1965) have generalized the slices method to non-circular sliding surfaces , taking into account at the same time the interslices forces. However, to make the problem statically determinated Janbu has fixed the line of thrust interslices and Morgenstern and Price have fixed the ratio between shear and normal force at the slices interfaces. Defining safety by a reduction factor F for cohesion c and friction tan ϕ, which brings the slope to the verge of failure, both these authors have to make a number of iterations to obtain F. Since F comes in the denominator in the constitutive equations sometimes difficulties arrise in the convergence of the process.

To have a solution fully valid both in statics and kinematic Martins et al. (1979 and 1981) have appealed to the use of mathematical programming (linear programming, LP). In their method sliding surfaces of any shape both external and internal can be taken into account. Furthermore, external forces in the soil or rock mass other than self weight are easily taken into account also.

All the methods refered above give a safety factor only. They

J. B. Martins (ed.), Numerical Methods in Geomechanics, 305–334.

tell nothing about the soil or jointed rock mass deformation prior to rupture. To obtain this information one has to resort to elastic or elasto-visco-plastic methods and use numerical techniques, v.g., finite elements. F.E. methods give not only the displacements before failure but also information on the overstressed areas of the soil or rock mass, providing in this way information for better design by either change of geometry and/or quality of the materials in the overstressed areas. However, on approaching collapse those methods may diverge (Zienkiewicz et al., 1977). Collapse load determination by MP also avoids the error accumulation which arises both on the "initial strain" and "initial load" F.E. methods (G. Maier, 1977).

2. SLOPE STABILITY ANALYSIS BY THE METHODS OF FELLENIUS AND BISHOP

The slices method has been invented by Fellenius (1927) and initially has been applied mainly in Sweden, reason by which is sometimes called the Swedish method.

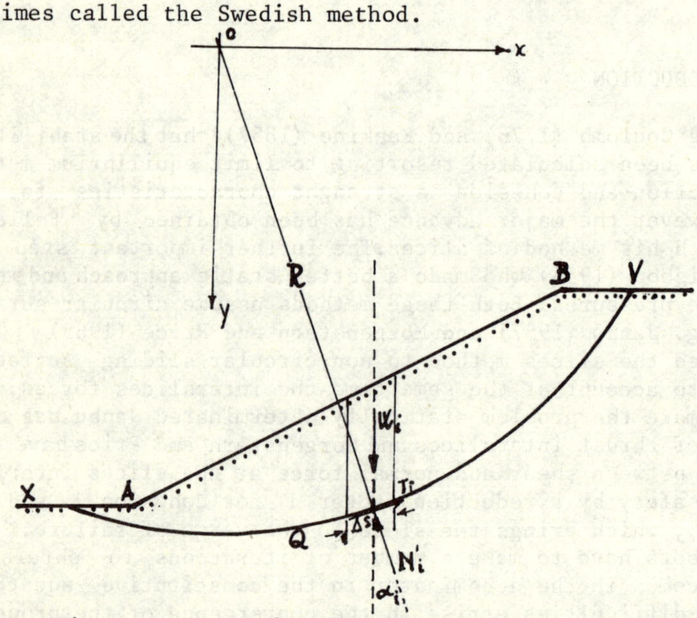

Figure 1. Fellenius Method of Slices

It assumes a circular surfaces of sliding (X Q V) and defines the factor of safety F as a reduction factor for cohesion and tan ϕ, such that the soil mass is brought to the verge of slide, i.e., to a limit equilibrium. To state this limit equilibrium the soil mass is divided into an arbitrary number n of vertical slices and the weight W_i of each slice is assumed to be in equilibrium with a normal efective force N_i and tangencial force T_i at the base of the slice. At the limit

$$T_i = \frac{c_i \, \Delta s_i}{F} + N_i \, \frac{tg\phi_i}{F} . \tag{2.1}$$

Writing down the equilibrium equations for the whole mass we have

$$\Sigma F_{ix} = \sum_{i=1}^{n} T_i \cos\alpha_i - \sum_{i=1}^{n} N_i' \sin\alpha_i = 0 \tag{2.2}$$

$$\Sigma F_{iy} = \sum_{i=1}^{n} W_i - \sum_{i=1}^{n} T_i \sin\alpha_i - \sum_{i=1}^{n} N_i' \cos\alpha_i = 0 \tag{2.3}$$

$$\Sigma M_{io} = \sum_{i=1}^{n} W_i x_i - \sum_{i=1}^{n} T_i R = 0 . \tag{2.4}$$

Since N_i' and T_i are obtained by decomposition of W_i in the normal and tangencial directions, eqs. (2.2) and (2.3) are automatically satisfied. Substituting T_i given by (2.1) and $x_i = R \sin\alpha_i$ into (2.4) we obtain

$$\sum_{i=1}^{n} W_i \sin\alpha_i = \frac{1}{F} \sum_{i=1}^{n} (c_i' \, \Delta s_i + N_i' \, tg\,\phi_i) . \tag{2.5}$$

Noting that $N_i' = W_i \cos\phi_i$, substituting into (2.5) and solving F, the classical formula for the factor of safety is obtained

$$F = \frac{\sum_{i=1}^{n} (c_i' \, \Delta s_i + (W_i \cos\alpha_i) \, tg\,\phi_i)}{\sum_{i=1}^{n} W_i \sin\alpha_i} \tag{2.6}$$

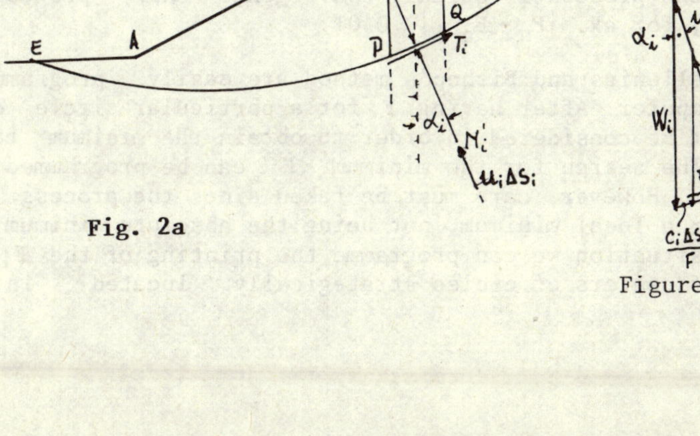

Fig. 2a

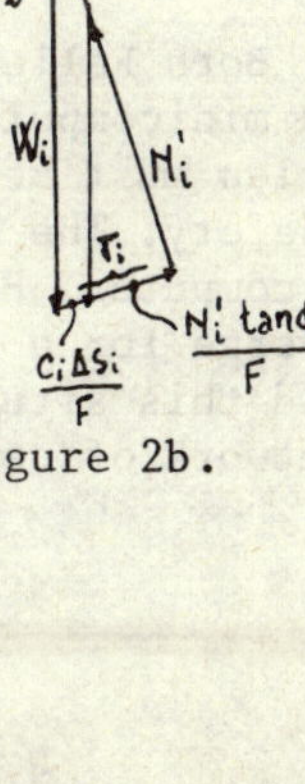

Figure 2b.

Later Bishop (1965) did generalize (2.6) to the case where pore pressures are set up at the base of each slice (Fig. 2a).

To estimate pore pressures Bishop assumed $u_i = \bar{B}\,(\frac{W_i}{\Delta x_i})$. (2.7)

Where $\bar{B}$ is a overall pore pressure coefficient depending on the nature and degree of saturation of the clayey soil.

In Figure 2b. projecting the polygon of forces on the vertical the following relationship is obtained

$$u_i\,\Delta s_i \cos\alpha_i + N'_i \cos\alpha_i + c'_i \times \frac{\Delta s_i}{F} \sin\alpha_i + \frac{N'_i}{F} \tan\phi_i \sin\alpha_i = W_i \qquad (2.8)$$

Solving in order to N'_i gives

$$N'_i = \frac{W_i - (u_i \cdot \cos\alpha_i + (c'_i / F)\sin\alpha_i) \cdot \Delta s_i}{\cos\alpha_i + \frac{\tan\phi_i}{F}\sin\alpha_i} . \qquad (2.9)$$

Substituting (2.9) into (2.5) and solving in order to F the classical Bishop's formula for the factor of safety is obtained:

$$F = \frac{1}{\sum^{n} W_i \sin\alpha_i} \sum_{i=1}^{n} \left[\{c'_i\,\Delta\alpha_i + \mathrm{tg}\,\phi_i \left[W_i\,(1-\bar{B})\right]\} \frac{1}{\cos\alpha_i + \frac{\tan\phi_i \sin\alpha_i}{F}} \right] \qquad (2.10)$$

The factor of safety is taken from (2.10) by iteration. Starting with some factor F_o (for ex. $F_o = 1.5$) and puting it in the right hand member of (2.10) a new F (F_1) is obtained. Puting F_1 in the place of F_o a new F is obtained (F_2) and so on. Experiences shows that the process is quickly convergent. The process is stoped when, for ex. $|F_n - F_{n-1}| \leq 0.01$.

Both Fellenius and Bishop's method are easily programmable in a minicomputer. After having F for a particular circle other circles must be considered in order to obtain the minimum factor of safety. The search for the minimum F can be programmed in the computer. However, care must be taken since the process may converge for a local minimum, not being the absolute minimum. To avoid this situation we can programme the printing of the F_i for a network of centers of cicles strategically located in the neighbourhood of the vertical that passes by the toe of the slope. the center of the cicle giving the absolute minimum usually is not far from that vertical.

3. NON-CIRCULAR SURFACES OF SLIDING

Many are the cases in pratice where circular surfaces of sliding are not the most critical. Figure 3. shows one of those cases.

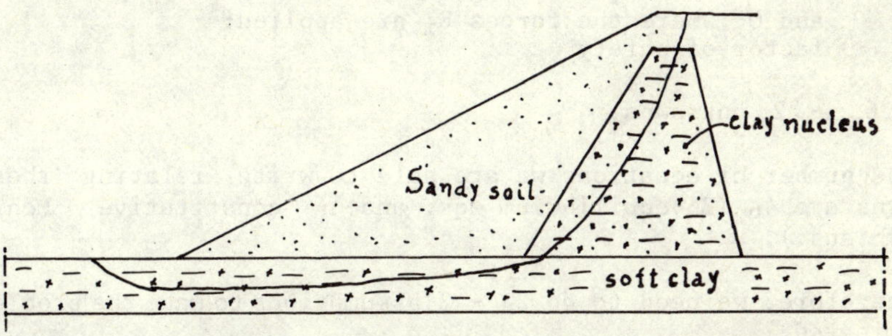

Fig. 3 Non-circular Surface of Sliding

Let us consider the sliding surface y(x) (figure 4a.) and divide the sliding mass into a number n of vertical slices, as

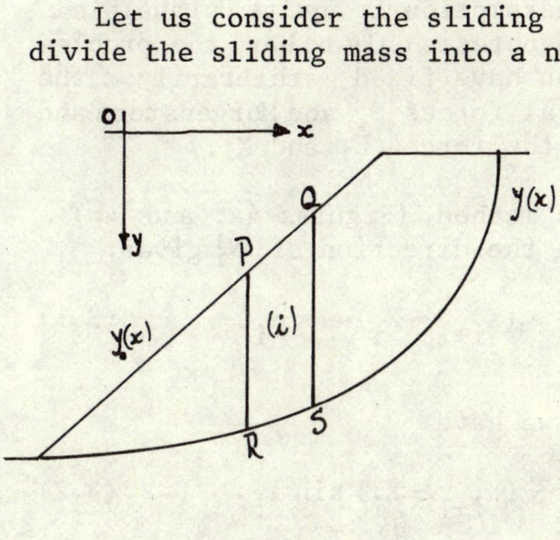

Figure 4a.

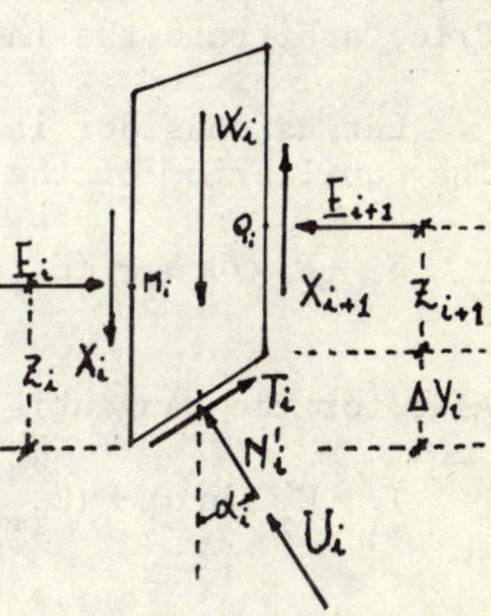

Figure 4b.

before. The forces that act on slice i are represented in Figure 4b. Those forces are: the normal forces $N_i = N_i' + U_i$ ($U_i = u_i\ \Delta s_i$ the pore pressure force on the base); the tangencial force T_i at the base. The normal forces E_i and E_{i+1} and the tangential forces X_i and X_{i+1} on the sides of the slice (Figure 4b.). Finally the weight of the slice W_i. We assume that there are no other external forces other than weight. However, if they exist we can take them into account. Let us see how many unknowns we have, n being the number of slices:

- n normal forces N'_i;
- n tangential forces T_i at the bases of the slices;
- n - 1 forces E_i on the vertical sides of the slices;
- n - 1 forces X_i in the vertical sides of the slices;
- n coordinates y_i (or x_i) defining the position of the points P_i where N'_i are applied;
- n - 1 coordinates Z_i defining the position of the ponts M_i and Q_i where the forces E_i are applied;
- 1 factor of safety

Total 6_n - 2 unknowns.

The number of equations we are able to write relating these unknowns are 4_n ($3n$ equilibrium eqs. and n constitutive Mohr-Coulomb eqs.).

Therefore, we need to do 2_n - 2 assumptions to make the problem determinated.

The assumption common to the various methods is that the points P_i of application of the forces N'_i are in the midle of the Δs_i of the slices. For thin slices this assumption is quite plausible. Therefore, we need to make n - 2 hypotesis only to have the problem determinated. To this purpose Janbu have fixed arbitrarily the points of application of the lateral forces E_i and Morgenstern and Price arbitrate the ratio between the forces E_i and X_i.

Let us consider in detail this method (Figures 4a. and 4b). The equilibrium of the slice i in the direction of N'_i gives:

$$N_i = W_i \cos\alpha_i - (E_{i+1} - E_i)\sin\alpha_i - (X_{i+1} - X_i)\cos\alpha_i . \qquad (3.1)$$

Also for the tangential direction we have

$$T_i = W_i \sin\alpha_i + (E_{i+1} - E_i)\cos\alpha_i - (X_{i+1} - X_i)\sin\alpha_i . \qquad (3.2)$$

Taking moments in relation to the midle point P_i we further have

$$-E_i\left(Z_i - \frac{(\Delta y_i)}{2}\right) + E_{i+1}\left(Z_{i+1} + \frac{(\Delta y_i)}{2}\right) + (X_{i+1} + X_i)\frac{\Delta x_i}{2} = 0 . \qquad (3.3)$$

In addition to these the constitutive equation of Mohr-Coulomb must be considered. For slice i we have

$$T_i = \frac{1}{F}\left[c_i'\,\Delta x_i \sin\alpha_i + (N_i - U_i)\tan\phi_i\right] \qquad (3.4)$$

where $U_i = U_i \times \Delta s_i$ is the resulting force from pore pressure at the base of the slice.

Puting (3.4) into (3.2), after having (3.1) into (3.4), we obtain

$$\frac{1}{F}\{c_i'\,\Delta x_i \sin\alpha_i + \left[W_i \cos\alpha_i - (E_{i+1} - E_i)\sin\alpha_i - (X_{i+1} - X_i)\right.$$

$$\left.\cos\alpha_i - U_i\right]\tan\phi_i = W_i \sin\alpha_i + (E_{i+1} - E_i)\cos x_i -$$

$$-(X_{i+1} - X_i)\,\mathrm{sen}\,\alpha_i\,. \qquad (3.5)$$

Equations (3.5) and (3.2) form a system of 2n of eqs. (n-1) unknowns E_i, (n-1) unknowns X_i, (n-1) unknowns Z_i and one unknown F. If we arbitrate, as Morgenstern and Price do, the ratio between X_i and E_i, to less then a common factor λ, we have the relationship:

$$X_i = \lambda f(x_i)E_i \qquad (3.6)$$

where f(x) is a "reasonable" function, which the authors take linear in principle and in pratice they take sometimes constant. Being reasonable to assume that the ratio X_i/E_i be greater in the zone of greater curvature of the sliding surface. According to Figure 4a., and taking into account that in the published cases $\lambda f(x_i)$ as an average near - 1/3, we could write

$$f(x) = \frac{1}{3} + \frac{2x - (x_1 + x_o)}{5(x_1 - x_o)} \qquad (3.7)$$

x_o being the smaller abcissa of the sliding mass and x_1 the larger value of x for that mass. As seen (3.7) gives $f(x) = 1/3$ at the midle point $x = (x_1 + x_o)/2$. In that case λ would come near -1 as in the examples published.

Putting (3.7) into (3.6) and the result into (3.5) and (3.3) we will have a system of 2n equations, (n-1) being the unknowns E_i, (n-1) the unknowns Z_i and the unknowns F and λ.

Let us now see how the problem could be solved.

Since the unknown F comes in the denominator in (3.5) the system of equations is not linear. Also due to the products $E_i Z_i$ and $E_{i+1} Z_{i+1}$ in (3.3) the system is not linear in that equations, either.

Therefore we start taking $F = 1.5$ and $\lambda = -1$, for example, and with these values stating the equilibrium of slice 1. To this end we solve (3.5) for $E_{i+1} = E_1$, putting $E_i = E_o = 0$. With E_i, acting in the slice 2 and stating the equilibrium of this slice, equation (3.5), we would obtain E_2; and so forth up to the slice 7 from which equilibrium the value E_7 is obtained. Putting the values of E_i into (3.3) the values of Z_i could be obtained. In this way we

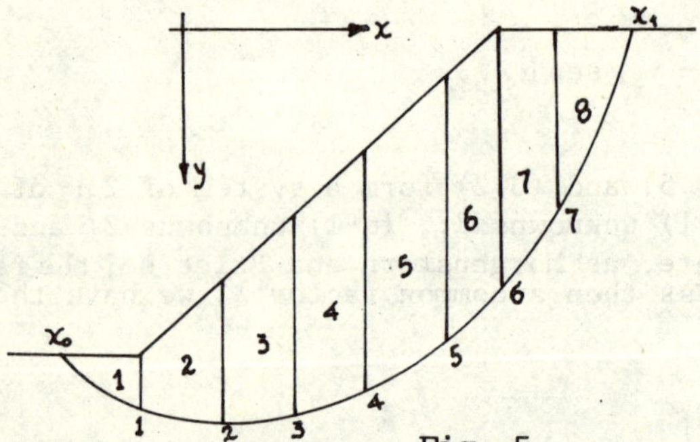

Fig. 5

would arrive at slice 8 having as unknowns F, λ. Having (3.6) into (3.5), putting $E_9 = 0$ both in (3.5) and (3.3) everything will be known except F and λ. Since F is in the denominator the equations will not be linear, and therefore the solution will not be unique. In practice, the solutions with $F > 0$ and $\lambda < 0$ will be chosen. With these new values for F and λ a new iteration should be performed starting again at slice 1. After a number of iterations (going up to 20) the values of F_n and λ_n in iteration n must be close to the values of F_{n-1} and λ_{n-1} of the proceeding iteration. For ex:

$$|F_n - F_{n-1}| \leq 0.001 \text{ and } |\lambda_n - \lambda_{n-1}| \leq 0.005.$$

If after 20 iterations the process does not converge, Morgenstern and Price (1967) advise to restart, putting $\lambda = 0$.

The method of Janbu also may lead to a divergent solution (Morgenstern and Price, 1965).

The lack of convergence may be due not only to the wrong choice of λ and F in the begining or in the non unique solutions for the last slice, but also due to the fact that sliding may occur not only at the external surface of sliding but also on the interslices surfaces. Therefore we might have

$$|X_i|\ \max \leq c'_i\ |y_o(x_i) - y(x_i)| + E'_i \tan\phi_i \tag{3.8}$$

where $E'_i = E_i - U_i$, U_i being the force resulting from pore pressures interslices.

Also tensile forces E_i must not be allowed and therefore if they occur we must make $E_i = X_i = 0$. If this provision is not taken, again divergence may arrive in the iterative process.

Assuming that convergence is assured, even so, a number of different sliding surfaces must be tried in order to have a minimum for F.

4. VARIATIONAL METHODS

To obtain the surface of sliding y(x) giving the minimum value for the factor of safety F, a number of authors (R. Bark and M. Garber (1977); Ramamurthy et al. (1977); Castillo and Revilla (1977) have resorted to variational methods (Figure 6.).

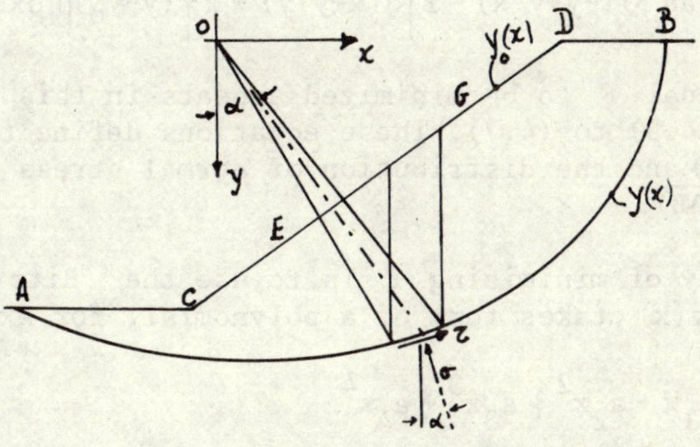

Figure 6. Variational Methods

The approaches of all those authors are not the same, but we can lump them using the basic principles of the calculus of variations.

Assuming that the geometry of the slope is represented by the function $y_o(x)$ and that the sliding surfaces is represented by the function f(x), γ being the unit weight of the soil, we may state the equilibrium equations of a slice as follows:

$$\int_{xA}^{xB} (\tau \cos\alpha - \sigma \sin\alpha)\, ds = 0 \tag{4.1}$$

$$\int_{xA}^{xB} (-\tau \sin\alpha - \sigma \cos\alpha)ds + \int_{xA}^{xB} \gamma\,(y - y_o)dx = 0 \qquad (4.2)$$

$$\int_{xA}^{xB} -(\tau \cos\alpha - \sigma\,\text{sen}\,\alpha)y\,ds + \int_{xA}^{xB}(\tau \sin\alpha + \gamma \cos\alpha)x\,ds -$$

$$-\int_{xA}^{xB} \gamma\,(y-y_o)x\,dx = 0\,. \qquad (4.3)$$

To these equations we must add the Mohr-Coulomb constitutive law

$$\tau = \frac{c}{F} + \sigma\,\frac{\tan\phi}{F}\,. \qquad (4.4)$$

Putting (4.4) into (4.1) to (4.3) we have

$$\int_{xA}^{xB} (c + \sigma\tan\phi - F\sigma y')dx = 0, \quad y' = \tan\alpha \qquad (4.5)$$

$$\int_{xA}^{xB} -(c + \sigma\tan\phi)y'dx + \int_{xA}^{xB} F\gamma(y-y_o - \sigma)dx = 0 \qquad (4.6)$$

$$\int_{xA}^{xB} \{(c + \sigma\tan\phi)(-y+y'x) - F\left[\sigma(x+y'y) + x\gamma(y-y_o)\right]\}\,dx\,. \qquad (4.7)$$

The functional F to be minimized appears in this case in all the equations (4.5) to (4.7). These equations define the sliding surface $y = y(x)$ and the distribution of normal stress $\sigma = \sigma(x)$ on that surface $(\overline{AB})$.

The best way of minimizing F is to use the Ritz method, assuming that y(x) takes form of a polynomial, for ex.,

$$y(x) = a_o + a_1x + a_2x^2 + a_3x^3 + a_4x^4. \qquad (4.8)$$

If we assume that the sliding surface would pass at points x_A and x_B only three of the constants $a_i (i = 0, 1, \ldots, 4)$, would be unknown.

Also, the normal stresses $\sigma(x)$ would be assumed to have a polynomial distribution

$$\sigma(x) = b_o + b_1x + b_2x^2 + b_3x^3 + b_4x^4. \qquad (4.9)$$

In this case, as boundary conditions only $\sigma(x)\Big|_{x = x_B} = 0$ (4.10) could be stated, giving, therefore, one of the constants b_j ($j = 0, 1, \ldots, 4$). In what concerns $y_o(x)$, the geometry of the slope, to avoid discontinuity in the derivatives we could take

$$y_o(x) = c_o + c_1 x + c_2 x^2 + c_3 x^3 + c_4 x^4 + c_5 x^5 \tag{4.11}$$

determinating the constants c_k ($k = 0, 1, \ldots, 5$) by the condition of $y_o(x)$ pass at the points A, B, C, D, E, G.

Putting (4.8), (4.9) and (4.11) into (4.5) to (4.7) and integrating, we would obtain

$$f_1 (a_i, b_j, F) = 0 \tag{4.12}$$

$$f_2 (a_i, b_j, F) = 0 \tag{4.13}$$

$$f_3 (a_i, b_j, F) = 0. \tag{4.14}$$

In (4.12) to (4.14) we need to find the constant a_i and b_j which minimize the implicit function F. To solve this problem the best is to take the function

$$Z = F \tag{4.15}$$

and minimize it subject to the constraints (4.12) to (4.14). This can be done using Lagrange multipliers γ_1, γ_2 and γ_3 and considering the function

$$G = F + \gamma_1 f_1 + \gamma_2 f_2 + \gamma_3 f_3 . \tag{4.16}$$

Stating the conditions of extremum, we have

$$\frac{\partial G}{\partial a_i} = 0 \rightarrow \gamma_1 \frac{\partial f_1}{\partial a_i} + \gamma_2 \frac{\partial f_2}{\partial a_i} + \gamma_3 \frac{\partial f_3}{\partial a_i} = 0 \tag{4.17}$$

$$\frac{\partial G}{\partial b_j} = 0 \rightarrow \gamma_1 \frac{\partial f_1}{\partial b_j} + \gamma_2 \frac{\partial f_2}{\partial b_j} + \gamma_3 \frac{\partial f_3}{\partial b_j} = 0 \tag{4.18}$$

$$\frac{\partial G}{\partial F} = 0 \rightarrow 1 + \gamma_1 \frac{\partial f_1}{\partial F} + \gamma_2 \frac{\partial f_2}{\partial F} + \gamma_3 \frac{\partial f_3}{\partial F} = 0 \tag{4.19}$$

$$\frac{\partial G}{\partial \gamma_1} = 0 \rightarrow f_1 = 0 \tag{4.20}$$

$$\frac{\partial G}{\partial \gamma_2} = 0 \rightarrow f_2 = 0 \qquad (4.21)$$

$$\frac{\partial G}{\partial \gamma_3} = 0 \rightarrow f_3 = 0 \,. \qquad (4.22)$$

In (4.17) to (4.22) we have 11 equations to 11 unknowns, 3 values a_i (assuming that $y(x)$ passes in $x = x_A$ and $x = x_B$, fixed); 4 values b_j, 1 value F and 3 values γ_k $(k = 1, 2, 3)$.

The solution of the above equations is not easy since they are not linear. Furthermore, if the soils are not homogeneous the problem results much more involved. Also the introduction of pore pressures produce new difficulties. Finally it should be noted that in any case this method still is a slices method although the slices here are infinitesimal.

5. ANALYSIS OF THE STABILITY OF SLOPES BY LINEAR PROGRAMMING (LP)

5.1. Introduction

Figures 7 and 8 represent two simplified examples of slopes where linear programming can be used for the determination of the factor of safety F.

We assume that the joints between blocs have any orientation and that the material filling the joints has friction ϕ and cohesion c, which can change from joint to joint. We assume also that the contact between blocs happens at a finite number of points. Therefore, the cohesion force at a point of contact between two blocs is the integral of the cohesion strenght in the midle

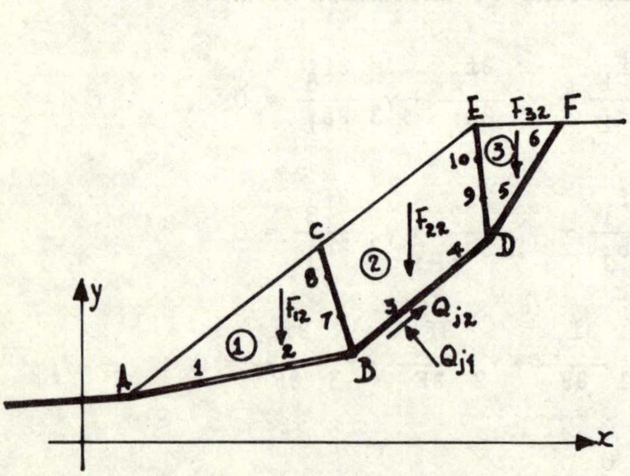

Figure 7. Jointed Rock Slope

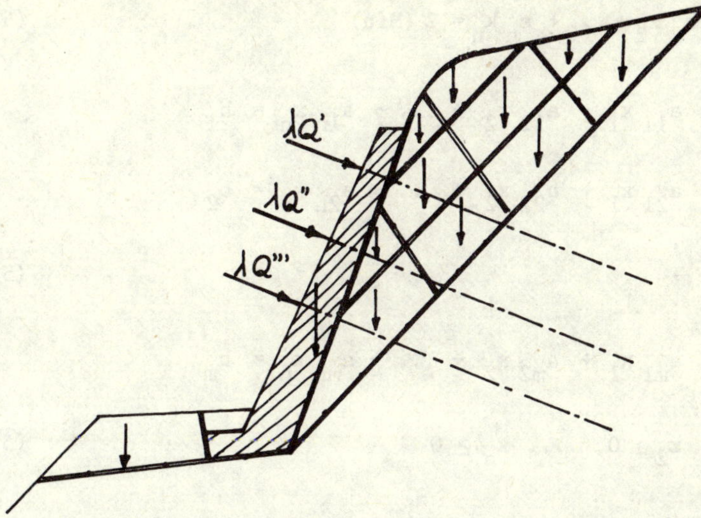

Figure 8. Anchored Wall

distance between that point and its neighborours. For example if the lenght of joint AB is ℓ and points 1 and 2 (Figure 7.) are equally spaced in AB, $C_1 = C_2 = (\ell/2) \cdot c$, where c is the unit cohesion.

Regarding the definition of factor of safety F we consider it as being a reduction factor for $\tan\phi$ and cohesion c, such that $\tan\theta = (\tan\phi)/F$ and $c' = c/F$ would lead the jointed rock mass or soil to the verge of failure. Since the variable F comes in the denominator in the final system of governing eqs., this definition implies a non linear programming problem. However, arbitrating initially a reasonable value for F, say $F_o = 1{,}5$, a maximum load factor λ applied to the external loads (in the case of Figure 7, the weights of the blocs) can be determinated in such a way that the soil rock mass reaches failure. If we find $\lambda > 1$ $(\lambda < 1)$ a second iteration must be done with $F > F_o$ $(F < F_o)$. After a few iterations we find $\lambda \simeq 1$. The corresponding F is the actual factor of safety. In the case of Figure 8. the factor of safety may be fixed before hand and the load factor λ would be applied to the anchoring cable loads only. In that case a sole iteration would be needed.

To obtain the maximum load factor in each iteration we must solve a LP problem. That can be done by the well known "Simplex" method.

In "standard form" the problem consists of determinating the minimum of a linear function, subjected to a number of linear constraints. I.e.

$$c_1 x_1 + c_2 x_2 + \ldots + c_n k_n = Z \text{ (Min)} \tag{5.1}$$

subject to
$$\begin{aligned} a_{11} x_1 + a_{12} x_2 + \ldots + a_{1n} x_n &= b_1 \\ a_{21} x_1 + a_{22} x_2 + \ldots + a_{2n} x_n &= b_2 \\ &\vdots \\ a_{m1} x_1 + a_{m2} x_2 + \ldots + a_{mn} x_n &= b_m \end{aligned} \tag{5.2}$$

$$x_1 \geq 0, \; x_2 \geq 0, \; \ldots, \; x_n \geq 0 \tag{5.3}$$

where, in general, $n > m$ and the c_i's, b_i's and a_{ij}'s are real constants and the x_i's are real numbers to be determined. We always assume $b_i \geq 0$. To reach this aim it may be necessary to multiply the i the equation by minus unity.

The relationships (5.1) to (5.4) may be writen under the compact form

$$\tilde{C}^T \cdot \tilde{x} = z \text{ (Min)} \tag{5.1)'$$

subject to $A \cdot \tilde{x} = \tilde{b}$ (5.2)'

$\tilde{x} \geq 0$ (5.3)'

The ~ on a letter means column vector, $\tilde{C}^T$ means a row vector and A^T the transpose of the matrix A.

It is clear that to reach the standard form (5.1) to (5.3) we must manipulate the equilibrium and constitutive equations for the jointed rock mass of Figure 7.

Let us consider the following notation:

The external forces that act on the bloc i ($i = 1, 2, \ldots, n_b$) are noted by F_{i1}, F_{i2} and $F_{i3} \cdot F_{i1}$, being the x component; F_{i2} the y component, assumed at the center of gravity of bloc i; and F_{i3} is the moment.

The internal forces are the normal and shear force at each contact point. At the contact point j, Q_{j1} would be the force normal to the joint and Q_{j2} would be the shear or tangential force.

Writing down the equilibrium equations would lead to

$$- A\tilde{Q} + \lambda\tilde{F} = 0. \tag{5.4}$$

In the case of Figure 7 the matrix A has the dimensions 9x20, $\tilde{Q}$ has 20x1 and $\tilde{F}$ has 9x1 elements. λ is the load factor.

To the (5.4) we must joint the constitutive eqs. expressing the Mohr-Coulomb's law. For point j we have

$$|Q_{j2}| \le C_j/F + Q_{j1}(\tan\phi)/F \,. \tag{5.5}$$

The relationship (5.5) gives rise to the inequalities

$$Q_{j2} - [(\tan\phi)/F]\, Q_{j1} \le C_j/F \qquad (5.5)'$$

$$- Q_{j2} - [(\tan\phi)/F]\, Q_{j1} \le C_j/F \,. \qquad (5.5)''$$

The constitutive equations in (5.5)' and (5.5)" form the "yield surface", which is represented in Figure 9. In this case this surface is open, but in general it is a convex polyhedron (or polygon, when forces have only two components).

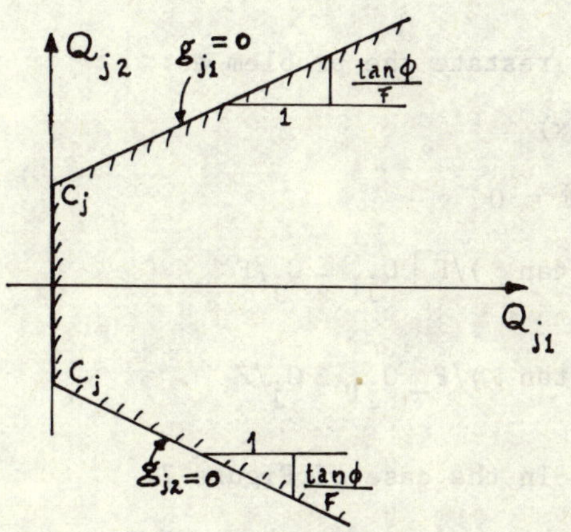

Figure 9.

The yield surface, which represents the plasticity conditions, if not linearized, i.e., not reduced to a polyhedron, would give rise to a non linear programming problem.

In general (5.5)' and (5.5)" may be writen:

$$g_{j\cdot k}\,(Q_{j1}, Q_{j2}\ \ldots) \leq 0 \tag{5.6}$$

where k represents the k face of the polyhedron. In the case of Figure 9.

$$g_{j1} = Q_{j2} - \left[(\tan\phi)/F\right] Q_{j1} - C_j/F, \tag{5.7}$$

$$g_{j2} = Q_{j2} - \left[(\tan\phi)/F\right] Q_{j1} - C_j/F. \tag{5.8}$$

The "feasible" domain at each internal contact point j is the set of pairs of forces Q_{j1} (normal) and Q_{j2} (shear) that are inside the shaded area of Figure 9.

Equilibrium equations (5.1) and inequalities (5.5)' and (5.5)", together with the objective function

$$\lambda = Z\,(\text{Max}) \text{ or } -\lambda = Z\,(\text{Min}) \tag{5.10}$$

form a linear programming problem, provided we fix F beforehand.

The matrix A of the equilibrium equations would be constructed easily taking account of the geometry of the joints, the coordinates of the points of contact and those of the centers of gravity of the blocs.

Therefore we may restate the problem as:

find $\lambda = Z(\text{Max})$ (5.10)

subject to $-A\tilde{Q} + \lambda\tilde{F} = 0$ (5.4)

and $+Q_{j2} - \left[(\tan\phi)/F\right] Q_{j1} \leq C_j/F$ (5.5)'

$-Q_{j2} - \left[(\tan\phi)/F\right] Q_{j1} \leq C_j/F$ (5.5)"

$j = 1, 2, \ldots, 10$, in the case of Figure 7.

For the moment we would consider Q_{j1} and Q_{j2} as variables unrestrited in sign.

According to the usual procedure in PL, the inequalities (5.5)' and (5.5)" will be converted into equalities introducting the slack variables R_{1j} and R_{2j} ($j = 1, 2, ..., 10$).

For the given example we would have 41 variables alltogether ($10\,Q_{j1} + 10\,Q_{j2} + \lambda + 20\,R_{nj}$). The number of eqs. after introduction of the slack variables will be 29 (9 equilibrium eqs. for the 3 blocs and 20 eqs. from (5.5)' and (5.5)").

Following the "simplex" method we must now put all the constraints, i.e.,

$$-A\tilde{Q} + \lambda\tilde{F} = 0 \tag{5.4}$$

$$\text{and } Q_{j2} - [(\tan\phi)/F]\,Q_{j1} + R_{j1} = C_j/F \tag{5.6}$$

$$-Q_{j2} - [(\tan\phi)/F]\,Q_{j1} + R_{j2} = C_j/F \tag{5.7}$$

$$R_{j1} \geq 0 \text{ and } R_{j2} \geq 0 \tag{5.8}$$

into "canonical form". This form implies that a "basic solution" of the eqs. is known, i.e., that a bloc of m x m coefficients in the matrix A in (5.2)' became a unit matrix of rank m, and that the new constant vector is non-negative ($b_i' \geq 0$). Assuming that we rearrange the variables x_i in (5.2) in vector $\tilde{x}$, the system (5.2) would become.

$$\begin{bmatrix} 1 & 0 & \dots & 0 & a'_{1,m+1} & \dots & a'_{1,n} \\ 0 & 1 & \dots & 0 & a'_{2,m+1} & \dots & a'_{2,n} \\ \cdot & & & & & & \\ \cdot & & & & & & \\ \cdot & & & & & & \\ 0 & 0 & \dots & 1 & a'_{m,m+1} & \dots & a'_{m,n} \end{bmatrix} \left\langle \begin{matrix} x_1 \\ x_2 \\ \cdot \\ \cdot \\ \cdot \\ x_n \end{matrix} \right\rangle = \left\langle \begin{matrix} b'_1 \\ b'_2 \\ \cdot \\ \cdot \\ \cdot \\ b'_n \end{matrix} \right\rangle . \tag{5.9}$$

Let us now try to reduce our problem to this form. To this aim we may remove a number of connections at the contact points in order to have an isostatic (or basic) stable system. In the case of Figure 7. we may remove, for example, all the internal connections (at points 7 to 10). This is equivalent to state $Q_{j1} = Q_{j2} = 0$ ($j = 7, 8, 9, 10$) in (5.4). To have a basic system we must state also that, for example, $Q_{22} = Q_{42} = Q_{62} = 0$ and $\lambda = 1$ in (5.4), starting from the "simplex" tableau corresponding to the eqs. (5.4) and (5.6) to (5.8). Operating on that tableau, doing "pivots" at the variables Q_{11}, Q_{12} and Q_{21} (bloc 1), Q_{31}, Q_{32} and Q_{41} (bloc 2), Q_{51}, Q_{52} and Q_{61} (bloc 3), putting $\lambda = 1$, we would obtain the following modified eqs.:

$$-\lambda = -1 \tag{5.10}$$

$$A'\tilde{Q} - \lambda\tilde{F}' = 0 \tag{5.11}$$

$$k_{pp}\tilde{R} = \tilde{b}' . \tag{5.12}$$

Under the form of tableaus we would start with tableau (5.1).

Pivoting on this tableau in order to make unit the coefficients of Q_{11}, Q_{12}, Q_{21}, (bloc 1), ..., Q_{51}, Q_{52}, Q_{61} (bloc 3), we would obtain a "feasible" solution. After those transformations tableau (5.1) would become tableau (5.2).

That tableau corresponds to eqs. (5.11) and (5.12).

To have tableau (5.2) in the canonical form we need to divide each of the last 20 eqs. by K_{pp} ($p = 1, 2, ..., 20$). Even after this operation the tableau would not yet in the canonical form since some variables would not be restricted in sign. This would be the case for the internal tangential forces at contact points $Q_{7,2}$ to $Q_{10,2}$. To make these variables non-negative, the best way (Luenberger, 1969) is to do a change of variables putting

$$Q'_{j2} = Q_{j2} + Q_0 \tag{5.13}$$

where Q_0 is an upper bound for the set of the Q_{j2}. Bringing (5.13) into (5.11) and (5.12), we would end with

$$-\lambda = Z\,(\text{Min}) \tag{5.10}$$

$$\text{subject to} \quad \begin{bmatrix} A'- & \tilde{F} & 0 \\ & & \\ 0 & 0 & K_{pp} \end{bmatrix} \begin{bmatrix} \tilde{Q}'- Q_0 \\ \lambda \\ \tilde{R} \end{bmatrix} = \begin{bmatrix} 0 \\ \\ b'' \end{bmatrix}$$

$$\text{i.e. subject to} \quad \begin{bmatrix} A'- & \tilde{F}' & 0 \\ & & \\ 0 & 0 & I \end{bmatrix} \begin{bmatrix} \tilde{Q}' \\ \lambda \\ \tilde{R} \end{bmatrix} = \begin{bmatrix} \tilde{b} \\ \\ b'' \end{bmatrix} \tag{5.14}$$

Tableau (5.1)

Q_{11}	Q_{12}	Q_{21}	Q_{22}	$\cdots$	Q_{kj}	$\cdots$	$Q_{10,1}$	$Q_{10,2}$	λ	R_{11}	R_{12}	$\cdots$	$R_{10,2}$		
$a_{11,11}$	$a_{11,12}$	$\cdots$			$a_{11,kj}$				$-F_{11}$	0	0	0	0	=	0
$a_{21,11}$	$a_{21,12}$								$-F_{21}$	0	0	0	0	=	0
$\vdots$									$\vdots$					$\vdots$	$\vdots$
$a_{33,11}$									$-F_{33}$	0	0	0	0	=	0
$-\tan\phi_1/F$	1	0	0	0	0	0	0	0	0	1	0	0	0	=	C_1/F
$-\tan\phi_1/F$	-1	0	0	0	0	0	0	0	0	0	1	0	0	=	C_1/F
$\vdots$														$\vdots$	$\vdots$
0	0	0	0	0	0	0	$-\tan\phi_{10}/F$	1	0	0	0	1	0	=	C_{10}/F
0	0	0	0	0	0	0	$-\tan\phi_{10}/F$	-1	0	0	0	0	1	=	C_{10}/F

Tableau (5.2)

Q_{11}	Q_{12}	Q_{21}	...	Q_{51}	Q_{52}	...	Q_{kj}	...	$Q_{10,1}$	$Q_{10,2}$	λ	R_{11}	R_{12}	...	$R_{10,2}$		
1	0	0		0	$a'_{11,52}$		$a'_{11,kj}$		$a'_{11;10,1}$	$a'_{11;10,2}$	$-F'_{11}$	0			0	=	0
0	1	0		0	$a'_{21,52}$		$a'_{21,kj}$		$a'_{21;10,1}$	$a'_{21;10,2}$	$-F'_{21}$	0			0	=	0
0	0	1		0	...		...		...	...	...	0			0	=	0
⋮	⋮	⋮		⋮			⋮		⋮	⋮	⋮	⋮			⋮	⋮	⋮
0	0	0		1	$a'_{33,52}$		$a'_{33,kj}$		$a'_{33;10,1}$	$a'_{33;10,2}$	$-F'_{33}$	0			0	=	0
0	0	0	0	0	0		0		0	0	0	K_{11}	0	0	0	=	b'_1
0	0	0	0	0	0		0		0	0	0	0	K_{22}	0	0	=	b'_2
0	0	0	0	0	0		0		0	0	0	0				=	...
0	0	0	0	0	0		0		0	0	0					=	
0	0	0	0	0	0		0		0	0	0	0	0	0	$K_{20,20}$	=	b'_{20}

In (5.14) the first 9x9 elements of A' would be a unit matrix I, according to the tableau (5.2).

Since the basic variables (the first 9 elements of vector Q') must be non-negative, so must be the vector $\tilde{b}$. If any element of $\tilde{b}$ is negative, we could increase the value of Q_o to make it positive. With eqs. (5.14) under canonical form we are ready to apply the usual "simplex" technique, which is part of the library of programmes of any computer.

Should we find $\lambda > 1$, (or $\lambda < 1$) the value of F should be increased (decreased) and the calculations repeated. However, the same basic solution for the equilibrium eqs. could be used.

6. THE DUAL PROBLEM AND ITS INTERPRETATION

It is well known from the mathematical programming theory that any "primal" problem, in our case LP1, has a "dual" (LP2). It can be shown that between the linear programming primal problem (LP1) and its dual (LP2) we have the following relationship:

Primal Problem (LP1)	Dual Problem (LP2)
Objective function = Z_1 (Max) (linear in x_i)	Objective function = Z_2 (Min) (linear in y_i)
Coefficients of the objective form	Constant terms in the eqs. and inequalities
Constant terms in the eqs. and inequalities	Coefficients of the objective form
Coefficient Matrix	Transpose of the corfficient Matrix
i^{th} inequality: $\leq$	Variable $y_i \geq 0$
i^{th} equation: =	Variable y_i unrestricted in sign
Variable $x_i \geq 0$	j^{th} inequality: $\geq$
Variable x_i unrestricted in sign	j^{th} equation: =

These relationship can be stated in the form of the following tableau for 5 variables x_i, one equation and two inequalities in LP1:

Tableau (6.1)

variables in LP1

constant terms in LP1 = = coefficients of the objective form in LP2

variables in LP2	x_1	x_2	$x_3 \geq 0$	$x_4 \geq 0$	$x_5 \geq 0$			
y_1	a_{11}	a_{12}	a_{13}	a_{14}	a_{15}	=	b_1	x_1, x_2: unrestricted in sign
$y_2 \geq 0$	a_{21}	a_{22}	a_{23}	a_{24}	a_{25}	≤	b_2	$x_3 \geq 0$, $x_4 \geq 0$, $x_5 \geq 0$.
$y_3 \geq 0$	a_{31}	a_{32}	a_{33}	a_{34}	a_{35}	≤	b_3	$y_2 \geq 0$, $y_3 \geq 0$;
	=	=	≥	≥	≥		=	y_1: unrestricted
	C_1	C_2	C_3	C_4	C_5	=	Min Max	

coefficients of the objective form in LP1 = = constant terms in the eqs. and inequalities in LP2.

The dual problem in our case consists in obtaining the set of external displacements $f_{i\ell}$, corresponding to the external forces $F_{i\ell}$, ($i = 1, 2, 3$, are the blocs and $\ell = 1, 2, 3$, are the components of the displacements; f_{i3} are the rotations of the blocs). And also obtaining the set of internal displacements q_{jh}, corresponding to the internal forces Q_{jh}. j ($= 1, 2, 3, \ldots, 10$) is the point of contact between blocs or between blocs and the external sliding surface and h is the component of the local displacement or force ($h = 1$ for normal and $h = 2$ for tangential component).

With this notation we can built a tableau for the relationships between LP1 and LP2 in our case.

Tableau (6.2) may be writen in matrix form as

$$-A\tilde{Q} + \lambda\tilde{F} = 0 \tag{6.1}$$

and $$D\tilde{Q} \leq \tilde{K} \tag{6.2}$$

$$\lambda = Z \text{ (Max)} \tag{6.3}$$

Where A is a (9x20) matrix derived from statics D is a (20x20) matrix and $\tilde{K}$ is a 20x1 vector both derived from yield conditions (Mohr-Coulomb law). $\tilde{Q}$ (20x1) is the internal and $\tilde{F}$ (9x1) is the external forces vector.

According to the rules to form the dual from the primal problem, applied to tableau (6.2) we obtain

Tableau (6.2)

	Q_{11}	Q_{12}	Q_{21}	Q_{22}	$\cdots$	$Q_{10,1}$	$Q_{10,2}$	λ		
f_{11}	$-a_{11,11}$	$-a_{11,12}$	$\cdots$	$\cdots$	$\cdots$		$-a_{11,10,2}$	$+F_{11}$	$=$	0
f_{12}	$-a_{12,11}$	$-a_{12,12}$	$\cdots$	$\cdots$	$\cdots$		$-a_{12,10,2}$	$+F_{1,2}$	$=$	0
f_{13}	$-a_{13,11}$	$-a_{13,12}$	$\cdots$	$\cdots$	$\cdots$		$\cdots$	$\cdots$		
$\vdots$	$\vdots$	$\vdots$								
f_{33}	$-a_{33,11}$	$-a_{33,12}$			$\cdots$		$-a_{33,10,2}$	$+F_{3,3}$	$=$	0
u_{11}	$-\frac{\tan\phi}{F}$	1							$\leq$	C_1/F
u_{12}	$-\frac{\tan\phi}{F}$	-1							$\leq$	C_1/F
$\vdots$									$\vdots$	$\vdots$
$u_{10,1}$						$-\frac{\tan\phi}{F}$	1		$\leq$	C_{10}/F
$u_{10,2}$						$-\frac{\tan\phi}{F}$	-1		$\leq$	C_{10}/F
	$=$	$=$	$=$			$=$	$=$	$=$		
	0	0	0			0	0	$+1$	$=$	Min Max

$$-A^T \cdot \tilde{f} + D^T\tilde{u} = 0 \tag{6.4}$$

and $\tilde{F}^T \cdot \tilde{f} = 1.$ (6.5)

Also the objective function will be $\tilde{K}^T \cdot \tilde{u} = \text{Min},\ \tilde{u} \geq 0.$ (6.6)

Since the work done by the external forces in the external displacements is equal to the work of the internal forces in the internal displacements (principle of virtual work), $A\tilde{Q} = \tilde{F}$ implies

$$A^T\tilde{f} = \tilde{q}. \tag{6.7}$$

Putting (6.7) into (6.4) we obtain

$$D\tilde{u} = \tilde{q}. \tag{6.8}$$

In the other hand if we try to solve the problem stated in (6.1) to (6.3) using the Kuhn - Tucker conditions, derived from Lagrange multipliers, the following normality condition must be satisfied:

$$g_{jk} \cdot u_{jk} = (D\tilde{Q} - \tilde{K})^T \cdot \tilde{u} = 0\,, \quad j = 1, 2, \ldots, 10; \quad k = 1, 2. \tag{6.9}$$

from (6.9) we notice that

$$\frac{\partial g_{jh}}{\partial Q_{jh}} = D \qquad (h = 1, 2). \tag{6.10}$$

Introducing (6.10) into (6.8), we obtain

$$\frac{\partial g_{jk}}{\partial Q_{jh}} \cdot u_{jk} = q_{jh}\,. \tag{6.8)'$$

Each plasticity factor u_{jk} is closely related to the corresponding face of the polihedron representing the yield surface in the space of the internal forces at each point j. In fact by (6.9) for u_{jk} to be different from zero, must be $g_{jk} = 0$, that is, the corresponding point must be on the yield surface on the face $g_{jk} = 0$. At that point the internal forces will be Q_{jh} $(h = 1, 2)$.

Looking at (6.8)' we see the well known normality rule; i.e., at each point of the yield surface the internal displacement (or the velocity of deformation) is normal to the surface, that is, has the direction of the gradient $\partial g_{jk}/\partial Q_{jh}$.

Concerning the eq. (6.5) it is clear that it represents the work done by the external forces. In (6.5) it appears as that work being fixed. However, it is understood that the work might be multiplied by any constant with appropriated adjustment of the other variables.

It remains now to interpret (6.6). Taking into account (6.9), we may write

$$[D\tilde{Q}]^T \cdot \tilde{u} = \text{Min} \qquad \text{or} \qquad [D\tilde{u}]^T \cdot \tilde{Q} = \text{Min}. \tag{6.6)'$$

Since $D\tilde{u} = \tilde{q}$, by (6.8), (6.6)' becomes

$$\tilde{q}^T \cdot \tilde{Q} = \text{Min.}$$

Therefore, (6.6) states that the virtual work done by the internal forces must be minimum.

Summing up we may say that the limit equilibrium of any jointed rock or soil mass can be analysed by one of two LP_S, with the same optimal value:

$$\text{Max } \lambda, \text{ subject to } -A\tilde{Q} + \lambda\tilde{F} = 0 \quad \text{and} \quad D\tilde{Q} \leq \tilde{K} \tag{6.12}$$

or

$$\text{Min } \tilde{K}^T \cdot \tilde{u}, \text{ subject to } \tilde{F} \cdot \tilde{f} = 1, \; -A^T\tilde{f} + D^T\tilde{u} = 0, \; \tilde{u} \geq 0. \tag{6.13}$$

A computer programm for the LP1 above has been written in FORTRAN, the flow chart of which is given next page.

7. ELASTO - PLASTIC STRESSES AND STRAINS

We have seen that, given the external loads $\tilde{F}$ on the set of blocs we assumed the sliding soil or rock mass to be divided into, given a set of "control" points on the joints between blocs and the constitutive laws (Coulomb) of the material filling the joints (ineq. (5.7) and (5.8)), we could find a load factor λ leading the set to collapse, the internal forces $\tilde{Q}$ (generalized stress), internal displacements $\tilde{q}$ (generalized strains) and external displacements $\tilde{f}$. To reach this aim we had to solve a LP and its dual LCP (linear complementary problem).

This procedure implicily ignores elastic stresses and strains. To take those into account we might use the Collonetti principle (Donato - 1968) which states that any elasto - plastic problem may be divided into two steps a) the determination of the stress and strain state on the assumption that linear elasticity is of unlimited validity; b) the determination, on the same elastic assumption, of a set of self stresses due to the plastic "disloca - tion" $\tilde{q}^P$, which corrects the stress state determined in a).

Both these calculations may be done by usual methods. In our case we might use, for ex., finite element methods and if the blocs were small enough, the blocs themselves could be taken as elements and the "control" points between slices as "nodes" of the F.E. methods.

According to Collonetti's principle the internal forces would be given by

Flow Chart

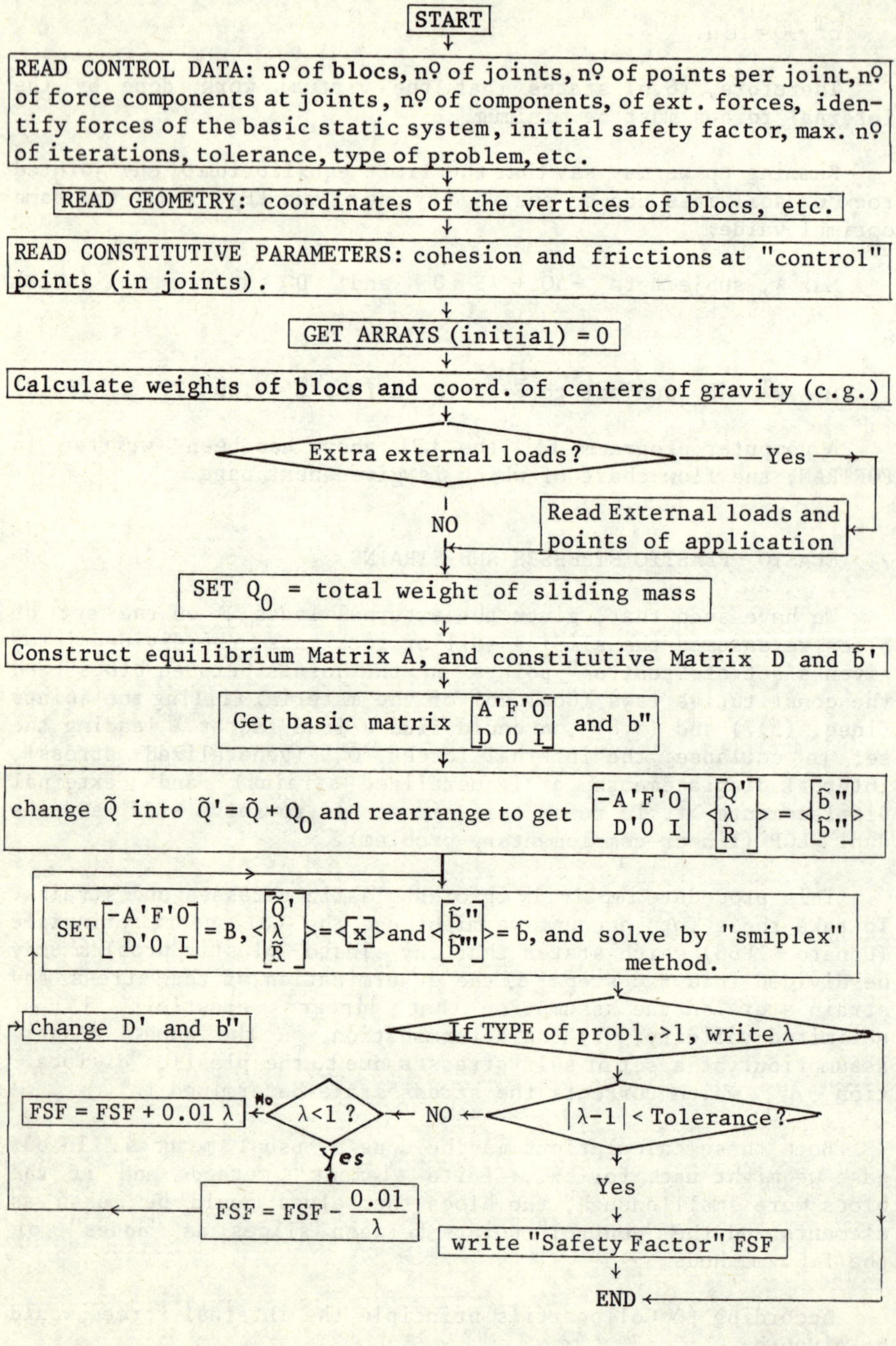
START
READ CONTROL DATA: nº of blocs, nº of joints, nº of points per joint, nº of force components at joints, nº of components, of ext. forces, identify forces of the basic static system, initial safety factor, max. nº of iterations, tolerance, type of problem, etc.
READ GEOMETRY: coordinates of the vertices of blocs, etc.
READ CONSTITUTIVE PARAMETERS: cohesion and frictions at "control" points (in joints).
GET ARRAYS (initial) = 0
Calculate weights of blocs and coord. of centers of gravity (c.g.)
Extra external loads?
Yes
Read External loads and points of application
NO
SET Q_0 = total weight of sliding mass
Construct equilibrium Matrix A, and constitutive Matrix D and $\tilde{b}'$
Get basic matrix $\begin{bmatrix} A'F'0 \\ D\ 0\ I \end{bmatrix}$ and b''
change $\tilde{Q}$ into $\tilde{Q}' = \tilde{Q} + Q_0$ and rearrange to get $\begin{bmatrix} -A'F'0 \\ D'0\ I \end{bmatrix} \left\langle \begin{matrix} Q' \\ \lambda \\ R \end{matrix} \right\rangle = \left\langle \begin{matrix} \tilde{b}'' \\ \tilde{b}'' \end{matrix} \right\rangle$
SET $\begin{bmatrix} -A'F'0 \\ D'0\ I \end{bmatrix} = B, \left\langle \begin{matrix} \tilde{Q}' \\ \lambda \\ \tilde{R} \end{matrix} \right\rangle = \langle x \rangle$ and $\left\langle \begin{matrix} \tilde{b}'' \\ \tilde{b}''' \end{matrix} \right\rangle = \tilde{b}$, and solve by "smiplex" method.
change D' and b''
If TYPE of probl.>1, write λ
FSF = FSF + 0.01 λ
No
λ<1 ?
NO
|λ-1| < Tolerance?
Yes
$FSF = FSF - \frac{0.01}{\lambda}$
Yes
write "Safety Factor" FSF
END

$$\tilde{Q} = \tilde{Q}^E + \tilde{Q}^s \qquad (7.1)$$

where $\tilde{Q}^E$ would be the internal forces for the linear elastic problem ignoring yield and

$$\tilde{Q}^s = Z \cdot \tilde{q}_p \qquad (7.2)$$

The matrix Z, representing the "influence" of dislocations on internal "stress" (generalized) would be obtained in a simple way in the case of linear structures such that of Figure 10. In the case of the plane strain problem, we are concerned with, Z could also be obtained but resorting to F.E. elastic solutions or other numerical solutions.

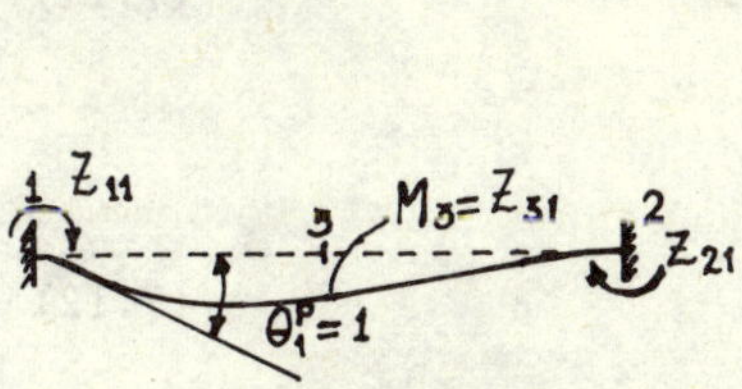

Figure 10.a

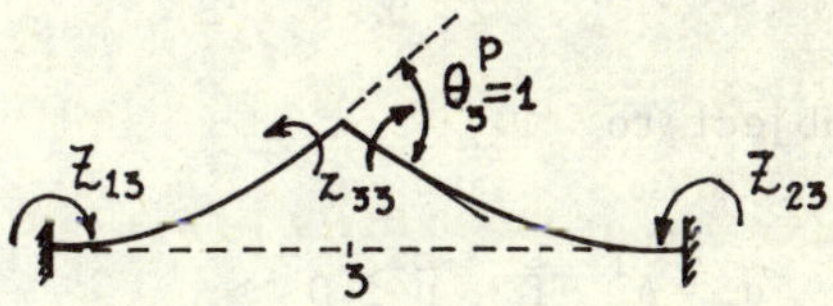

Figure 10.b

The elasto - plastic problem could be put from the begining in the following way, and solved by quadratic programming.

Given the vector of the independent external forces $\tilde{F}$; the bloc diagonal matrix C (S) with the flexibilities (stiffnesses) of the elements of the structure the equilibrium matrix A (5.4), find the structural holonomic response of the structure ($\tilde{Q}$, $\tilde{q}$, $\tilde{f}$).

$\tilde{Q}$ the internal forces (generalized stresses);
$\tilde{q}$ the corresponding generalized strains;

$$\tilde{q} = \tilde{q}_p + \tilde{q}_E{}' \qquad (7.3)$$

$\tilde{q}_p$ = plastic generalized strains; $\tilde{q}_E$ = elastic generalized strains;

$$\tilde{q}_E = C\tilde{Q} \quad \text{(Hooke's law).} \qquad (7.4)$$

$$\tilde{q}_p = D\tilde{u} \quad \text{(normality rule)} \qquad (7.5) = (6.8)$$

$$D\tilde{Q} - \tilde{K} \leq 0 \quad \text{(yield conditions} \qquad (7.6) = (6.3)$$

$$(D\tilde{Q} - \tilde{K}) \cdot \tilde{u} = 0 \quad \text{(holonomy, no-local-unloading conditions)} \qquad (7.7) = (6.9)$$

$$\tilde{u} \geq 0 \qquad (7.8)$$

Maier (1970) showed the above problem may be solved by one of the following quadratic forms:

Find the minimum of objective function

$$\psi(\tilde{Q}) = \frac{1}{2}\tilde{Q}^T C \tilde{Q} \tag{7.9}$$

subject to

$$A\tilde{Q} = \tilde{F}, \quad D\tilde{Q} - \tilde{K} \leq 0; \tag{7.10}$$

or find the minimum of the objective function

$$\phi(\tilde{f}, \tilde{u}) = \frac{1}{2}\tilde{q}_E^T \cdot S\tilde{q}_E - \tilde{F}^T \cdot \tilde{f} + \tilde{K}\tilde{u} \tag{7.11}$$

subject to

$$\tilde{q}_E = A^T \cdot \tilde{f}, \quad \tilde{u} \geq 0 \tag{7.12}$$

$\tilde{q}_E$ may be eliminated from (7.11) after substitution of (7.12):

$$\phi(\tilde{f}, \tilde{u}) = \frac{1}{2}\tilde{f}^T A \cdot S \cdot A^T \tilde{f} - \tilde{F}^T \cdot \tilde{f} + \tilde{K}\tilde{u}. \tag{7.13}$$

The bloc diagonal matrix with the stiffnesses of the elements may be obtained, in our plane strain problem, by means of interpolation or shape functions for the displacement field inside each element, as in the usual way in F.E. methods(e.g., Zienckiewicz 1971).

It should be noticed, however, that we did not put nodes on the vertices of the elements; we did put them on the joints only. This is because our "control" nodes are not the elements themselves as suggested elsewhere (Fonseca, 1980). In this way we avoid the ambiguity of having elements with common nodes, some plastified and some not.

8. CONCLUSIONS

a) The stability of a mass of soil or of rock with joints of any orientation and filled with any kind of materials and any set of external forces can be analysed by linear programming.

b) The stability and the elasto - plastic displacements can be obtained by quadratic programming, associated to the usual F.E. method.

9. ACKNOWLEDGEMENTS

This work have been sponsored by NATO (Research Grant Nº 044.80)

REFERENCES

(1) Backer R. and M. Garber, "Variational Approach to Slope Stability" Proc. 9th I.C.S.M.F.E., Tokyo, 1977, 2, pp. 2 - 12.

(2) Bishop, A.W., "The Use of the Slip Circle in the Stability Analysis of Slopes", Gèotechnique, Vol. 5 (1955), 1 (March), pp. 7 - 17.

(3) Castillo, E. and J. Revilla, "The Calculus of Variations and the Stability of Slopes", Proc. 9th. I.C.S.M.F.E., Tokyo, 1977, 2, pp. 25 - 30.

(4) Coulomb, C.A., "Éssai sur une Application des Règles des Maximes et Minimis à quelques Problèmes de Statique Relatifs à l' Architecture", Mem. Acad. Roy. près. divers Savants, Vol. 7, Paris (1776).

(5) Donato, O.D., "Extension to Continua of Some Minimum Theorems in Elastoplastic Theory", Meccanica, J. IATAM (AIMETA), 3, Nº 4, Dec. 1968, 259 - 264.

(6) Fellenius, W., (1936), "Calculation of the Stability of Earth Dams", Trans. 2nd. Cong. Large Dams, 4, pp. 445.

(7) Fonseca, A.M.A. da, "Plastic Analysis and Synthesis of Plates and Shells by Mathematical Programming", Thesis, I. College, Dep. Civil Engn., 1980.

(8) Janbu, N. "Earth Pressure and Bearing Capacity by Generalized Procedure of Slices", Int. Conf. Soils Mechanics; 2, 207 - 212, London, 1957.

(9) Luenberger, D.G., "Introduction to Linear and Nonlinear Programming", Addison - Wesley Dub. Cº., Reading, Mass., 1972.

(10) Maier, G., "On a Matrix Structural Theory of Piecewise Linear Elastoplacity with Interacting Yield Planes", Meccanica, 5, nº 1, March, 1970, 54 - 66.

(11) Maier, G., "Future Directions in Engineering Plasticity" in Engineering Plasticity by Mathematical Programming, Proc. NATO, ASI, Univ. of Waterloo, Canada, Aug. 2 - 12, 1977, Eds. M.Z. Cohn, G. Maier and D.E. Grierson, Publ. Pergamon Press, pp. 631 - 648.

(12) Martins, J.B., and A.C. Matos (1979) "A Survey of the Methods to

Calculate Safety Against Collapse in Soil and Rock Masses, 3rd. Int. Conf. Num. Meth. in Geomechanics, (1), 339 - 345, Aachen, 2/6 April, 1979.

(13) Martins, J.B., E.B. Reis and A.C. Matos, "New Methods of Analysis for the Stability of Slopes" X. Int. Conf. Soil Mech., F. E., Stockholm, June 1981.

(14) Morgenstern, N.R. and V.E. Price, "A Method for Solving the Equations of Stability of General Slip Surfaces" Computer J., V.9., Nº 4, Feb., 1967.

(15) Morgenstern, N.R. and V.E. Price, "The Analysis of the Stability of General Slip Surfaces" Geotechnique, March, 1965, 79 - 93.

(16) Ramamurthy, T.C., G.P. Narayan and V.P. Bhatkar, "Variational Methodh for Slope Stability Analysis", Proc. 9th. I.C.S.M.F.E., Tokyo, 1977, 139 - 142.

(17) Rankine, W.J.M., "On Stability of Loose Earth", Proc. Phil.Trans. Roy. Soc., London, Vol. 147, 1857, French Trad. 1874.

(18) Zienckiewicz, O.C., "The Finite Element Method in the Engineering Science", McGraw Hill, London, 1971.

(19) Zienckiewicz, O.C. and C. Humpheson, (1977) "Viscoplasticity a generalized model for Description of Soil Behavior" in "Numerical Methods in Geotechnical Engineering", C.S. Desai and J. T. Christian (Eds.), McGraw Hill Book, pp. 116 - 142.

CHAPTER IV

BOUNDARY ELEMENT TECHNIQUES

FUNDAMENTALS OF BOUNDARY ELEMENTS

C. A. Brebbia

Department of Civil Engineering, The University,
Southampton, U.K.

1. INTRODUCTION

The numerical techniques of analysis used in engineering are based on the approximate solution of a set of equations describing a physical problem. The most widely known approximate method is Finite Differences which approximates the governing equations of the problem using local expressions for the variables, generally based on truncated Taylor series. The technique can be interpreted as a special case of the more general weighted residual method (Brebbia [1]).

The Finite Element method has attracted the attention of the analysts largely due to its feature of dividing the continuum into a series of elements, which can be associated with physical parts with well-determined properties. The existing literature on Finite Elements is, by now, very extensive and encompasses structural[2] as well as fluid flow[3] and other types of problems. The method can sometimes be based on variational principles or, more generally, on weighted residual expressions. The great interest that the method attracted stimulated an impressive amount of work on computational techniques and efficient engineering software and originated research into basic physical principles such as variational techniques and weighted residual formulations.

The basic mathematical techniques used in approximate methods can, however, be traced to pre-computer times. They involve different ways of solving the governing equations of the problem, i.e. Galerkin, collocation, least-squares, line techniques, matrix progression or transfer matrices, the combina-

J. B. Martins (ed.), Numerical Methods in Geomechanics, 337–354.

tion of different techniques, etc. The methods reappeared in the Finite Element literature under different names, such as Galerking Finite Elements, Finite Element strip-method, some integration schemes, etc. Another important development in approximate analysis was the development of mixed principles and the realization that physical problems can be expressed and solved in many different ways in accordance with the part or equations of the problem that we need to approximate. The mixed formulations can be traced to Reissner [6] and more specifically for Finite Elements to Pian [7]. An excellent exposition of mixed methods in structural analysis can be found in the book by Washizu [8].

Integral equation techniques were, until recently, considered to be a different type of analytical method, somewhat unrelated to approximate methods such as Finite Differences or Finite Elements. They became popular in Europe through the work of a series of Russian authors such as Muskhelishvili [9], Mikhlin [10], Kupradze [11] and Smirnov [12], but were not very popular with engineers. A predecessor of some of this work was Kellog [13] who applied integral equations for the solution of Laplace's type problems. Integral equation techniques were mainly used in some fluid mechanics and general potential problems and known as the 'source' method, which is now classified as an 'indirect' method of analysis; i.e. the unknowns are not the physical variables of the problem. Work in this method continued during the 1960's and 1970's in the pioneering work of Jaswon [14] and Symm [15], Massonnet [16], Hess [17] and others.

It is difficult to point out precisely who was the first to propose the direct Boundary Integral method of analysis. It is found in a different way in Kupradze's book [11]. It seems fair, however, from the engineering point of view to consider that the method originated in the work of Cruse and Rizzo [18] in elastostatics.

Since the early 1960's a research group at Southampton University started working on the applications of integral equations to solve stress analysis problems. Hadid's [19] partly based on Hajdin's work [20] was published in 1964 and dealth with the use of integral equations in shall analysis. Unfortunately, the presentation of the problem, the difficulty of defining the appropriate Green's functions and the parallel emergence of the Finite Element method all contributed to minimizing the importance of this work. Another important thesis was due to Tomlin [21] who solved stress analysis as well as seepage problems in layer soils. In 1973 the group produced another significant thesis, Watson [22], dealing with three-dimensional elastostatics problems. By then, recent developments in Finite Elements has started to find their way into the formulation of Boundary Integral equations. The idea of using general curved elements originated in an idea by C. A. Brebbia and is contained in another Southampton University thesis presented by Lachat [23] in 1975. Still the question of how to effectively relate the boundary integral equations to other approximate techniques was unresolved. This was done by Brebbia who published a series of papers on the relationship of different approximate methods, culminating in 1978 with the first book [1] for which the title 'Boundary Elements' was used. More recently this work has been expanded to encompass time dependent and non-linear problems [24]. Two important International Conferences were held at Southampton University in 1978 and 1980 and another at Irvine, California in 1981. The edited Proceedings of these Conferences - the only ones so far on this topic - are now standard references [25, 26, 27].

One of the most important recent applications of Boundary Elements has been for the solution of time dependent problems. This has now opened the way for solving more complex problems such as wave propagation, Navier-Stokes flow, elastodynamics, etc. The first Boundary Integral solution for the diffusion equation was due to Rizzo and Shippy [28],who used the Laplace's transform to remove the time dependence. This approach was discussed by Liggett and Liu [29] and some of its disadvantages were pointed out. Chang et al [30] in 1973 solved the problem in the time domain using space and time piecewise constant values for the variables. This approach has been extended by Wrobel and Brebbia [31] who applied to solve complex temperature and other problems. Brebbia and Walker [24] proposed a step by step Finite Difference scheme for the time derivatives. The use of a time dependent fundamental solution, implies that one does not need to perform any finite differences on the time derivative and produces an accurate and efficient solution specially when higher order interpolation functions are employed [32, 33].

The first applications of Boundary Elements in elastodynamics can be traced to Rizzo and Cruse [34, 35] and the technique is now widly accepted for solving soil-structure interaction problems. Alarcón et al [36] used the method to study the response of inbedded formulation to forced vibrations. Cruse's work has been of fundamental importance for the development of Boundary Elements as he has contributed to its application not only in elastostatics and elastodynamics, but also in fracture mechanics [37] where Boundary Elements are much more accurate than Finite Elements.

The Boundary Element method generally offers the advantage over Finite Difference or Finite Elements of working with smaller systems of equations. Because of this, the technique is of interest for solving a range of non-linear problems as well as linear ones. The first application of Boundary Elements for plasticity was presented in 1971 by Swedlow and Cruse [38] in a purely theoretical paper which laid the foundations of all subsequent work. However, they did not present the expressions for initial stresses or strains which are needed to solve the stepwise plastic analysis. These expressions were presented by Mendelson [39] in 1973 for two and three-dimensional bodies although the plane strain formulation was correctly presented by Mukherjee [40] only in 1977. None of the above authors gave the proper expression for strains and stresses at internal points which are required to efficiently solve the plasticity problem. It was Bui [41] who, in 1978, published the correct expressions for strains and stresses for three-dimensional bodies and finally, in 1979, Telles and Brebbia [42, 43] presented the complete formulation for Boundary Elements applied to two and three-dimensional problems.

It may be convenient in many applications to combine Boundary Elements with Finite Elements, as some subregions of the problem may be more amenable to solution using the latter. This may be achieved in either of two ways (1) by considering the whole problem using an equivalent B.E.M. approach or (2) by converting the B.E.M. subregions into equivalent F.E.M. The two approaches are described in detail in Brebbia and Georgiu [44]. As the F.E.M. is very well established, with computer codes readily available, representing the B.E. region by an equivalent finite element matrix formulation seems most attractive. The formulation of this 'equivalent' matrix presents, however, certain problems; at sharp geometric discontinuities, there are also discontinuities of surface tractions which require special attention and the equivalent F.E. matrix formed is most inherently symmetric, unlike the classical F.E. approach.

A technique which overcomes these problems and provides an acceptable F.E. type formulation using the B.E.M. method has been suggested by Georgiu and Brebbia [45]. The technique is based on an idea advanced by Chaudonneret [46] in 1978. An early paper involving another way of symmetrizing the matrices has been published by Zienkiewicz et al [47]. This work has been extended by a recent paper by Mustoe et al [48] who presented an interesting way of overcoming some of the difficulties due to the non-symmetric way in which the integrations are carried out. It involves weighting the Somigliana's identity with the actual shape functions used for the boundary variables. This, in effect, means that instead of applying a point source to form each equation, a distributed source on either side of the node is applied and a numerical integration process which serves the 'influence' of a series of 'sources' as opposed to a single one has to be applied. Unfortunately, this significantly increases the number of integrations involved.

The Boundary Element method is still in a state of early development and many other applications of the technique to non-linear problems are currently being carried out. The Boundary Element method is a technique which offers important advantages over 'domain' type techniques and, consequently, is becoming very popular in engineering practice. One of the most interesting features of the method is the much smaller systems of equations and the considerable reduction generally achieved in the data required to run a problem. In addition the numerical accuracy of Boundary Elements is generally greater than that of Finite Elements. These advantages are more marked in two and three-dimensional problems. The method is also well suited to problem solving with infinite domains such as those occurring in geomechanics, for which the classical domain techniques are unsuitable.

This lecture introduces some basic concepts involved in Boundary Elements approximations, stressing the relationship between the fundamental concepts involved in B.E. and the classical approximation techniques.

2. THE BOUNDARY ELEMENT METHOD AS A WEIGHTED RESIDUAL TECHNIQUE

The engineering analyst usually tries to find an approximate solution to a mathematical problem expressed in function of a differential equation or set of equations. The equation or system of equations is applicable over a certain domain and has a set of boundary conditions. Let us write the equation as,

$$L(u_o) = 0, \qquad \text{in } \Omega \text{ domain} \qquad 2.1$$

with boundary conditions,

$$S(u_o) = s, \text{ essential boundary condition on } \Gamma_1$$
$$\qquad 2.2$$
$$G(u_o) = g, \text{ natural boundary condition on } \Gamma_2$$

$$\text{Total boundary } \Gamma = \Gamma_1 + \Gamma_2$$

One can then propose a discretization of the continuum into elements or cells and assume that over each of these subregions the u_o unknown varies in accordance with certain predetermined shapes or modes. Notice that for the case of Boundary Elements only the surface of the domain needs to be discretized. These shape functions belong to a set of linearly independent functions with unknown coefficients, whose values are determined once the resulting algebraic system of equations is solved.

Let us write the approximations for u_o as a set of linearly independent functions ϕ_i multiplied by a series of unknown α_i coefficients i.e.

$$u_o \simeq u = \alpha_1\phi_1 + \alpha_2\phi_2 + \alpha_3\phi_3 \cdots \qquad 2.3$$

Introducing this approximation to the u_o function into Equations 2.1 and 2.2 will produce certain errors in the Ω domain and on the Γ boundary. The aim of the different types of approximate techniques is to make this error as small as possible. The errors can be made zero at certain points, collocation, finite differences, etc. or they can be distributed over a subregion (collocation by subregions, Galerkin, lest square, cell-collocation, Finite Elements, etc.). If the dimensions of the problems are reduced by one, then one may only be required to distribute the errors on the boundary of the region under consideration. The way in which the error is distributed produces different types of weighted residual techniques and hence approximate methods.

To understand how the distribution can be carried out let us consider a function w which is to be our weighting function and is defined by a set of known linearly independent functions ψ_i multiplied by arbitrary β_i coefficients, i.e.

$$w = \beta_1\psi_1 + \beta_2\psi_2 + \beta_3\psi_3 + \ldots \qquad 2.4$$

Consider in what follows that L is a linear operator for simplicity and that all boundary conditions are identically satisfied by the approximate function u, Equation 2.3. In this case we only need to weight Equation 2.1 by the w function and we can now identify the following techniques.

1) Method of Point Collocation

In this case we simply collocate the L(u) equation at a series of points 'i' and make it equal to zero at these points. We can represent this operation by the following integral expression,

$$\int_\Omega L(u)\, \Delta_i \, d\Omega = 0 \qquad \text{for all } i \qquad 2.5$$

where Δ_i is a Dirac delta function whose integral is equal to 1 at the point 'i' and zero everywhere else. Hence w can be written as,

$$w = \beta_1\Delta_1 + \beta_2\Delta_2 + \beta_3\Delta_3 + \ldots \qquad 2.6$$

2) Method of Moments

The weighting function w is now assumed to be such that moments of the L(u) operator can be taken, i.e.

$$\int_\Omega L(u)\ \psi_i\ d\Omega = 0 \qquad 2.7$$

where ψ_i is defined for a two-dimensional problem for instance, by

$$w = \beta_1 1 + \beta_2 x + \beta_3 y + \beta_4 xy + \ldots \qquad 2.8$$

3) Method of Collocation by Subregions

In this case the integral of the error over a subregion cell or subdomain is set equal to zero. The finite difference technique can be interpreted as a subregion or cell collocation technique Lau and Brebbia [49], for which the value of u is defined over a cell, i.e. the part of the domain connecting the centre point to the neighbouring points, and then we collocate the L(u) operator using Equation 2.5. The main difference with Equation 2.5 is that the u functions of the cell-collocation method are valid only over a part or cell of the Ω domain. In addition to the Dirac delta we can propose over our subregion a simple function such as 1 or other more complex ψ_i weighting functions.

4) Generalized Weighting

In this case the functions ϕ_i and ψ_i belong to two different sets. The ϕ_i satisfy the boundary conditions of the problem and the ψ_i sometimes but not necessarily satisfy the homogeneous version of those conditons. This gives the following weighted residual statement.

$$\int_\Omega L(u)\ \psi_i\ d\Omega = 0 \qquad 2.9$$

5) Galerkin Method

If the weighting functions are chosen to be equal to those in the approximating set, i.e. $\psi_i = \phi_i$, we have a special version of Equation 2.9 called the Galerkin method. Here

$$\int_{\Omega} L(u)\ p_i\ d\Omega = 0 \qquad 2.10$$

6) Least-square Method

The least-square technique is somewhat different to the previous ones. It consists of defining the square of the error function, i.e.

$$\varepsilon^2 = (L(u))^2 \qquad 2.11$$

and minimizing it with respect to the α_i coefficients of u. This gives for a linear operator L,

$$\int_{\Omega} L(u)\ L(\psi_i)\ d\Omega = 0 \qquad 2.12$$

Hence the weighting functions here are not the original ψ_i functions as in Galerkin's but the expressions found by applying the L( ) operator to them.

7) Weak Formulations

Let us assume that the L( ) operator is symmetric and can be decomposed as follows:

$$\int_{\Omega} L(u)\ w\ d\Omega = -\int_{\Omega} D(u)\ D(w)\ d\Omega + \int_{\Gamma} G(u)\ S(w)\ d\Gamma \qquad 2.13$$

where G and S indicate the natural and essential boundary condition operators. Consider for simplicity that the w functions are such that they satisfy the homogeneous version of the essential b.c., i.e. $S(w) \equiv 0$. This gives,

$$\int_{\Omega} L(u)\ w\ d\Omega = -\int_{\Omega} D(u)\ D(w)\ d\Omega \qquad 2.14$$

Notice that the order of derivation for u has decreased and the derivatives of w have correspondingly increased. D( ) is an operator of lower order than L( ).

The classical Rayleigh-Ritz technique can be interpreted as a particular case of Equation 2.14 for which $\psi_i = \phi_i$. Hence the stationarity condition for a Rayleigh-Ritz function ϕ_i can be written as,

$$\int_\Omega D(u)\, D(\phi_i)\, d\Omega = 0 \qquad \text{for all } i \tag{2.15}$$

It is now apparent that the least-square solution, Equation 2.12, implies a higher order weighting than Rayleigh-Ritz Equation 2.15. As it is to be expected, least-square solutions tend to satisfy the governing equation operator more closely than Rayleigh-Ritz. Both methods have the property of giving symmetric matrices for the above type of symmetric operator.

8) Boundary Solutions

In these cases the ϕ_i or the ψ_i set of functions identically satisfy the operator L but not the boundary conditions which are then approximated using any of the above discussed methods.

Consider that the functions of the ϕ_i set satisfy the governing equations but that ψ_i may not do so. The problem consists now of approximating boundary conditions 2.2, i.e.

$$\begin{aligned} S(u) &= s, \text{ on } \Gamma_1 \\ G(u) &= g, \text{ on } \Gamma_2 \end{aligned} \tag{2.16}$$

This can be done by substituting the approximation for u into Equation 2.16 and weighting the boundary error functions as follows:

$$\begin{aligned} &\int_{\Gamma_1} (S(u)-s)\, G(w)\, d\Gamma = 0 \\ &\int_{\Gamma_2} (G(u)-g)\, S(w)\, d\Gamma = 0 \end{aligned} \tag{2.17}$$

Equations 2.17 allow us to find a set of algebraic equations from which the unknown α_i coefficients can be obtained. Notice that the solution implies only boundary integrations, i.e. the order of the problem has been reduced by one.

Another possibility is to propose weighting functions w which satisfy the governing equations, i.e. $L(w) \equiv 0$. We can start with

$$\int_{\Omega} L(u)\ w\ d\Omega = 0 \qquad 2.18$$

and integrate by parts until the L( ) operator passes to the w function. This gives,

$$\int_{\Omega} L(u)\ w\ d\Omega = \int_{\Omega} u\ L(w)\ d\Omega + \int_{\Gamma} G(u)\ S(w)\ d\Gamma - \int_{\Gamma} S(u)\ G(w)\ d\Gamma \qquad 2.19$$

or if $L(w) \equiv 0$,

$$\int_{\Omega} L(u)\ w\ d\Omega = \int_{\Gamma} G(u)\ S(w)\ d\Gamma - \int_{\Gamma} S(u)\ G(w)\ d\Gamma \qquad 2.20$$

We can also write Equation 2.17 using the following conditions; $G(u) = g$ on Γ_2, $S(u) = s$ on Γ_1. This gives,

$$\int_{\Gamma_1} G(u)\ S(w)\ d\Gamma + \int_{\Gamma_2} g\ S(w)\ d\Gamma = \int_{\Gamma_2} S(u)\ G(w)\ d\Gamma + \int_{\Gamma_1} s\ G(w)\ d\Gamma \qquad 2.21$$

Equation 2.21 is a generalization of Equation 2.17.

We can use the above ideas to generalize the weighted residual expressions. Let us propose a more general statement corresponding to approximation of the boundary conditions as well as the domain equation, i.e.

$$\int_{\Omega} L(u)\ w\ d\Omega = \int_{\Gamma_2} (G(u)-g)\ S(w)\ d\Gamma - \int_{\Gamma_1} (S(u)-s)\ G(w)\ d\Gamma \qquad 2.22$$

This gives, after integrating by parts,

$$\int_\Omega D(u)\, D(w)\, d\Omega = \int_{\Gamma_1} (S(u)-s)\, G(w)\, d\Gamma + \int_{\Gamma_2} g\, s(w)\, d\Gamma + \int_{\Gamma_1} G(u)\, S(w)\, d\Gamma \tag{2.23}$$

Integrating again we obtain the inverse relationship i.e.,

$$\int_\Omega u\, L(w)\, d\Omega = \int_{\Gamma_1} s\, G(w)\, d\Gamma + \int_{\Gamma_2} S(u)\, G(w)\, d\Gamma - \int_{\Gamma_2} g\, S(w)\, d\Gamma - \int_{\Gamma_1} G(u)\, S(w)\, d\Gamma \tag{2.24}$$

9) Singular Functions

In certain boundary problems it is convenient to choose weighting functions which are singular and represent the solution of Equation 2.1 without any boundary conditions, i.e. the solution of,

$$L(w^*) = \Delta_i \tag{2.25}$$

where w^* is this special type of singular function. It is identified with an asterisk to denote this and differentiate it from w. Δ_i is a Delta Dirac function whose integral is equal to 1 at the point under consideration and zero everywhere else. In this case Equation 2.19 becomes,

$$u_i + \int_\Gamma G(u)\, S(w^*)\, d\Gamma = \int_\Gamma S(u)\, G(w^*)\, d\Gamma \tag{2.26}$$

Taking this expression to the boundary produces a coefficient 'c' in the limit i.e.,

$$c_i\, u_i + \int_\Gamma G(u)\, S(w^*)\, d\Gamma = \int_\Gamma S(u)\, G(w^*)\, d\Gamma \tag{2.27}$$

This expression reduces to Somigliana's identity and other similar integral relationships depending on the definition of the inner products and the type of physical problem under consideration.

Another possibility when using these w^* type functions is to consider that they are distributed over part of the boundary rather than concentrated at node 'i'. This distribution of the identity with the distribution shape function proposed. Singularity means, in effect, that to form each equation, a distributed source on either side of the 'i' source node is applied. Some authors [48] have proposed to carry out this distribution using a numerical integration scheme. Unfortunately, this significantly increases the number of integrations required and is too expensive for practical applications, although it can be a useful technique to symmetrize B.E. matrices. Notice that if we have the fundamental solution for a distributed source between two points the double integrals would not be needed.

Finally, notice that we can also propose least-square boundary solutions to symmetrize the problem, starting with,

$$\begin{aligned} S(u)-s &= 0 \qquad \text{on } \Gamma_1 \\ G(u)-g &= 0 \qquad \text{on } \Gamma_2 \end{aligned} \tag{2.28}$$

The conditions to satisfy are now

$$\begin{aligned} \int_{\Gamma_1} (S(u)-s)\, S(w) &= 0 \\ \int_{\Gamma_2} (G(u)-g)\, G(w) &= 0 \end{aligned} \tag{2.29}$$

for a given w function which satisfies $L(w) \equiv 0$. The above formulation does not seem to have been applied in practice.

3. CONCLUSIONS

This lecture presents the fundamentals of Boundary Integral solutions and their relationships to other approximate methods of analysis. Weighted residuals have been used to classify and systematize the domain and boundary solutions. This approach appears to be the most appropriate way of looking at the approximate techniques in order to simplify their presentation and understand the relationship between the different methods. The final classification has been schematically presented in Figure 1, where the techniques are divided into two groups, those for which the same basis functions are used for u and w and those for which the basis functions are

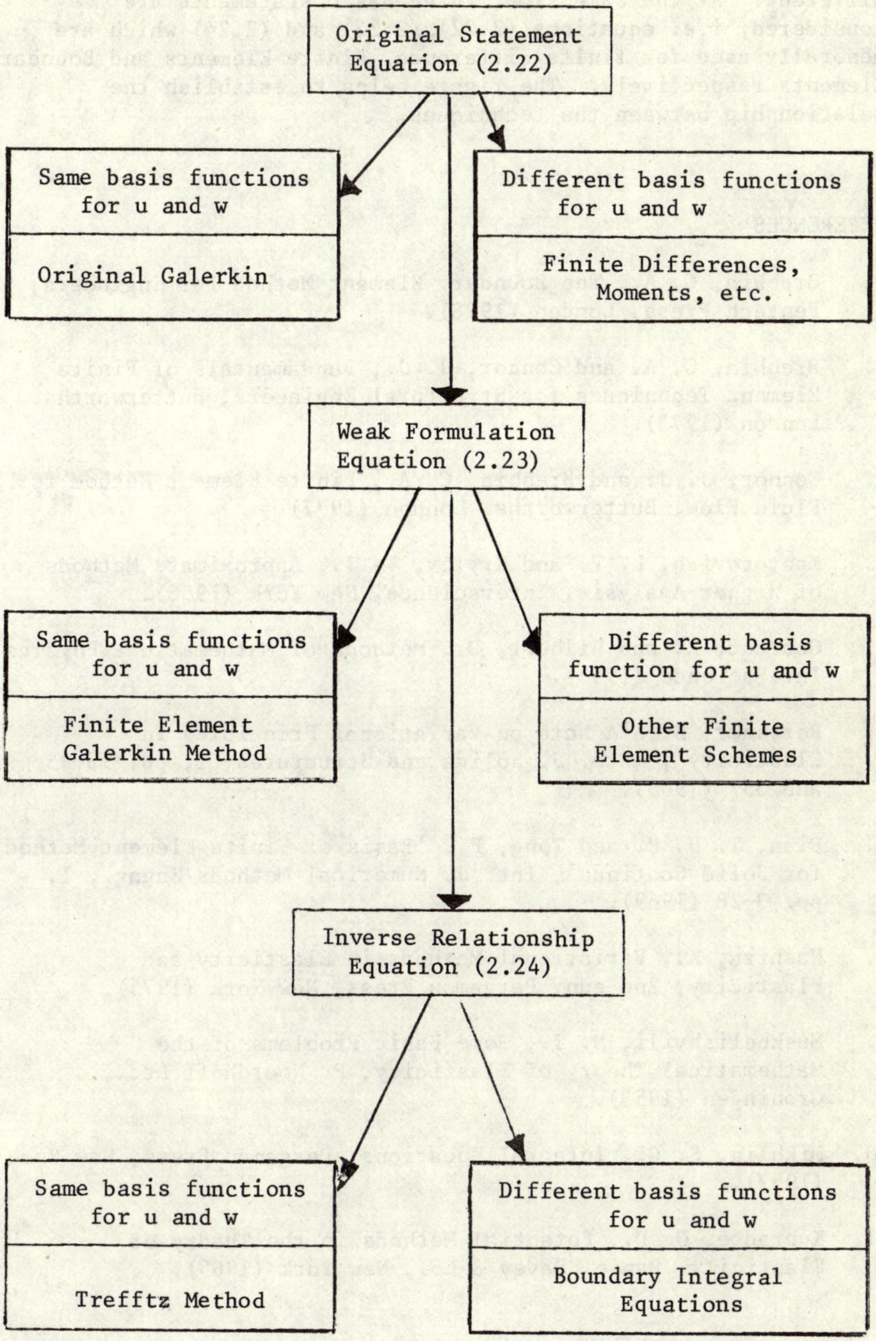
Original Statement
Equation (2.22)
Same basis functions
for u and w
Original Galerkin
Different basis functions
for u and w
Finite Differences,
Moments, etc.
Weak Formulation
Equation (2.23)
Same basis functions
for u and w
Finite Element
Galerkin Method
Different basis
function for u and w
Other Finite
Element Schemes
Inverse Relationship
Equation (2.24)
Same basis functions
for u and w
Trefftz Method
Different basis functions
for u and w
Boundary Integral
Equations

Figure 1

different. At the same time, three basic statements are considered, i.e. equations (2.22), (2.23) and (2.24) which are generally used for finite difference, Finite Elements and Boundary Elements respectively. The figure helps to establish the relationship between the techniques.

REFERENCES

1. Brebbia, C. A., The Boundary Element Method for Engineers, Pentech Press, London (1978).

2. Brebbia, C. A. and Connor, J. J., Fundamentals of Finite Element Techniques for Structural Engineers, Butterworths, London (1973).

3. Connor, J. J. and Brebbia, C. A., Finite Element Method for Fluid Flow, Butterworths, London (1977).

4. Kantorovich, L. V. and Krylov, V. I., Approximate Methods of Higher Analysis, Interscience, New York (1958).

5. Courant, R. and Hilbert, D., Methods of Mathematical Physics, Interscience (1953).

6. Reissner, E., 'A Note on Variational Principles in Elasticity', Into. J. Solids and Structures, 1, pp. 93-95 and 357 (1965).

7. Pian, T. H. H. and Tong, P., 'Basis of Finite Element Method for Solid Continua', Int. J. Numerical Methods Engng., 1, pp. 3-28 (1969).

8. Washizu, K., Variational Methods in Elasticity and Plasticity, 2nd edn, Pergamon Press, New York (1975).

9. Muskhelishvili, N. I., Some Basic Problems of the Mathematical Theory of Elasticity, P. Noordhoff Ltd., Groningen (1953).

10. Mikhlin, S. G., Integral Equations, Pergamon Press, New York (1957).

11. Kupradze, O. D., Potential Methods in the Theory of Elasticity, Daniel Davey & Co., New York (1965).

12. Smirnov, V. J., 'Integral Equations and Partial Differential Equations', in A Course in Higher Mathematics, Vol. IV, Addison-Wesley (1964).

13. Kellogg, P. D., Foundations of Potential Theory, Dover, New York (1953).

14. Jaswon, M. A. 'Integral Equation Methods in Potential Theory, I' Proc. R. Soc., Ser. A, p. 273 (1963)

15. Symm, G. T., 'Integral Equation Methods in Potential Theory II' Proc. R. Soc., Ser. A, p. 275 (1963).

16. Massonnet, C. E., 'Numerical Use of Integral Procedures in Stress Analysis', Stress Analysis, Zienkiewicz, O. C. and Holister, G. S. (eds) Wiley (1966).

17. Hess, J. L. and Smith, A. M. O., 'Calculation of Potential Flow about Arbitrary Bodies'. Progress in Aeronautical Sciences Vol. 8, Kuchemann, D. (Ed). Pergamon Press (1967).

18. Cruse, T. A. and Rizzo, F. J., 'A Direct Formulation and Numerical Solution of the General Transient Elasto-Dynamic Problem I', J. Math. Analysis Applic., 22 (1968).

19. Hadid, H. A., An Analytical and Experimental Investigation into the Bending Theory of Elastic Conoidal Shells, Ph.D. Thesis, University of Southampton (1964).

20. Hajdin, N., 'A Method for Numerical Solutions of Boundary Value Problems and its Application to Certain Problems in the Theory of Elasticity', Belgrade University Publication (1958).

21. Tomlin, G. R., 'Numerical Analysis of Continuum Problems in Zoned Anisotropic Media', Ph.D. Thesis, University of Southampton (1972).

22. Watson, J., Analysis of Thick Shells with Holes by Using Integral Equation Method, Ph.D. Thesis, University of Southampton (1973).

23. Lachat, J. C., A Further Development of the Boundary Integral Technique for Elastostatics, Ph.D. Thesis, University of Southampton (1975).

24. Brebbia, C. A. and Walker, S., The Boundary Element Technique in Engineering, Newnes-Butterworths, London (1979).

25. Brebbia, C. A. (Ed.). Recent Advances in Boundary Element Methods, Proc. 1st Int. Conference Boundary Element Methods, Southampton University, 1978, Pentech Press, London (1978).

26. Brebbia, C. A. (Ed.). New Developments in Boundary Element Methods, Proc. 2nd Int. Conference Boundary Element Methods, Southampton University, 1980, C.M.L. Publications, Butterworths, London, (1980).

27. Brebbia, C. A. (Ed.). Boundary Element Methods. Proc. 3rd Int. Conference Boundary Element Methods, Irvine, California, 1981. C.M.L. Publications - Springer-Verlag, Berlin, (1981).

28. Rizzo, F. J. and Shippy, D. J. 'A Method of Solution for Certain Problems of Transient Heat Conduction', AIAA J., 8, No. 11, pp. 2004-2009 (1970).

29. Liggett, J. A. and Liu, P. L., 'Unsteady Flow in Confined Aquifers: A Comparison of Two Boundary Integral Methods', Water Resources Res., 15, No. 4, pp. 861-866 (1979)

30. Chang, Y. P., Kang, C. S. and Chen D. J., 'The Use of Fundamental Green's Function for the Solution of Problems of Heat Conduction in Anisotropic Media', Int. J. Heat Mass Transfer, 16 pp. 1905-1918 (1973).

31. Wrobel, L. C. and Brebbia, C. A., 'The Boundary Element Method for Steady-State and Transient Heat Conduction' 1st Int. Conf. Numerical Methods in Thermal Problems, Swansea, Taylor, C. and Morgan, K. (Eds.), Pineridge Press, (1979).

32. Wrobel, L. C. and Brebbia, C. A., 'A Formulation of the Boundary Element Method for Axisymmetric Transient Heat Conduction', Int. J. Heat Mass Transfer.

33. Wrobel, L. C. and Brebbia, C. A., 'Boundary Elements in Thermal Problems', Numerical Methods in Heat Transfer, Lewis, R. W. (Ed.), Wiley, Chichester, England (1981).

34. Cruse, T. A. and Rizzo, F. J., 'A Direct Formulation and Numerical Solution of the General Elastodynamic Problem, I', J. Math. Analysis Applic., 22 (1968).

35. Cruse, T. A. and Rizzo, F. J., 'A Direct Formulation and Numerical Solution of the General Transient Elastodynamic Problem, II', J. Math. Analysis Applic., 22 (1968).

36. Alarcon, E. et al. 'Elastodynamics'. Chapter in the Book 'Progress in Boundary Element Methods' (Ed. C. A. Brebbia). Pentech Press, London, Halstead Press, N.Y., 1981.

37. Cruse, T. A. and Vanduren, W., 'Three-Dimensional Elastic Stress Analysis of a Fracture Specimen with Edge Crack', Int. J. Fract. Mech., 7, pp. 1-15 (1971).

38. Swedlow, J. L. and Cruse, T. A., 'Formulation of Boundary Integral Equations for Three-Dimensional Elasto-Plastic Flow', Int. J. Solids Struct., 7, pp. 1673-1683, (1971).

39. Mendelson, A., 'Boundary Integral Methods in Elasticity and Plasticity', Report No. NASA TN D-7418 (1973).

40. Mukherjee, S., 'Corrected Boundary Integral Equation in Planar Thermoelasticity', Int. J. Solids Struct., 13, pp. 331-335 (1977).

41. Bui, H. D., 'Some Remarks about the Formulation of Three-Dimensional Thermoelastoplastic Problems by Integral Equations', Int. J. Solids Struct., No. 14, pp. 935-939 (1978)

42. Telles, J. C. F. and Brebbia, C. A., 'On the Application of the Boundary Element Method to Plasticity', Appl. Math. Modelling, 3, pp. 446-470 (1979).

43. Telles J. C. F. and Brebbia, C. A., 'The Boundary Element Method in Plasticity', in New Developments in Boundary Element Methods, Brebbia, C. A. (Ed.), C.M.L. Publications, England (1980).

44. Brebbia, C. A. and Georgiu, P., 'Combination of Boundary and Finite Elements in Elastostatics', Applied Math. Modelling, 3, pp. 213-220 (1978).

45. Georgiu, P. and Brebbia, C. A., 'On the Combination of Boundary and Finite Element Solution'. Applied Math. Modelling. To be published.

46. Chaudonneret, M., 'On the Discontinuity of the Stress Vector in the Boundary Integral Equation Method for Elastic Analysis', 'Recent Advances in Boundary Element Methods', Brebbia, C. A. (Ed.), Pentech Press, London (1978).

47. Zienkiewicz, O. C., Kelly, D. W. and Bettess, P., 'The Coupling of the Finite Element Method and Boundary Solution Procedures', Int. J. Num. Meth. Engng., 11, 355-375.

48. Mustoe, G., 'A Symmetric Direct Boundary Integral Equation Method for Two-Dimensional Elastostatics', Paper presented at the 2nd Int. Symposium on Boundary Element Methods, Southampton University, March (1980).

49. Lau, P. and Brebbia, C. A., 'The Cell Collocation Method in Continuum Mechanics', Int. J. Mech. Sci., 20, pp. 83-85 (1978).

BOUNDARY ELEMENTS IN GROUNDWATER FLOW PROBLEMS

L. C. Wrobel and C. A. Brebbia

Computational Mechanics Centre, 125 High Street,
Southampton, SO1 OAA England.

ABSTRACT

The present work deals with applications of the Boundary Element Method to several groundwater flow problems. Steady and transient cases are discussed; two-dimensional, axisymmetric and fully three-dimensional problems considered; it is also shown how features like orthotropy and anisotropy, piecewise homogeneous regions and semi-infinite regions can be included in the analysis. Several numerical examples are studied in order to show the wide range of groundwater flow problems which can be efficiently solved using the Boundary Element Method.

INTRODUCTION

The Boundary Element Method (B.E.M.) is now firmly established as an important alternative technique to the prevailing numerical methods in continuum mechanics. Much attention is being given to its development over the last few years, as can be seen by the number of recently published books and proceedings of international conferences on the subject ([1] to [11]).

The technique basically consists of the transformation of the partial differential equation describing the behaviour of the unknown inside and on the boundary of the domain into an integral equation relating only boundary values, and the numerical solution of this equation. If values at internal points are required, they are calculated afterwards from the boundary data. Since all numerical approximations take place only at the boundaries, the dimensionality of the problem is reduced by one

J. B. Martins (ed.), Numerical Methods in Geomechanics, 355–374.

and smaller systems of equations obtained in comparison with those achieved through differential methods.

The present paper attempts to review some applications of the B.E.M. to groundwater flow problems. Steady and transient cases are discussed; two-dimensional, axisymmetric and fully three-dimensional problems considered; it is shown how orthotropy and anisotropy can be introduced into the analysis, as well as regions with different material properties. Several numerical examples are included in order to show the wide range of groundwater flow problems which can be efficiently solved using the B.E.M.

BASIC FORMULATION

In this work, we consider saturated groundwater flow problems satisfying Darcy's law. The principle of conservation of mass requires that the divergence of the seepage velocity equals zero,

$$\frac{\partial v_1}{\partial x_1} + \frac{\partial v_2}{\partial x_2} + \frac{\partial v_3}{\partial x_3} = 0 \qquad (1)$$

For an anisotropic soil, since

$$v_i = -k_{ij}\frac{\partial u}{\partial x_j} \qquad i,j = 1,2,3 \qquad (2)$$

where k_{ij} is the permeability tensor and u the piezometric head (or velocity potential), equation (1) becomes

$$k_{ij}\frac{\partial^2 u}{\partial x_i \partial x_j} = 0 \qquad (3)$$

in which the summation convention for repeated indices is implied.

If the medium is orthotropic, i.e. the principal axes of k can be used as the coordinate axes, equation (3) reduces to

$$k_{11}\frac{\partial^2 u}{\partial x_1^2} + k_{22}\frac{\partial^2 u}{\partial x_2^2} + k_{33}\frac{\partial^2 u}{\partial x_3^2} = 0 \qquad (4)$$

where k_{ii} is the permeability of the medium in the direction of orthotropy i.

A simpler expression is obtained for the case when the medium is isotropic,

$$k \nabla^2 u = 0 \tag{5}$$

i.e. the governing equation becomes a Laplace equation for the piezometric head.

The boundary conditions of the problem may be of two types,

$$u = \bar{u} \quad \text{on } \Gamma_1$$
$$q = k_{ij} \frac{\partial u}{\partial x_j} n_i = \bar{q} \quad \text{on } \Gamma_2 \tag{6}$$

where n_i are the direction cosines of the outward normal to the boundary, $\bar{u}$ and $\bar{q}$ are prescribed values.

For simplicity, let us first discuss the isotropic case. Using Green's third identity [1], [12] or the weighted residual technique [2], [3], the problem represented by equation (5) with boundary conditions (6) can be recast into a boundary integral equation of the form,

$$c(P)\, u(P) + \int_\Gamma u(S)\, q^*(P,S)\, d\Gamma(S) = \int_\Gamma q(S)\, u^*(P,S)\, d\Gamma(S) \tag{7}$$

where the coefficient c is a function of the solid angle of the boundary at the point P [2]. The function u^* is the fundamental solution to equation (5) and q^* its derivative with respect to the outward normal at S. The fundamental solution to Laplace's equation is well known [1], [2], [12]

$$u^*(P,S) = \frac{1}{4\pi k\; r(P,S)} \quad \text{in 3-D} \tag{8}$$

$$u^*(P,S) = \frac{1}{2\pi k} \ln \frac{1}{r(P,S)} \quad \text{in 2-D} \tag{9}$$

where r(P,S) is the distance between P and S (figure 1).

The fundamental solution for axisymmetric problems can be obtained by integrating the three-dimensional one around a ring [13],

$$u^*(P,S) = \frac{K(m)}{\pi k \{[R(P)+R(S)]^2 + [Z(P)-Z(S)]^2\}^{\frac{1}{2}}} \tag{10}$$

in which K is the complete elliptic integral of the first kind and

$$m = \frac{4R(P)\ R(S)}{[R(P)+R(S)]^2 + [Z(P)-Z(S)]^2} \tag{11}$$

Note that, unlike the two and three-dimensional cases, the axisymmetric fundamental solution cannot be written as simply a function of the distance between two points, but it also depends on the distance of the points to the axis of revolution.

Equation (7) can be solved through a numerical approach which generally consists of the following steps (see, for instance, [1], [2], [3]):

a) The boundary Γ is discretised into a series of elements over which the function u and its normal derivative q are assumed to vary according to interpolation functions. The geometry of these elements can be modelled using straight lines, circular arcs, parabolas, etc. (for the two-dimensional and axisymmetric cases), flat or curved triangles or quadrilaterals (for the three-dimensional case);

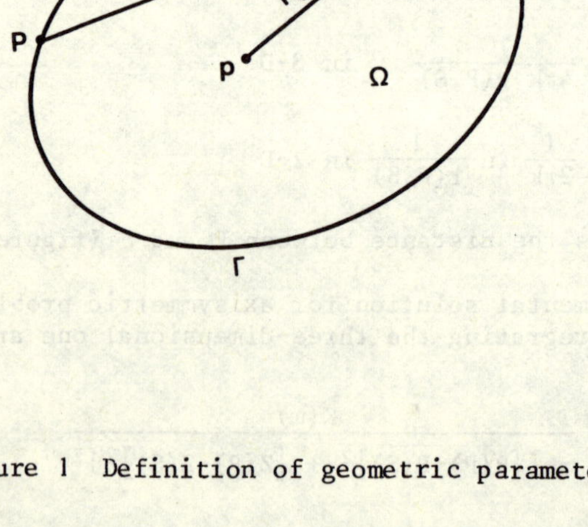

Figure 1 Definition of geometric parameters

b) By using the method of collocation, the discretised equation is applied to a number of particular nodes within each element where values of u and q are associated;

c) The integrals over each element are carried out by using, in general, a numerical quadrature scheme;

d) By imposing the prescribed boundary conditions of the problem, a system of linear algebraic equations is obtained. The solution of this system, which can be effected using direct or iterative methods, produces the remaining boundary data.

If the medium over which the problem is defined is orthotropic, the governing equation (4) can also be transformed into an integral equation of the type (7). The two-dimensional fundamental solution to the problem, for instance, then becomes [2],

$$u^*(P,S) = \frac{1}{2\pi(k_{11}\ k_{22})^{\frac{1}{2}}} \ln \frac{1}{r(P,S)} \tag{12}$$

where

$$r(P,S) = \left\{\frac{1}{k_{11}} [x_1(P)-x_1(S)]^2 + \frac{1}{k_{22}} [x_2(P)-x_2(S)]^2\right\} \tag{13}$$

and

$$q^*(P,S) = k_{11} \frac{\partial u^*(P,S)}{\partial x_1(S)} n_1(S) + k_{22} \frac{\partial u^*(P,S)}{\partial x_2(S)} n_2(S) \tag{14}$$

Similarly, for an anisotropic medium, we have [14]

$$u^*(P,S) = \frac{1}{2\pi|k_{ij}|^{\frac{1}{2}}} \ln \frac{1}{r(P,S)} \tag{15}$$

where $|k_{ij}|$ is the determinant of the medium permeability coefficients matrix and

$$r(P,S) = \left\{\frac{1}{k_{11}} [x_1(P)-x_1(S)]^2 + \frac{2}{k_{12}} [x_1(P)-x_1(S)][x_2(P)-x_2(S)]\right.$$

$$\left. + \frac{1}{k_{22}} [x_2(P)-x_2(S)]^2\right\}^{\frac{1}{2}} \tag{16}$$

$$q^*(P,S) = \left[k_{11} \frac{\partial u^*(P,S)}{\partial x_1(S)} + k_{12} \frac{\partial u^*(P,S)}{\partial x_2(S)}\right] n_1(S)$$

$$+ \left[k_{12} \frac{\partial u^*(P,S)}{\partial x_1(S)} + k_{22} \frac{\partial u^*(P,S)}{\partial x_2(S)}\right] n_2(S) \quad (17)$$

If the problem under consideration is defined over a region which is only piecewise homogeneous, the numerical procedures previously described can be applied to each homogeneous sub-region as they were separated from the others. The final system of equations for the whole region is obtained by adding the set of equations for each sub-region together with compatibility and equilibrium conditions between their interfaces [2].

The fundamental solutions we have considered so far can be recognised as Green's functions for infinite regions. In some practical problems, the problem region may be confined in some regular way and it may be more convenient to find a fundamental solution specific to the region [3].

As an example, let us derive the fundamental solution for a semi-infinite region. Figure 2 represents a problem of groundwater flow round a tunnel which runs under a river. In a problem of this nature, it is preferable to remove the infinite boundary $\bar{\Gamma}$. By choosing a fundamental solution which identically satisfies the boundary condition on the surface $\bar{\Gamma}$ we shall not need to discretise this surface, thus considerably reducing the amount of numerical work involved in the solution of the problem.

Consider a source of intensity $\sigma(S)$ at a point $S \in \Gamma$ (figure 2). The potential generated by this source will somehow be reflected at the surface $\bar{\Gamma}$, depending on the boundary condition applied there. In order to represent this reflection, we shall introduce an image source of intensity $\sigma(S')$ at a point S' symmetrically located with respect to $\bar{\Gamma}$. Thus, the potential at any point p will be the superposition of the ones generated by both sources, i.e.

$$u(p) = \sigma(S)\, u^*(p,S) + \sigma(S')\, u^*(p,S') \quad (18)$$

where u^* is the infinite space fundamental solution.

Applying the boundary condition $u = 0$ at the surface $\bar{\Gamma}$, we obtain

$$\sigma(S)\, u^*(p,S) + \sigma(S')\, u^*(p,S') = 0 \quad \text{on } x_2(p) = 0 \quad (19)$$

For two-dimensional problems, condition (19) implies that

$$\sigma(S)\ \ell n \frac{1}{\{[x_1(p)-x_1(S)]^2 + [-x_2(S)]^2\}^{\frac{1}{2}}}$$

$$+ \sigma(S')\ \ell n \frac{1}{\{[x_1(p)-x_1(S)]^2 + [x_2(S)]^2\}^{\frac{1}{2}}} = 0 \tag{20}$$

which gives

$$\sigma(S) = -\sigma(S') \tag{21}$$

Since, by definition, the fundamental solution is equivalent to the potential generated by unit sources, the fundamental solution for the semi-infinite region with zero potential at the interface is simply

$$u^*(p,s) = \frac{1}{2\pi k}\ \ell n \left\{\frac{[x_1(p)-x_1(s)]^2 + [x_2(p)-x_2(s)]^2}{[x_1(p)-x_1(s)]^2 + [x_2(p)+x_2(s)]^2}\right\}^{\frac{1}{2}} \tag{22}$$

If the boundary condition at the interface $\bar{\Gamma}$ is that of zero normal flux, i.e.

$$\sigma(S)\ q^*(p,S) + \sigma(S')\ q^*(p,S') = 0 \quad \text{on } x_2(p) = 0 \tag{23}$$

implying the condition

$$-\sigma(S) \frac{x_2(S)}{[x_1(p)-x_1(S)]^2 + [-x_2(S)]^2} + \sigma(S') \frac{x_2(S)}{[x_1(p)-x_1(S)]^2 + [x_2(S)]^2}$$

$$= 0 \tag{24}$$

which gives,

$$\sigma(S) = \sigma(S') \tag{25}$$

then the fundamental solution of the problem is

$$u^*(p,s) = \frac{1}{2\pi k}\ \ell n\ (\{[x_1(p)-x_1(s)]^2 + [x_2(p)-x_2(s)]^2\}^{\frac{1}{2}}\ \{[x_1(p)-x_1(s)]^2$$

$$+ [x_2(p)+x_2(s)]^2\}^{\frac{1}{2}}) \tag{26}$$

Fundamental solutions for other problems such as parallel layered regions can be constructed in the same way, as well as three-dimensional fundamental solutions.

The B.E.M. can also be applied to study transient groundwater flow problems, such as unsteady flows in confined aquifers. Such flows are described by a diffusion equation. Considering, for simplicity, that the medium is homogeneous and isotropic and there is no recharge or withdrawal of fluid inside the aquifer, the governing equation becomes,

$$T \nabla^2 u = \lambda \frac{\partial u}{\partial t} \qquad (27)$$

where T is the transmissivity, λ is the storage coefficient and u the hydraulic head.

The boundary conditions of the problem are of the type (6) and, since the problem is now time-dependent, some initial conditions at $t = t_o$ must also be given,

$$u = u_o \quad \text{in } \Omega \qquad (28)$$

The problem described by equation (27) with boundary conditions (6) and initial conditions (28) can be recast into a space-time boundary integral equation of the form [15], [16], [17]

$$c(P)\, u(P,t_F) + \alpha \int_{t_o}^{t_F} \int_{\Gamma} u(S,t)\, q^*(P,S,t_F,t)\, d\Gamma(S)\, dt$$

$$= \alpha \int_{t_o}^{t_F} \int_{\Gamma} q(S,t)\, u^*(P,S,t_F,t)\, d\Gamma(S)\, dt +$$

$$+ \int_{\Omega} u_o(s,t_o)\, u^*(P,s,t_F,t_o)\, d\Omega(s) \qquad (29)$$

where $\alpha = T/\lambda$. The fundamental solutions to equation (27) are as follows [18], [19], [20],

$$u^*(P,S,t_F,t) = \frac{1}{8(\pi\alpha\tau)^{3/2}} \exp\left[-\frac{r(P,S)}{4\alpha\tau}\right] \qquad (3\text{-D})$$

$$u^*(P,S,t_F,t) = \frac{1}{4\pi\alpha\tau} \exp\left[-\frac{r(P,S)}{4\alpha\tau}\right] \qquad (2\text{-D})$$

$$u^*(P,S,t_F,t) = \frac{\pi}{4(\pi\alpha\tau)^{3/2}} \exp\left[-\frac{R^2(P)+R^2(S)+[Z(P)-Z(S)]^2}{4\alpha\tau}\right]$$

$$\times \quad I_o\left[\frac{R(P)R(S)}{2\alpha\tau}\right] \qquad \text{(axisymmetric)}$$

where $\tau = t_F - t$ and I_o is the modified Bessel function of the first kind of order zero.

For the solution of equation (29), a time-stepping technique is employed. This permits the time integrals in the equation to be carried out analytically. The resulting boundary integral equation is solved numerically for each time step. Two different time-marching schemes can be adopted: the first treats each time step as a new problem and so, at the end of each step, computes values of the function u at a sufficient number of internal points in order to use them as pseudo-initial values for the next step; in the other, the time integration process always starts at time t_o and so, despite the increasing number of intermediate steps as the time progresses, values of u at internal points need not be recomputed. Furthermore, if u_o satisfies Laplace's equation, the domain integral in equation (29) can be transformed into equivalent boundary integrals. A comprehensive discussion of both time-marching schemes, including all the necessary procedures for their numerical implementation and a comparison of their accuracy and computer efficiency, can be found in [17].

EXAMPLES OF APPLICATION

Groundwater Flow Round a Tunnel [17]

The first example analysed was the one represented in figure 2, that is, a two-dimensional problem of groundwater flow round a tunnel with permeable invert. As seen in the previous section, if the medium is homogeneous and isotropic the problem is reduced to that of Laplace's equation for the groundwater pressure u. The boundary conditions of the problem are,

$$u = d \qquad \text{on } \overline{\Gamma}$$

$$u = 0 \qquad \text{on } \Gamma_1$$

$$q = -\cos\theta \qquad \text{on } \Gamma_2$$

where d is the depth of the river and θ is the angle measured from the vertical (see figure 2). The surface Γ_1 is the permeable invert of the tunnel where we assume that water flows in freely and Γ_2 is the impermeable part of the tunnel lining, where the condition of no flow across the surface holds. Notice that for a point at infinity we have the condition $u = d-x_2$.

The problem can be reformulated by subtracting out the solution at infinity. The groundwater pressure u is divided into two parts,

$$u = u_1 + u_2$$

where $u_1 = d-x_2$ satisfies the infinity condition. Then we have that u_2 tends to zero at infinity and furthermore, that $\nabla^2 u_2 = 0$ such that the problem can now be solved in terms of u_2. The boundary conditions for u_2 become then,

$$\begin{aligned} u_2 &= 0 && \text{on } \bar{\Gamma} \\ u_2 &= -(d+a-r_t\cos\theta) && \text{on } \Gamma_1 \\ q_2 &= 0 && \text{on } \Gamma_2 \end{aligned} \qquad (30)$$

where r_t is the radius of the tunnel and a is the distance of the centre of the tunnel below the bottom of the river. The numerical values adopted for the parameters are $d = 60$, $a = 30$, $r_t = 3.5$ and $A\hat{O}B = 3\pi/4$.

The same computer program that uses the infinite space fundamental solution can be applied to this problem by specifying antisymmetry with respect to the x_1-axis, thus also satisfying condition (21). The solution of the new problem, i.e. Laplace's equation in the infinite domain bounded internally by Γ and its image Γ', with boundary conditions $u_2 = -u_1$ on Γ_1, $u_2 = u_1$ on Γ'_1, $q_2 = 0$ on Γ_2 and Γ'_2, is identical in the lower half-space to the solution of the original problem with boundary conditions (30). By taking into account the symmetry with respect to the x_2-axis, only one half of the tunnel surface needs to be discretised.

Results for the function u_2 at some boundary points are presented in table 1 for three different discretisations using linear boundary elements and compared to a finite element solution [21] obtained by discretising the whole semi-infinite region into 152 triangular finite elements plus some infinite elements (see figure 3). The discrepancy between both solutions is due to the coarseness of the finite element mesh around the tunnel (see figure 3), which does not take into account properly the discontinuity of the radial flux at the point B.

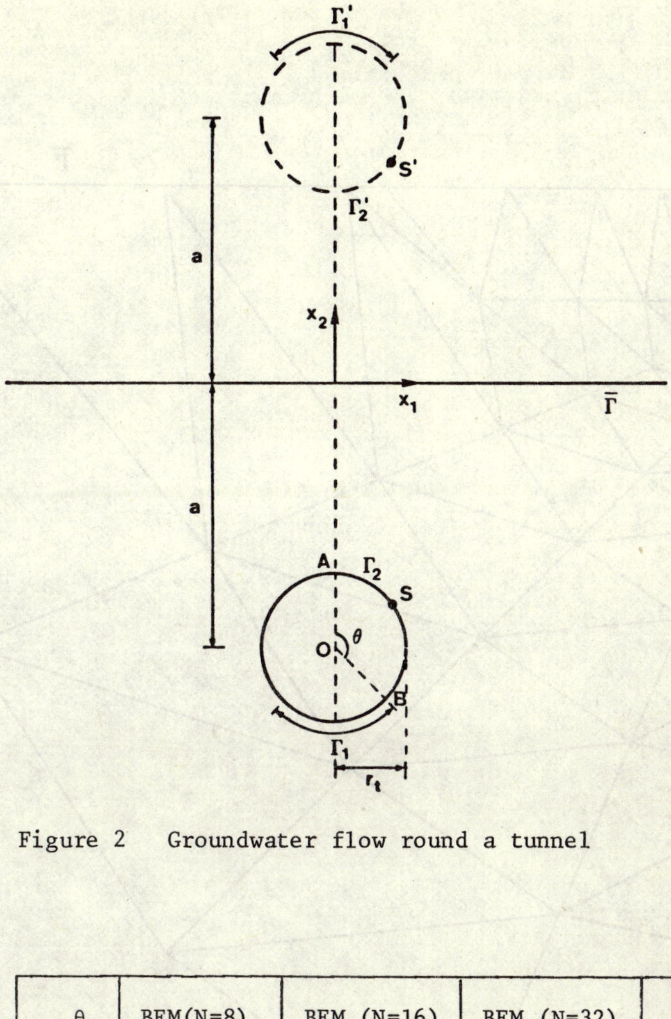

Figure 2 Groundwater flow round a tunnel

θ	BEM(N=8)	BEM (N=16)	BEM (N=32)	FEM
0	30.51	29.08	28.87	33.44
π/8	31.52	30.03	29.83	34.51
π/4	34.60	32.87	32.73	37.76
3π/8	39.96	38.05	37.85	43.46
π/2	48.24	45.96	45.72	52.34
5π/8	61.43	58.42	58.10	66.46

Table 1 - Values of u_2 (-) on tunnel surface

Figure 3 Finite element mesh for groundwater flow round a tunnel

Flow under Dam with Two Different Orthotropic Strata [22]

The geometry of the problem can be seen in figure 4. The coefficients of permeability of the soil in the lower layer are $k_{11} = 0.25 \times 10^{-5}$m/s and $k_{22} = 0.075 \times 10^{-5}$m/s; the principal axes of permeability of the upper stratum makes an angle of 45° with the horizontal and the coefficients of permeability are $k_{11} = 4. \times 10^{-5}$m/s and $k_{22} = 1. \times 10^{-5}$m/s. The dam retains 20m of water upstream and has 5m of tailwater downstream.

The pressure distribution on the dam base, the equipotential lines and the flow velocities at various points under the dam are plotted in figure 4. The results were obtained by discretising the boundary of the two different regions (including their interface) using 74 constant boundary elements, as shown in the figure.

Flow in Axisymmetric Region [13]

The following example is a finite hollow cylinder $a<R<b$, $-\ell<Z<\ell$ (figure 5) subject to the following boundary conditions,

$u = 0$ at $R = b$ and $Z = \pm\ell$

$q = 0$ at $R = a$, $-\ell<Z<-L$

$q = 0$ at $R = a$, $L<Z<\ell$

$q = V$ at $R = a$, $-L<Z<L$

This would correspond to the case of flow through a porous medium for an axisymmetric specimen. An analytical solution of this problem can be found by employing an infinite series expansion in terms of Bessel functions [19].

Due to the symmetry with respect to the R-axis, only one half of the cylinder boundary needed to be discretised into 14 equal linear elements. The numerical values assumed were $a = 2$, $b = 6$, $\ell = 5$, $L = 2$, $k = 1$ and $V = 1$. Results are compared in table 2, showing good agreement.

Flow in Three-Dimensional Region [17]

Another simple problem was considered in this example of flow through a unit cubic block of a porous medium with Dirichlet boundary conditions as follows (see figure 6),

$u = 1$ at $x_1 = 0.5$

$u = 2$ at $x_1 = -0.5$

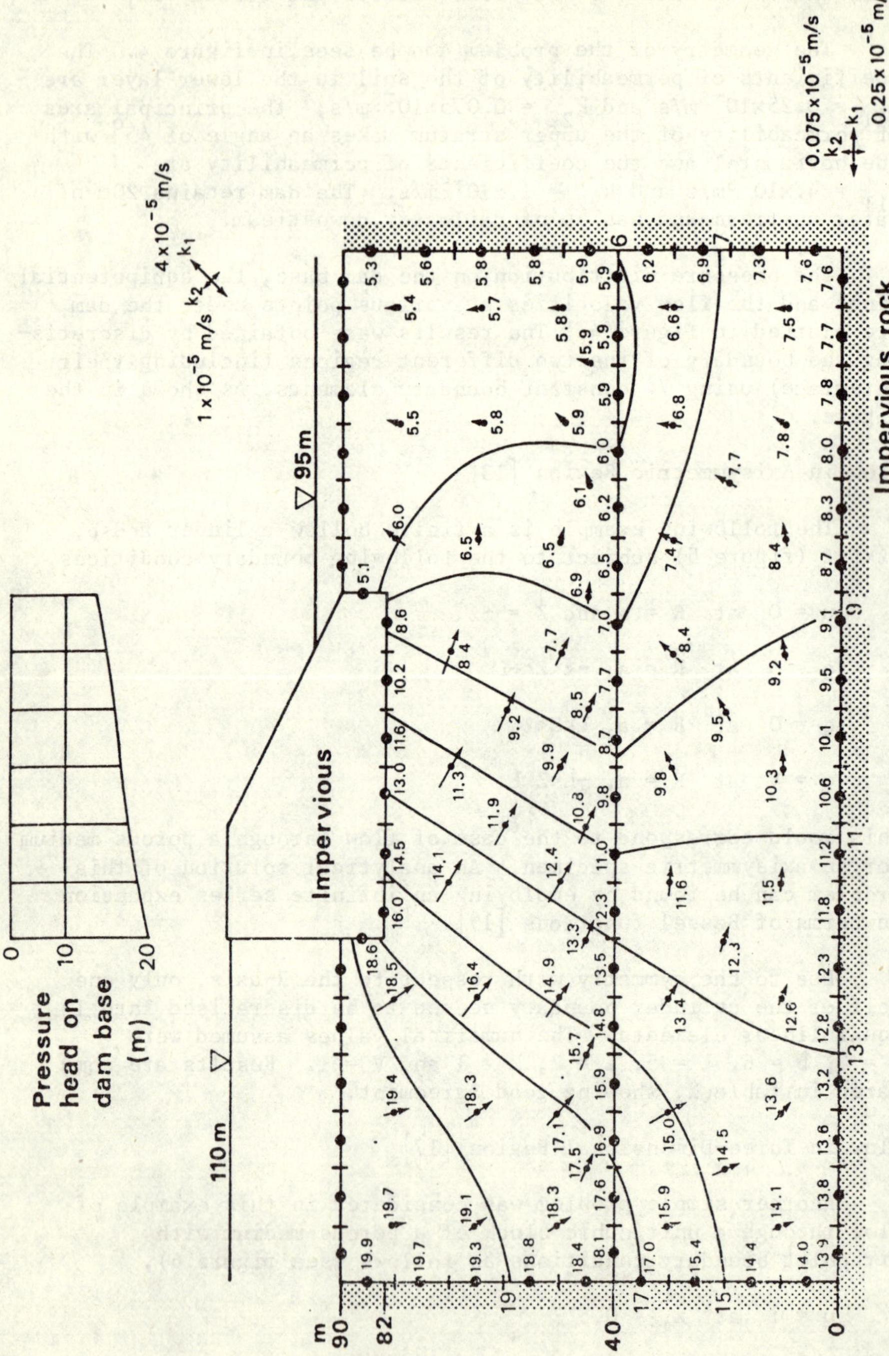

Figure 4 Seepage flow in two orthotropic soils

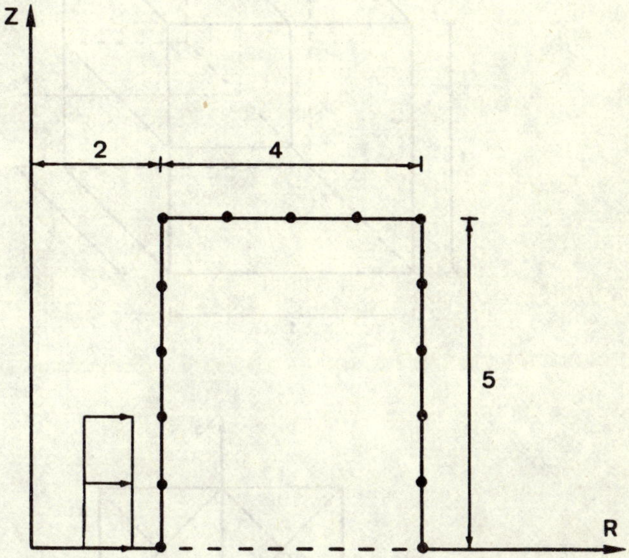

Figure 5 Discretisation of hollow cylinder

R	Z	BEM	ANALYTICAL
3.0	4.0	0.141	0.141
3.0	3.0	0.320	0.319
3.0	2.0	0.556	0.556
3.0	1.0	0.760	0.761
3.0	0.0	0.831	0.831
5.0	4.0	0.043	0.043
5.0	3.0	0.088	0.088
5.0	2.0	0.133	0.133
5.0	1.0	0.167	0.167
5.0	0.0	0.180	0.180

Table 2 - Values of u at internal points

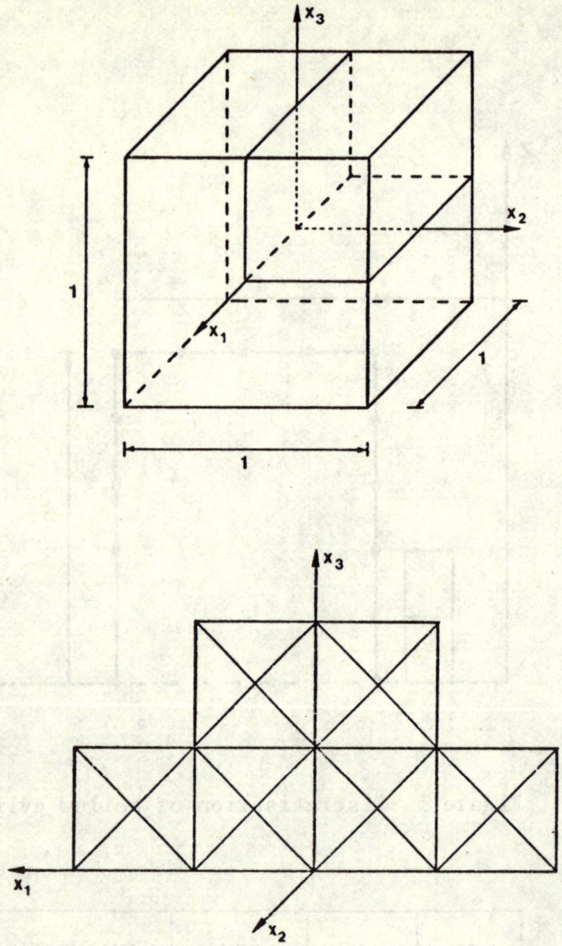

Figure 6 Unit cube: a) Geometry; b) Discretisation

x_1	BEM (12 elements)	BEM (24 elements)	ANALYTICAL
-0.375	1.637	1.472	1.430
-0.250	1.044	0.979	0.967
-0.125	0.678	0.661	0.659
0.	0.500	0.500	0.500
0.125	0.478	0.472	0.472
0.250	0.597	0.566	0.560
0.375	0.855	0.770	0.748

Table 3 - Values of u along the x_1-axis

$u = 0$ at $x_2 = \pm 0.5$

$u = 0$ at $x_3 = \pm 0.5$

Due to symmetry with respect to the planes $x_1 - x_2$ and $x_1 - x_3$, only one quarter of the actual cube needed to be analysed. Two different meshes of constant triangular boundary elements were employed, the finer of which is shown in figure 6. For simplicity, the parameter k was assumed to be unity. Results are presented in table 3, compared with an analytical solution of the problem [19].

Unsteady Flow in Confined Aquifer [23]

This example consists of finding the variation of the hydraulic head with time in a rectangular block of a porous medium. For simplicity, we assume that both the storage and transmissivity coefficients are equal to unity. Initially, the hydraulic head is zero everywhere. The boundary conditions of the problem are as follows,

$u = 1$ at $x_1 = \pm L$

$q = 0$ at $x_2 = \pm \ell$

Due to symmetry, only one quarter of the region was discretised into 9 constant boundary elements. The numerical values adopted for the cross-section were $\ell = 4$, $L = 5$. Results are shown in figures 7 and 8, compared with an available analytical solution [19].

CONCLUSIONS

In this paper, several applications of the B.E.M. to groundwater flow problems have been presented. More realistic geotechnical problems including piecewise homogeneous media, orthotropy and anisotropy were analysed. It was also shown that, for problems involving semi-infinite regions, specific fundamental solutions which identically satisfy the boundary conditions at the interface can considerably reduce the amount of numerical work involved in the solution of the problem.

The solution of time-dependent problems such as unsteady flows in confined aquifers was also discussed. Although a time-stepping technique is usually employed, the fundamental solution itself is time-dependent and this allows for large time steps to be adopted.

In conclusion, the Boundary Element Method is an efficient

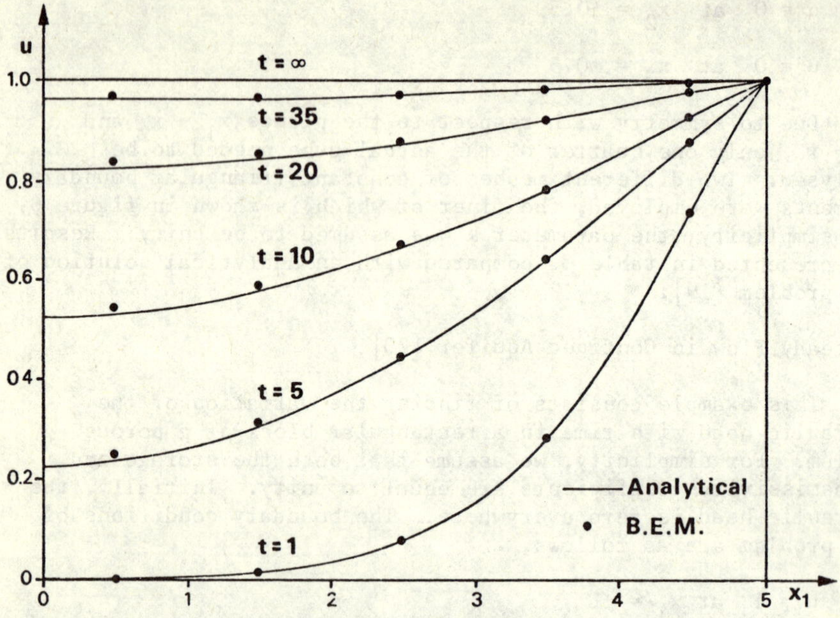

Figure 7 Values of u along the x_1-axis

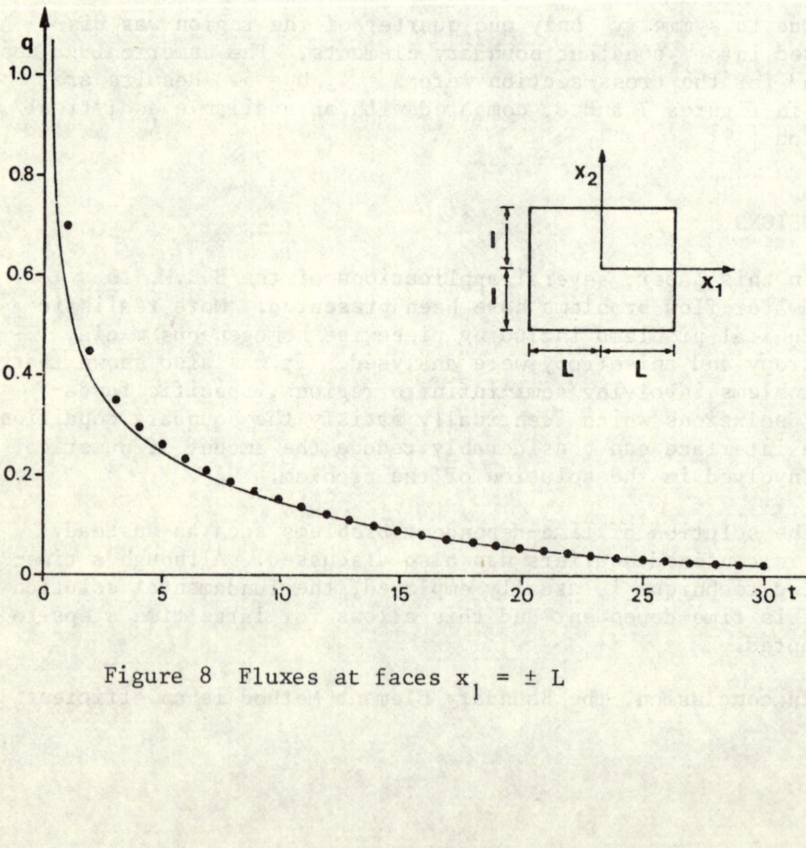

Figure 8 Fluxes at faces $x_1 = \pm L$

and powerful technique for solving many geotechnical problems of groundwater flow and engineers should be made aware of its potentialities.

REFERENCES

1. M. A. Jaswon and G. T. Symm, Integral Equation Methods in Potential Theory and Elastostatics, Academic Press, London, 1977.

2. C. A. Brebbia, The Boundary Element Method for Engineers, Pentech Press, London, 1978.

3. C. A. Brebbia and S. Walker, Boundary Element Techniques in Engineering, Newnes-Butterworths, London, 1980.

4. T. A. Cruse and F. J. Rizzo (eds.), Boundary Integral Equation Method: Computational Applications in Applied Mechanics, ASME, AMD-11, New York, 1975.

5. T. A. Cruse, J. C. Lachat, F. J. Rizzo and R. P. Shaw (eds.), First Int. Symp. on Innovative Numerical Analysis in Applied Engineering Science, CETIM, Versailles, 1977.

6. C. A. Brebbia (ed.), Recent Advances in Boundary Element Methods, Pentech Press, London 1978.

7. P. K. Banerjee and R. Butterfield (eds.), Developments in Boundary Element Methods, Applied Science Publishers, London, 1979.

8. C. A. Brebbia (ed.), New Developments in Boundary Element Methods, C.M.L. Publications, Southampton, 1980.

9. R. P. Shaw et al. (eds.), Innovative Numerical Analysis for the Engineering Sciences, University Press of Virginia, Charlotesville, 1980.

10. C. A. Brebbia (ed.), Progress in Boundary Elements, Vol. 1, Pentech Press, London, 1981.

11. C. A. Brebbia (ed.), Boundary Element Methods, Springer-Verlag, Heidelberg, 1981.

12. O. D. Kellogg, Foundations of Potential Theory, Springer-Verlag, Berlin, 1929.

13. L. C. Wrobel and C. A. Brebbia, Axisymmetric Potential Problems, in [8].

14. Y. P. Chang, C. S. Kang and D. J. Chen, The use of Fundamental Green's Functions for the Solution of Problems of Heat Conduction in Anisotropic Media, Int. J. Heat Mass Transfer, Vol. 16, pp. 1905-1918, 1973.

15. L. C. Wrobel and C. A. Brebbia, Boundary Elements in Thermal Problems, Chapter 5 in Numerical Methods in Heat Transfer, R. W. Lewis, K. Morgan and O. C. Zienkiewicz (eds.), J. Wiley, Chichester, 1981.

16. L. C. Wrobel and C. A. Brebbia, Time-dependent Potential Problems, Chapter 6 in [10].

17. L. C. Wrobel, Potential and Viscous Flow Problems using the Boundary Element Method, Ph.D. Thesis, Southampton University, 1981.

18. P. M. Morse and H. Feshbach, Methods of Theoretical Physics, McGraw-Hill, New York, 1953.

19. H. S. Carslaw and J. C. Jaeger, Conduction of Heat in Solids, 2nd edn, Clarendon Press, Oxford, 1959.

20. L. C. Wrobel and C. A. Brebbia, A Formulation of the Boundary Element Method for Axisymmetric Transient Heat Conduction, Int. J. Heat Mass Transfer, Vol. 24, pp. 843-850, 1981.

21. W. L. Wood, On the Finite Element Solution of an Exterior Boundary-value Problem, Int. J. Num. Meth. Engng, Vol. 10, pp. 885-891, 1976.

22. O. V. Chang, Boundary Elements Applied to Seepage Problems in Zoned Anisotropic Soils, M.Sc. Thesis, Southampton University, 1979.

23. C. A. Brebbia and L. C. Wrobel, Steady and Unsteady Potential Problems using the Boundary Element Method, Chapter 1 in Recent Advances in Numerical Methods in Fluids, C. Taylor and K. Morgan (eds.), Pineridge Press, Swansea, 1980.

BOUNDARY ELEMENT FORMULATION TO SOLVE NO-TENSION PROBLEMS IN GEOMECHANICS

W. S. Venturini and C. A. Brebbia

Department of Civil Engineering, University of Southampton, Southampton, SO9 5NH, U.K.

1. INTRODUCTION

The Boundary Element method for stress analysis has developed considerably since its beginning in 1967 for plane elastostatics[1]. The technique has been applied to a large number of problems in engineering such as plasticity[2,3], fracture mechanics[4], viscoelasticity[5] and others[6].

Another problem of practical importance in geomechanics is the modelling of no-tension materials. This problem has been solved using Finite Element Methods by Valliappan[7], applying an iterative process which consists of computing at each step an equivalent initial stress field. Recently, Venturini and Brebbia[8] have applied a similar technique using the Boundary Element Method.

Stress analysis in geomechanics are ideally suited for Boundary Element solutions, usually requiring a very small number of nodes by comparison with Finite Element Solutions. As only the surface of the continuum needs to be discretized, problems extending to infinite can be described by a very small number of elements or unknowns. In addition, the boundary conditions of the infinite domain are properly defined.

Study of no-tension material in geomechanics is important to predict regions of loosenings over tunnel roofs and other underground structures. This loosening is usually prevented by recovering the tunnel with a thin layer of shotcrete, or by prestressing the material surrounding the tunnel. It is also important to be able to predict no-tension zones around pressure

J. B. Martins (ed.), Numerical Methods in Geomechanics, 375–388.

tunnels. The material in this region is affected by the excavation process and unable to sustain any tensile stress, resulting in an increase of stress in the lining.

The Boundary Element solution presented in this paper can be applied to solve all these geomechanical problems. Numerical results have been validated by comparing them against known Finite Element and theoretical solutions. They point out the advantages of using Boundary Elements for the solution of no-tension material problems in Geomechanics.

2. THE BOUNDARY ELEMENT METHOD

Consider a plane body (Fig. 1) described by the domain Ω bounded by surface Γ. The boundary integral equation for displacements and stresses at any internal point can be written as shown in references [3, 8].

$$u_\ell = - \int_\Gamma p^*_{\ell k} u_k \, d\Gamma + \int_\Gamma u^*_{\ell k} p_k \, d\Gamma + \int_\Omega u^*_{\ell k} b_k \, d\Omega + \int_\Omega \varepsilon^*_{\ell mk} \sigma^o_{mk} \, d\Omega \qquad 2.1$$

and

$$\sigma_{ij} = - \int_\Gamma S_{ijk} u_k \, d\Gamma + \int_\Gamma D_{ijk} p_k \, d\Gamma + \int_\Omega D_{ijk} b_k \, d\Omega + \int_\Omega F_{ijmk} \sigma^o_{mk} \, d\Omega - \frac{1}{8(1-\nu)} \left[2\sigma^o_{ij} + (1-4\nu)\sigma^o_{kk} \delta_{ij} \right] \qquad 2.2$$

where u_k and p_k are the displacement and traction components applied on the boundary Γ; b_k and σ_{mk} are body forces and initial stress field in the domain Ω. $p^*_{\ell k}$, $u^*_{\ell k}$, $\varepsilon^*_{\ell mk}$, S_{ijk}, D_{ijk}, F_{ijmk}, are tensor components related to the elastic fundamental solution, and δ_{ij} is the Kronecker delta. For details related to these tensor components see references [3,6,8].

The equation (2.1) can also be applied for points on the boundary by introducing a term $c_{\ell m}$, i.e.

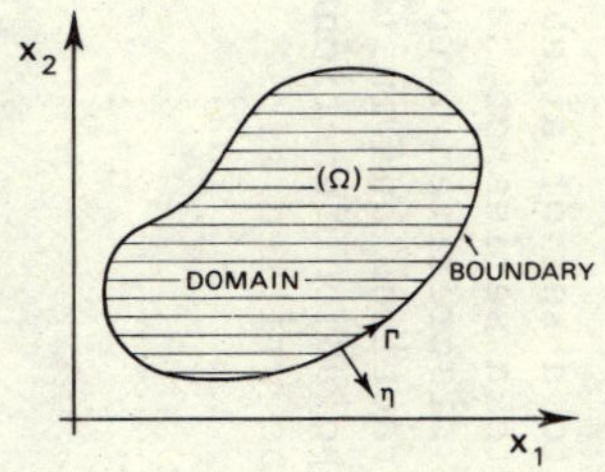

Figure 1 Definitions

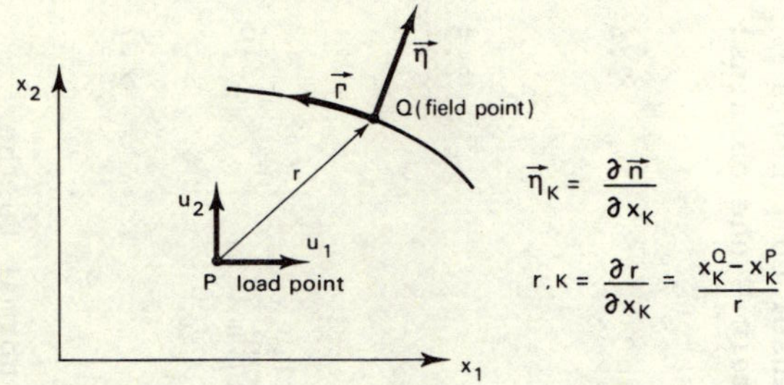

Figure 2a Geometric Parameters

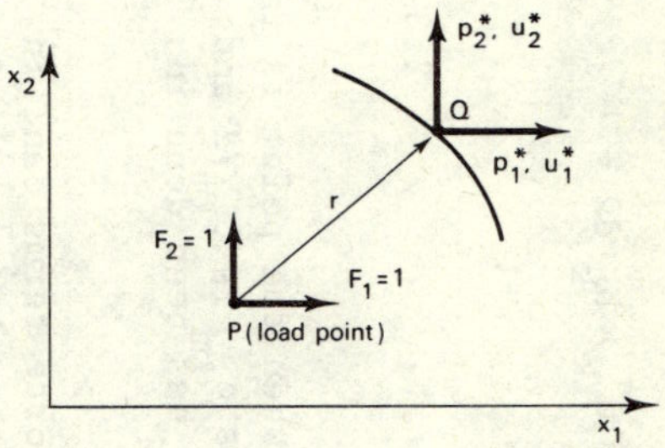

Figure 2b Fundamental solution

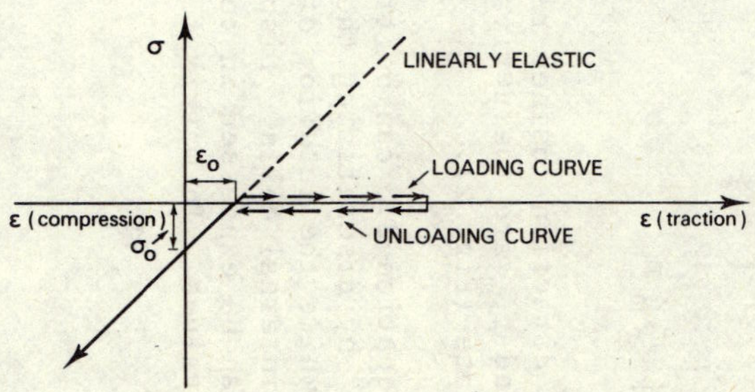

Figure 3 No–tension as a non–path dependent problem

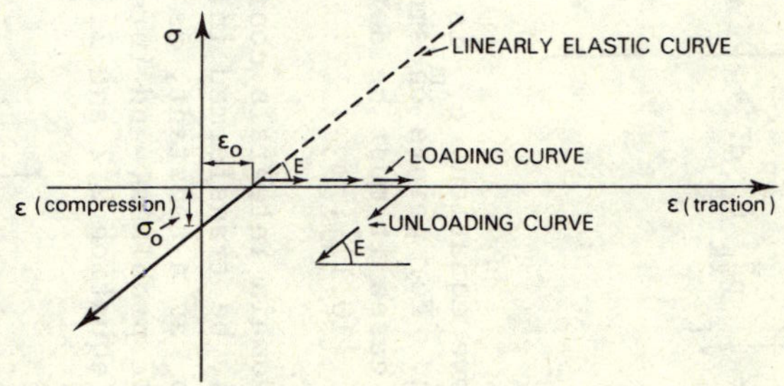

Figure 4 No–tension as a path dependent problem

$$u_m \, c_{\ell m} = - \int_\Gamma p^*_{\ell k} \, u_k \, d\Gamma + \int_\Gamma u^*_{\ell k} \, p_k \, d\Gamma + \int_\Omega u^*_{\ell k} \, b_k \, d\Omega + $$

$$+ \int_\Omega \varepsilon^*_{\ell mk} \, \sigma^o_{mk} \, d\Omega \qquad 2.3$$

in the above equation, $c_{\ell m}$ is equal to $\delta_{\ell m}$ when the point is in the domain. For points on smooth boundaries $c_{\ell m}$ is $\delta_{\ell m}/2$ and for other cases it must be determined as has been done in references [10,11].

The domain integrals containing body force terms can, in many cases, be transformed into boundary integrals [12,13]. Considering b_k as a constant, as it is frequently the case in geomechanic problems, and integrating the tensor components $u^*_{\ell k}$ and D_{ijk} (equations 2.2 and 2.3) over the domain Ω one obtains;

$$\int_\Omega u^*_{\ell k} \, b_k \, d\Omega = \int_\Gamma V_{\ell k} \, b_k \, d\Gamma \qquad 2.4$$

$$\int D_{ijk} \, b_k \, d\Omega = \int W_{ijk} \, b_k \, d\Gamma \qquad 2.5$$

where

$$V_{\ell k} = \frac{1}{16\Pi G(1-\nu)} \left\{ -(3-4\nu)(\ell n r - \tfrac{1}{2}) \, \delta_{\ell k} + r,_\ell r,_k \right\} \eta_m \, r,_m \qquad 2.6$$

$$W_{ijk} = D_{ijk} \, r,_m \, \eta_m \qquad 2.7$$

η_m are the direction cosines related to the normal to the boundary and $r,_m$ are the derivatives of r with respect to the x_m coordinate (Fig. 2).

The equation 2.3 can be transformed into a set of algebric equations[10] by discretizing the boundary into a series of elements, while the interior of the body is discretized into a number of internal cells. Displacements, tractions and initial stresses can be expressed in terms of interpolation functions and nodal values. In matrix form they can be written as;

$$\underset{\sim}{u} = \underset{\sim}{\phi}^T \underset{\sim}{u}^N$$

$$\text{on } \Gamma \qquad 2.8$$

$$\underset{\sim}{p} = \underset{\sim}{\phi}^T \underset{\sim}{p}^N$$

and

$$\underset{\sim}{\sigma}^o = \underset{\sim}{\psi}^T \underset{\sim}{\sigma}^{oN} \quad \text{in } \Omega \qquad 2.9$$

Where $\underset{\sim}{u}^N$, $\underset{\sim}{p}^N$ $\underset{\sim}{\sigma}^{oN}$ are nodal values for displacements, tractions and initial stresses respectively, and $\underset{\sim}{\phi}^T$ and $\underset{\sim}{\psi}^T$ are interpolation functions such as those used in Finite Elements[14].

Equation (2.3) can now be written in a discretized form by expressing each integral as a summation of element or cell integrals as is explained in detail in reference [10]. The matrix forms of (2.3) and (2.2) can be written as,

$$\underset{\sim}{C}\,\underset{\sim}{U} + \underset{\sim}{\hat{H}}\,\underset{\sim}{U} = \underset{\sim}{G}\,\underset{\sim}{P} + \underset{\sim}{V}\,\underset{\sim}{b} + \underset{\sim}{Q}\,\underset{\sim}{\sigma}^o \qquad 2.10$$

and

$$\underset{\sim}{\sigma} = \underset{\sim}{G}'\,\underset{\sim}{p} - \underset{\sim}{H}'\,\underset{\sim}{u} + \underset{\sim}{V}'\,\underset{\sim}{b} + (\underset{\sim}{Q}' + \underset{\sim}{E}')\underset{\sim}{\sigma}^o \qquad 2.11$$

In both equations the matrices $\hat{H}$, B, V, Q, G', H', V', and Q' are obtained by integrating the correspondent tensors and the interpolation functions over all elements and cells. $\underset{\sim}{E}'$ represents the independent term in equation 2.2.

Equation 2.10 needs to be applied to boundary nodes only, then one can write.

$$\underset{\sim}{H}\,\underset{\sim}{u} = \underset{\sim}{G}\,\underset{\sim}{p} + \underset{\sim}{V}\,\underset{\sim}{b} + \underset{\sim}{Q}\,\underset{\sim}{\sigma}^o \qquad 2.12$$

where

$$\underset{\sim}{H} = \underset{\sim}{\hat{H}} + \underset{\sim}{C} \qquad 2.13$$

For a well-posed problem, a sufficient number of tractions and boundary displacements need to be prescribed. Reordering the system to keep all unknowns in the left-hand side vector $\underset{\sim}{x}$, the above equation can be written;

$$\underset{\sim}{A}\,\underset{\sim}{x} = \underset{\sim}{f} + \underset{\sim}{Q}\,\underset{\sim}{\sigma}^o \qquad 2.14$$

and similarly the equation 2.11 becomes,

$$\underset{\sim}{\sigma} = -\underset{\sim}{A}' \underset{\sim}{x} + \underset{\sim}{f}' + (\underset{\sim}{Q}'+\underset{\sim}{E}')\underset{\sim}{\sigma}^o \qquad 2.15$$

Multiplying the equation 2.14 by A^{-1} results;

$$\underset{\sim}{x} = \underset{\sim}{m} + \underset{\sim}{R}\, \underset{\sim}{\sigma}^o \qquad 2.16$$

where

$$\underset{\sim}{m} = \underset{\sim}{A}^{-1} \underset{\sim}{f} \qquad 2.17$$

and

$$\underset{\sim}{R} = \underset{\sim}{A}^{-1} \underset{\sim}{Q} \qquad 2.18$$

substituting 2.15 into 2.11 gives,

$$\underset{\sim}{\sigma} = \underset{\sim}{n} + \underset{\sim}{S}\, \underset{\sim}{\sigma}^o \qquad 2.19$$

in which

$$\underset{\sim}{S} = \underset{\sim}{Q}' + \underset{\sim}{\bar{E}}' - \underset{\sim}{A}' \underset{\sim}{R} \qquad 2.20$$

and

$$\underset{\sim}{n} = \underset{\sim}{f}' - \underset{\sim}{A}' \underset{\sim}{m} \qquad 2.21$$

Notice that the matrix E' in equation 2.11 was replaced by $\underset{\sim}{\bar{E}}'$ to allow one to use the iterative process in the next section. In this case $\underset{\sim}{\bar{E}}'$ can be written as,

$$\underset{\sim}{\bar{E}}' = \underset{\sim}{E}' + \underset{\sim}{I} \qquad 2.22$$

where $\underset{\sim}{I}$ is the unit matrix.

Equation 2.19 applies to internal stresses, but the boundary stresses can also be included as shown in reference [8] by using the interpolation functions for displacements on the boundary.

The above development applies to homogeneous bodies. In many geomechanical applications one needs to solve problems with different materials. It can be done by treating each material region independently and joining them together applying compatibility and equilibrium conditions[15].

3. ANALYSIS OF NO-TENSION MATERIAL

In many practical applications, tunnels can be built only by excavation and no structural lining is required to garantee the cavity stability. Careful analysis of the stress in the rock medium is needed. The assumption that the material is linearly elastic is generally unsafe, even when the structure is in a sound rock, because the tensile strength of this material is usually very small.

For the last 15 years, many criteria have been developed trying to model the real behaviour of the rock. One of those more successfully applied in rock mechanics has been the no-tension material criterion. It has been applied to solve innumerous problems in geomechanics, using Finite Element formulations.

In tunnels, the determination of no-tension zones can be a good approximation for designing the temporary support required to prevent loosening over the roof. Usually a thin layer of shotcrete, with rock bolts for additional support if necessary, applied to the rock face immediately after blasting prevents loosening, reduces decompression to a certain degree and transforms the surrounding rock into a self-supporting arch.

The method is also suitable for a more accurate evaluation of the stresses in the rock and in the lining in water pressure tunnels. Usually, in this case, the rock surrounding the tunnel can not withstand any tensile stresses due to the rock excavation process.

The scheme to obtain the final no-tension solution consists of applying the "stress transfer method" which transfers any tensile stresses arising at any point to the compressive zones. Using the Boundary Element Method, the transferring stress process can be easily done introducing an initial stress field as shown in equations 2.16 and 2.19.

The no-tension concept can be better defined by considering the material behaviour as shown in Figures 3 and 4. In the first case, the problem is assumed to be elastic and the load can be applied in one step. For the second case, the problem is considered path dependent and the load should be applied in increments small enough to achieve the required accuracy. The two cases will be considered here, and similar iterative procedures are adopted for them to reach the no-tension solution. The iterative process consists of the following steps:

i) The problem is solved elastically with isotropic properties. The vector $\sigma_{e\ell}$, which contrains the elastic stresses due to the load, is determined. These stresses are added to any original rock stresses, σ_{to}, present before the construction

$$\underset{\sim}{\sigma} = \underset{\sim}{\sigma}_{to} + \underset{\sim}{\sigma}_{e\ell} \qquad 3.1$$

The vector $\underset{\sim}{\sigma}_m$ which gives the linear elastic path during the process (dashed line Fig. 3) is assumed to be equal to $\underset{\sim}{\sigma}$.

ii) A new true stress vector $\underset{\sim}{\sigma}_t$ is obtained assigning zero values to any point where $\underset{\sim}{\sigma}_m$ is in the no-tension part of the curve of figure 3. This must be done using principal stresses.

iii) The initial stress vector, $\underset{\sim}{\sigma}_i$, can be found at this stage by the following expression:

$$\underset{\sim}{\sigma}_i = \underset{\sim}{\sigma}_{e\ell} - (\underset{\sim}{\sigma}_t - \underset{\sim}{\sigma}_{to}) \qquad 3.2$$

iv) The elastic stress vector, now defined in function of initial stress, can be determined by,

$$\underset{\sim}{\sigma}_{e\ell} = \underset{\sim}{S}\, \underset{\sim}{\sigma}_i \qquad 3.3$$

and recalculating the vectors $\underset{\sim}{\sigma}$ and $\underset{\sim}{\sigma}_m$ one has,

$$\underset{\sim}{\sigma} = \underset{\sim}{\sigma}_t + \underset{\sim}{\sigma}_{e\ell} \qquad 3.4$$

$$\underset{\sim}{\sigma}_m = \underset{\sim}{\sigma}_m + \underset{\sim}{\sigma}_{e\ell} \qquad 3.5$$

v) At this stage, the convergence criterion must be applied. If the initial stress vector values are small with respect to our specified tollerance, the iterative process can stop. If the process should be continued, let

$$\underset{\sim}{\sigma}_{to} = \underset{\sim}{\sigma}_t \qquad 3.6$$

and repeat all items from ii).

For the path dependent criterion described before, the iterative process would be the same with $\underset{\sim}{\sigma}_m = \underset{\sim}{\sigma}_{to}$ when each new load increment is applied.

4. APPLICATIONS

Two examples are presented in this section. The first is a steep valley, previously solved using Finite Element Methods[7]. The other example is a circular lined tunnel in rock, considering the region closed to the lining as a no-tension zone. The

The final results will be compared with the theoretical solution[16].

Steep Valley

The case of a steep valley was solved by Villiappan[7] using Finite Elements and no-tension material. The load applied to this structure is only due to the removal of residual compressive stresses which are taken to be equal to $-\gamma y$ and $-\gamma K_o y$ in vertical and horizontal directions respectively (Fig. 5).

The sloped valley boundary is discretized into a series of straight elements over which tractions corresponding to the residual stresses are applied. In Figure 6 the mesh used to solve the problem is presented. The convergence of the results was tested by using a finer mesh which produced practically the same values. Notice that for this problem the domain tends to infinite and no displacement boundary conditions have to be prescribed.

Figure 7 shows the elastic initial solution and Figure 8 the final no-tension results in function of the principal stresses. The results agree well with a published Finite Element solution[7], for which a 247 elements and 144 nodes mesh was used. The Boundary Element solution only required 24 nodes and 23 Boundary Elements. For integrating the initial stress term

This example was solved using the two no-tension solution criteria previously discussed (see Figs. 3 and 4) giving practically the same results. It was also found during the solution that the discretization of the curved boundary into straight elements produces small tranctions forces which cannot be completely eliminated.

Lined Tunnel

This example consists of a lined tunnel with internal pressure in an infinite rock medium. The only load considered in this case is 100tf/m internal pressure.

The internal tunnel diameter is equal to 400cm. The 30cm thick reinforced concrete lining is assumed to be able to sustain any tensile stresses and its Poisson's ratio is 0.15. The rock medium is considered to be divided into two distinct regions. A no-tension region ring 115cm thick closed to the lining is assumed fissured, while any other points outside that region are considered capable of sustaining tensile stresses. The rock Poisson's ratio is 0.2 and the ratio between the concrete and rock elastic modulus is considered equal to 2. The original stresses present in the rock mass were not included in the

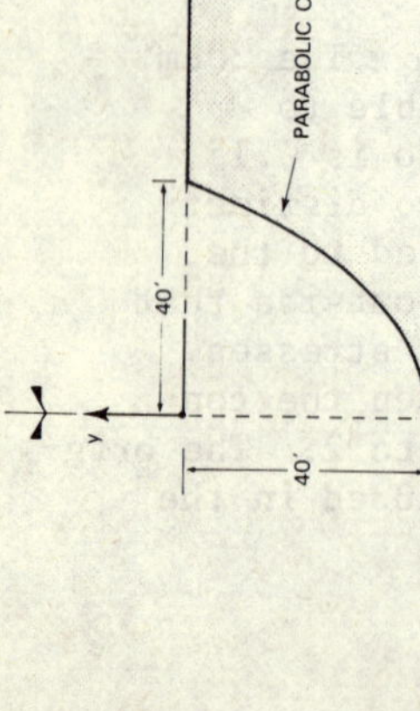

Figure 5 Valley to be excavated

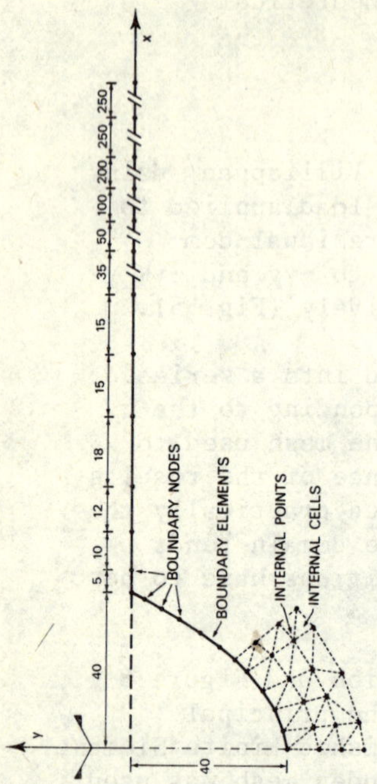

Figure 6 Steep valley. Discretization

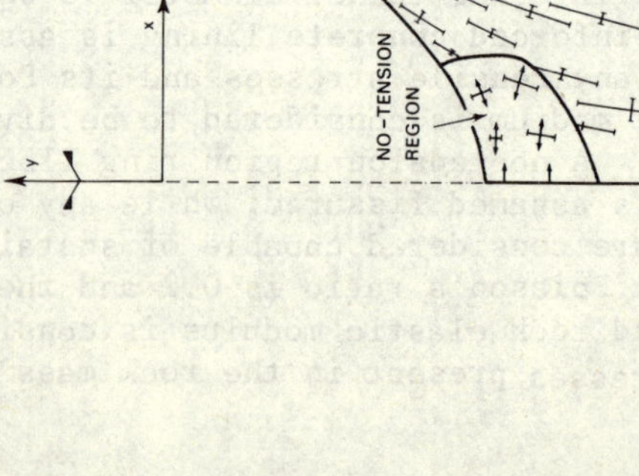

Figure 7 Steep valley. Elastic solution

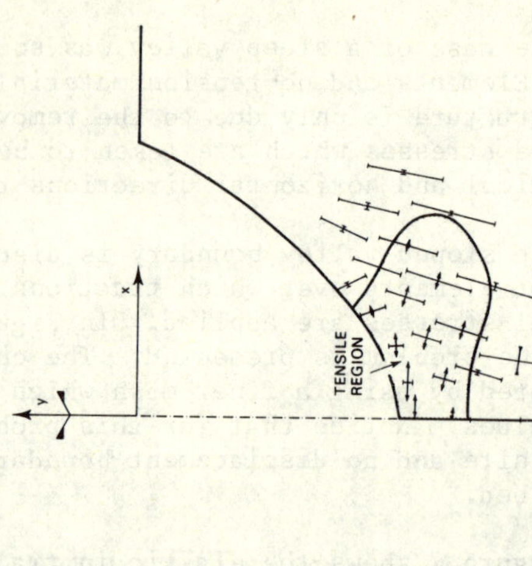

Figure 8 Steep valley. No-tension solution

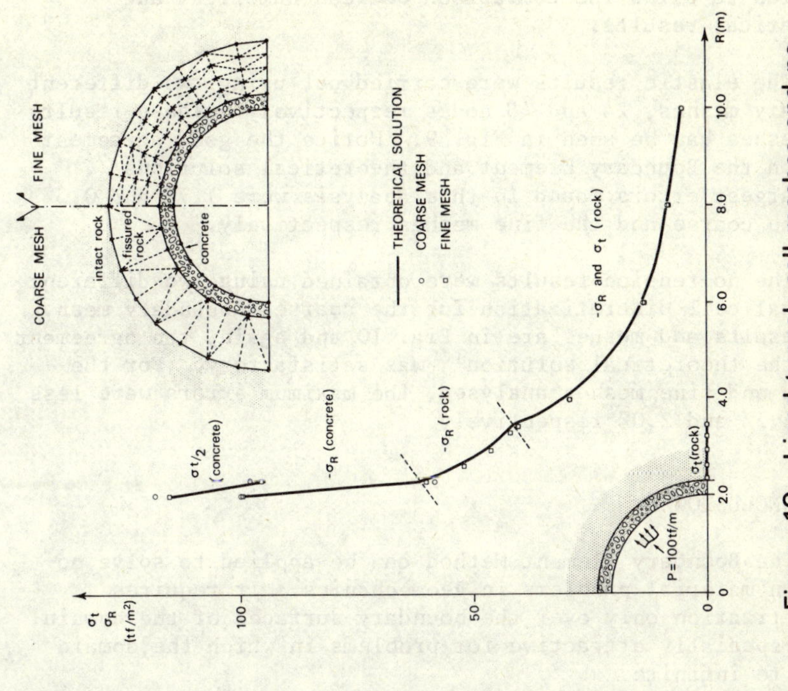

Figure 10 Lined tunnel; cell meshes and no-tension results

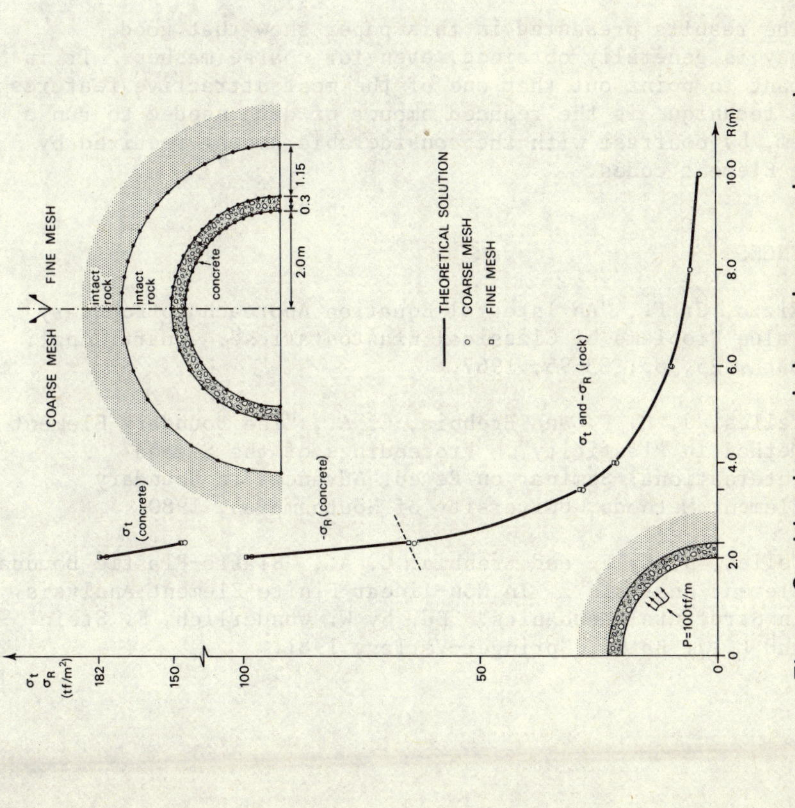

Figure 9 Lined tunnel; boundary meshes and elastic results

solution to allow the comparison between numerical and theoretical results.

The elastic results were carried out using two different boundary meshes, 24 and 48 nodes respectively. These results and meshes can be seen in Fig. 9. Notice the good agreement between the Boundary Element and theoretical solutions[17]. The largest errors found in this analysis were 1.7% and 0.3% for the coarse and the fine meshes respectively.

The no-tension results were obtained using two different internal cell discretization for the coarsest boundary mesh. The results and meshes are in Fig. 10 and again, the agreement with the theoretical solution[16] was satisfactory. For the coarse and fine meshes analysed, the maximum errors were less than 4.5% and 2.0% respectively.

5. CONCLUSIONS

The Boundary Element Method can be applied to solve no-tension material problems in geomechanics. It requires discretization only over the boundary surfaces of the domain. It is specially attractive for problems in which the domain tends to infinite.

The results presented in this paper show that good accuracy is generally obtained, even for coarse meshes. It is important to point out that one of the most attractive features of the technique is the reduced amount of data needed to run a problem, by contrast with the considerable amount required by Finite Element codes.

REFERENCES

1. Rizzo, J. F., "An Integral Equation Approach to Boundary Value Problems of Classical Elastostatics". Quart. Appl. Math. 25, pp. 83-95, 1967.

2. Telles, J. C. F. and Brebbia, C. A., "The Boundary Element Method in Plasticity". Proceedings of the Second International Seminar on Recent Advances in Boundary Element Methods, University of Southampton, 1980

3. Telles, J. C. F. and Brebbia, C. A., "Elasto-Plastic Boundary Element Analysis". In Non-linear Finite Element Analysis in Structural Mechanics. Ed. by W. Wunderlich, E. Stein and L. J. Bathe, Springer-Verlag, 1981.

4. Cruse, T. A. and Vanburen, W., "Three-Dimensional Elastic Stress Analysis of a Fracture Specimen with an Edge Crack", Int. J. Fract. Mech., 7. pp. 1-15, 1971.

5. Rizzo, F. J. and Shippy, D. J., "An Application of the Correspondence Principles of Linear Viscoelastic Theory". SIAM, J. Appl. Math., 21, pp. 321-330, 1971.

6. Brebbia, C. A. and Walker, S., "Boundary Element Techniques in Engineering". Newnes-Butterworths, 1980.

7. Valliappan, S., "Non-linear Stress Analysis of Two-Dimensional Problems with Special Reference to Rock and Soil Mechanics". Ph.D. Thesis, University of Wales, Swansea, 1968.

8. Venturini, W. S. and Brebbia, C. A., "The Boundary Element Method for the Solution of No-tension Materials". Proceedings of the Third International Seminar on Recent Advances in Boundary Element Methods, to be held at Irvine, California. 1981.

9. Rabicewicz, L. V., "The New Australian Tunnelling Method". Water Power, November, December, 1964 and January 1965.

10. Brebbia, C. A., "The Boundary Element Method for Engineering". Pentech Press, 1978.

11. Cruse, T. A., "Mathematical Foundations of the Boundary Integral Equation Method in Solid Mechanics". Air Force Office of Scientific Research, Report No. AFOSR-7R-10021, 1977.

12. Stippes, M. and Rizzo, F. J., "A Note on the Body Force Integral of Classical Elastostatics". ZAMP. Vol. 28, 1977.

13. Rizzo, F. J. and Shippy, D. J., "An Advanced Boundary Integral Equation Method for the Three-Dimensional Thermoelasticity", Int. J. Num. Meth. Eng. Vol. 11, 1977.

14. Brebbia, C. A. and Connor, J. J., "Fundamentals of Finite Element Techniques for Structural Engineers". Butterworths, 1973.

15. Lachat, J. C., "A Further Development of the Boundary Integral Technique for Elastostatics". Ph.D. Thesis, Southampton University, 1975.

16. Camargo, W. M., "Projeto de Túneis em Maciço Rochoso sob Pressão Hydrostática Interna". Thesis presented to University of São Paulo, 1968.

17. Jaeger, C., "Rock Mechanics and Engineering". Cambridge University Press, 1972.

ELASTOPLASTIC ANALYSIS USING BOUNDARY ELEMENTS

J. C. F. Telles and C. A. Brebbia

Department of Civil Engineering, University of Southampton, Southampton, SO9 5NH, U.K.

INTRODUCTION

This paper presents some recent developments on the application of boundary elements to plasticity on the lines previously presented by the authors (Telles and Brebbia, 1979, 1980a, 1980b).

The application of boundary elements (or boundary integral equations) to nonlinear material problems started with Riccardella (1973) who presented a valid plasticity formulation but used a not entirely successful numerical procedure. Later on Mukherjee and Kumar (1978) produced a formulation based on original constitutive relations which involved plasticity as well. All the implementations were then restricted to piecewise constant interpolation for the plastic strains.

In 1979 Telles and Brebbia produced the complete and correct formulation for two and three dimensional bodies, including the implementation of higher order interpolation functions for the plastic strains. In 1980 (a and b) they applied this formulation to study two-dimensional plasticity problems showing how different numerical procedures could be properly used; namely initial strain, initial stress and fictitious tractions and body forces. Emphasis was given to the accuracy and efficiency of the numerical formulation, firmly establishing the potentialities of boundary elements in plasticity.

One of the advantages of the implementation which was not

J. B. Martins (ed.), Numerical Methods in Geomechanics, 389–406.

pointed out in the past is that it overcomes the well-known disadvantage of finite elements first reported by Nagtegaal et al (1974); i.e. the impossibility of representing properly the limit load due to the severe constraints created by the incompressibility of the plastic strains.

In the present paper the plasticity formulation is extended to solve half-plane problems using the elastic fundamental solution of Melan's problem. This author (Melan, 1932) presented the stresses due to point loads within the semi-infinite plane. Telles and Brebbia (1981) extended this work to compute displacements, strains and surface tractions, presenting the complete fundamental solution needed in generalized boundary elements. This solution is the one used in the examples shown here.

In addition to the known advanges of the Kelvin elastoplastic formulation, the main benefit of the half-plane implementation is that boundary elements are not needed over the traction-free surface of the semi-plane. This property is of great significance in problems connected with geomechanics and has not been presented before for plasticity.

BASIC EQUATIONS

The incremental equilibrium equations for three-dimensional bodies in terms of displacement rates ($\dot{u}_i$) and plastic strain rates ($\dot{\varepsilon}^p_{ij}$), are

$$\dot{u}_{j,\ell\ell} + \frac{1}{1-2\nu}\dot{u}_{\ell,\ell j} = 2\left(\dot{\varepsilon}^p_{ij,i} + \frac{\nu}{1-2\nu}\dot{\varepsilon}^p_{kk,j}\right) - \frac{\dot{b}_j}{G} \quad (1)$$

where $\dot{b}_j$ represents the body force rate components, ν is Poisson's ratio and G is the shear modulus. Time derivatives (or simply increments) are indicated by a dot and space derivatives by a comma.

The traction boundary conditions for equation (1) can be written as

$$\dot{p}_i + 2G\left(\dot{\varepsilon}^p_{ij}n_j + \frac{\nu}{1-2\nu}\dot{\varepsilon}^p_{kk}n_i\right) = \frac{2G\nu}{1-2\nu}\dot{u}_{\ell,\ell}n_i + G(\dot{u}_{i,j} + \dot{u}_{j,i})n_j \quad (2)$$

where $\dot{p}_i$ represents the traction rate components and n_j stands for the direction cosines of the outward normal to the boundary of the body.

If plastic strains are considered as initial strains, the application of Hooke's law to the elastic part of the total strain rate tensor leads to the following expression for the stress rate components,

$$\dot{\sigma}_{ij} = 2G(\dot{\varepsilon}_{ij} - \dot{\varepsilon}^{p}_{ij}) + \frac{2G\nu}{1-2\nu}(\dot{\varepsilon}_{kk} - \dot{\varepsilon}^{p}_{kk})\,\delta_{ij} \tag{3}$$

in which δ_{ij} is the Kronecker delta.

Alternatively, equation (3) can be written in terms of initial stresses,

$$\dot{\sigma}_{ij} = 2G\dot{\varepsilon}_{ij} + \frac{2G\nu}{1-2\nu}\dot{\varepsilon}_{kk}\delta_{ij} - \dot{\sigma}^{p}_{ij} \tag{4}$$

where $\dot{\sigma}^{p}_{ij}$ stands for the components of the "initial stresses" given by

$$\dot{\sigma}^{p}_{ij} = 2G\dot{\varepsilon}^{p}_{ij} + \frac{2G\nu}{1-2\nu}\dot{\varepsilon}^{p}_{kk}\delta_{ij} \tag{5}$$

Expressions (1) to (5) are valid for plane strain ($i,j,\ell = 1,2$; $k = 1,2,3$) and also for plane stress ($i,j,k,\ell = 1,2$) if ν is replaced by $\overline{\nu} = \nu/(1+\nu)$.

HALF-PLANE FUNDAMENTAL SOLUTION

The complete fundamental solution for half-plane problems can be obtained by adding to the well-known two-dimensional Kelvin fundamental solution a complementary part. This provides satisfaction of the traction-free condition over the surface of the half-plane. In order to save space, only the complementary part of the solution will be discussed, but the complete fundamental solution is obtained by the following relation,

$$(\ )^* = (\ )^k + (\ )^c \tag{6}$$

where $(\ )^k$ and $(\ )^c$ stand for Kelvin part and complementary part respectively.

In the present work the traction-free surface of the half-plane is assumed to be represented by the axis $x_1 = 0$.

With reference to figure 1, the complementary expressions for u^*_{ij} and p^*_{ij} corresponding to the displacements and tractions in j direction due to a unit point load applied in i direction, have been presented elsewhere (Telles and Brebbia 1981a,b).

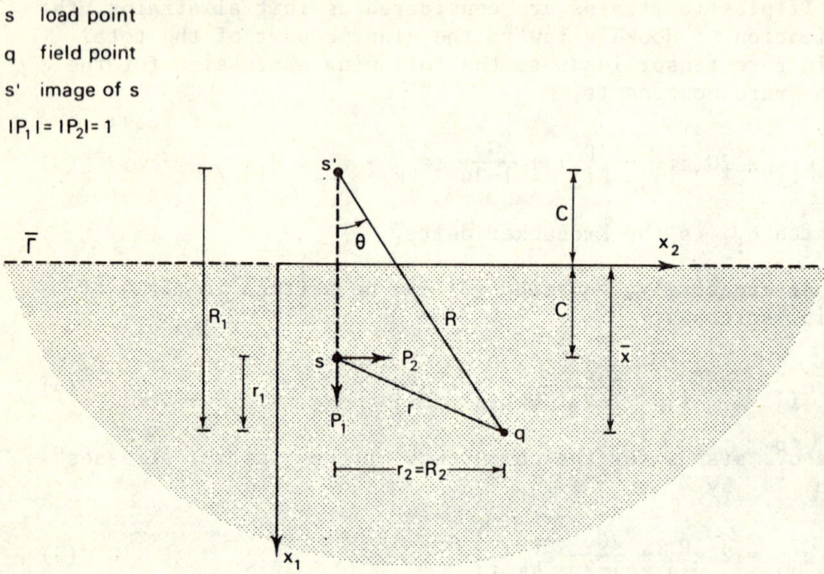

Figure 1. Unit point loads applied within the half-plane

It is interesting to note that such complementary part of the fundamental solution does not present any singularities within the domain $x_1 \geq 0$ when $c > 0$. Therefore, many expressions already obtained for the Kelvin elasto-plastic implementation, can still be used in the present formulation.

BOUNDARY ELEMENT FORMULATION

The extension of the elastic half-plane boundary element formulation (Telles and Brebbia, 1981(a)) to elastoplastic problems, follows the same procedure as for the Kelvin implementation.

If we consider the plastic strains to be incompressible, the starting equation for the initial strain formulation is given by,

$$c_{ij} \dot{u}_j = \int_{\Gamma} u^*_{ij} \dot{p}_j \, d\Gamma - \int_{\Gamma'} p^*_{ij} \dot{u}_j \, d\Gamma + \int_{\Omega} u^*_{ij} \dot{b}_j \, d\Omega$$

$$+ \int_{\Omega} \hat{\sigma}^*_{jki} \dot{\varepsilon}^p_{jk} \, d\Omega \qquad (7)$$

where Γ represents the total boundary of the body, Γ' the part of Γ in which $x_1 > 0$ and Ω stands for the domain of the body that has to be always located within the half-plane $x_1 \geq 0$. Note that the second boundary integral is performed over Γ' because the fundamental solution is zero on the surface of the half-plane and hence the integral vanishes there.

For plane strain problems the complementary part of the tensor that multiplies the plastic strain rates is (Mukherjee, 1977),

$$\hat{\sigma}^c_{jki} = \sigma^c_{jki} - \nu \, \sigma^c_{\ell\ell i} \, \delta_{jk} \qquad (8)$$

whereas for plane stress

$$\hat{\sigma}^c_{jki} = \sigma^c_{jki} \qquad (9)$$

Equation (7) is valid for any location of the load point ($s \in \Gamma$ or $s \in \Omega$), provided c_{ij} and the boundary integral over Γ' are properly interpreted as known from the previous elastic application of the half-plane fundamental solution.

By suitably modifying the plastic strain rate integral in equation (7), an initial stress formulation without the condition of incompressibility of the plastic strains (Telles and Brebbia, 1980b) can be equally obtained,

$$c_{ij} \dot{u}_j = \int_{\Gamma} u^*_{ij} \dot{p}_j \, d\Gamma - \int_{\Gamma'} p^*_{ij} \dot{u}_j \, d\Gamma + \int_{\Omega} u^*_{ij} \dot{b}_j \, d\Omega$$

$$+ \int_{\Omega} \varepsilon^*_{jki} \dot{\sigma}^p_{jk} \, d\Omega \qquad (10)$$

where $\dot{\sigma}^p_{jk}$ was given in expression (5).

It is interesting to note that in contrast with the initial strain formulation, equation (10) can now be used for plane strain and plane stress problems with the replacement of ν by $\bar{\nu}$ being the only modification.

STRESSES AT INTERNAL POINTS

It is well-known that one of the advantages of boundary elements over "domain type" techniques in elastic analysis, is that stresses at internal points can be accurately computed by using the appropriate integral equation. The elastoplastic implementation of the technique is by all means no exception, and from the beginning the present authors have used the correct integral expression for computing the internal values of the stress rate tensor (Telles and Brebbia, 1979). In spite of this, some boundary element researchers have found the internal stresses by computing displacements at internal points, and differentiating them numerically as it is done in finite differences or finite elements (Cathie, 1980). This procedure possesses some drawbacks which are going to be commented on in other sections of this paper.

In order to accurately compute the stress rates at internal points, the derivatives of (7) are combined to produce the expressions for the total strain rate and then substituted into equation (3). Here one notices that, due to the non singular nature of the complementary tensors, the derivatives of the plastic strain rate integral create exactly the same singularities obtained for the single Kelvin implementation. Hence, for plane strains one has,

$$\dot{\sigma}_{ij} = \int_{\Gamma} u^*_{ijk}\, \dot{p}_k\, d\Gamma - \int_{\Gamma'} p^*_{ijk}\, \dot{u}_k\, d\Gamma + \int_{\Omega} u^*_{ijk}\, \dot{b}_k\, d\Omega + \int_{\Omega} \hat{\sigma}^*_{ijk\ell}\, \dot{\varepsilon}^p_{k\ell}\, d\Omega + f_{ij}(\dot{\varepsilon}^p) \quad (11)$$

in which the plastic strain rate integral is to be computed in the principal value sense and f_{ij} is the same independent term obtained for the Kelvin formulation. i.e.,

$$f_{ij} = -\frac{G}{4(1-\nu)}\left[2\, \dot{\varepsilon}^p_{ij} + (1-4\nu)\dot{\varepsilon}^p_{\ell\ell}\delta_{ij}\right] \quad (12)$$

in addition,

$$\hat{\sigma}^c_{ijk\ell} = \sigma^c_{ijk\ell} - \nu\, \sigma^c_{ijmm}\, \delta_{k\ell} \quad (13)$$

and

$$\sigma^c_{ijk\ell} = G\left(\frac{\partial\sigma^c_{k\ell i}}{\partial x_j} + \frac{\partial\sigma^c_{k\ell j}}{\partial x_i}\right) + \frac{2G\nu}{1-2\nu}\frac{\partial\sigma^c_{k\ell m}}{\partial x_m}\delta_{ij} \tag{14}$$

where the derivatives are taken with reference to the coordinates of the load point. These derivatives together with u^c_{ijk} and p^c_{ijk} are given in (Telles and Brebbia, 1981a).

An interesting feature of the half-plane implementation is that if the problem to be analysed satisfies the traction-free condition ($p_k = 0$) over some part of the boundary $\Gamma - \Gamma'$, stresses at points located along this part of the boundary can be computed as if they were internal points. In order to validate equation (11) for such cases, only the expression of f_{ij} needs to be modified to take into consideration the limiting case $c = 0$. This expression can be easily obtained as follows; let us assume a semi-circular free body, of radius ρ, whose straight boundary is contained by the surface of the half-plane. If body forces are not considered, the application of a uniform plastic strain field ($\overline{\varepsilon}^p_{ij}$) to this body, will produce only displacements, internal stresses and tractions remain zero throughout the process.

The application of equation (11) to represent the stresses at the centre of the semi-circle leads directly to,

$$\int_{\Gamma'} p^*_{ijk}\,\overline{u}_k\,d\Gamma = \overline{\varepsilon}^p_{k\ell}\int_\Omega \hat{\sigma}^*_{ijk\ell}\,d\Omega + f_{ij}(\overline{\varepsilon}^p) \tag{15}$$

moreover, from the condition of existence of the principal value (Mikhlin, 1962), one can prove that,

$$\overline{\varepsilon}^p_{k\ell}\int_\Omega \hat{\sigma}^*_{ijk\ell}\,d\Omega = 0 \tag{16}$$

hence,

$$f_{ij}(\overline{\varepsilon}^p) = \int_{\Gamma'} p^*_{ijk}\,\overline{u}_k\,d\Gamma \tag{17}$$

where the relevant boundary displacements (neglecting rigid boundary movements) can be computed by (Telles and Brebbia, 1979),

$$\bar{u}_i = \rho(\bar{\varepsilon}^p_{ij} - \nu\,\bar{\varepsilon}^p_{kk}\,\delta_{ij})n_j \tag{18}$$

in which n_i represents the direction cosines of the outward normal to the curved boundary.

Equation (17) therefore provides the required expression for f_{ij} when $c = 0$,

$$f_{11} = f_{12} = 0$$

$$f_{22} = -\frac{G}{2(1-\nu)}(\dot{\varepsilon}^p_{22} - \dot{\varepsilon}^p_{11}) \tag{19}$$

For the initial stress formulation the procedure is entirely similar and the equation equivalent to (11) is of the following form,

$$\dot{\sigma}_{ij} = \int_\Gamma u^*_{ijk}\,\dot{p}_k\,d\Gamma - \int_{\Gamma'} p^*_{ijk}\,\dot{u}_k\,d\Gamma + \int_\Omega u^*_{ijk}\,\dot{b}_k\,d\Omega + \int_\Omega \varepsilon^*_{ijk\ell}\,\dot{\sigma}^p_{k\ell}\,d\Omega + g_{ij}(\dot{\sigma}^p) \tag{20}$$

where $\varepsilon^c_{ijk\ell}$ is obtained from (20) by substituting σ^c_{jki} by ε^c_{jki},

$$g_{ij} = -\frac{1}{8(1-\nu)}\left[2\dot{\sigma}^p_{ij} + (1-4\nu)\dot{\sigma}^p_{\ell\ell}\,\delta_{ij}\right] \quad \text{for } c > 0 \tag{21}$$

and

$$g_{11} = g_{12} = 0$$

$$g_{22} = -\frac{1}{4(1-\nu)}\left[\dot{\sigma}^p_{22} - \dot{\sigma}^p_{11}\right] \quad \text{for } c = 0 \tag{22}$$

Plane stress problems can be handled by equations (11) and (20) if Poisson's ratio is modified as before, $\hat{\sigma}^*_{ijk\ell} = \sigma^*_{ijk\ell}$ and expression (12) is replaced by

$$f_{ij} = -\frac{G}{4(1-\bar{\nu})}\left[2\dot{\varepsilon}^p_{ij} + \dot{\varepsilon}^p_{\ell\ell}\,\delta_{ij}\right] \tag{23}$$

Notice that expressions (19), (21) and (22) still remain valid.

Although body forces have been considered in the equations presented so far, in the following sections this term will not be included for simplicity.

NUMERICAL IMPLEMENTATION

For numerical implementation of the equations presented in the previous sections, linear boundary elements and linear triangular cells were used in much the same way as described by Telles and Brebbia (1980a, b) for the initial strain and initial stress formulations. It is worth mentioning that integration of the complementary part of the expressions presents no singularities when $c > 0$. Consequently, simple quadrature formulae can be used for the boundary and domain integrals. The limiting case $c = 0$ can be shown to produce singularities of the same order as those present in Kelvin's, therefore the two parts of the fundamental solution are added up and integrals are properly computed using the same integration scheme normally employed for the Kelvin part.

In addition, computation of stress rates at boundary nodes ($s \in \Gamma'$ and $s \in$ non traction-free part of $\Gamma - \Gamma'$) is included into the implementation through simple expressions which do not require any integration and involve the interpolated traction and displacement rates over the boundary elements.

DIFFERENT SOLUTION TECHNIQUES

At present, the initial strain formulation has been implemented with the von Mises criterion and employs Mendelson's successive elastic solutions method (Mendelson, 1968). This solution technique, also called "elastic predictor-radial corrector method" by Schreyer et al. (1979), has proved to be very efficient and stable with reference to the load increment size. Therefore it is recommended for von Mises material problems. The initial stress implementation, on the other hand, is more general and can handle four different yield criteria (Mises, Tresca, Mohr-Coulomb and Drucker-Prager), with two different iterative routines. The first is a pure incremental technique, comparable to what was used by Zienkiewicz et al. (1969) for finite elements. The second deals with accumulated values of the initial stresses, in a similar fashion to the initial strain implementation.

A common feature of the alternative implementations is that they are all incremental-iterative processes, capable of performing iterations by using a single recursive expression relating stresses to the plastic strains and the initial elastic solution. In what follows, it will be seen that the equivalent relations are not so straightforward when internal displacements are interpolated for computation of the internal stresses.

Recalling the procedure presented by Cathie (1980), one notices that in order to implement linear interpolation functions for the initial stresses, triangular cells with quadratic interpolation for the displacements must be used. Consequently, the number of internal points to which the cell is connected (also called internal points by the author), is doubled to six. By differentiating the interpolated displacements, a linear variation of the total strains is obtained and used to calculate the initial stresses at three so-called stress points. All operations are performed in incremental form.

For comparing the number of operations during one iteration cycle, one can consider, for simplicity, the case of unique cell located within the body. Following Cathie's procedure, the computation of internal displacement increments at the six internal points would be carried out by using a 12 x 9 matrix (called $\tilde{H}_2$ in the reference), which multiplies the increments of initial stress at the three stress points. On top of this the derivatives still have to be performed to obtain the corresponding strains.

For the pure incremental procedure of the initial stress formulation, the equivalent operation will only involve the product of a 9 x 9 matrix by the initial stress increments, leading directly to the increments of elastic stress. By elastic stresses one means the result from the application of Hooke's law to the total strains, which already saves one operation when applying the elastoplastic constitutive relations (Telles and Brebbia, 1980b). Note that for quadrilateral cells the same argument still applies. Furthermore, as the number of cells increases, the difference between the two procedures becomes even more pronounced.

In conclusion, the above demonstrates that for accurate and efficient elastoplastic BE implementation, where the influence of the plastic term should be interpolated at least in linear piecewise form, computation of internal stresses by using the proper integral equation should always be preferred.

EXAMPLES

In the present section, the results of some geomechanics applications of the technique described are compared with numerical and analytical solutions presented in the literature.

Strip Footing

In this example, the plane strain analysis of a flexible strip footing under uniform loading is presented. The finite soil stratum is discretized taking full advantage of both, symmetry and free-surface condition, using the reduced number of 14 boundary elements and 42 internal points as shown in figure 2.

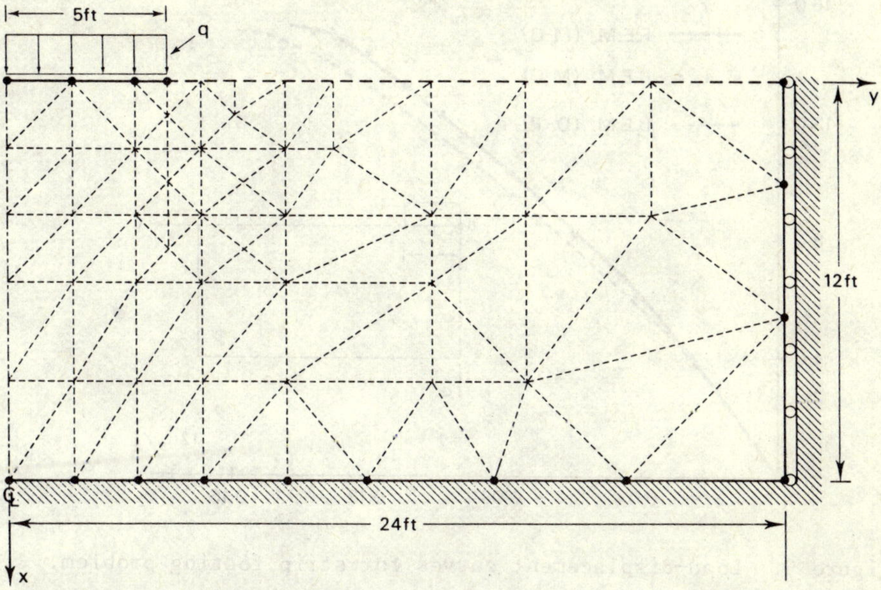

Figure 2. Strip footing on elastoplastic soil. Discretization used for B.E. results

The soil was considered to be a perfectly plastic material, obeying the associated Mohr-Coulomb (M-C) criterion, with

$E = 30000$ psi
$c = 10$ psi (cohesion)
$\phi = 20^{o}$ (angle of internal friction)
$\nu = 0.3$

An alternative solution was also carried out using the associated Drucker-Prager (D-P) yield criterion (extended von Mises) given by,

$$\frac{3\tan\phi}{(9+12\tan^2\phi)^{\frac{1}{2}}}\ \bar{p} + \sqrt{J_2} - \frac{3c}{(9+12\tan^2\phi)^{\frac{1}{2}}} = 0$$

where $\bar{p}$ is the hydrostatic component of the stress tensor and J_2 is the second invariant of the deviatoric stresses.

Ground surface displacements are presented in figure 3. Also included is the equivalent M-C finite element solution obtained by Zienkiewicz et al. (1975) using quadratic isopara-

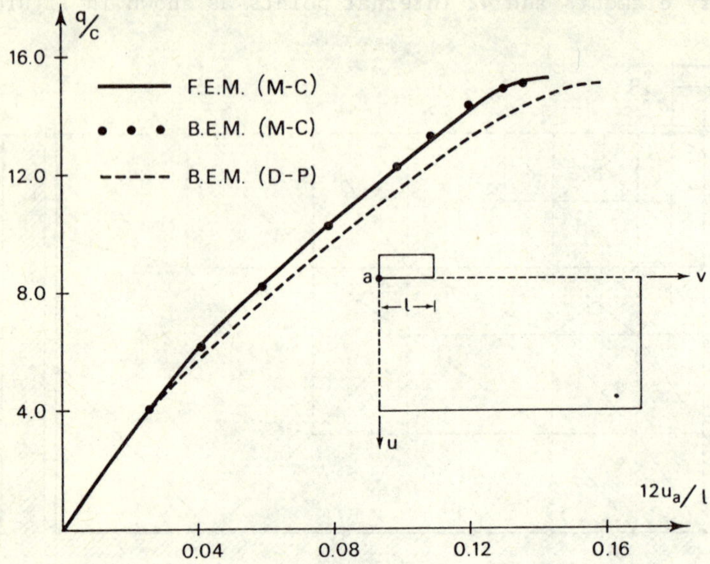

Figure 3. Load-displacement curves for strip footing problem.

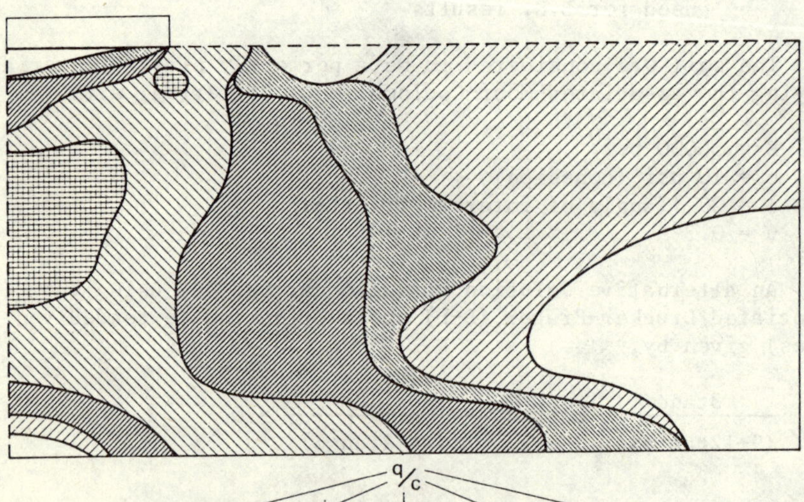

Figure 4. Spread of plastic zones at different load levels. Mohr-Coulomb yield criterion.

metric elements with 121 nodal points. The collapse loads achieved by the boundary element and finite element techniques (M-C) are $q/c = 14.9$ and $q/c = 15.1$ respectively, which agree well with the Prandtl solution (Chen, 1975) $q/c = 14.8$. As for the D-P results, it is seen that although the displacements were larger, the maximum load obtained was still not far from the previous ones.

Zones of yielding defined by the M-C solution are shown in figure 4. These zones compare well with the reported finite element computations.

Shallow Tunnel

In a recent paper the present authors have applied the elastoplastic boundary element technique to solve the problem of a deep circular excavation of radius r' in an infinite medium. The great advantages of the technique were then pointed out when comparing results with different finite element solutions. Herein, an analogous problem is studied by considering the tunnel to be shallow, located within the semi-infinite domain and with its centre at a depth of 5r'.

The rock-like material was assumed to follow the D-P yield criterion presented in the last example, with the following characteristics

$E = 500$ ksi
$c = 0.28$ ksi
$\phi = 30^o$
$\nu = 0.2$

In order to produce a more realistic analysis, the semi-infinite medium was assumed to be initially under the in-situ linearly varying stress field given by the formulae,

$$\sigma_v = \overline{\sigma}_v + \gamma h \qquad \text{(vertical stress)}$$

$$\sigma_h = 0.4\sigma_v \qquad \text{(horizontal stresses)}$$

where $\overline{\sigma}_v$ is a uniform pressure that may be due to an overburden of water or very weak material, γ is the unit weight of the rock and h is the distance from the ground surface.

To simulate the stress state adopted for the deep tunnel problem ($\sigma_v = 1$ ksi) at the depth of the excavation axis, the following values were chosen,

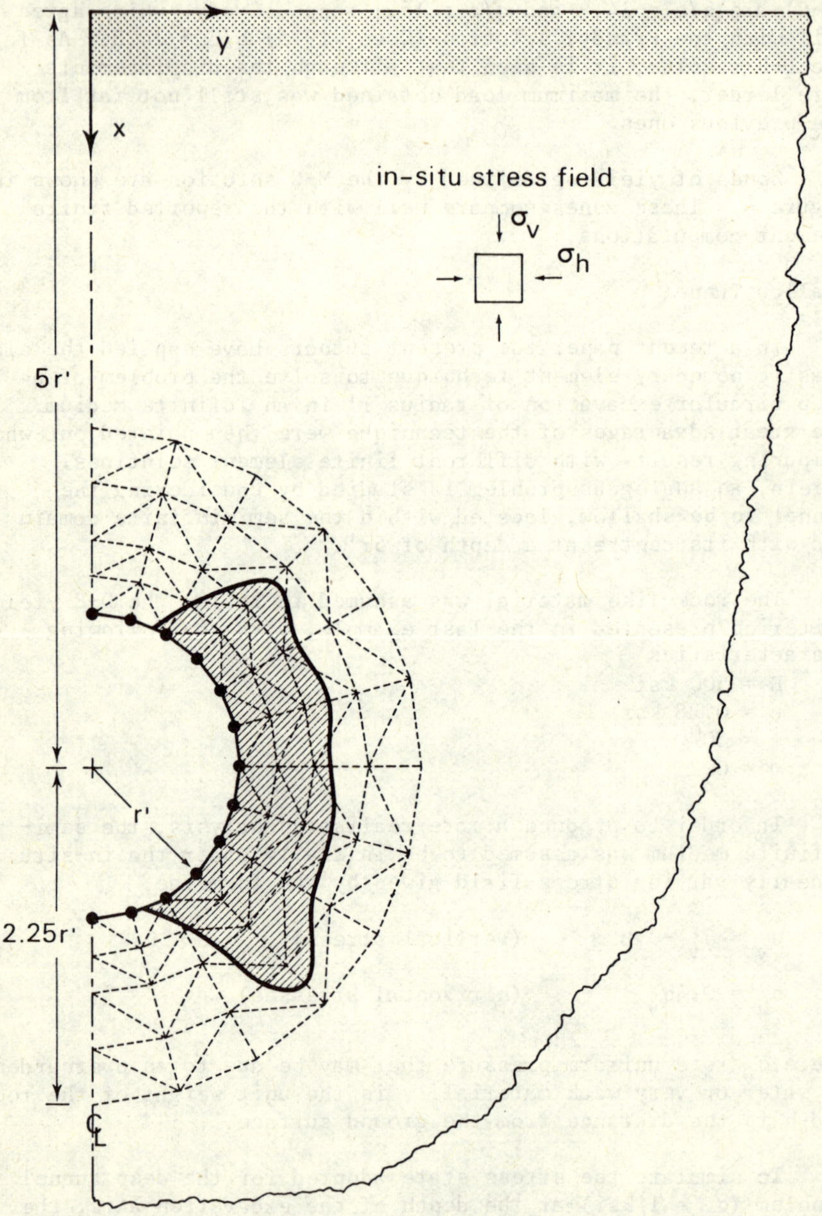

Figure 5. Shallow circular tunnel problem. Discretization used for B.E. results and total spread of plastic zone.

$\overline{\sigma}_v = 0.3$ ksi
$\gamma = 8.9 \times 10^{-2}$ lb/in^3
$r' = 131$ ft

The plane strain analysis was carried out by applying increments of external loads, corresponding to the relaxation of the in-situ stresses, over the boundary of the cavity. The discretization employed is depicted in figure 5 where the total extent of the plastic zone is also shown.

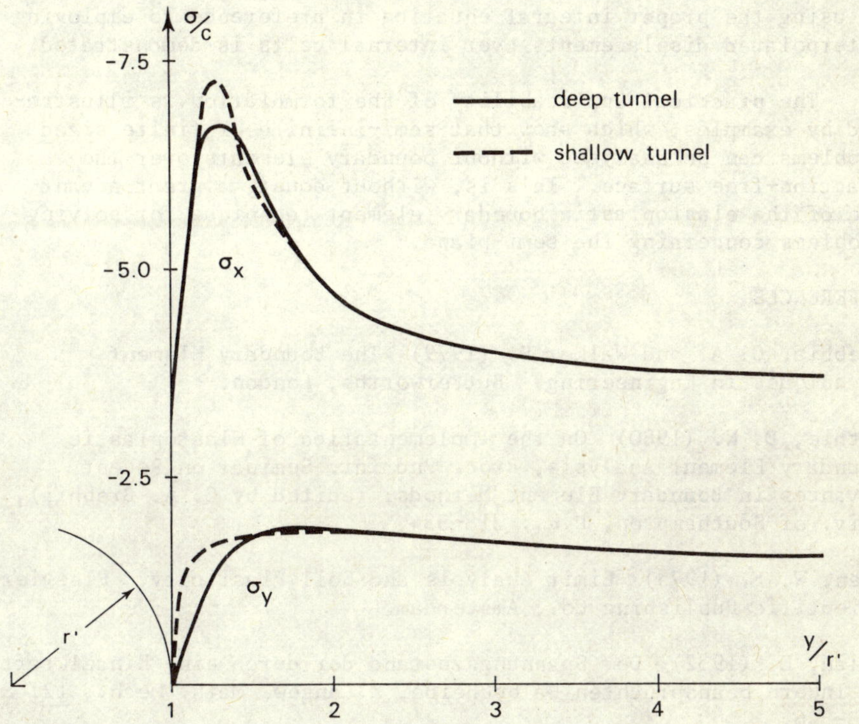

Figure 6. Final stresses along the horizontal section through the medium

Final stresses along the horizontal section are presented in figure 6 with the equivalent results from the deep tunnel case (Telles and Brebbia, 1980b) included for comparison. It should be noted that stress values outside the internal cell region were computed at simple internal points.

The above example clearly indicates the powerfulness of the half-plane implementation. Problems of this sort can only be satisfactorily solved by using this technique, which requires neither ground surface nor outer boundary discretization.

CONCLUSIONS

In this paper the complete formulation for the half-plane boundary element technique applied to plasticity has been presented. The correct integral equations for stresses at internal points and stresses at the traction-free surface were derived following the same procedures previously established by the authors.

The efficiency of computing stresses at internal points by using the proper integral equation in preference to employing interpolated displacements over internal cells is demonstrated.

The practical applicability of the formulation is illustrated by examples, which show that semi-infinite or finite sized problems can be analysed without boundary elements over the traction-free surface. This is, without doubt, a great advantage of the elastoplastic boundary element technique for solving problems concerning the semi-plane.

REFERENCES

Brebbia, C. A. and Walker S. (1979) The Boundary Element Techniques in Engineering. Butterworths, London.

Cathie, D. N. (1980) On the Implementation of Elastoplastic Boundary Element Analysis, Proc. 2nd Int. Seminar on Recent Advances in Boundary Element Methods, (edited by C. A. Brebbia), Univ. of Southampton, U.K.: 318-334.

Chen, W. F. (1975) Limit Analysis and Soil Plasticity. Elsevier Scientific Publishing Co., Amsterdam.

Melan, E. (1932) Der Spannungszustand der durch eine Einzelkraft in Innern beanspruchten Halbscheibe, Z. Angew. Math. Mech., 12: 343-346.

Mendelson, A. (1968) Plasticity: Theory and Application. MacMillan, New York and London.

Mikhlin, S. G. (1962) Singular Integral Equations. Amer. Math. Soc. Trans., Series 1, 10: 84-197.

Mukherjee, S. (1977) Corrected Boundary Integral Equation in Planar Thermoelastoplasticity, Int. J. Solids Structures, 13: 331-335.

Mukherjee, S. and Kumar, V. (1978) Numerical Analysis of Time-Dependent Inelastic Deformation in Metallic Media Using the Boundary Integral Equation Method, trans. ASME, J. Appl. Mech., 45: 785-790.

Nagtegaal, J. C., Parks, D. M. and Rice, J. R. (1974) On Numerically Accurate Finite Element Solutions in the Fully Plastic Range, Comp. Meth. Appl. Mech. Engng., 4: 153-177.

Prager, W. and Hodge, P. G. (1968) Theory of Perfectly Plastic Solids. Dover, New York.

Riccardella, P. C. (1973) An Implementation of the Boundary Integral Technique for Planar Problems in Elasticity and Elasto-plasticity. Report No. SM-73-10, Dept. Mech. Engng., Carnegie Mellon Univ., Pittsburgh.

Schreyer, H. L., Kulak, R. F. and Kramer, J. M. (1979) Accurate Numerical Solutions for Elastic-Plastic Problems, trans. ASME, J. Pressure Vessel Technology, 101: 226-234.

Telles, J. C. F. and Brebbia, C. A. (1979) On the Application of the Boundary Element Method to Plasticity, Appl. Math. Modelling, 3: 466-470.

Telles, J. C. F. and Brebbia, C. A. (1980a) The Boundary Element Method in Plasticity, Proc. 2nd Int. Seminar on Recent Advances in Boundary Element Methods, (edited by C. A. Brebbia), Univ. of Southampton, U.K.: 295-317.

Telles, J. C. F. and Brebbia C. A. (1980b) Elastoplastic Boundary Element Analysis, Proc. U.S.-Europe Workshop on Non-linear Finite Element Analysis in Structural Mechanics, (edited by Wunderlich et al). Ruhr-University Bochum, W. Germany: 403-434.

Telles, J. C. F. and Brebbia, C. A. (1981b) New Developments in Elasto Plastic Analysis, in Boundary Element Methods, Springer-Verlag, Berlin, 1981.

Telles, J. C. F. and Brebbia, C. A. (1981a) Boundary Element Solution for Half-Plane Problems, Int. J. Solids Structures, in press.

Zienkiewicz, O. C., Valliappan, S. and King, I. P. (1969) Elasto-plastic Solutions of Engineering Problems, Initial Stress Finite Element Approach, Int. J. Num. Meth. Engng., 1: 75-100.

Zienkiewicz, O. C., Humpheson, C. and Lewis, R. W. (1975) Associated and Non-Associated Visco-Plasticity and Plasticity in Soil Mechanics, Géotechnique, 25: 671-689.

CHAPTER V

SEISMIC PROBLEMS

SEISMIC SITE AND SOIL-STRUCTURE INTERACTION ANALYSIS

John Lysmer

University of California
Berkeley, California, U.S.A.

ABSTRACT

The relationship between free-field motions and interaction motions is discussed. Conclusions are reached regarding the likely spatial and temporal variations of free-field motions at shallow depth. Several finite element methods for soil-structure analysis are discussed including the Flexible Volume Method.
This method, which employs substructuring, makes it possible to obtain solutions to a broad range of three-dimensional soil-structure and structure-soil-structure interaction problems.

1. INTRODUCTION

Seismic soil response problems and soil-structure interaction problems are closely interrelated and much can be learned about the nature of soil-structure interaction by studying the wave fields which are generated locally in a site during an earthquake. In fact, as will be discussed below, the seismic motion of an embedded structure depends mainly on the free-field motions of the soil volume displaced by the structure. Thus, in order to perform a proper soil-structure interaction analysis, it is of utmost importance to understand which type of motions can realistically exist within shallow depths in a site and, also, how to specify such motions for design purposes. Once these problems have been resolved it is relatively easy, although sometimes costly and time-consuming, to perform a soil-structure interaction analysis; even for complicated wave fields in two or three dimensions. The results of such calculations will often show that the additional motions in the displaced soil volume are small

J. B. Martins (ed.), Numerical Methods in Geomechanics, 409–454.

compared to the free-field motions of the same volume. Thus it may be argued that the proper solution of the site response problem is more important than the solution of the interaction problem. Nevertheless, for important structures, such as nuclear plants and off-shore facilities, the interaction problem must be solved and a large number of sophisticated computer codes have been developed for this purpose. Some of these codes, including a new substructure code, SASSI, will be discussed. This code, which is based on the Flexible Volume Method, can solve the general three-dimensional structure-soil-structure interaction problem for structures on horiztonally layered sites.

2. DEFINITION OF THE SITE AND INTERACTION PROBLEMS

The basic problem of soil-structure interaction involves the determination of the motions of one or more structures at a given site from a knowledge of a given motion (the control motion) at a specified point (the control point) of the site prior to construction. A complete soil-structure interaction analysis must necessarily consist of two distinct parts; a site response analysis and an interaction analysis.

The site response analysis involves the determination of the temporal and spatial variations of the free-field motions. The interaction analysis involves the determination of the motions of a structure placed in the above seismic environment. These are different types of problems but each needs to be addressed to determine a solution to the soil-structure interaction problem.

Each of the above problem types can in principle be formulated in terms of continuum models or discretized models, and it is not possible here to describe all of the possible forms of equations of motions which have been proposed. It is, however, useful for a better understanding of the nature of and the connection between the two problem types to consider the equations of motion for the three linear models shown in Fig. 1. The models are identical in the sense that all are of the finite element type and all are spanned by the same finite element mesh. Also, all masses and stiffnesses are the same, except that the structural part of the model shown in Fig. 1(b) has no stiffness and mass, and that for this model the structural nodes above ground level are assumed to be fixed in space.

Since the fixed nodes have no influence on the motion of the ground, Fig. 1(b) represents a free-field site response problem. It has the equation of motion

$$[M_f]\{\ddot{u}_f\} + [C_f]\{\dot{u}_f\} + [K_f]\{u_f\} = \{Q_f\} \quad (1)$$

where $[M_f]$, $[C_f]$, $[K_f]$ are the mass, damping, and stiffness matrix, respectively, for the free field, and $\{u_f\}$ is a vector containing the nodal point displacements. Since the source of excitation is outside the model the load vector $\{Q_f\}$ has non-zero elements on the external boundary only. Particular solutions to the equation of motion, Eq. (1), can be obtained by site response analysis as described below and it will here be assumed that a free-field solution is available. Thus $\{u_f\}$ and $\{Q_f\}$ are known.

Figure 1(a) represents the corresponding interaction problem. The total displacements can be written

$$\{u\} = \{u_f\} + \{u_i\} \tag{2}$$

where $\{u_f\}$ are the known free-field displacements and $\{u_i\}$ are the interaction displacements. Assuming that the external boundary is very far away from the structure the equation of motion for the interaction problem is

$$[M]\{\ddot{u}\} + [C]\{\dot{u}\} + [K]\{u\} = \{Q_f\} \tag{3}$$

where $\{Q_f\}$ is the same load vector as in Eq. (1) and [M], [C], and [K] are the total mass, damping, and stiffness matrices, respectively. Substitution of Eqs. (1) and (2) into Eq. (3) yields

$$[M]\{\ddot{u}_i\} + [C]\{\dot{u}_i\} + [K]\{u_i\} = \{Q_i\} \tag{4}$$

where

$$\{Q_i\} = ([M_f]-[M])\{\ddot{u}_f\} + ([C_f]-[C])\{\dot{u}_f\} + ([K_f]-[K])\{u_f\} \tag{5}$$

The load vector, $\{Q_i\}$, in Eq. (5), can be computed from the known free-field displacements. It depends only on the difference in properties between the structure and the excavated soil. Thus $\{Q_i\}$ has non-zero elements only at the structure and Eq. (4) is the equation of motion for the problem illustrated by Fig. 1(c). This problem is well-posed and can be solved for the interaction displacements, $\{u_i\}$. The total displacements for the soil-structure interaction problem can be found by superposition as indicated by Eq. (2).

Equations (2) and (4) remain valid even as the distance to the boundary goes to infinity and the mesh size shrinks to infinitesimal dimensions. Hence, the above formulation can be extended to continuum mechanics and three dimensions.

The above formulations reveal three important characteristics of the soil-structure interaction phenomenon:

1. The only free field ground motions which are of importance for the interaction phenomena are those within the volume to be excavated for the embedded part of the structure.

2. For embedded structures the amount of interaction depends on the difference in mass and stiffness between the structure and the volume of excavated soil, see Eqs. (4) and (5). This implies that embedment of structures will usually lead to smaller interaction effects.

3. Soil-structure interaction analysis implies the use of superposition, see Eq. (2). Thus true nonlinear analyses may not be possible.

The first observation has far reaching consequences; especially for embedded structures on relatively soft sites since, for such sites, both theory and observation indicate that the free-field motions vary significantly within normal foundation depths.

3. THE SITE RESPONSE PROBLEM

Site response problems involve the determination of the temporal and spatial variation of motions within a site. In principle, these motions can be determined from a large model which includes the source of earthquake. However, in practice the source parameters and the regional geology cannot be determined in sufficient detail to solve this problem with a high degree of accuracy in the frequency range of interest for design. Thus, current methods of site response analysis attempt to predict the above variation of motions locally from a single specified control motion at some control point within the site. This problem is mathematically ill-posed and unique solutions can be obtained only by the introduction of assumptions regarding the geometry of the site and the nature of the wave field causing the control motion. In practice, consistent solutions can be obtained only for horizontally layered sites. Possible wave patterns include: vertically propagating or inclined plane body waves, and horizontally propagating surface waves. Only the case of vertically propagating shear waves can currently be solved by truly nonlinear methods.

3.1. Control Motion and Control Point

The inherent problem in site response analysis is the choice of wave field to be used in the analysis and it is therefore natural to classify and discuss the different available methods according to the type of wave field assumed. However, before doing so it should be mentioned that the choice of an appropriate control motion and control point is just as, and in most cases, more important than the choice of wave field.

The control motion should be chosen with due respect to observed relations between earthquake magnitude, epicentral distance, maximum acceleration, duration, frequency content, see Idriss (1978) and Ref. 1, and the site-dependent characteristics established by Hayashi et al., (1971), Seed et al. (1976a, 1976b), Faccioli (1978), and Ref. 1.

Except for the obvious case in which the control motion is an observed record at the control point, the preferable control point is a point either at the ground surface or, as discussed below, at an assumed rock outcrop at the depth of which the motion is specified. This is so because most of our data base of strong motion earthquake records from which the control motion had to be estimated was obtained at surface stations and, even more important, because the frequency content of motions at points below the ground surface is strongly influenced by reflections at the free surface, the major effect being the suppression of certain frequencies by wave interference. Thus the specification that the control motion at depth should be a broad-band spectrum or a motion recorded at another site or depth may result in completely unrealistic computed motions for the site.

3.2. Choice of Wave Type

With the control motion and control point fixed, the solution to the site response problem depends entirely on the nature of the wave field producing the ground surface motions. This wave field may consist of many components including Rayleigh waves, Love waves, plane body waves inclined at different angles, and some other wave types such as spherical and cylindrical waves which are usually not considered.

At the present time seismologists cannot advise engineers in sufficient detail on the relative contents of the different possible wave forms which make up the surface motion. Thus, even though soil-structure interaction analyses can be performed for an arbitrary wave field, in practice, such analyses cannot be made due to lack of data on the characteristics of the wave field involved.

Under these conditions, a typical engineering approach is to make analyses for extreme cases of possible wave fields, e.g., for a motion represented by all Rayleigh waves or for a motion represented entirely by a system of body waves, and to determine the influence of the motion specification on the results of the analysis. If the differences are small, then precise specification of the components of the wave field is considered unnecessary. If the differences are large, then increasing efforts must be made to determine even a crude assessment of the relative components of different wave types, or alternatively, conservative choices of wave components may have to be made for different parts of the analysis. It is important therefore to examine the characteristics of the different wave forms which might contribute to the surface control motion; especially at shallow depths.

3.3. Horizontally Propagating Waves

For horizontally layered sites it is relatively easy to set up linear methods of analysis for horizontally propagating waves. However, many possible choices of wave patterns exist (inclined body waves at different angles of incidence, different modes of surface waves, etc.) and it is currently impossible to determine from available seismological data the exact contributions of each wave type to earthquake motions near the surface in the frequency range of interest to earthquake engineers. However, some estimates have been made (Trifunac and Brune, 1970; Randall, 1971; Chandra, 1972; Nair and Emery, 1975; Liang and Duke, 1977; and Toki, 1977). These estimates involve considerable uncertainties however; in view of these uncertainties analysis of site response for motions represented only by horizontally propagating waves is mainly an academic exercise based on assumed data. Nevertheless, as will be discussed below, some practical conclusions can be drawn from such analyses.

The free-field motions caused by horizontally propagating waves will be discussed in three parts: Surface waves are discussed immediately below, inclined body waves are discussed after the section on vertically propagating waves and, finally, the three wave types are discussed together in a section on motions at shallow depths.

3.3.1. Surface Waves. Rayleigh waves in a perfect elastic half-space are well-known and the theory for these are given in standard textbooks, e.g., Richart et al., (1970). However, for obvious reasons soil dynamics analysts are much more interested in surface waves in multi-layered systems, Thomson (1950), Haskell (1953), Ewing et al. (1957). Two types of waves may occur in such systems: Love waves, in which the motions are horizontal and perpendicular to the direction, X, of wave propagation, and Rayleigh waves which involve both vertical and horizontal motions in the vertical XZ-plane. For plane harmonic waves the displacement fields are of the form:

Love waves: $$u_y = \sum_{s=1}^{\infty} L_s \cdot h_s(z) \cdot e^{i(\omega t - c_s x)} \qquad (6)$$

$$\text{Rayleigh waves:} \quad \left\{ \begin{aligned} u_x &= \sum_{s=1}^{\infty} R_s \cdot f_s(z) \cdot e^{i(\omega t - k_s x)} \\ u_z &= \sum_{s=1}^{\infty} R_s \cdot g_s(z) \cdot e^{i(\omega t - k_s x)} \end{aligned} \right\} \quad (7)$$

where ω and t are the frequency and time, respectively, and L_s and R_s are unknown mode participation factors. The infinite sets of wave numbers, c_s and k_s, and mode shapes, $h_s(z)$, $f_s(z)$, and $g_s(z)$, may in principle be determined by methods developed by Thompson (1950) and Haskell (1953). The wave numbers are directly related to the phase velocities of the different wave modes through $V_L = \omega/c$ and $V_R = \omega/k$. Thus the fundamental problem of site response analysis with surface waves is to determine the infinite set of factors L_s and R_s from a single given amplitude of the control point. This is clearly an ill-posed problem and solutions can only be obtained by further assumptions, the most common of which is to assume that only the fundamental Rayleigh or Love mode, corresponding to s = 1, exists. For undamped systems the frequency-dependent phase velocity and mode shape can be found by the Thomson-Haskell method and the amplitude L_1 or R_1 may be determined from the control motion.

Continuum analyses are possible for the case of viscoelastic layers over an undamped half-space (Ewing et al. (1957), Boncheva (1977)). However, for this case it is more practical to first discretize the semi-finite system by the use of finite elements as proposed by Lysmer (1970) and Waas (1972) for Rayleigh waves and Waas (1972), Lysmer and Waas (1972) for Love waves. Only Rayleigh waves will be discussed here. The theoretical model is shown in Fig. 2. It involves the assumption of a linear variation of displacements between layer interfaces and the existence of a stationary rigid base at some finite depth. If this depth is chosen to be considerably larger than the wave length of the Rayleigh waves of interest, a half-space is simulated by this model.

For an N-layer system these assumptions reduce the equation of motion for the layered system to a quadratic eigenvalue problem:

$$([A]k^2 + [B]k + [C] - \omega^2[M])\{v\} = \{0\} \quad (8)$$

where [A], [B], [C] and [M] are simple 2N x 2N matrices which can be formed from the stiffnesses, damping ratios and mass densities of the layered system, and {v} is an eigenvector (mode shape) which contains the 2N displacement amplitudes of the layer interfaces. The mode shape represents the functions $f_s(z)$

and $g_s(z)$ in Eq. (7). For a given frequency, ω, Eq. (8) can be solved by methods developed by Waas (1972). The solution consists of 2N possible wave numbers, k_s, and associated mode shapes $\{v\}_s$ and, in analogy with Eq. (7), the general solution to the equation of motion may be expressed in the form:

$$\{u\} = \sum_{s=1}^{2N} R_s\{v\}_s \cdot e^{i(\omega t - k_s x)} \tag{9}$$

For a damped system all the wave numbers will be complex with negative imaginary parts. Hence, Eq. (9) can also be written

$$\{u\} = \sum_{s=1}^{2N} e^{im(k_s)x} \cdot R_s\{v\}_s \cdot e^{i(\omega t - Re(s_s)x)} \tag{10}$$

which represents a system of generalized Rayleigh waves (modes) which propagate in the positive x-direction, each with its own mode shape, $\{v\}_s$, phase velocity, $\omega/Re(k_s)$, and decay factor, $\exp[2\pi Im(k_s)/Re(k_s)]$, per wave length, $\lambda_s = 2/Re(k_s)$. Experience with the method has shown that most of the Rayleigh modes decay extremely rapidly in the frequency range of interest to earthquake engineers and only a few terms of Eq. (9) need therefore be considered. If it is assumed that only the fundamental mode (defined as the mode with the largest value of $Re(k_s)$) is present, Eq. (9) reduces to

$$\{u\} = R_1 \cdot \{v\} e^{i(\omega t - k_1 x)} \tag{11}$$

and the mode participation factor, R_1, can be determined at each frequency from the amplitude of the control motion at say the surface at x = 0. Transient motions can be handled by Fast Fourier Transform techniques. This requires that the eigenvalue problem in Eq. (8) be solved for each frequency of the transform.

The above method has been used by Chen et al. (1981) to determine possible Rayleigh wave motions in several sites. A typical example of these computations is shown in Fig. 3. The site consisted of 128 ft of medium dense sand overlying a rock formation. A horizontal control motion with the maximum acceleration 0.25 g was specified at the ground surface (distance 0). The response spectrum of the control motion is shown in Fig. 3 (top curve in upper left graph). The computational model consisted of 18 layers in the sand plus 10 thicker layers in the rock.

The equivalent linear method was used to simulate nonlinear effects, i.e. the stiffness and damping ratio in each sand layer were adjusted iteratively to be compatible with the shear stress amplitude developed in the layer. As a result of these iterations the shear wave velocities in the sand profile varied from 621 fps near the surface to 786 fps at the rock interface. Since the average low strain (microseismic) shear wave velocity was about 1150 fps this indicates quite strong nonlinear effects at the excitation level specified. The damping ratio converged to an average value of about 10% with slightly lower values near the ground surface.

Figure 3 shows surface response spectra for a Rayleigh wave field propagating from left to right. It is readily apparent from these spectra that the high frequency components of the wave field are rapidly damped out as a result of the relatively high damping ratio in the sand and in fact, at a distance of a few hundred feet, virtually all motions with frequencies higher than 2 Hz have decayed to insignificant values. Similar results have been obtained for other soil profiles. Since most soil deposits extend horizontally for thousands of feet, it is thus unrealistic to expect that a significant part of the high frequency components of strong motions in such deposits could result from horizontal propagation of fundamental Rayleigh waves. Similar results have been obtained for Love waves in soil sites.

The above theoretical observation is confirmed by the fact that, as far as the writer can ascertain, the seismological literature contains no reports of observation of strong Rayleigh waves at frequencies higher than 2 Hz.

The only practical conclusion which can be reached from the above observations is that fundamental surface waves need not be considered in the seismic design of relatively stiff structures of small lateral extent on soil sites.

Such motions may, however, be important for structures with very low natural frequencies, say below 1 Hz, and for structures on rock where the much lower damping and higher wave velocities could lead to lower rates of attentuation of high frequency surface waves in the horizontal direction. However, even in rock surface waves have not been observed in the frequency range above 2 Hz. Furthermore, soil-structure interaction effects are likely to be small for structures built on rock sites.

In summary, fundamental surface waves are not likely to be of significance for the design of relatively stiff strucutres of small lateral extent, say nuclear power plants.

3.3.2. Higher-order Surface Waves. The fundamental Rayleigh mode defined above is, as shown by Lysmer (1970), identical to the fundamental mode considered by seismologists in layered systems overlying a deformable half space. The rest of the terms of Eq. (10) represent higher-order Rayleigh modes (in the terminology of seismologists) and body waves. These modes will have longer wavelength and will propagate faster than the fundamental mode which by definition has the shortest wavelength and thus the lowest phase velocity. While most of these higher modes can be neglected, since they decay rapidly in the direction of wave propagation, others may decay less rapidly than the fundamental mode. This phenomenon occurs only at relatively high frequencies on sites with a marked increase in stiffness with depth; say a sand profile over rock. These low-decay modes could conceivably contribute significantly to the motion at a surface control point. However, studies by Chen et al. (1981) have shown that such modes, when they occur, cause near surface motions which are similar to those caused by vertical or slightly inclined body waves. That this is so is not surprising when one considers the propagation mechanism of these modes. The very facts that the waves travel at high velocities and decay slowly indicate that the major part of the energy propagation occurs in deeper layers with high body wave velocities and low damping. This immediately implies that insignificant amounts of energy are propagated horizontally in the softer surface layers or, in other words, that the higher frequency motions in the surface layers are maintained through nearly vertical energy propagation through a mechanism similar to that of slightly inclined body waves. The result is that the upper parts of mode shapes of surface waves (i.e. the variation of displacements with depth) are virtually identical to those found in analyzing vertically propagating or slightly inclined body waves.

Thus, in practical calculations the effects of higher-order surface wave components can be considered by assuming a certain content of slightly inclined or vertical body waves in the control motion.

3.4. Vertically Propagating Waves

The great majority of methods for site response analysis use Kanai's (1952) assumption of vertically propagating shear waves. This assumption leads to simple one-dimensional mathematical models for horizontally layered systems and has, partly because of the similarity between motions caused by different wave fields, led to remarkable success in predicting the major features of site response during earthquakes, especially since the analytical procedure was modified by introduction of the equivalent linear method by Seed and Idriss (1969).

Linear site response problems with vertically propagating waves can be solved by a multitude of numerical techniques which are described in texts on soil dynamics, e.g., Desai and Christian (1976). The most efficient method for computing free-field motions from a specified surface control motion appears to be the complex response method used in the program SHAKE, Schnabel et al. (1972). With these methods it is currently possible to analyze any layered viscoelastic soil system overlaying a visco-elastic half-space. The control motion can be specified at the ground surface, at any depth in the soil deposit or as an outcrop motion. Non-linear effects can be approximated by the equivalent linear method.

Recent efforts have been directed towards the development of true non-linear methods of analysis. Several methods have been proposed for performing non-linear total stress analysis of site response problems with vertically propagating shear waves. The most important of these are: The method of characteristics, Streeter et al. (1974), Idriss et al. (1976), Taylor and Larkin (1978); the finite difference method, Joyner (1977); and implicit integration schemes, Martin (1975). In addition several methods of effective stress analysis have been proposed, Ghaboussi and Dikmen (1978), Zienkiewicz et al. (1978), Finn et al. (1977), Liou et al. (1977), and Martin and Seed (1979), which can predict the pore pressure build-up in saturated sands during seismic excitation.

Comparative studies of ground motion characteristics computed by the equivalent linear method and non-linear methods show relatively small differences except where motions are very strong and soils relatively weak. Thus the development of non-linear analysis techniques has further confirmed the fact that equivalent linear methods are sufficiently accurate for virtually all practical purposes in evaluating site response for soil-structure interaction analyses.

3.4.1. Inclined Body Waves. Some energy may be arriving at the control point in the form of non-vertically propagating body waves. This includes the effect of the high-order surface wave modes discussed in a previous section.

The response of horizontally layered sites to plane harmonic body waves arriving at a specified incident angle through an underlying elastic half-space has been investigated by several researchers. The fundamental work was done by Thomson (1950) and Haskell (1960, 1962) who developed an efficient matrix method for computing the frequency-dependent transmission coefficients in a layered continuum for incident SH, SV and P-waves. Efficient computer codes for the Thomson-Haskell method were developed by Hannon (1964) and Teng (1967). Silva (1976) extended the

Thomson-Haskell method to include damping in the soil layers. With this method it is possible to solve linear or equivalent linear site response problems with inclined body waves for systems consisting of viscoelastic soil layers overlying a uniform undamped half-space, provided the incident angle in the half-space is known. Since no damping is included in the half-space the resulting surface motions do not decay in the horizontal direction. More recently, Chen et al. (1981) have developed a method which includes damping in the underlying half space.

Analyses of surface response to inclined body waves have also been made by Joyner et al. (1976), who determined the transfer functions from bedrock to the soil surface for a soil deposit 186 m in depth for shear waves propagating at various angles to the vertical. The results of this study are shown in Fig. 4, and it is apparent that for angles of incidence up to 60°, there is a negligible difference between the motions computed for inclined waves and for vertically propagating waves.

It is reasonable to conclude therefore that the variation of horizontal motions with depth within a soil deposit are for all practical purposes the same, whether they are computed for vertical or inclined directions of propagation within the depth range of interest to engineers. On this basis it is appropriate to use analyses for vertically propagating waves because of their greater simplicity and availability of solutions.

3.5. Motions of Shallow Depth

As discussed earlier, only the motions within a relatively shallow depth (the projected depth of embedment) of the free field will influence the motions of structures. The same discussion also indicates that both the spatial and temporal variation of the free-field motions within this depth are of importance in evaluating soil-structure interaction effects. It is therefore appropriate to discuss in more detail how the amplitude and frequency content of free-field motions vary with depth and in particular how they vary near the free ground surface.

As will be shown in the following, the presence of the strong discontinuity represented by the free ground surface imposes predictable and observable limitations on how horizontal amplitudes and the frequency content of motions vary with depth near the ground surface.

The potential effects of the free ground surface on the amplitude and frequency content of waves at various depths in a uniform deposit is shown in Figs. 5 and 6. Both figures show the variation of amplitude with the dimensionless depth z/λ_s in in a perfect half-space, where $\lambda_s = V_s/f$ is the wavelength of shear waves at the frequency, f[Hz], considered.

Figure 5 corresponds to the case of vertically propagating shear waves and Fig. 6 corresponds to the case of horizontally propagating Rayleigh waves.

The two types of wave fields are obviously quite different. It is remarkable, however, that both the shear wave field and the Rayleigh wave field produce monotonically decreasing horizontal displacements within the approximate depth

$$z \simeq \frac{1}{4}\lambda_s \quad \text{to} \quad \frac{1}{5}\lambda_s \simeq \frac{V_s}{4f} \quad \text{to} \quad \frac{V_s}{5f} \tag{12}$$

and that all horizontal displacements vanish at this depth. A similar phenomenon occurs for inclined shear waves and for layered soil systems where V_s in Eq. (12) can be replaced by the average shear wave velocity, $\overline{V}_s$ above the depth z. As can be seen from the dotted curve shown in Fig. 5 the existence of material damping does not change the substance of these observations.

Two important conclusions can be drawn from these analyses:

(1) Any horizontal motion computed (or observed) at the depth z must be deficient of components of the frequency

$$f \simeq \frac{\overline{V}_s}{4z} \quad \text{to} \quad \frac{\overline{V}_s}{5z} \tag{13}$$

i.e., its response spectrum will have a dip at the approximate frequency f, which incidentally is equal to the fixed base natural frequency of the soil column above the depth z. Thus, the only level at which a smooth spectrum can exist is at a free surface, and specifying a control motion with a smooth spectrum at any other depth will, as experience has shown, lead to completely unrealistic results. This free surface can be the actual ground surface or a real or imaginary outcrop; however, a smooth spectrum cannot exist within a soil deposit, whether the motions be due to near-vertically propagating body waves or to horizontally propagating Rayleigh waves.

(2) In a deposit with uniform properties, seismic motions will decrease with depth below the ground surface at least down to the depth

$$z = \frac{\bar{V}_s}{5f_{max}} \tag{14}$$

where f_{max} is the highest frequency present in the motion.

This follows directly from Figs. 5 and 6 which show that all components decrease in amplitude within the above depth. Because of variations in soil characteristics with depth this predicted reduction will often extend to depths greater than those indicated by Eq. (14). For a typical soil site, with say $\bar{V}_s$ = 1000 fps, and a seismic environment with say f_{max} = 20 Hz, the above decrease will occur within the upper 10 ft (or deeper if the predominant frequency of the control motion is lower) of the site. Thus in view of the discussion in connection with Eq. (4), even relatively shallow embedment may significantly influence the seismic response of structures on soft sites and both the embedment and the reduction in the amplitude of the seismic environment with depth should be considered in a rational interaction analysis.

Substitution of realistic values of V_s and f into Eq. (12) will show that z is typically larger than 20 ft for soil sites and 60 ft for rock sites. Thus typical structures experience only the upper part (z/λ_s < 0.2) of the motions shown in Figs. 5 and 6. In this "shallow" depth range horizontal motions produced by any seismic environment with the same horizontal surface control motion are quite similar. It is therefore to be expected that the horizontal motions produced at points below the ground surface during earthquakes will be relatively independent of the type of wave field producing the motion.

The above observations were made for motions in a uniform half-space. For layered systems, the stiffness of which usually increases with depth, calculations have shown that the similarity between motions produced at shallow depth by different types of wave fields is even more pronounced. This is so because, just as for the higher-order surface waves discussed above, the upper parts of the mode shapes of fundamental surface waves in layered systems tend toward the mode shapes produced by vertically propagating or slightly inclined body waves.

An interesting example of these effects in a 600 feet deep soil deposit overlying a rigid half-space is shown by the analytical results presented in Figs. 7 and 8. To study the response at different depths in this deposit, analyses were made using vertically propagating shear wave theory for 15 different excitation records. In eight of the analyses, existing records obtained on deep soil deposits were scaled to have a peak acceleration of 0.20 g and considered to be developed at the ground surface. The distribution of acceleration with depth and the frequency characteristics of the motions developed at depths of 40 and 76 ft were then determined by deconvolution analyses.

For the same soil deposit, a second study was made in which seven records representative of rock motions were used as base excitation and the base motions were scaled in each case to produce a peak acceleration of 0.20 g at the ground surface. In spite of the fact that specification of control motions at depth is not recommended see above; for this case there was surprisingly little difference in the computed distribution of motions whether the excitation was applied at the ground surface or whether it was applied at the base of the soil deposit. The results of the two sets of studies were analyzed statistically to determine the mean acceleration distribution separately for the deconvolution analyses and for the base input analyses. The results of this analysis are shown in Fig. 7. On the whole the results are remarkably similar, all showing a marked drop in peak acceleration within the upper 60 ft.

The response spectra for the motions developed at depths of 40 and 76 ft were also computed and analyzed statistically for the two different groups. The 84 percentile spectra for surface motions, motions at 40 ft depth and motions at 76 ft depth for the deconvolution analyses are shown in Fig. 8. It may be seen that while the spectrum for the surface motions is of the broad band type, the spectrum for motions at a depth of 40 ft contains a marked suppression of frequencies corresponding to a period of about 0.18 second while that for motions at a depth of 76 ft shows a marked suppression of frequencies corresponding to a period of about 0.3 second. The fixed base natural periods of this deposit for 40 ft of soil and 76 ft of soil were about 0.18 and 0.3 seconds respectively. Similar results were obtained for the base excitation analyses. Thus it may be seen that the frequency suppression effect, as predicted by Eq. (13), is mainly a feature of the geometry and material characteristics of the deposit and depends only slightly on the time history of the motion involved.

3.6. <u>Summary and Conclusions</u>

At the outset of this section it was shown that any analysis of soil-structure interaction must necessarily be based on a knowledge of the seismic environment to which the structure will be subjected. This requires an understanding of the spatial distribution of motions in the ground surrounding and underlying the structure. In the light of the discussion of this subject presented in the preceding pages it seems reasonable to draw the following conclusions concerning the role of the seismic environment in soil-structure interaction analyses.

1. On rock sites structures are likely to be founded near the surface. For such sites, earthquake motions may consist of an unknown mixture of Rayleigh waves, Love waves and near-vertically propagating body waves. Because of the low damping in the rock, attenuation of Rayleigh or Love waves may be small within the general area of the site but the contribution of these types of waves to the total ground motion, within the frequencies of interest for nuclear power plants, will nevertheless be small. The presence of such waves will tend to increase the rocking and torsional excitation on the base of the structure due to out of phase effects as the waves pass across the base. Thus structures located on rock should be analyzed for these motions to determine the potential severity of their contributions to the total response of the structure. However, the greater part of the response can be considered to result from vertically propagating body waves.

2. For soil sites, structures are likely to be embedded at some depth (say 20 to 80 ft) below the ground surface. The effects of fundamental Rayleigh and Love waves need not be considered at such sites in most designs because the high frequency components of these waves (greater than 1 Hz) will have been damped out by the soil if it extends to any significant distance (say 1000 ft) around the location of the structure. Higher order Rayleigh modes can be simulated by inclined body waves. Thus the main source of excitation will be inclined body waves and for all practical purposes, these can be analyzed as if they propagated in a vertical direction. However in soil deposits there will be an important variation in motion characteristics with depth and this should be considered in the analysis if meaningful results are to be obtained. The assumption of uniform motions in the upper layers of a soil deposit is inconsistent with the physical nature of wave mechanics and observations in the field and can only lead to misleading results unless the specified control motion is intended to take the natural variations in motion characteristics into account in some way. Without knowing something about the variations in motion it is difficult to see how this can be done realistically without introducing an unwarranted degree of conservatism into the soil-structure analysis procedure.

It should not be construed from the above statements that the assumption of vertically propagating waves at soil sites is appropriate for all types of structures. The long-period components of horizontally propagating waves may be extremely important for the design of buried pipelines, tunnel linings and earth retaining structures. However, except for increased stresses in the walls of buried or embedded structures the change in the stress field due to these waves appears to have little effect on the overall horizontal motions of such structures.

The propagating nature of the displacement field may also induce additional displacements and stresses in long above-ground structures such as bridges, Bogdanoff et al. (1965), Johnson and Galletly (1972), Abdel-Ghaffar and Trifunac (1976); and rocking and torsional motions in long period single structures, Wong (1975), Scanlan (1976) and Wong and Luco (1976).

Finally, and perhaps most important, control motions should be chosen with due respect to site conditions and, if a broad band design spectrum is used the control point should be located at the ground surface or, alternatively, at an imaginary outcrop, where it could conceivably exist, and not at some arbitrary depth below the ground surface where the boundary conditions resulting simply from the existence of a ground surface preclude this possibility.

The above conclusions have been drawn mainly on theoretical grounds. However, despite the fact that no concerted effort has been made to date to obtain field data to confirm the above conclusion, a substantial body of field data does in fact exist. Thus, observations by, e.g. Joyner et al. (1976) and Ohsaki and Higawara (1970) confirm that near surface seismic motions do in fact decrease with depth and Hays et al. (1979) and Gazetas and Bianchini (1979) have presented data which confirm the depth dependent frequency suppression.

4. SOIL-STRUCTURE INTERACTION

The topic of soil-structure interaction analysis is very large and it is not possible herein to discuss in detail all of the sophisticated methods which have been proposed. Instead the writer will concentrate on a few rather general methods all of which can handle embedded foundations. Readers interested in further details and simplified methods for surface foundations are referred to several recent state-of-the-art reports on the topic, Refs. (2), (3) and Johnson (1981).

The methods discussed herein are all of the finite element type since only methods which discretize the structure and its foundation can handle the complicated geometry of real sites and embedded structures. The methods are essentially linear. However, nonlinearities can be approximated by equivalent linear methods.

The assumption of linearity is dictated by the superposition stated in Eq. (2). With the current definition of the seismic soil-structure interaction problem this superposition will occur, in some form or another, in any rational mathematical formulation of the problem. This superposition presents a basic dilemma of seismic soil-structure interaction analysis. Its corollary is that even if perfect non-linear constitutive laws were available for the soil and perfect non-linear numerical procedures were available for evaluating the free-field motions, $\{u_f\}$, and the interaction motions, $\{u_i\}$, it would not be legal to add these motions to obtain the total motions. Thus, from a purely theoretical standpoint it is not possible to perform a non-linear seismic soil-structure interaction analysis. This, of course, should not be construed to mean that approximate non-linear solutions cannot be obtained by the introduction of reasonable engineering assumptions, e.g. by specifying free-field motions on the boundaries of the model shown in Fig. 1(a) it may be possible to obtain an approximate non-linear solution for the motion of the structure without the use of superposition.

Another limitation on the methods to be discussed is that they all refer to horizontally layered sites. This restriction is necessitated by the fact that, currently, the site response problem can only be solved for such sites.

4.1. Complete Methods

Complete methods are characterized by employing models which include both the structure and part of the site. Since the discretization must be small compared to the wavelength of the seismic waves to be modeled, these methods lead to very large finite element models and their use is for practical purposes limited to two-dimensional and axisymmetric situations. Some of the methods employ special transmitting boundaries to reduce or eliminate reflections from the boundaries of the necessarily finite model for the semi-infinite site.

Figure 9 shows a typical computational scheme for a complete interaction analysis. The scheme, used by Seed and Idriss (1973), and Lysmer et al. (1975), is limited to problems involving vertically propagating body waves. As for all methods of soil-structure interaction analysis the site response problem is solved first. This involves a deconvolution of the surface control motion to some level below the ground surface where it can be assumed that the presence of the structure will not influence the ground motion, see Fig. 9(a). In the second step the computed base motion is used as a specified boundary motion for a finite element analysis of the soil-structure system shown in Fig. 9(b).

The finite element model used must be very wide or must be equipped with transmitting lateral boundaries in order to avoid reflections, Lysmer et al. (1975), Seed et al. (1975). However, as will be discussed below in connection with Fig. 11, numerous computations have shown that for typical seismic motions only insignificant reflections will occur from the lower rigid boundary on which the control motion is specified. This is so because the interaction wave field consists mainly of surface waves which do not penetrate deeply into the ground. Hence, the depth to the rigid base can often be chosen surprisingly small.

The computational scheme indicated in Fig. 9 may in principle be used to analyze any three-dimensional soil-structure interaction problem involving vertically propagating waves and a horizontally layered site. However, well-known restrictions on computer cost and capacity limit our current ability in treating the third dimension rigorously except for the case of axial symmetry of the structure, Kausel and Roesset (1974), Berger (1976), Berger et al. (1975a, 1975b), Tajimi and Shimomura (1976), Shimizu et al. (1977). An approximate method for three-dimensional analysis has been presented by Hwang et al. (1975). The three-dimensional effect is achieved by placing Lysmer-Kuhlemeyer (1969) dashpots at all nodal points of the plane-strain finite element slice analyzed as shown in Fig. 10. These dashpots are only active on the interaction velocities and simulate the propagation of shear waves in the direction perpendicular to the plane of analysis. Structure-soil-structure interaction can be handled by this method for several structures in the plane of analysis.

The theory discussed in connection with Fig. 1 and Eqs. (1) to (5) constitutes a complete method which can handle arbitrary incident wave fields. This method was used by Gomez-Masso et al. (1978, 1980) to solve a number of soil-structure interaction problems involving transient vertically propagating waves and fundamental Rayleigh waves in soil and rock sites. A typical model for the interaction part of the analysis is shown in Fig. 11. The model was equipped with transmitting boundaries as indicated in the figure and also with viscous boundaries as indicated in Fig. 10 to account for three-dimensional effects. The results obtained showed that the motions of the structure were independent of whether or not a viscous boundary was used at the base of the model, indicating, as mentioned above, that most of the interaction displacements consist of surface waves which dissipate through the lateral boundaries rather than the base of the model. Control computations using the procedure illustrated by Fig. 9 and vertically propagating waves indicated, as expected, identical results for the two methods.

The site response problem was solved by the methods discussed in Section 3. Typical Rayleigh wave mode shapes for the rock site are shown in Fig. 12. These mode shapes were determined from a site response model in which the depth varied inversely with frequency, according to Chen et al. (1981), in order to simulate the existence of a half space in the lower part of the site. The Rayleigh wave field is strongly dispersive, i.e. each of the modes shown in Fig. 12 propagates horizontally with its own phase velocity.

For the structure shown in Fig. 11 the computed responses due to Rayleigh waves and vertically propagating body waves with the same horizontal control motion at the ground surface were found to be surprisingly similar. Typical response spectra for the rock site case are shown in Fig. 13. In this figure the full curve corresponds to Rayleigh wave excitation and the broken curve (S+P Wave) to a combined shear wave and P-wave excitation which produces the same horizontal and vertical motions at the control point as the Rayleigh wave field. The results for a soil site showed even greater similarity. These results would seem to confirm the assertion made above that the motions of embedded structures are relatively independent of the type of wave field causing the motion. Furthermore, in evaluating the differences indicated by Fig. 13 it should be remembered that a surface motion consisting of pure Rayleigh waves is unlikely to exist on any site. Most likely, at frequencies above 2 Hz, the Rayleigh wave component will amount to only a few percent of the total surface motion.

4.2. Substructure Methods

The mathematical problems involved in modeling the essentially semi-infinite soil mass and a desire to break complicated soil-structure interaction problems into more manageable parts have led to a large number of methods in which the soil mass is treated as a continuum and the structure as a discretized model. The half-space is analyzed first, usually in the frequency domain, and the impedance and scattering properties at the soil-structure interface are established. In the second step these properties are used as boundary conditions in a dynamic analysis of the structure with a loading which depends on the free-field motions. In recent years several substructure methods have appeared in which the half-space solution is obtained by finite element analysis with transmitting boundaries.

For the case of structures founded at the ground surface substructuring does indeed lead to simple and very effective methods of soil-structure interaction analysis. However, most real structures are embedded and the effect of this embedment, which can be quite strong even for shallow depths of embedment, is not easily considered by substructure methods. The basic difficulty is that, because the free-field motions vary considerably with depth, especially in softer materials, it is difficult to specify the distribution of the free-field forces on the embedded part of the structure.

Substructure methods for embedded foundations have been proposed by several authors. They fall into three distinct groups according to the number of degrees-of-freedom at which the half-space and the structure interacts.

4.2.1. Rigid Boundary Methods. Kausel and Roesset (1974) have proposed a 3-step method for the case of rigid embedded foundations. This procedure is illustrated in Fig. 14. The first step is a scattering analysis in which the site includes a rigid massless foundation with the same shape as the actual foundation. The solution to this problem produces a set of rigid body accelerations, $\ddot{u}_i$, for points in the structure. The second step is a foundation vibration (impedance) analysis the solution to which produces the impedance matrix for the foundation and with this the springs and dashpots to be used in the last step of the analysis where the loading on the structure is computed from the rigid body motions obtained from Step 1. The total displacements follow by superposition of the motions from Steps 1 and 3. Unfortunately, rigorous solutions to the first step are difficult to obtain except by finite element methods and since the second step also requires a finite element analysis the complete 3-step method is hardly competitive with the complete finite element methods discussed above. However, with approximate solutions to Step 1, Kausel et al. (1978), and Step 2, Novak and Beredugo (1972), Novak (1974), Kausel and Roesset (1975), the method does lead to very economical analysis for the case of vertically propagating waves.

A different, but similar, formulation of the rigid embedded foundation interaction problem has been proposed by Luco et al. (1975) and used by Day (1977, 1978) to determine the response of structures on cylindrical embedded foundations subjected to inclined SH-waves. The method is similar to the 3-step method discussed above in that it usually requires two finite element analyses of the models shown in Fig. 14(a) and Fig. 14(b). The first analysis (the scattering problem) involves determining the forces which the free-field motions exert on the footing when this is held fixed in space. The second analysis (the impedance problem) is identical to that of the above prcoedure. The third step is similar to Step 3 in Fig. 14(c). The method can, as opposed to the Kausel-Roesset method, be adapted for structure-soil-structure interaction analysis.

A disadvantage of the rigid boundary methods is of course that they assume a rigid basement. Thus, unless it can be shown in each case that this assumption is valid, these methods provide no means of evaluating the actual deformations and stresses in the embedded part of the structure.

4.2.2. Flexible Boundary Methods. If the flexibility of the embedded part of the structure is to be considered a flexible boundary method must be used. The basic elements of a theory for a rigorous finite element flexible boundary method have been developed by Gutierrez (1976) and Gutierrez and Chopra (1977, 1978).

As with the methods discussed above, the complete solution of a soil-structure interaction problem by the flexible boundary method requires first the elevation of a site response (scattering) problem similar to the one in Fig. 14(a) to determine the motion of the now flexible boundary, and second the solution of an impedance problem similar to the one stated in Fig. 14(b). The latter problem now involves more degrees-of-freedom and leads to a larger impedance matrix. The third step involves an analysis of the structure alone and is only slightly more complicated than the problem shown in Fig. 14(c).

Unfortunately, as of this time, no satisfactory method has emerged for the solution of the scattering part of the problem. Thus the method has only been used for the analysis of surface structures, excited by vertically propagating body waves, Chopra and Gutierrez (1973) and Aydinoglu and Cakiroglu (1977).

4.2.3. Flexible Volume Methods. Both the scattering and impedance problems for the above methods can be greatly simplified if more common degrees-of-freedom between the half-space and the structure are included in the interaction problem. Let the soil and the structure be partitioned not at the interface but according to the partitioning shown in Fig. 15. In this partitioning the structure, Fig. 15(c) consists of the superstructure plus the basement minus the excavated soil, and the foundation consists of the original site, Fig. 15(b), i.e., the soil to be excavated is retained with the foundation. Interaction between the structure and the foundation occurs at all basement nodes.

The equation of motion for the complete problem, Fig. 15(a) is:

$$[M]\{\ddot{\hat{u}}\} + [K]\{u\} = \{\hat{Q}_b\} \tag{15}$$

where [M] and [K] are the total mass and stiffness matrices, respectively. $\{\hat{u}\}$ is the vector of nodal point displacements and $\{\hat{Q}_b\}$ are the external forces. Since the source of excitement is outside the model, $\{\hat{Q}_b\}$ has non-zero elements only at the degrees of freedom corresponding to boundary nodes, which are assumed to be far away from the structure.

For harmonic excitation of the frequency ω the load and displacement vectors can be written

$$\{\hat{Q}_b\} = \{\hat{Q}_b\}e^{i\omega t} \quad \text{and} \quad \{\hat{u}\} = \{\hat{u}\}e^{i\omega t} \tag{16}$$

where $\{Q_b\}$ and $\{u\}$ now contain complex force and displacement amplitudes. Hence, for each frequency the equation of motion takes the form

$$[C]\{u\} = \{Q_b\} \quad , \quad [C] = [K] - \omega^2[M] \tag{17}$$

where [C] is a complex, frequency dependent stiffness matrix. Damping can be introduced by forming [K] from complex moduli.

Introducing the following subscripts which refer to degrees of freedom (DOF) associated with different nodes:

Subscript	Nodes	
s	superstructure	
i	basement	See also Fig. 15
f	excavated soil	
b	external boundary	
g	remaining soil	

the equation of motion for the foundation in Fig. 15(b) can be written

$$\begin{bmatrix} C_{ff} & C_{fg} & C_{fb} \\ C_{gf} & C_{gg} & C_{gb} \\ C_{bf} & C_{bg} & C_{bb} \end{bmatrix} \begin{Bmatrix} u_f \\ u_g \\ u_b \end{Bmatrix} = \begin{Bmatrix} Q_f \\ 0 \\ Q_b \end{Bmatrix} \tag{18}$$

where $\{Q_f\}$ are the interaction forces from the structure in Fig. 15(c). Similarly, the equation of motion for the structure in Fig. 15(c) can be written:

$$\begin{bmatrix} C_{ss} & C_{si} \\ C_{is} & (C_{ii} - C_{ff}) \end{bmatrix} \begin{Bmatrix} u_s \\ u_f \end{Bmatrix} = \begin{Bmatrix} 0 \\ -Q_f \end{Bmatrix} \tag{19}$$

where compatibility of displacements ($u_i = u_f$) and equilibrium ($Q_i + Q_f = 0$) has been enforced. The term ($C_{ii} - C_{ff}$) simply indicates the stated partitioning according to which the stiffness and mass of the excavated soil is subtracted from the stiffness of the structure.

Assuming that the external boundary is very far away from the structure the equation of motion for the free field (Fig. 15(b)) can be written:

$$\begin{bmatrix} C_{ff} & C_{fg} & C_{fb} \\ C_{gf} & C_{gg} & C_{gb} \\ C_{bf} & C_{bg} & C_{bb} \end{bmatrix} \begin{Bmatrix} u'_f \\ u'_g \\ u'_b \end{Bmatrix} = \begin{Bmatrix} 0 \\ 0 \\ Q_b \end{Bmatrix} \tag{20}$$

where $\{u'\}$ are the free-field motions (the solution to the site response problem). Subtraction of Eq. (20) from Eq. (18) yields:

$$\left[\begin{array}{cc|c} C_{ff} & C_{fg} & C_{fb} \\ \hline C_{gf} & C_{gg} & C_{gb} \\ C_{bf} & C_{bg} & C_{bb} \end{array}\right] \left\{\begin{array}{c} r_f \\ \hline r_g \\ r_b \end{array}\right\} = \left\{\begin{array}{c} Q_f \\ \hline 0 \\ 0 \end{array}\right\} \tag{21}$$

where $\{r\} = \{u\} - \{u'\}$ are the interaction displacements. By partitioning Eq. (21) as indicated $\{r_g\}$ and $\{r_b\}$ can be eliminated and $\{Q_f\}$ can be expressed in the form:

$$\{Q_f\} = [X_f]\{r_f\} = [X_f](\{u_f\} - \{u'_f\}) \tag{22}$$

The frequency dependent matrix $[X_f]$ is the impedance matrix corresponding to the nodal points indicated by heavy dots in the foundation (free field) model, Fig. 15(b). It can be determined as the inverse of the flexibility matrix for this model, i.e. from the solution for point loads in a uniformly layered system.

Substitution of Eq. (22) into Eq. (19) yields:

$$\begin{bmatrix} C_{ss} & C_{si} \\ C_{is} & (C_{ii} - C_{ff} + X_f) \end{bmatrix} \begin{Bmatrix} u_s \\ u_f \end{Bmatrix} = \begin{Bmatrix} 0 \\ [X_f]\{u'_f\} \end{Bmatrix} \tag{23}$$

from which the final motions of the structure can be determined. Except for the term C_{ff} which corresponds to the excavated soil this equation is similar to that used in the general substructure method proposed by Gutierrez (1976).

Thus the solution of the soil-structure interaction problem has been reduced to three steps:

1. Solve the site response problem
   to determine the free field motions $\{u'_f\}$ within the embedded part of the structure, see Fig. 16(a).

2. Solve the impedance problem
   to determine the matrix $[X_f]$, see Fig. 16(f).

3. Solve the structural problem
   This involves forming the complex stiffness matrices and load vector in Eq. (23) and solving this equation.

For horizontally layered sites the site response problem can be solved by the methods described in Section 3. Current state-of-the-art would allow the consideration of any superposition in three dimensions of plane vertical and inclined body waves and surface waves in horizontally layered visco-elastic systems with nonlinearities considered by the equivalent linear method.

The scattering problem has been eliminated, see Fig. 16, since it is identical to the site response problem. And the impedance problem is simpler than any of the problems shown in Figs. 16(c) and 16(e) because of the more regular surface boundary. It can be solved as the inverse of the flexibility problem which involves finding the displacements of the interaction nodes for unit loads placed at each node in succession. Each of these subproblems can be reduced to an axisymmetric problem, which has to be solved for only one node per layer interface.

Because of the above simplifications the flexible volume method is more effective than any of the other substructure methods mentioned above for solving soil-structure interaction problems involving one or more embedded foundations. A finite element computer code, SASSI, which implements the method in both two and three dimensions has recently been developed at the University of California at Berkeley, Lysmer et al. (1981).

Figure 17 illustrates a typical example of a three-dimensional soil-structure interaction problem which can be solved by the SASSI code. The model represents an off-shore tripod structure which stands on the surface of a deep sand deposit. The elements shown with broken lines are not part of the actual structure but are included in the model as a fictitious basement. This makes it possible to consider the confining effects of the weight of the structure and non-linear effects (by the equivalent linear method) within the sand volume spanned by the added elements. Several other three-dimensional cases have been solved. However, it is not possible within the space of this paper to present results and the reader is referred to Lysmer et al. (1981) for further details.

4.3. Summary and Conclusions

It has already been concluded in Section 3.6 that for most practical purposes the appropriate seismic environment for a seismic interaction analysis is a system of vertically propagating waves defined by a control motion at the ground surface.

For isolated surface structures the interaction analysis is most simply performed by a substructure method. For embedded structures and for multi-structure systems, however, the choice is not so simple. It has often been argued that substructure methods are simpler and cheaper to perform. However, if basement flexibility is to be considered, substructuring is anything but simple and it might be cheaper to perform a complete analysis.

A major advantage of complete methods over the substructure methods is that the analysis is performed at the actual stress level for all soil elements and that it is therefore possible with these methods to consider the spatial effects of nonlinearity, say by the equivalent linear method.

A disadvantage of the complete methods is that more computer time and storage are required to perform the analysis of the larger models involved. However, as pointed out by Wight (1977), experience has shown that while the complete methods are computer-intensive, the substructure methods, even the approximate rigid base methods, are manpower-intensive and the total dollar cost of analysis will be about the same for the two methods.

Among the substructure methods the flexible volume method seems to have several advantages over the rigid base and flexible boundary methods. In particular it is the only method which can currently handle general three-dimensional geometries and wave fields, and structure-structure interaction between flexible embedded foundations.

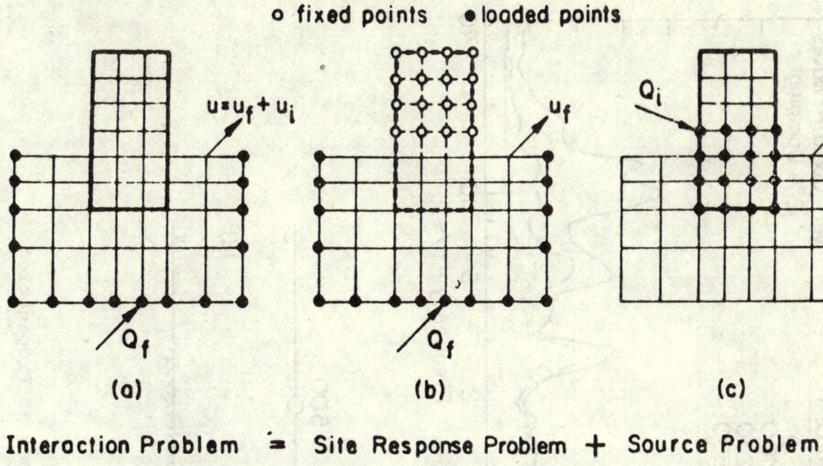

Fig. 1. Superposition Theorem for Interaction Problems

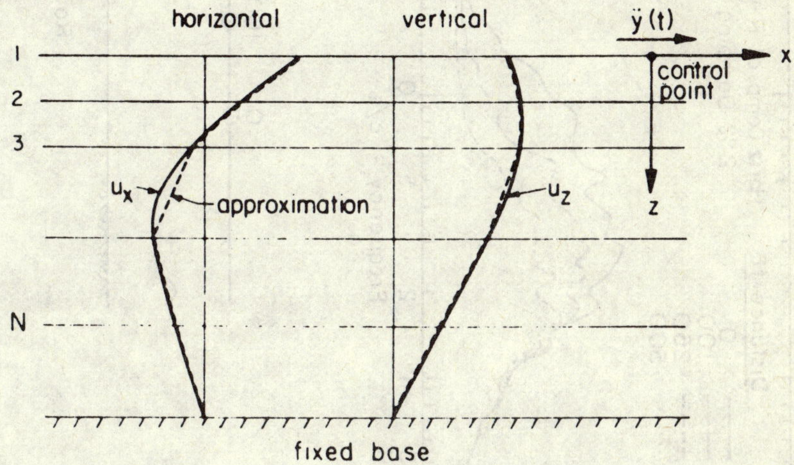

Fig. 2. Typical Soil Profile and Rayleigh Wave Mode Shape

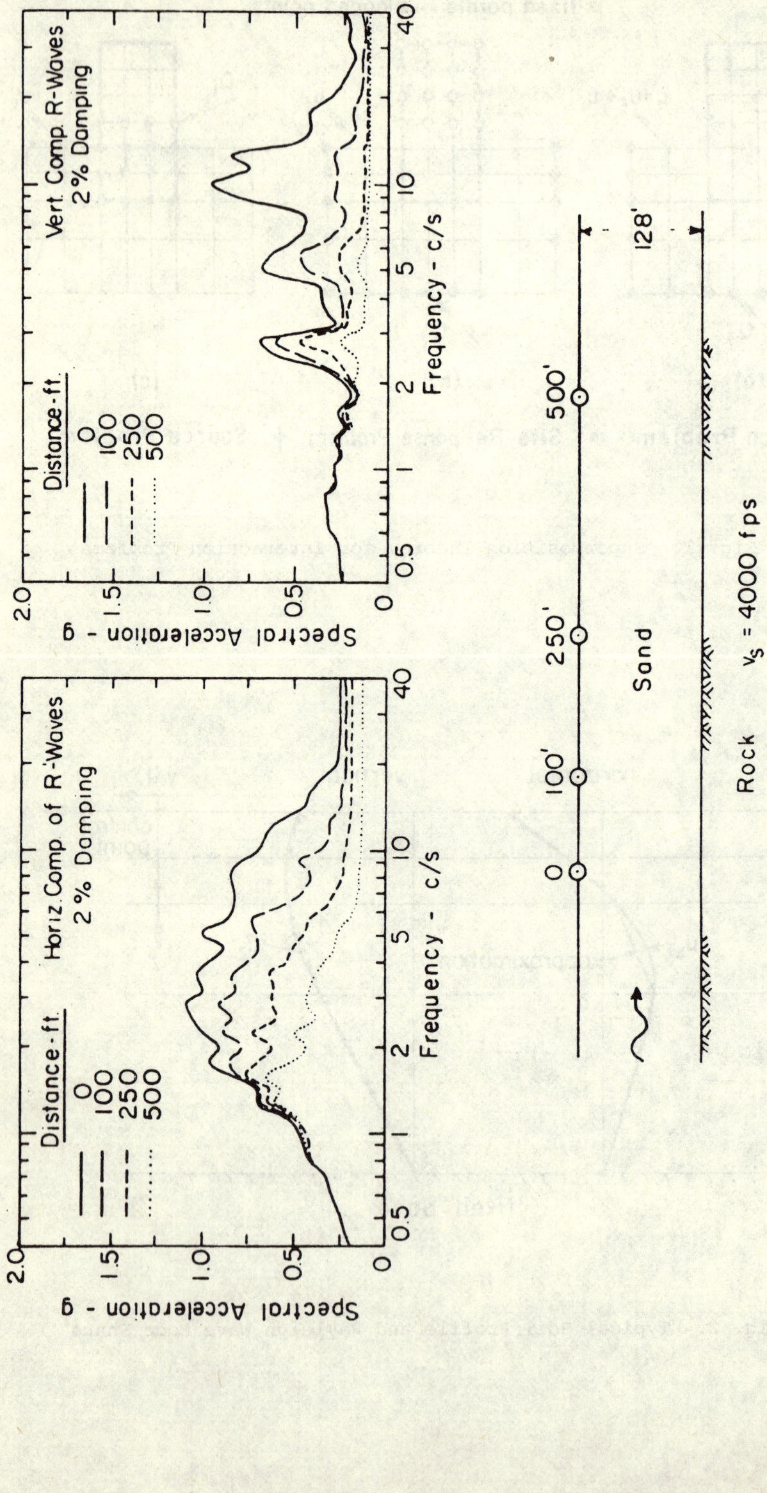

Fig. 3. Attenuation of Rayleigh Waves with Distance of Propagation

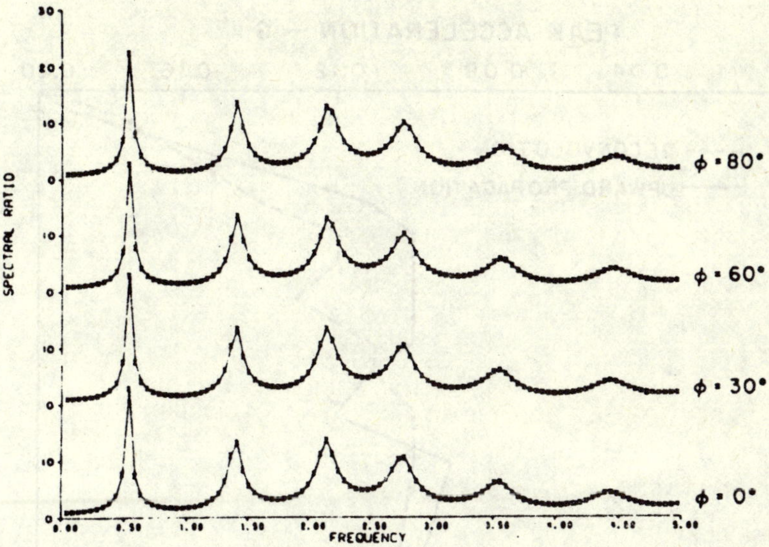

(c) Computed Horizontal Spectral Ratios Between Surface and 186 m

Fig. 4. Influence of Angle of Shear Wave Incidence on Computed Surface Response

(after Joyner et al., 1976)

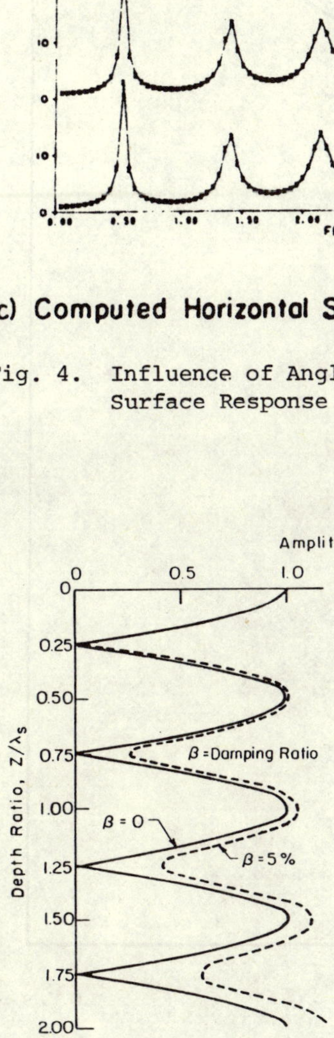

Fig. 5. Shear Waves

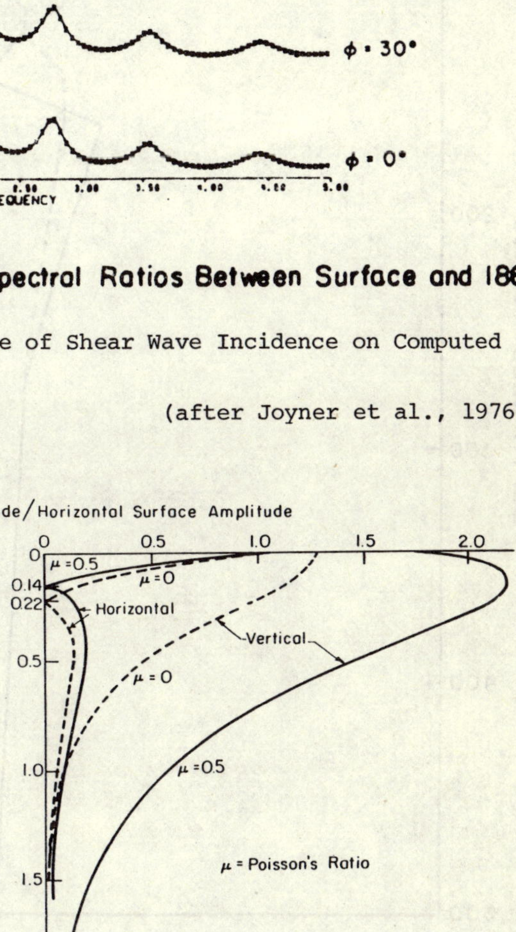

Fig. 6. Rayleigh Waves

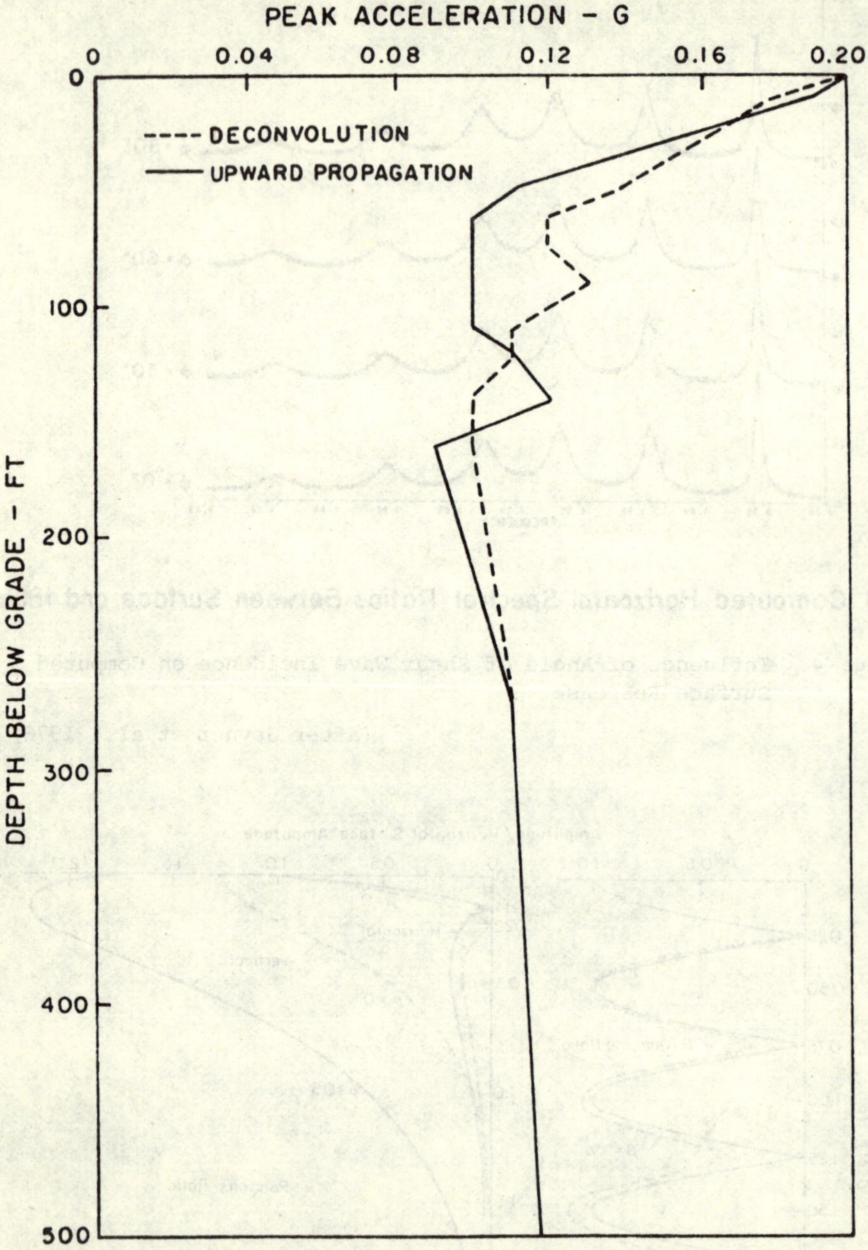

Fig. 7. Variation of Mean Peak Acceleration with Depth

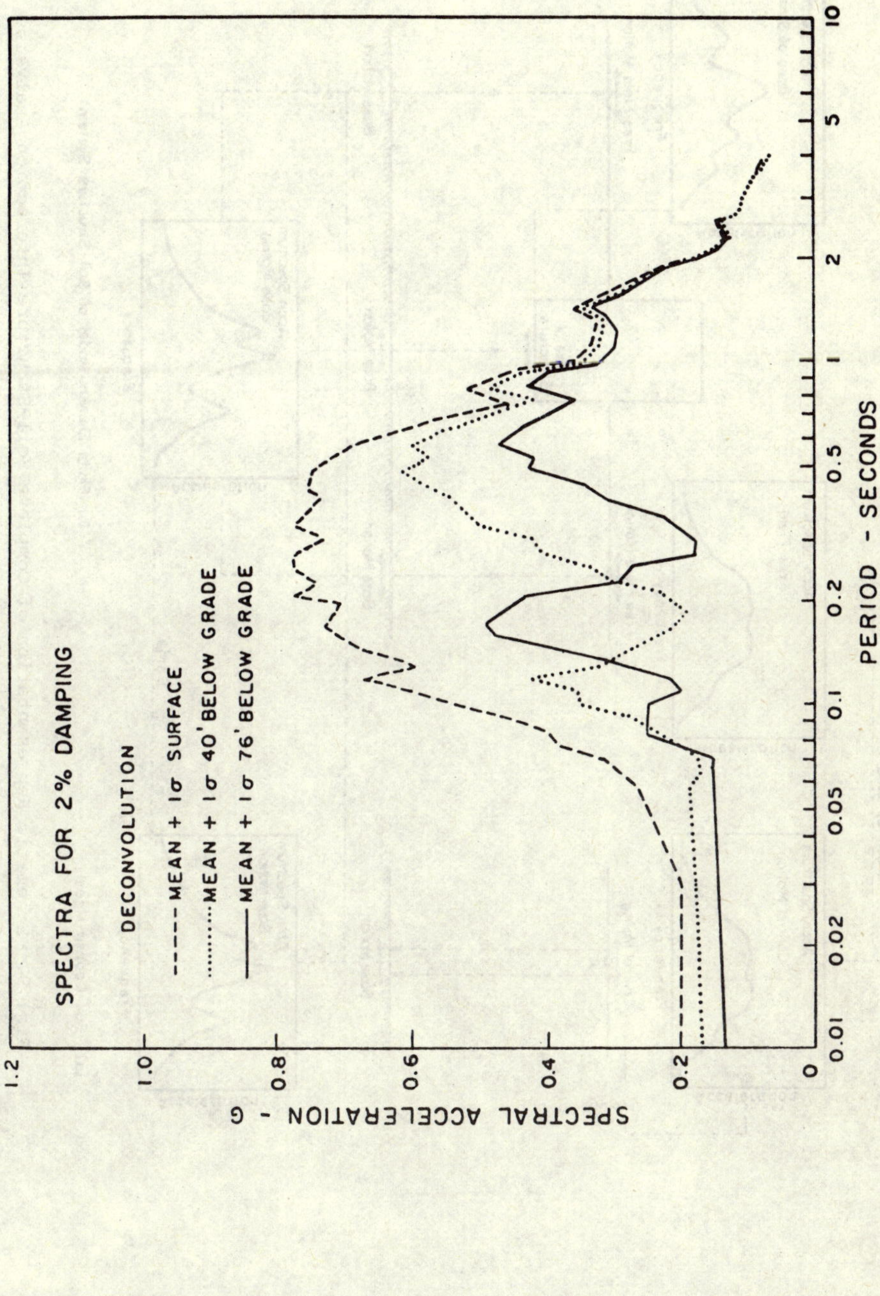

Fig. 8. Computed Spectral Shapes at Different Depths in Soil Deposit

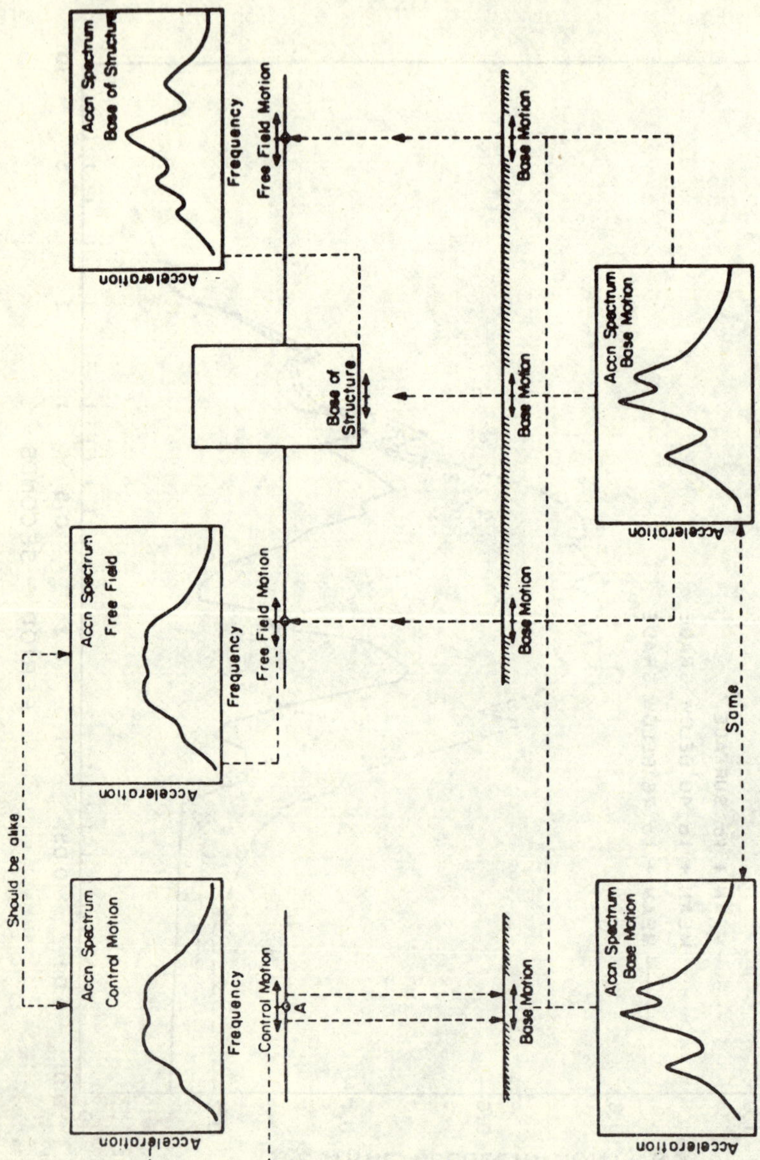

Fig. 9. Schematic Representation of Complete Soil-Structure Interaction Analysis

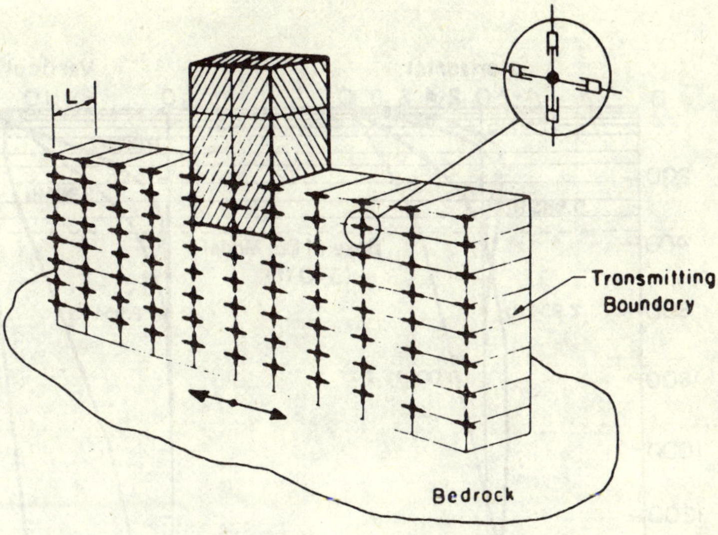

Fig. 10. Schematic View of a Simplified 3-D Model

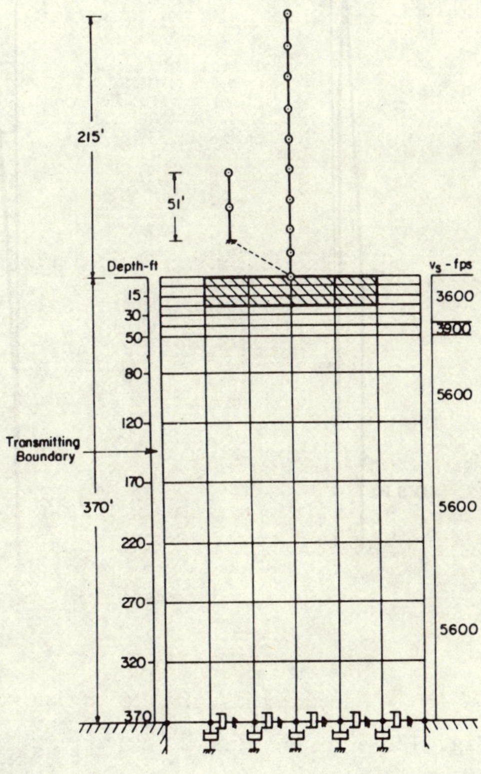

Fig. 11. Finite Element Model

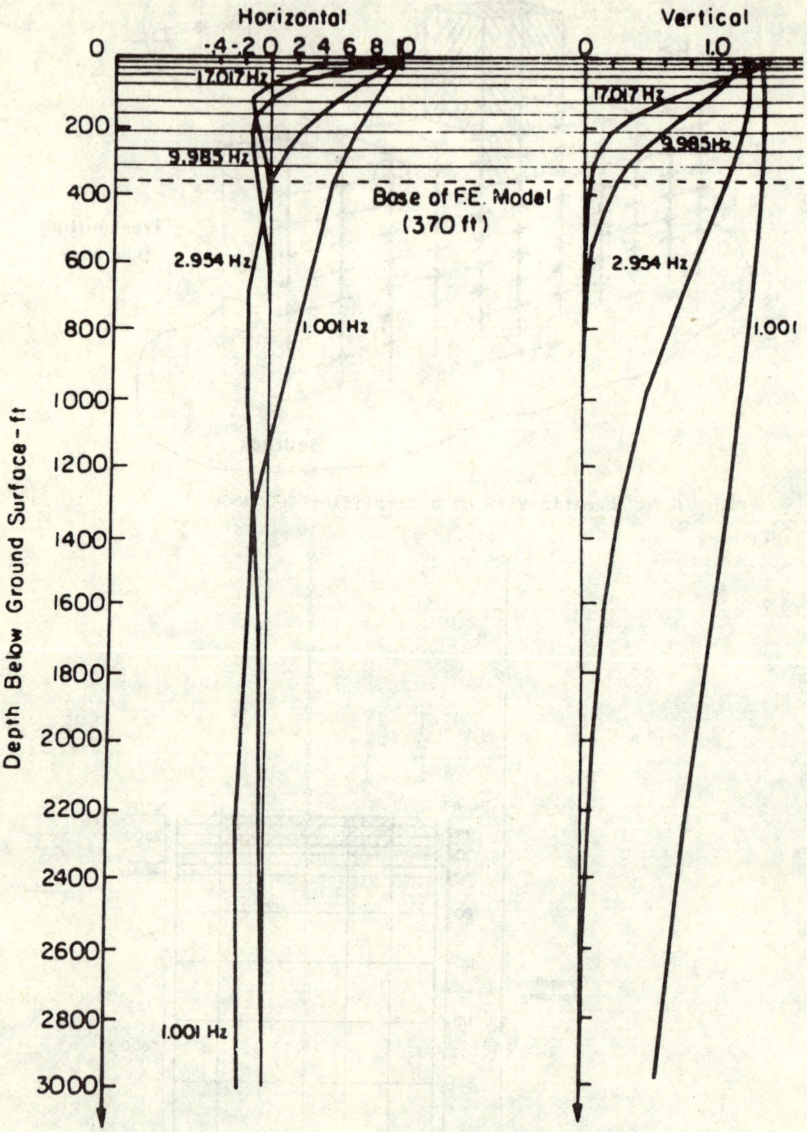
Horizontal
Vertical
-.4 -.2 0 .2 .4 .6 .8 1.0
0 1.0
17.017 Hz
9.985 Hz
2.954 Hz
1.001 Hz
1.001
Base of F.E. Model (370 ft)
Depth Below Ground Surface - ft
0
200
400
600
800
1000
1200
1400
1600
1800
2000
2200
2400
2600
2800
3000

Fig. 12.

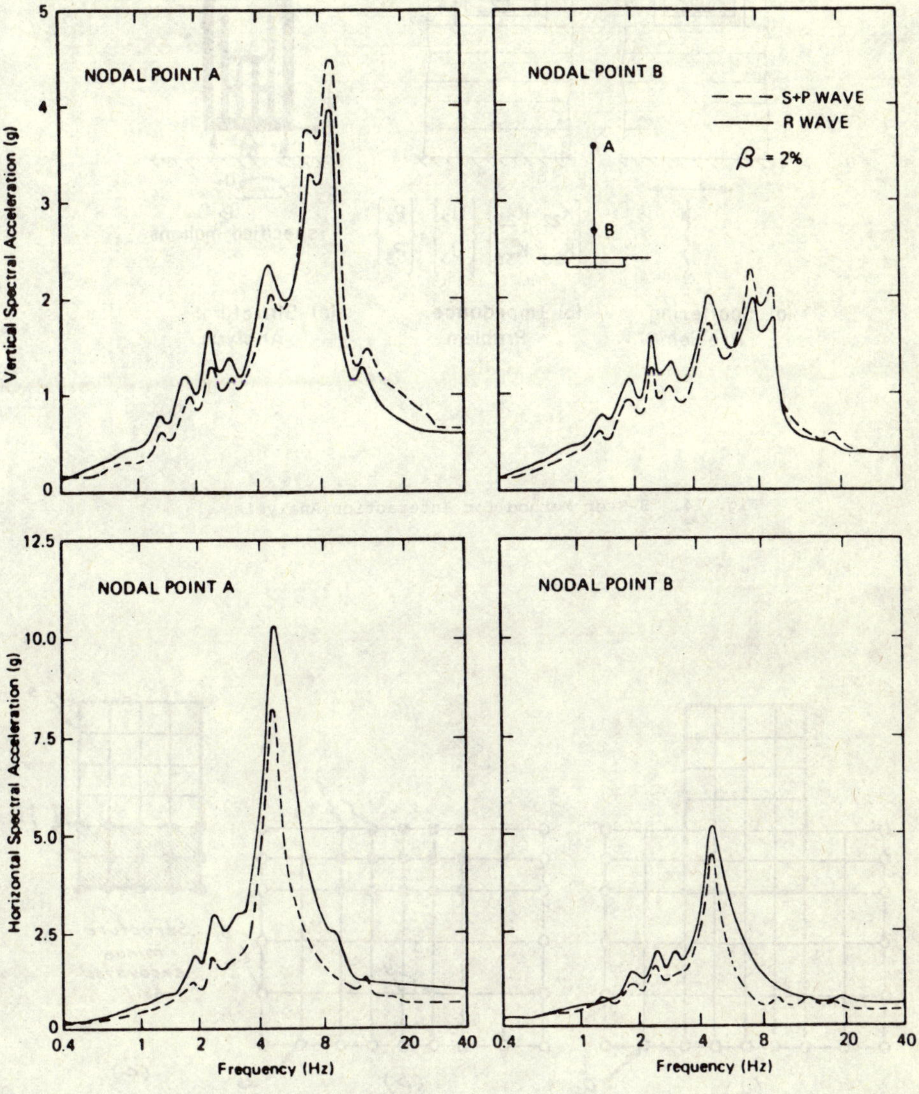

Fig. 13. Comparison of Response Spectra at Nodal Points A and B (Rock Site)

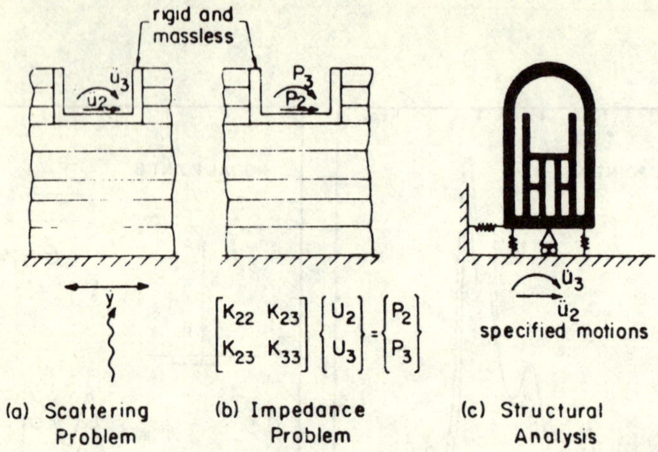

Fig. 14. 3-Step Method for Interaction Analysis

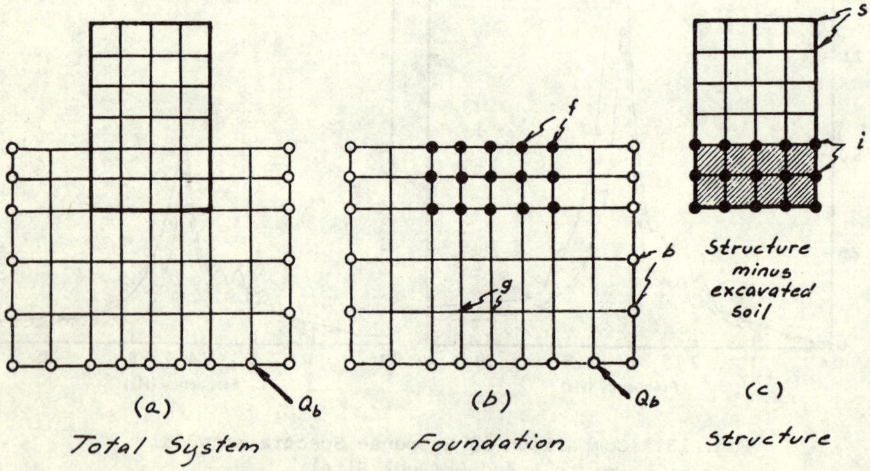

Fig. 15. Substructuring for the Flexible Volume Method

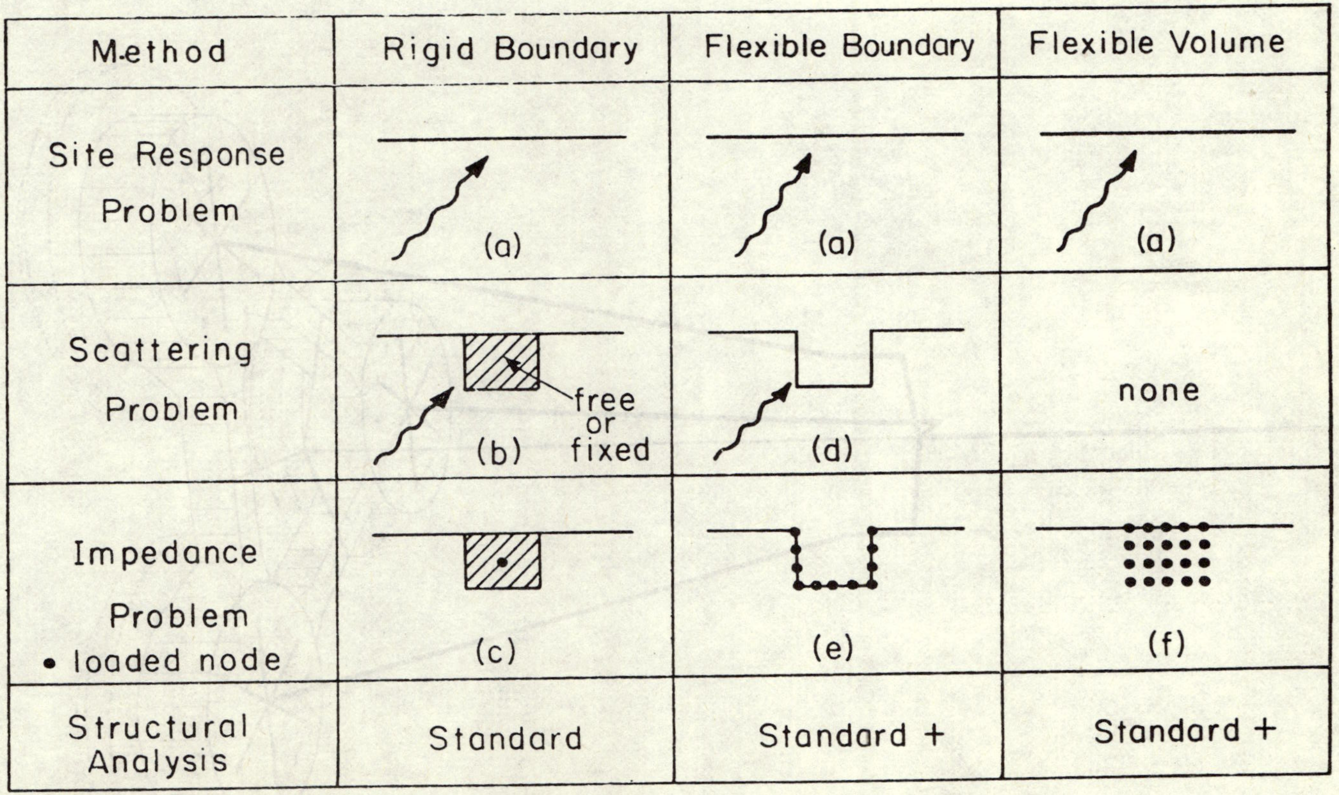

Fig. 16: Summary of Substructure Methods

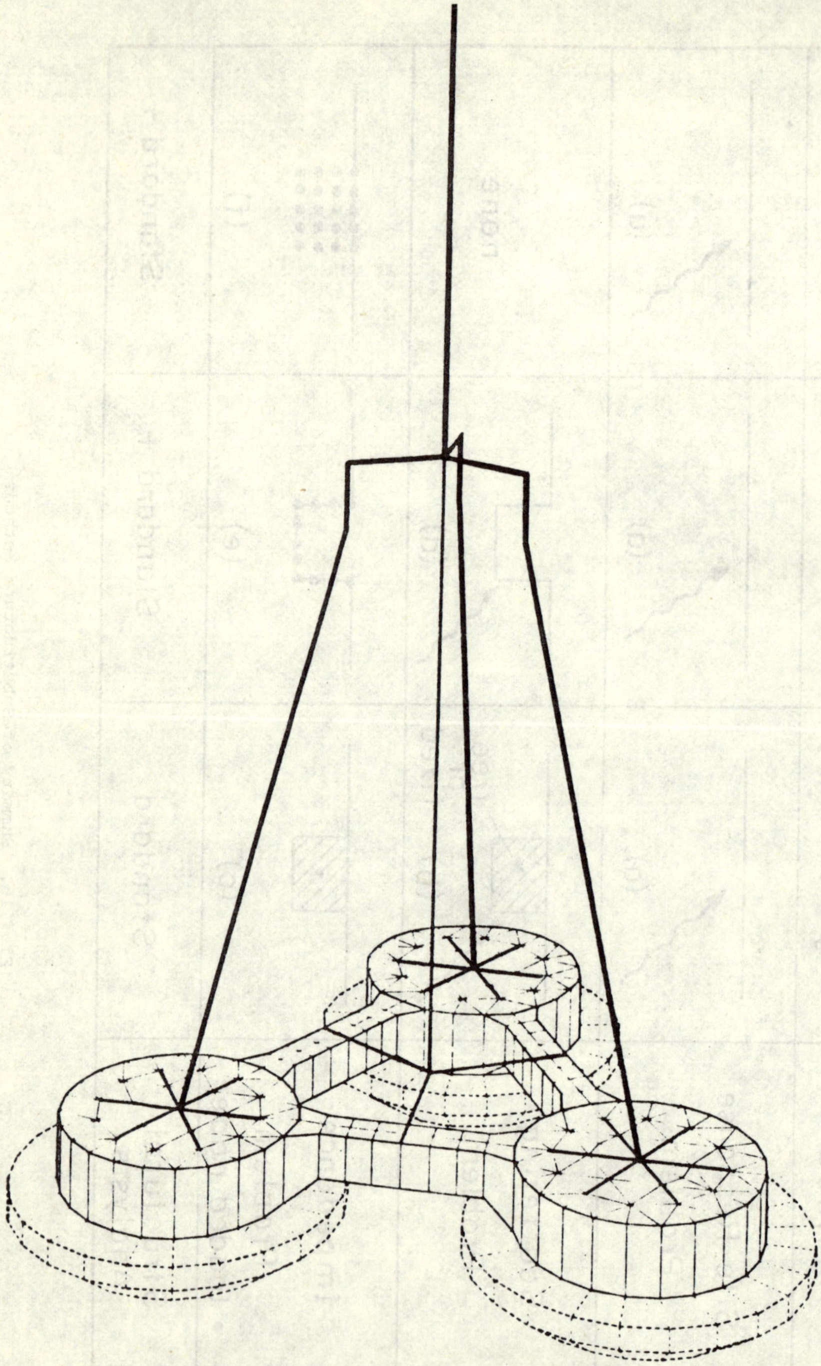

Fig. 17. View of Finite Element Model

REFERENCES

1. "Statistical Analysis of Earthquake Ground Motion Parameters," Report No. NUREG/CR-1175 prepared for the U.S. Nuclear Regulatory Commission by Shannon & Wilson, Inc. and Agbabian Associates, December, 1979.

2. "Analyses for Soil-Structure Interaction Effects for Nuclear Power Plants," Report by the Ad Hoc Group on Soil-Structure Interaction of the Committee on Nuclear Structures and Materials of the Structural Division of ASCE, I. M. Idriss, Chairman, American Society of Civil Engineers, 1979.

3. "Seismic Input and Soil Structure Interaction," Report No. NUREG/CR-0693 prepared for the U.S. Nuclear Regulatory Commission by D'Appolonia Consulting Engineers, February, 1979.

4. Abdel-Ghaffar, A. M. and Trifunac, M. D. (1976) "Antiplane Drafting Soil-Bridge Interaction for Incident Plane SH-Waves (draft)," California Institute of Technology, Pasadena, California, 1976.

5. Aydinoglu, M. N. and Cakiroglu, A. (1977) "Dynamic Interaction Between Soil and a Group of Buildings," Proc. 6th World Conference on Earthquake Engineering, Vol. 4, pp. 133-138, New Delhi, India, January, 1977.

6. Berger, E. (1976) "Seismic Response of Axisymmetric Soil-Structure Systems," Ph.D. Dissertation, University of California, Berkeley, June, 1976.

7. Berger, E., Lysmer, J. and Seed, H. B. (1975a) "ALUSH - A Computer Program for Seismic Response Analysis of Axisymmetric Soil-Structure Systems," Earthquake Engineering Research Center, University of California, Berkeley, unpublished.

8. Bogdanoff, J. L., Goldberg, J. E. and Schiff, A. J. (1965) "The Effect of Ground Transmission Time on the Response of Long Structures," Bull. Seis. Soc. Am., Vol. 55, pp. 627-640, June, 1965.

9. Boncheva, H. (1977) "Soil Amplification Factor of Surface Waves," Proc. 6th World Conf. Earthq. Engrg., Vol. 6, pp. 6-189, New Delhi, 1977.

10. Chandra, U. (1972) "Angles of Incidence of S-Waves," Bull. Seis. Soc. Am., Vol. 62, No. 4, pp. 903-915, 1972.

11. Chen, J.-C., Lysmer, J. and Seed, H. B. (1981) "Analysis of Local Variations in Free Field Seismic Ground Motion," Report No. EERC 81/03, Earthquake Engineering Research Center, University of California, Berkeley, January, 1981.

12. Chopra, A. K. and Gutierrez, J. A. (1973) "Earthquake Analysis of Multi-story Buildings including Foundation Interaction," Report No. EERC 73/13, Earthquake Engineering Research Center, University of California, Berkeley, June, 1973.

13. Day, S. M. (1977) "Finite Element Analysis of Seismic Scattering Problems," Ph.D. Dissertation, University of California, San Diego, 1977.

14. Day, S. M. (1978) "Seismic Response of Embedded Foundations," Preprint No. 3450, ASCE Convention and Exposition, Chicago, October, 1978.

15. Desai, C. S. and Christian, J. T. (1976) "Numerical Methods in Geotechnical Engineering," McGraw-Hill Book Company, 783 pp., 1976.

16. Ewing, W. M., Jardetzky, W. S. and Press, F. (1957) "Elastic Waves in Layered Media," McGraw-Hill Book Company, 1957.

17. Faccioli, E. (1978) "Response Spectra for Soft Soil Site," Proc. ASCE Specialty Conference, Earthquake Engineering and Soil Dynamics, Vol. 1, pp. 441-456, Pasadena, California, June, 1978.

18. Finn, W. D. L., Lee, K. W. and Martin, G. R. (1977) "An Effective Stress Model for Liquefaction," Journal Geotech. Engrg. Div., ASCE, Vol. 103, No. GT6, pp. 517-533, June, 1977.

19. Gazetas, G. and Bianchini, G. (1979) "Field Evaluation of Body and Surface-Wave Soil-Amplification Theories," Proc. 2nd U.S. Nat. Conf. on Earthq. Engrg., August, 1979.

20. Ghaboussi, J. and Dikmen, S. U. (1978) "Liquefaction Analysis of Horizontally Layered Sands," Journal Geotech. Engrg. Div., ASCE, Vol. 104, No. GT3, pp. 341-356, March, 1978.

21. Gomez-Masso, A. J. (1978) "Soil Structure Interaction in an Arbitrary Seismic Environment," Ph.D. Dissertation, University of Texas at Austin, May, 1978.

22. Gomez-Masso, A., Lysmer, J., Chen, J.-C. and Seed, H. B. (1981) "Soil-Structure Interaction with Rayleigh Waves." This reference is part of Ref. 2, pp. V-1 to V-36.

23. Gutierrez, J. A. (1976) "A Substructure Method for Earthquake Analysis of Structure-Soil Interaction," Report No. EERC 76/9, Earthquake Engineering Research Center, University of California, Berkeley, April, 1976.

24. Gutierrez, J. A. and Chopra, A. K. (1977) "Evaluation of Methods for Earthquake Analysis of Structure-Soil Interaction," Proc. 6th World Conference on Earthquake Engineering, Vol. 4, pp. 1-6, New Delhi, India, January, 1977.

25. Gutierrez, J. A. and Chopra, A. K. (1978) "A Substructure Method for Earthquake Analysis of Structures Including Structure-Soil Interaction," J. Earthq. Engrg. Struc. Dyn., Vol. 6, pp. 51-69, 1978.

26. Hannon, W. J. (1964) "An Application of the Haskell-Thomson Matrix Method to the Synthesis of the Surface Motion due to Dilatational Waves," Bull. Seis. Soc. Am., Vol. 54, pp. 2067-2079, December, 1964.

27. Haskell, N. A. (1953) "The Dispersion of Surface Waves in Multi-Layered Media," Bull. Seis. Soc. Am., Vol. 43, No. 1, pp. 17-34, February, 1953.

28. Haskell, N. A. (1960) "Crustal Reflection of Plane SH Waves," Journal Geophysical Research, Vol. 65, No. 13, pp. 4147-4150, December, 1960.

29. Haskell, N. A. (1962) "Crustal Reflection of Plane P and SV Waves," Journal Geophysical Research, Vol. 67, No. 12, pp. 4751-4767, November, 1962.

30. Hayashi, S., Tsuchida, H. and Kurata, E. (1971) "Average Response Spectra for Various Subsoil Conditions," Third Joint Meeting, U.S. - Japan Panel on Wind and Seismic Effects, UJNR, Tokyo, May 10-12, 1971.

31. Hays, W. W., Rogers, A. M. and King, K. W. (1979) "Empirical Data About Local Ground Response," Proc. 2nd U.S. Nat. Conf. on Earthq. Engrg., Earthq. Engrg. Res. Inst., Stanford University, August, 1979.

32. Hwang, R., Lysmer, J. and Berger, E. (1975) "A Simplified Three-Dimensional Soil-Structure Interaction Study," Second ASCE Specialty Conference on Structural Design of Nuclear Plant Facilities, New Orleans, December, 1975.

33. Idriss, I. M. (1978) "Characteristics of Earthquake Motions --A State-of-the Art Review," Proc. ASCE Geotechnical Engineering Division Specialty Conference on Earthquake Engineering and Soil Dynamics, Vol. III, Pasadena, California, June, 1978.

34. Idriss, I. M., Dobry, R., Doyle, E. H. and Singh, R. D. (1976) "Behavior of Soft Clays Under Earthquake Loading Conditions," Proc. Offshore Technology Conference, Paper No. OTC 2671, Dallas, May, 1976.

35. Johnson, J. J. (1981) "Soil-Structure Interaction: The Status of Current Analysis Methods and Research," Seismic Safety Margins Research Program, Report No. NUREG/CR-1780, Lawrence Livermore Laboratory, January 1981.

36. Johnson, N. E. and Galletley, R. D. (1972) "The Comparison of Response of a Highway Bridge to Uniform and Moving Ground Excitation," Shock and Vibration Bulletin, Vol. 42, p. 2, January, 1972.

37. Joyner, W. B. (1977) "A FORTRAN Program for Calculating Nonlinear Seismic Ground Response," Open File Report No. 77-671, U.S. Geological Survey, Menlo Park, California, 1977.

38. Joyner, W. B., Warrick, R. E. and Oliver, A. A. (1976) "Analysis of Seismograms from a Downhole Array in Sediments Near San Francisco Bay," Bull. Seis. Soc. Am., Vol. 66, No. 3, pp. 937-958, June, 1976.

39. Kanai, K. (1952) "Relation Between the Nature of Surface Layer and the Amplitudes of Earthquake Motions," Bull. Earthquake Research Institute, Vol. 30, pp. 31-37, Tokyo, 1952.

40. Kausel, E. and Roesset, J. M. (1974) "Soil-Structure-Interaction Problems for Nuclear Containment Structures," ASCE Power Division Specialty Conference, Denver, Colorado, August, 1974.

41. Kausel, E. and Roesset, J. M. (1975) "Dynamic Stiffness of Circular Foundations," J. Engrg. Mech. Div., ASCE, Vol. 101, No. EM6, pp. 771-785, December, 1975.

42. Kausel, E., Roesset, J. M. and Waas, G. (1975) "Dynamic Analysis of Footings on Layered Media," J. Engrg. Mech. Div., ASCE, Vol. 101, No. EM5, pp. 679-693, October, 1975.

43. Liang, G. C. and Duke, C. M. (1977) "Separation of Body and Surface Waves in Strong Motion Records," Proc. 6th World Conference on Earthquake Engineering, Vol. 2, pp. 215-220, New Delhi, India, January, 1977.

44. Liou, C. P., Streeter, V. L. and Richart, F. E., Jr. (1977) "Numerical Model for Liquefaction," Journal Geotech. Engrg. Div., ASCE, Vol. 103, No. GT6, pp. 589-606, June, 1977.

45. Luco, J. E., Wong, H. L. and Trifunac, M. D. (1975) "A Note on the Dynamic Response of Rigid Embedded Foundations," Earthq. Engrg. and Struct. Dyn., Vol. 4, pp. 119-128, 1975.

46. Lysmer, J. (1970) "Lumped Mass Method for Rayleigh Waves," Bull. Seismological Society of America, Vol. 60, No. 1, pp. 89-104, February, 1970.

47. Lysmer, J. and Kuhlemeyer, R. L. (1969) "Finite Dynamic Model for Infinite Media," J. Engrg. Mech. Div., ASCE, Vol. 95, No. EM4, pp. 859-877, August, 1969.

48. Lysmer, J. and Waas, G. (1972) "Shear Waves in Plane Infinite Structures," Journal Engrg. Mech. Div., ASCE, Vol. 98, No. EM1, pp. 85-105, February, 1972.

49. Lysmer, J., Udaka, T., Tsai, C.-F. and Seed, H. B. (1975) "FLUSH - A Computer Program for Approximate 3-D Analysis of Soil-Structure Interaction Problems," Report No. EERC 75-30, Earthquake Engineering Research Center, University of California, Berkeley, November, 1975.

50. Lysmer, J., Tabatabaie-Raissi, M., Tajirian, F., Vahdani, S., Ostadan, F. "SASSI - A System for Analysis of Soil-Structure Interaction," Report No. UCB/GT/81-02, Geotechnical Engineering, Department of Civil Engineering, University of California at Berkeley, April, 1981.

51. Martin, P. P. (1975) "Non-Linear Methods for Dynamic Analysis of Ground Response," Ph.D. Thesis, University of California, Berkeley, June, 1975.

52. Martin, P. P. and Seed, H. B. (1979) "Simplified Procedure for Effective Stress Analysis of Ground Response," Journal Geotech. Engrg. Div., ASCE, Vol. 105, No. GT6, pp. 739-758, June, 1979.

53. Nair, G. P. and Emery, J. J. (1975) "Spatial Variations in Seismic Motions," Second Canadian Conference on Earthquake Engineering, Preprint No. 5, Hamilton, Ontario, June, 1975.

54. Ohsaki, Y. and Hagiwara, T. (1970) "On Effects of Soils and Foundations upon Earthquake Inputs to Buildings," Research Paper No. 41, Building Research Institute, Ministry of Construction, Japan.

55. Randall, M. J. (1971) "Revised Travel Time Table for S," Geophys. J., Vol. 22, pp. 229-234, 1971.

56. Richart, F. E., Jr., Hall, J. R., Jr. and Woods, R. D. (1970) "Vibrations of Soils and Foundations," Prentice-Hall, Inc., 1970.

57. Scanlan, R. H. (1976) "Seismic Wave Effects on Soil-Structure Interaction," Earthquake Engineering and Structural Dynamics, Vol. 4, pp. 379-388, June, 1976.

58. Schnabel, P. B., Lysmer, J. and Seed, H. B. (1972) "SHAKE - A Computer Program for Earthquake Response Analysis of Horizontally Layered Sites," Report No. EERC 72-12, Earthquake Engineering Research Center, University of California, Berkeley, December, 1972.

59. Seed, H. B. and Idriss, I. M. (1969) "The Influence of Soil Conditions on Ground Motions During Earthquakes," Journal Soil Mech. Found. Div., ASCE, Vol. 94, No. SM1, pp. 99-137, January, 1969.

60. Seed, H. B. and Idriss, I. M. (1973) "Soil-Structure Interaction of Massive Embedded Structures During Earthquakes," Fifth World Conference on Earthquake Engineering, Paper No. 233, Rome, 1973.

61. Seed, H. B., Lysmer, J. and Hwang, R. (1975) "Soil-Structure Interaction Analysis for Seismic Response," J. Geot. Engrg. Div., ASCE, Vol. 101, No. GT5, pp. 439-457, May, 1975.

62. Seed, H. B., Murarka, R., Lysmer, J. and Idriss, I. M. (1976b) "Relationships Between Maximum Acceleration, Maximum Velocity, Distance from Source and Local Site Conditions for Moderately Strong Earthquakes," Bull. Seis. Soc. Am., Vol. 66, No. 4, pp. 1323-1342, August, 1976.

63. Seed, H. B., Ugas, C. and Lysmer, J. (1976a) "Site Dependent Spectra for Earthquake Resistant Design," Bull. Seis. Soc. Am., Vol. 66, No. 1, pp. 221-243, February, 1976.

64. Shimizu, N. Yanamoto, S., Koori, Y. and Minowa, N. (1977) "Three-Dimensional Dynamic Analysis of Soil-Structure System by Thin Layer Element Method," Trans. A.I.J., Tokyo, Part 1: No. 253, March 1977, Part 2: No. 254, April 1977, Part 3: to be published.

65. Silva, W. (1976) "Body Waves in a Layered Anelastic Solid," Bull. Seis. Soc. Am., Vol. 66, No. 5, pp. 1539-1554, October, 1976.

66. Streeter, V. L., Wylie, E. B. and Richart, F. E., Jr. (1974) "Soil Motion Computations by Characteristics Method," Journal Geotech. Engrg. Div., ASCE, Vol. 100, No. GT3, pp. 247-263, March, 1974.

67. Tajimi, H. and Shimomura, Y. (1976) "Dynamic Analysis of Soil-Structure Interaction by Thin Layer Element Method," Trans. A.I.J., Tokyo, No. 243, May, 1976.

68. Taylor, P. W. and Larkin, T. J. (1978) "Seismic Site Response of Nonlinear Soil Media," Journal Geotech. Engrg. Div., ASCE, Vol. 104, No. GT3, pp. 369-383, March, 1978.

69. Teng, T.-L. (1967) "Reflection and Transmission from a Plane Layered Core-Mantle Boundary," Bull. Seis. Soc. Am., Vol. 57, No. 3, pp. 477-499, June, 1967.

70. Thomson, W. T. (1950) "Transmission of Elastic Waves Through a Stratified Soil Medium," Journal Appl. Physics, Vol. 21, February, 1950.

71. Toki, K. (1977) "Disintegration of Accelerograms into Surface and Body Waves," Proc. 6th World Conference on Earthquake Engineering, Vol. 2, pp. 209-214, New Delhi, India, January, 1977.

72. Trifunac, M. D. and Brune, J. N. (1970) "Complexity of Energy Release During the Imperial Valley, California Earthquake of 1940," Bull. Seis. Soc. Am., Vol. 60, No. 1, pp. 137-160, 1970.

73. Waas, G. (1972) "Analysis Method for Footing Vibrations through Layered Media," Ph.D. Dissertation, University of California, Berkeley, December, 1972; also published under the title "Earth Vibration Effects and Abatement for Military Facilities," Report 3, U.S. Army Engineer WES, Vicksburg, Miss., September, 1972.

74. Wight, L. H. (1977) "Soil-Structure Interaction in Nuclear Power Plants: A Comparison of Methods," Proc. 6th World Conference on Earthquake Engineering, Vol. 4, pp. 91-96, New Delhi, India, January, 1977.

75. Wong, H. L. (1975) "Dynamic Soil/Structure Interaction," Report No. EERC 75-01, Ph.D. Dissertation, California Institute of Technology, Pasadena, California, May, 1975.

76. Wong, H. L. and Luco, J. E. (1976) "Dynamic Response of Rigid Foundations of Arbitrary Shape," Earthquake Engineering and Structural Dynamics, Vol. 4, pp. 579-587, 1976.

77. Zienkiewicz, O. C., Chang, C. T. and Hinton, E. (1978) "Non-linear Seismic Response and Liquefaction," Submitted for publication in the Int. J. Numerical and Analytical Methods in Geomechanics, 1978.

LIQUEFACTION OF SANDY SOILS

José Folque

Laboratório Nacional de Engenharia Civil - Lisboa, Portugal

SYNOPSIS - A review of the most recent states-of-the-art allows a comprehensive revision of the questions involved in the major problems of liquefaction. Exact and simplified procedures of determining the liquefaction potential are revised. The question of induction and dissipation of pore pressure is dealt with.

1. INTRODUCTION

The experimental and quantitative studies of liquefaction phenomena of sandy soils, related with cyclic forces, started some twenty years ago. In this comparatively short period the proliferation of published studies was intense, hundreds of references being present in the files. A systematic review of the bibliography on the matter becomes in this way extremely difficult. Fortunately some Authors have produced excellent states-of-the--art and this circumstance makes easy a critical review of the subject.

To benefit from the most recent states-of-the-art and, having them as a basis, to review the modern trends and points of view, is a task that seems of interest. Simultaneously some comments are presented regarding controverted aspects of the question.

2. THE PHENOMENON "LIQUEFACTION"

The liquefaction of loose sands under the action of static forces is a well known phenomenon, described in conventional Soil Mechanics. Its study, made by Casagrande (1936) in systematic and well based terms, still does not seem to need revision. The same cannot be said concerning the liquefaction caused by vibratory forces: some aspects of its causes and

J. B. Martins (ed.), Numerical Methods in Geomechanics, 455–469.

mechanisms require further clarification.

Reflecting the existing ambiguities, even in terminologic aspects the question is not settled (Cf. Marcuson, 1978). Seed, at the initial section of a state-of-the-art (1976), felt the necessity of raising terminologics questions. This Author proposed the following definitions.

"a. Liquefaction - denotes a condition where a soil will undergo continued deformation at a constant low residual stress or with no residual resistance, due to the build-up and maintenance of high pore water pressures which reduce the effective confining pressure to a very low value; pore-pressure build-up leading to true liquefaction of this type may be due to static or cyclic stress applications".

"b. Initial liquefaction - denotes a condition where, during the course of cyclic stress applications, the residual pore water pressure on completion of any full stress cycle becomes equal to the applied confining pressure; the development of initial liquefaction has no implications concerning the magnitude of the deformations which the soil might subsequently undergo; however, it defines a condition which is a useful basis for assessing various possible forms of subsequent soil behavior".

"c. Initial liquefaction with limited strain potential, cyclic mobility or cyclic liquefaction - denotes a condition in which cyclic stress applications develop a condition of initial liquefaction and subsequent cyclic stress applications cause limited strains to develop either because of the remaining resistance of the soil to deformation or because the soil dilates, the pore pressure drops, and the soil stabilizes under the applied load".

It must be pointed out that for c. Casagrande and Castro prefer the term "cyclic mobility". Seed, Arango and Chan (1975) used for c. the term "Initial liquefaction with limited strain potential".

In order to explain the occurrence of liquefaction Seed (1976) assumes the mechanism that hereafter is summed up and commented.

The state of stress and stress changes at play in the liquefaction phenomena associated with seismic actions are those represented in fig. 1; in the element A, initially subjected to an anisotropic state of stress, tangential reversible cyclic stresses are superimposed, predominating the magnitude of the horizontal stress changes.

In fig. 2 is plotted the variation of the voids ratio against the effective mean stress (or maximum principal stress in a laterally confined element). At point A stress cycles tend to produce decrease in soil volume; if drainage occurs the volume change becomes effective with the corresponding variation in the voids ratio Δe ; if there is no drainage pore pressure builds up avoiding the volume change: what happens is equivalent to an expansion of the soil following BC, an expansion that corresponds to the decrease in effective stress associated with the pore pressure increase When the soil is loose sand, true liquefaction takes place and subsequent applications of cyclic stresses cause incessant deformations with a resistance corresponding to a very low effective stress, σ'_f . When the soil is

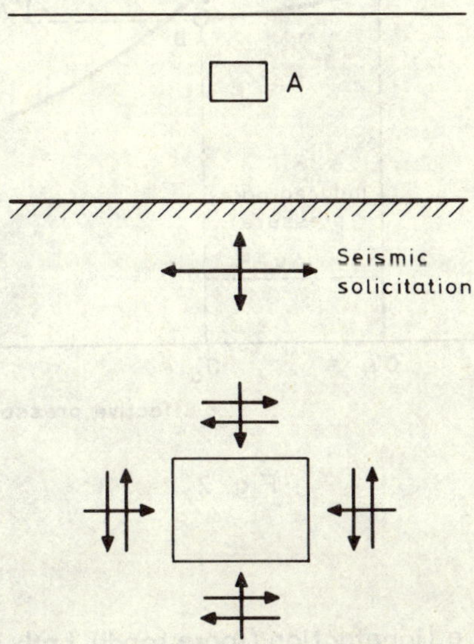

Fig. 1

dense sand subsequent cyclic stresses cause dilatancy with substantial reduc-tion in Δu and, consequently, the deformation ceases owing to the fact that the soil can support these additional stresses; these facts justify the designation of "initial liquefaction with limited strain potential". Seed does not point to the fact that the validity of the proposed mechanism is restricted to a given circumstance: the predominance of compression stresses in the initial stages of cyclic applications. In fact, if initially tangential stresses predominate, dilatancy can occur "ab initio" and thus initial li-quefaction does not arise. Further clarification seems necessary on this matter.

Castro and Poulos comment on liquefaction with out presenting physical reasons for the observed phenomena, which are exposed in a simple factual way. Considering an element of volume of loose sand - point C, fig. 3 if a monotonic or cyclic stress is applied pore pressures increase and the soil exhibits true liquefaction. If the soil is a dense sand - point D, Fig. 3 - - a monotonic application of stress causes a decrease in pore pressure which becomes negative; a cyclic application of stress induces irreversible defor-mations, positive pore pressures being developed: cyclic mobility arises.

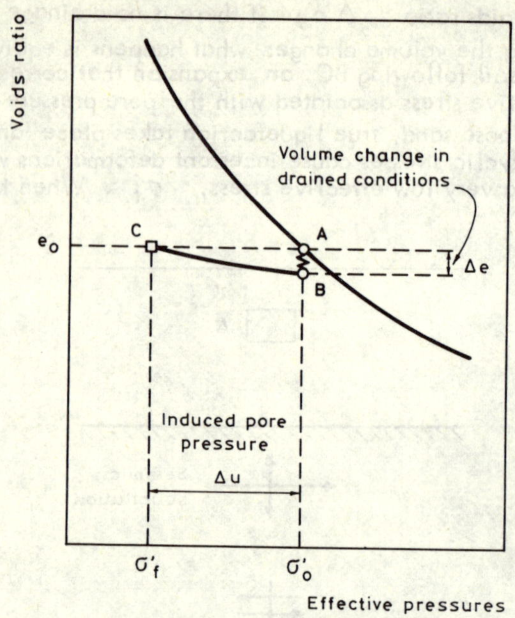

Fig. 2

In the case of true liquefaction (loose sands) both isotropic and deviatoric stresses cause increase in pore pressure. Finn, Lee and Martin (1976) presented a proposal of analytical prediction of pore pressure magnitude. The experimental requirements for the application of the method consist in the determination of the constitutive equations of the soil under study.

3. THE SUSCEPTIBILITY TO LIQUEFACTION OF SANDY SOILS

To evaluate the susceptibility to liquefaction of sandy soils Seed (1979) considers two fundamental types of methods:

a) - methods in which the stresses induced by earthquakes are calculated and compared with the stresses required to cause cyclic mobility or liquefaction of representative samples in the laboratory;

b) - empiric methods based on the known field performance of sand deposits correlated with some measure of the characteristics exhibited "in situ".

For the application of type a) methods laboratory cyclic tests are required. Their results are usually presented under the form of the relation "cyclic stress/effective overburden stress" (cyclic stress ratio) versus the number of cycles that causes liquefaction or excessive deformation.

For the computation of the stresses induced by earthquakes there are two basic alternatives: the adoption of simplified methods or the computation, by using dynamic methods of analysis, of the response of the soil layer to a given sequence of pulses (design earthquake) acting in the bed-rock. In these methods it is assumed that a series of regular pulses is equivalent to the sequence of irregular pulses caused by the earthquake. The soil mass is discretized in finite elements.

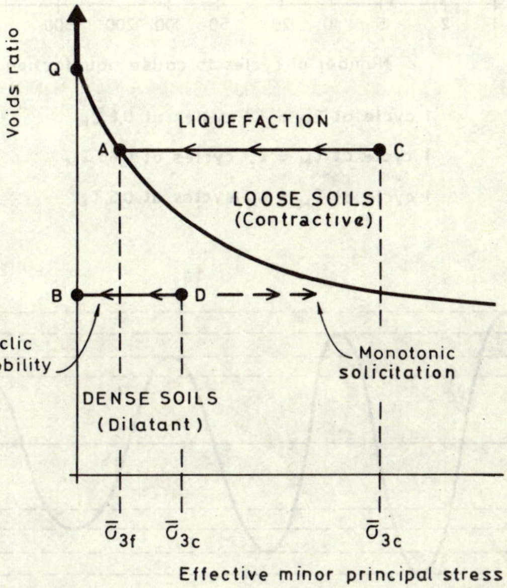

Fig. 3

As regards the number of uniform pulses equivalent to the sequence of actual pulses contained in the earthquake spectrum, in fig. 4 is presented a method for its computation. In the example shown in fig. 4 a) it can be seen that the effect of a pulse of magnitude τ_M is equivalent to the effect of 1.4 pulses at 0.65 τ_M and so on. Taking an irregular sequence, for instance the one shown in fig. 4 b), it can be seen that the irregular sequence of 4 pulses with maximum value τ_M is equivalent to a sequence of 5.1 cycles with the uniform value of 0.65 τ_M. (Seed-1976)

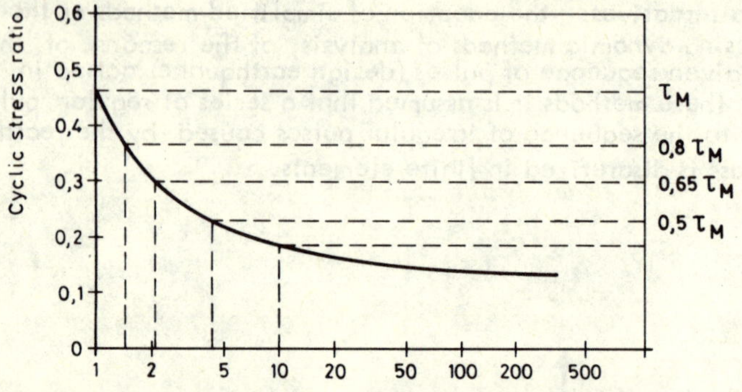

1 cycle at τ_M = 1,4 cycles at 0,8 τ_M

1 cycle at τ_M = 2,1 cycles at 0,65 τ_M

1 cycle at τ_M = 4,2 cycles at 0,5 τ_M

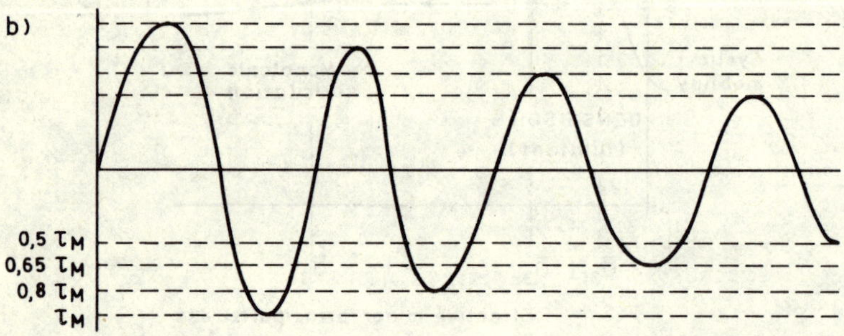

EQUIVALENCE: IRREGULAR SEQUENCE →

1 at τ_M ⟶ 2,1 at 0,65 τ_M → 2,1 + 1,5 + 1 + 0,5 = 5,1 at 0,65 τ_M

1 at 0,8 τ_M ⟶ $\frac{2,1}{1,4}$ = 1,5 at 0,65 τ_M

1 at 0,65 τ_M ⟶ 1 at 0,65 τ_M

1 at 0,5 τ_M ⟶ $\frac{2,1}{4,2}$ = 0,5 at 0,65 τ_M

Fig. 4

Among the simplified methods the most widely used is the one proposed by Seed and Idriss (1971). The expression that gives the cyclic stress induced at a given point of the mass is

$$\tau = 0.65 \frac{\gamma h}{g} a_{max} \cdot r_d$$

where

γ - unit weight of the soil mass

h - depth of the point under consideration

a_{max} - maximum acceleration induced by the earthquake

r_d - factor, representing the effect of depth, with values smalle than unity, taking the mean value of 0.9 at the depth of 12 m and decreasing to a mean value of 0.5 at the depth of 30 m.

The following step in the study is the analysis of cyclic laboratory test results in order to predict the behaviour of the soil for the stresses and number of pulses computed as summed up above.

Concerning the application of empiric methods based on the field performance of sand deposits it must be pointed out that it was de Nigata seism (1964) that supplied the most important information on this matter. Extensive surveys were performed correlating in situ characteristics, mainly

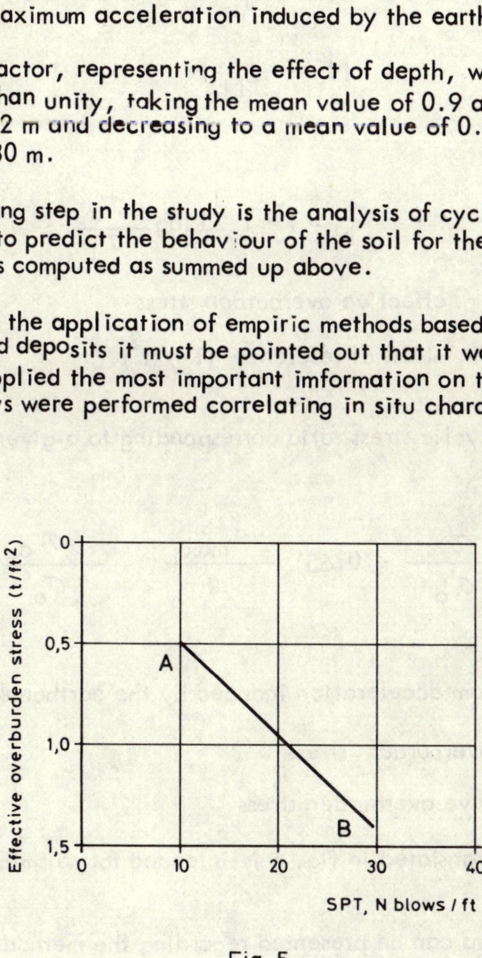

Fig. 5

assessed by means of SPT results, with the soil behaviour as regards liquefaction. Based on the results of these surveys Ohsaki (1966) presented a very simple proposal, which is reproduced in fig. 5: for situations at the right of line AB liquefaction dones not occur; on the contrary, for situations at the left of line AB the occurrence of liquefaction is very probable. Commenting this proposal it seems that an immediate objection can be raised concerning the fact that it is doubtful whether the results can be applied to a different place and for seisms of other magnitudes. Later on additional results, from different places, were gathered and, based on them, Seed and Peacock presented the proposal of fig. 6: a correlation is established between the cyclic stress ratio at the liquefaction threshold and the SPT results. It must be stressed that value N_1 of the SPT results is affeccted by the classical correction factor proposed by Gibbs and Holz, that is to say:

$$N_1 = C_N . N$$

where

$$C_N = 1 - 1.25 \log \frac{\sigma'_o}{\sigma'_1}$$

σ_o' - effective overburden stress

σ_1' - stress of reference (1 t/ft^2)

The cyclic stress ratio corresponding to a given seism is given by

$$\frac{\tau}{\sigma'_o} = 0.65 \frac{a_{max}}{g} \frac{\sigma_o}{\sigma_o'} r_d$$

were

a_{max} - maximum acceleration induced by the earthquake

σ_o - total overburden stress

σ'_o - effective overburden stress

The proposal translated in fig. 6 is intended for seismic of magnitude around 7.5.

Some criticisms can be presented regarding the method and Seed (1976) points out the following ones:

- the upper part of curve AB is extrapolated

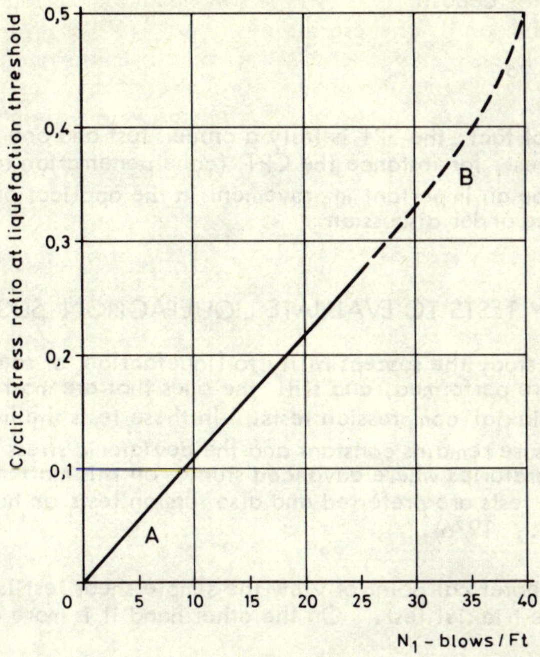

Fig. 6

- in the method some important circumstances are ignored, namely the duration of the earthquake

- the SPT is a crude test that can be misleading as regards the sand properties that govern the liquefaction behaviour.

The firs point underlines the need of numerous complementary observations in order to substantiate the basis of the correlation curve.

The second point refers to simplifications that are common in empiric methods and do not invalidate them if the basic correlations are sound.

As regards the adaptability of SPT results to study the liquefaction susceptibility of sands Seed (1976) shows that a lot of properties affect in the same way the SPT results and the resistance to liquefaction, namely:

- relative density

- fabric

- aging of the deposit

- value of k_o

But, as a matter of fact, the SPT is truly a crude test and one wonders why a more accurate test, for instance the CPT (cone penetration test) is not used. It would be an important improvement in the application of empiric methods as the one under discussion.

4. LABORATORY TESTS TO EVALUATE LIQUEFACTION SUSCEPTIBILITY

In order to study the susceptibility to liquefaction of sandy soils the first tests that were performed, and still the ones that are most frequently used, were the triaxial compression tests. In these tests the hydrostatic confinement pressure remains constant and the deviatoric stress is pulsed. Recently, in laboratories where advanced studies on this matter are carried out, simple shear tests are preferred and also torsion tests on hollow cylinders (Cho and al., 1976).

From the theoretical point of view the simple shear test is really more adequate than the triaxial test. On the other hand it is more difficult to perform.

Concerning the infractions to similarity between the conditions in the field and the conditions reproduced in the triaxial test, one can point out:

- non-uniformity in the state of stress owing to irregularities caused by the plates at the ends of the sample;

- rotation of 90° in the direction of the maximum principal stress during each load cycle;

- constraints in the state of strain, "necking" of the samples being possible, which invalidates the results.

Simple shear tests reproduce better the in situ conditions. Especially well adapted to a good simulation of the actual conditions is the shaking table test in which samples with a big length in comparison with their thickness are used; the singularities occuring at the extremities are comparatively small, due to the high length/thickness ratio.

The results of triaxial tests are usually presented under the form of the ratio $\sigma_{ad}/2\ \sigma_3$ between the maximum cyclic stress and the confinement stress. Concerning simple shear tests their results are usually presented under the form of the ratio τ_h/σ'_o between the maximum tangential stress applied in cycles on the horizontal plan and the effective normal stress acting on the same plan. It is obvious that the two ratios do not presente the same value. If during the triaxial test substantial singularities in the stresses near the extremities of the sample did not occur, the two ratios can be correlated:

$$\tau_h / \sigma'_o = C_r \quad \sigma_{ad}/2 \; \sigma_3$$

A substantial amount of experimental results led to the following values of C_r

k_o = 0.4 .... C_r = 0.65 to 0.7

k_o = 1.0 .... C_r = 1.0

As an additional contribution to the problem the criticism of Cho and al. (1976) is noteworthy. For these Authors the only reliable laboratory test is the torsion test on hollow cylinders. They do not say how to obtain samples with this shape which fulfil the condition of being structurally representative of actual sand masses.

Another question with interest is raised by Castro (1976) who shows doubts regarding the possibility of in situ occurrence of initial liquefaction; initial liquefaction would be a phenomenon occurring only in laboratory tests owing to marked redistributions of the voids ratio inside the samples. Casagrande (1979) presents comments and criticism focussing similar aspects.

5. FACTORS AFFECTING THE IN SITU LIQUEFACTION OF SANDS

It is usual to try to predict the behaviour concerning liquefaction in terms of the relative density. Frequently one finds results, considered valid for all sands, of the ratio τ / σ'_o corresponding to the liquefaction threshold versus relative density. This means that no sand can exhibit liquefaction when a given relative density is attained. It must be pointed out how fallacious is such an assessment. The relative density quantifies the state of compacity, within the range of compacity variability, of a given sand. But in can not be assumed that a given relative density corresponds to the same behaviour for all the sands: for instance, a given uniform sand with relative density 0.8 can exhibit poorer behaviour than a well graded sand with relative density 0.5

It has been difficult to give up the use of the relative density to base judgements of this type. Nowadays some researchers still use the relative density to justify ill-grounded conclusions. Nevertheless, the trend is to reduce the significance of the relative density to its true capability.

It is today well recognized that the compacity is not the unique factor that affects the liquefaction susceptibility. The following additional factors can be pointed out:

- mineralogic composition
- fabric
- age of the deposit
- previous history of stresses, with special emphasis on the action of seismis
- value of the horizontal pressure, $\sigma_v k_o$
- overconsolidation

The effect of overconsolidation has been carefully studied by Ishihara and al. (1977).

Reflecting on the above circumstances one wonders if the tests carried out on samples reconstructed at the laboratory by compaction lead to valid results. It seems obvious that, in order to contemplate the effects of the listed factors, truly representative, undisturbed samples are needed. Ishihara and al. (1979) comment on this subject. The studies carried out by Townsend (1978) and Pollard (1979) to identify the phenomena that influence, the results of laboratory tests which may not exist in situ also present interest.

The difficulties of obtaining undisturbed samples of cohesionless soils raise the question of the incongruence between the refinement of the most sophisticated tests available today and the poor quality of the samples. It seems, in practical terms and having in mind the disturbances of the samples, that the refined test techniques are only of academic concern.

6. INDUCTION AND DISSIPATION OF PORE PRESSURES DUE TO CYCLIC LOADS

A recent development in the field of the behaviour of sand deposits under earthquake action is the study of induction and dissipation of pore pressures during and after the seism.

Knowing the deformability moduli of the soil for loading and unloading processes and its dilatancy (volume change induced by deviatoric stresses) it is possible to compute the pore pressures induced by cyclic forces, which implies the establishment of the constitutive laws of the material. Assuming additionally that the water flow is unidirectional (upwards) or radial, and knowing the coefficient of permeability, it is possible to compute the rate of dissipation of the induced pore pressure at a given elevation and its process of transfer to adjacent layers.

If the layer over the layer where liquefaction occurs is very compressible and its permeability is low, the upward flow can induce propagation of liquefaction to that layer, which will arise some hours after the seism.

If the permeability of the soil is high the rate of dissipation of pore pressures can follow the rate of induction and liquefaction ceases at the end of the earthquake.

An important practical consequence of the studies carried out on this matter is the use of gravel piles in order to contribute to a rapid dissipation of the induced pore pressures.

7. RHEOLOGICAL CONSIDERATIONS ON THE MATTER

In the structural constitution of the soil there exist elastic (hookean), viscous (newtonians) and plastic elements (elements of St. Venant). First of all let us consider loose soils: it seems that viscous elements are strong, but the elastic and plastic elemens are rather week. Cyclic applications of load cause irreversible yields in the plastic elements with associated transferences of stresses to the viscous elements. It must be pointed out that these viscous elements correspond mainly to the free pore water. In this way, in loose soils, pore pressures suffer progressive build up.

In dense sands the plastic and elastic elements are stronger. The application of low cyclic loads cause reversible deformations: in this way pore pressure build up does not arise. High loads cause the fracture of the elastic and plastic elements. Due to the fact that these elements owe their existence to a given small distance between particles, the fractures cause dilatant deformations of the soils. The pressure in the pore water exhibits in this way a process of decrease.

8. FINAL REMARKS

The studies concerning the liquefaction of sands have attained an elaborated and refined state of development, both in analytical and experimental aspects.

A somewhat intrincated situation has been reached: the acquired knowledge shows that the mechanisms involved are affected by factors that are very variable and difficult to determine in actual soil masses. The comments of Peck (1979 on this matter are very interesting. These circumstances lead to serious doubts concerning the efficacity, for practical purposes, of the studies based on laboratory tests (Cf. Mulilis and al. -

-1977 and Mori and al. - 1978). It seems very difficult to obtain representative samples in order to carry out laboratory tests that duplicate the field situation. On the other hand, even in the adoption of an earthquake spectrum for design purposes, there are very arbitrary decisions that must be taken. It may be concluded that a marked discrepancy exists between the refinement of the analytical methods and test techniques and, on the other hand, the crude approaches resulting from the poor samples and the uncertainty related with design earthquakes.

REFERENCES

- Casagrande, 1937, Characteristics of cohesionless soils, Journal of Boston Society of Civil Engineers, January

- Casagrande, 1979, Liquefaction and cyclic mobility of soils - a critical review, Harvard Soil Mech. Series, nº 88, January

- Castro and Poulos, 1976, Factors affecting liquefaction and cyclic mobility, in "Liquefaction problems in Geot. Eng.", ASCE annual Convention

- Cho, Rizzo and Humphreys, 1976, Saturated sands and cyclic dynamic tests, in "Liquefaction problems in Geot. Eng.", ASCE annual Convention

- Finn, Lee and Martin, 1976, An effective stress model for liquefaction, problems in Geot. Eng.", ASCE annual Convention

- Ishihara, Sodekawa and Tanaka, 1978, Effects to overconsolidation on liquefaction characteristics of sands containing fines, ASTM Spec. Pub. STP 654

- Ishihara, Silver and Kitagawa, 1979, Cyclic strength of undisturbed samples obtained by a piston sampler, Soils and Foundations, 19 nº 3

- Marcuson, 1978, Definiton of terms related to liquefaction, J. Geot. Div., ASCE 104, GT9

- Mori, Seed and Chan, 1978, influence of sample disturbance on sand response to cyclic loading, J. Geot. Div. ASCE 104 GT3

- Mulilis, Seed, Chan, Mitchel and Arulanandan, 1977, Effects of sample preparation on sand liquefaction, J. Geot. Div. ASCE 103 GT2

- Ohsaki, 1966, Niigata earthquake (1964) - Building damages and soil conditions, Soils and Foundations, 6 nº 2

- Peck, 1979, Liquefaction potential, science versus practice, J. Geot. Div. ASCE 105 GT3

- Pollard, Rate and time effects in cyclic triaxial testing, Cornell University, D. Thesis

- Seed and Idriss, 1971 Simplified procedure for evaluating soil liquefaction potential, J. Soil Mec. Div. ASCE 97, SM9

- Seed, Arango and Chan, 1975 Evaluation of soil liquefaction potential during earthquakes, Report EERC 75-28, University of California, Berkeley

- Seed, 1976, Evaluation of soil liquefaction effects on level ground during earthquakes, in "Liquefaction problems in Geot, Eng." ASCE annual Convention

- Seed, 1979, Soil liquefaction and cyclic mobility evaluation for level ground during earthquakes, J. Geot. Div. ASCE 105, GT2

- Townsend, 1978, A review of factors affecting cyclic triaxial tests, ASTM Spec. Pub. STP 654

CHAPTER VI

STATISTICAL METHODS

STATISTICAL AND PROBABILISTIC METHODS IN GEOMECHANICS

Achintya Haldar

School of Civil Engineering
Georgia Institute of Technology
Atlanta, Georgia 30332, U.S.A.

The application of statistical and probabilistic methods in geomechanics is reviewed here. Considerable progress has been made in this area, and it is explained here with the help of examples. Discussions are also made on how decision theory concepts can be used to solve some problems by considering the risk as well as the consequences in case of a design failure. The safety factor concept is also reviewed in great detail. Some of the commonly used techniques in uncertainty analysis are also discussed. Some cautions required in any probabilistic study are mentioned, along with the possible direction of some future work in this area.

1. INTRODUCTION

Statistical and probabilistic methods are being used to different degrees in the different areas of civil engineering. Although the classical statistical and probabilistic theories are well established, their application to civil engineering problems are recent. Thus, the question can be raised of whether statistical and probabilistic methods are necessary to solve geomechanics problems.

Quantitative methods of modeling and evaluating geometrical and material properties, the type, intensity and duration of the loading, and the behavior of the system under the applied loads are the tools of modern engineering. Some of these methods have become quite elaborate and include sophisticated mathematical modeling and analysis, computer simulation, and optimization techniques. However, regardless of the level of sophistication

J. B. Martins (ed.), Numerical Methods in Geomechanics, 473–504.

of the models (including experimental laboratory models), they are predicted on idealized assumptions or conditions using limited information; hence, information derived from these quantitative methods may or may not closely reflect reality. In a typical geomechanics problem, the types of loads that are encountered are dead load, live load, wind load, temperature load, seismic load, wave load, etc. In a design problem, none of these loads can be predicted with certainty. Moreover, identification of the most critical load combination could be extremely difficult. It is also well established that soil deposits are generally nonhomogeneous, and consequently soil and foundation engineering are characterized by numerous uncertainties (1,2,3,4,5). These uncertainties may arise from insufficient information on the subsoil condition due to inadequate sampling, limitations in the sampling techniques, the validity of several empirical relationships used, discrepancies in the laboratory and field loading conditions, lack of uniformity in the testing procedures, the mathematical model used to describe the problem and from several other sources.

In the development of engineering designs, decisions are often required regardless of the state of completeness of quality of information, and thus must be formulated under conditions of uncertainty, in the sense that the consequences of a given decision can not be determined with complete confidence (6). In other words, there is a finite amount of risk associated with every engineering design. For any engineering design, if there is more than one failure mode, it is essential that the risk associated with each mode be very similar; otherwise, the design will be deficient in terms of safety, serviceability and economic factors. A conventional deterministic design procedure may not address the problem adequately.

Moreover, it has been observed that the approaches taken by geomechanics people and the classical geotechnical people to solve a problem are incompatible. Geomechanics people often complain that the classical geotechnical approaches are often quite crude and empirical in nature. On the other hand, the geotechnical people state that their experience shows that their approaches work. The application of probabilistic concepts could narrow this gap considerably.

From the above discussion, it is clear that the application of probability concepts in geomechanics problems is not only desirable but necessary for a systematic treatment of the problem. If properly used, probability concepts may help to quantify experience or judgmental information; this will supplement the information obtained from the data sets whenever necessary. Thus, the use of probabilistic concepts is a very powerful design tool that can be used to complement the conventional design methods.

Like all other available methods, the probabilistic approach is not perfect in every respect. In practice, there is generally not enough data to justify or ascertain a particular probability distribution. Information on extreme conditions is rarely available. It is also recognized that the probability of failure is sensitive to the assumed distributions and some other assumptions, particularly for low risks of perhaps $<<10^{-3}$ (7). However, Bolotin (8) notes that reliability analysis is useful in a comparative rather than an absolute sense. Hence, for those problems with a risk level greater than 10^{-3}, perhaps a rough estimate of risk may suffice. For the present reliability analysis, the primary objective is to systematically analyze the uncertainties associated with the problem under consideration and to obtain a consistent basis for comparison between design alternatives. It should be stressed that the description of statistical information and estimation of statistics are not the only applications of probability theory; in fact, the much more significant role of probability concepts lies in the unified framework they provide for the quantitative analysis of uncertainty and assessment of associated risk, and in the formulation of trade-off studies relative to decision making, planning and design.

In the following sections, the concept of probability in geomechanics will be reviewed.

2. RANDOMNESS AND RANDOM VARIALBES

Randomness in a parameter means that more than one outcome is possible; in other words the actual outcome is (to some degree) unpredictable. The possible outcomes are usually a range of measured or observed values; moreover, within this range certain values may occur more frequently than others. Mathematical representation of a random variable is thus a primary task in any probabilistic formulation.

In a probabilistic design procedure, a prediction of the future is made using information from the past, including experiences and judgment whenever possible. Thus, it is necessary to collect all relevant information from the past for this purpose. A typical flow chart of the steps involved is shown in Fig. 1. The information collected will constitute the sample space. The randomness characteristics can be described graphically in the form of a histogram or frequency diagram as shown in Fig. 2. (6, 9). For a more general representation of the randomness, the frequency diagram can be fitted to some theoretical probability density functions $f_X(x)$. By integrating the probability density function thus obtained, a probability distribution function, $F_X(x)$ can be obtained. However, to

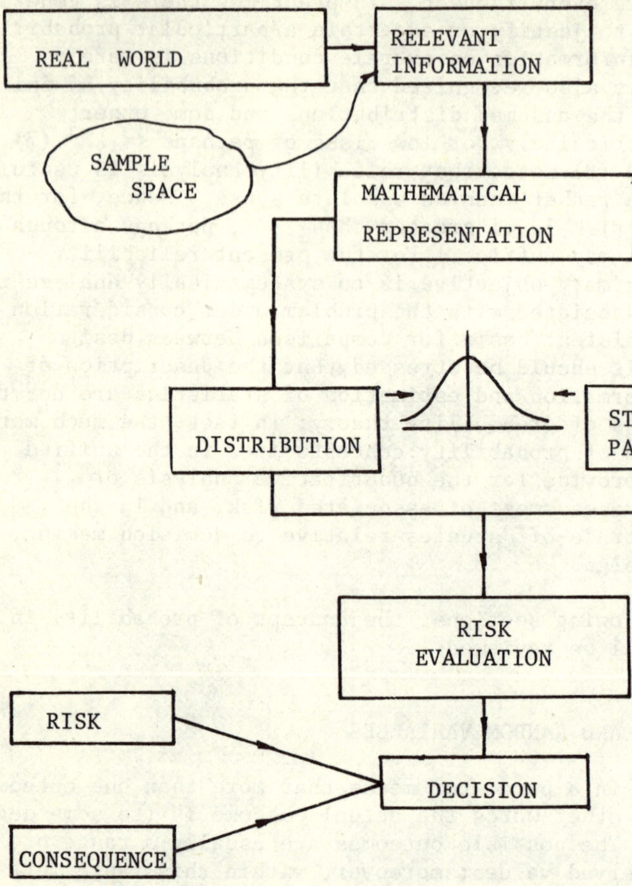

Figure 1. Steps in a Probabilistic Study

describe the probability density function uniquely, some parameters of the distribution need to be estimated. The estimation of these parameters, which are called statistics, is itself a major part of the uncertainty analysis. When the randomness in the load and the resistance parameters are modeled in terms of distributions and the corresponding statistics, the risk involved in the design can be estimated. This will be discussed in detail later. In this way, the risk associated with each design alternative can be evaluated. The information on risk and the corresponding consequences in case of failure can be combined using the decision analysis framework to obtain the so-called "best" solution to the problem under consideration.

Some of the elements shown in Fig. 1 need some elaboration here.

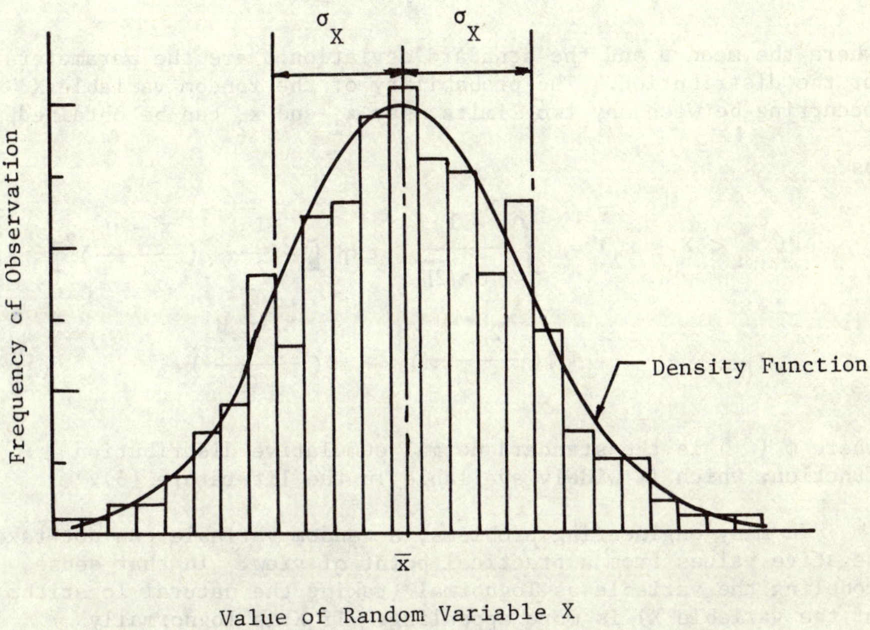

Figure 2. A Typical Histogram or Frequency Diagram

3. COMMON DISTRIBUTIONS

A random variable usually can be described mathematically by a distribution. A random variable may be discrete or continuous depending upon the underlying sample space. Most commonly used discrete random variables are described by the Binomial distribution, Poisson distribution, Geometric distribution Hypergeometric distribution, Hyperbinomial distribution, etc. Continuous random variables are generally described by the Normal distribution, Lognormal distribution, Exponential distribution, Gamma distribution, Beta distribution, Chi-square distribution, Student's t distribution, etc. However, for specific applications other distributions may prove to be more appropriate and useful, including the triangular, uniform trapezoidal, etc., as long as they satisfy all the properties of density functions.

Perhaps the best-known and most widely used probability

distribution is the normal distribution, also known as the Gaussian distribution. The normal distribution has a probability density function given by

$$f_X(x) = \frac{1}{\sigma\sqrt{2\Pi}} \exp\left[-\frac{1}{2}\left(\frac{x-\mu}{\sigma}\right)^2\right] \; ; \quad -\infty < x < \infty \quad (1)$$

where the mean μ and the standard deviation σ are the parameters of the distribution. The probability of the random variable X occurring between any two limits, say x_1 and x_2 can be obtained as

$$P(x_1 < X \le x_2) = \int_{x_1}^{x_2} \frac{1}{\sigma\sqrt{2\Pi}} \exp\left[-\frac{1}{2}\left(\frac{x-\mu}{\sigma}\right)^2\right] dx$$

$$= \Phi\left(\frac{x_2-\mu}{\sigma}\right) - \Phi\left(\frac{x_1-\mu}{\sigma}\right) \quad (2)$$

where ϕ ( ) is the standard normal cumulative distribution function, which is widely available in the literature (6).

In many engineering problems, a random variable can not take negative values from a practical point of view. In that sense, modeling the variable as lognormal (taking the natural logarithm of the variable X) is more appealing. If X is lognormally distributed, then

$$P(x_1 < X \le x_2) = \int_{x_1}^{x_2} \frac{1}{\sqrt{2\Pi}\,\xi\, x} \exp\left[-\frac{1}{2}\left(\frac{\ln x-\lambda}{\xi}\right)^2\right] dx$$

$$= \Phi\left(\frac{\ln x_2-\lambda}{\xi}\right) - \Phi\left(\frac{\ln x_1-\lambda}{\xi}\right) \quad (3)$$

where λ and ζ are the parameters of the lognormal distribution and can be estimated as

$$\lambda = E(\ln x) = \ln \mu - \frac{1}{2}\xi^2 \quad (4)$$

$$\xi^2 = \mathrm{Var}(\ln x) = \ln\left(1 + \frac{\sigma^2}{\mu^2}\right) \quad (5)$$

For more about normal, lognormal and other distributions, interested readers are referred to any standard textbook on probability and statistics (6, 9).

3.1. Extreme Value Distributions

In many civil engineering problems, concern often centers on the largest or smallest of a number of random variables. The success or failure of a system may rest solely on the maximum load effect or the minimum resistance effect, not simply the typical values. Earthquakes, floods, waves, wind and floor loadings are all random variables whose largest value in a sequence may be critical. For capacity, too, the weakest failure mode may depend on the extremes. Three specific cases, which have been studied in some detail and applied widely, need some mention here. They are Type I: Distribution of largest value, Type I: Distribution of smallest value, Type II: Limiting distribution of the largest value of many independent indentically distributed random variables, and Type III: Distribution limited in the tail of interest. For details of these distributions and their applications, readers are referred to Reference 9.

4. ESTIMATION OF PARAMETERS

For the appropriate probabilistic description of a random variable, some parameters or statistics need to be evaluated or estimated on the basis of a set of observed data obtained from the population. Among the most important statistical parameters are the mean value μ, which denotes the average or expected value of the random variable and the standard deviation σ, which denotes the dispersion of a random variable with respect to the mean value. Another commonly used parameter is the variance, which is the square of the standard deviation. The mean value and the variance can also be interpreted as, respectively, the centroidal distance and the central moment of inertia of the density function. For most engineering problems, the absolute dispersion about the mean value may not be as important as the ratio of the degree of dispersion to the mean value. Hence, the coefficient of variation (COV) is often preferred. COV is the ratio of the standard deviation and the mean value. Mean, variance, standard deviation and COV can be mathematically represented as

$$\mu = \int_{-\infty}^{\infty} x \, f_X(x) \, dx \tag{6}$$

$$Var(X) = \int_{-\infty}^{\infty} (x - \mu)^2 \, f_X(x) \, dx \tag{7}$$

$$\sigma = \sqrt{Var(X)} \tag{8}$$

$$COV = \sigma / \mu \tag{9}$$

Different methods for estimating these parameters are available; among these are the method of moments and the method of maximum likelihood. These are also sometimes called the point estimators. It may be desirable to mention here that there are certain properties that are desirable in a point estimator. They are unbiasedness, consistency, efficiency and sufficiency. In the method of moment, the parameters of the distribution are estimated from the mean and variance (and higher moments, if necessary) of the random variable. In some cases such as in the case of normal distribution, the moments, i.e., mean and variance, represent the parameters of the distribution. In the method of maximum likelihood the parameters of the distribution are estimated directly. The principle behind this method is as follows: for the given observations or sample values $x_1, x_2, --- x_n$, the estimated value of the parameter is the value most likely to produce these observed values.

5. DISTRIBUTION MODELS

To develop a probabilistic model, the underlying distribution of a random variable as well as the statistics needs to be known. In the previous section, a detailed discussion was made on the estimation of statistics. The methods available to empirically determine the distribution model will be discussed in this section.

In practice the choice of the probability distribution may be dictated by mathematical convenience. In many cases, the functional form of the required probability distribution may not be easy to determine, or more than one distribution may fit the available data. Under certain circumstances, the basis or properties of the physical process may suggest the form of the required distribution.

The required probability distribution may be determined empirically based entirely on the available observed data. A frequency diagram for the set of data can be constructed and a distribution model can be selected by visual comparison as shown in Fig. 2. Alternatively, the data may be plotted on probability papers prepared for specific distributions (6, 10). If the data plots approximately on a straight line on a probability paper, the distribution corresponding to this paper may be an appropriate distribution model.

When the distribution model is obtained from the aforementioned methods, or when two or more distributions appear to be plausible probability distribution models, statistical tests known as goodness-of-fit tests for distribution can be carried out to verify the distribution model. Two such tests that are commonly used for this purpose are the chi-square (χ^2) and the Kolmogorov - Smirnov (K-S) tests.

5.1. Chi-square Test for Distribution

In the chi-square goodness-of-fit test, observed data are divided into intervals. Observed frequencies n_1, n_2 --- n_k of k intervals of the random variable are compared with the corresponding frequencies e_1, e_2, --- e_k from an assumed theoretical distribution. It can be shown (11) that the quantity

$$\sum_{i=1}^{k} \frac{(n_i - e_i)^2}{e_i} \tag{10}$$

approaches the chi-square distribution with f=k-1 degrees of freedom as $n \to \infty$. The number of degrees of freedom f is a parameter of the chi-square distribution. For every unknown parameter of the theoretical distribution, the degrees of freedom should be reduced by one. On this basis, if an assumed distribution yields

$$\sum_{i=1}^{k} \frac{(n_i - e_i)^2}{e_i} < C_{1-\alpha,f} \tag{11}$$

where $C_{1-\alpha,f}$ is the value of the appropriate χ^2_f distribution at the cumulative probability $(1 - \alpha)$, the assumed theoretical distribution is an acceptable model, at the significance level α. Otherwise, the assumed distribution can be rejected at the α significance level.

5.2. Kolmogorov - Smirnov K-S Test for Distribution

The K-S test compares the experimental cumulative frequency and an assumed theoretical distribution function. The difference between the two distribution functions can be denoted by

$$D_n = \max_x \left| F(x) - S_n(x) \right| \tag{12}$$

where F(X) is the theoretical distribution function and $S_n(x)$ can be obtained by arranging the data in increasing order and using the following relationship:

$$S_n(x) = \begin{cases} 0 & ; \quad x < x_1 \\ \dfrac{k}{n} & ; \quad x_k \le x < x_{k+1} \\ 1 & ; \quad x \ge x_n \end{cases} \tag{13}$$

If $$P(D_n \leq D_n^\alpha) = 1 - \alpha \tag{14}$$

where D_n^α values at various significance levels α can be obtained from a standard mathematical Table (6), the theoretical distribution can be accepted with α significance level.

The advantage of the K-S test over the chi-square test is that it is not necessary to divide the data into intervals; thus the error associated with the number or size of the interval does not appear to be important.

6. UNCERTAINTY ANALYSIS

The discussion made in Sections 4 and 5 appears to be very simple. However, in real problems the uncertainty analysis could be quite involved. Some of the common cases encountered in practical problems are described in this section. For simplicity of discussion, they are divided into several subsections.

6.1. Single Random Variable

The uncertainty associated with a single random variable can be estimated with the help of a histogram or frequency diagram, density function, mean, variance, standard deviation, COV, etc. They are described in Sections 4 and 5 and will not be repeated here.

6.2. Function of Single Random Variable

In many geomechanics problems, the uncertainty associated with one random variable needs to be estimated indirectly from the information on uncertainty on another random variable. For example, X is a random variable and its density function representing the randomness is known. Y is another random variable whose uncertainty is of great concern in a particular problem. However, there is no direct information available on Y. But Y is functionally related to X as

$$Y = g(X) \tag{15}$$

Knowing the functional relationship g and the uncertainty associated with X, the uncertainty in Y in terms of density function can be shown to be

$$f_Y(y) = f_x(g^{-1}) \left| \frac{dg^{-1}}{dy} \right| \tag{16}$$

where g^{-1} represents $g^{-1}(y)$, the inverse function of g.

The following example taken from Ang and Tang (6) will indicate the steps involved in this procedure.

The freeboard height of earth dams can be estimated from the consideration of wind tide and wave height. The wind tide, in feet, above still-water level can be expressed as

$$Z = \frac{F}{1400\ d} V^2 \qquad (17)$$

where,

V= wind speed in miles per hour
F= fetch or length of water surface over which the wind blows, in feet
d= average depth of lake along the fetch, in feet.

If the wind speed V has an exponential distribution with mean speed v_o that is

$$f_V(v) = \frac{1}{v_o} e^{-\frac{v}{v_o}} \quad ; \quad v \geq 0 \qquad (18)$$
$$= 0 \quad ; \quad v < 0$$

the distribution of Z using Eq. 16 can be shown to be

$$f_Z(z) = \frac{1}{2v_o\sqrt{az}} \exp\left(-\frac{1}{v_o}\sqrt{\frac{z}{a}}\right) \quad ; \quad z \geq 0 \qquad (19)$$
$$= 0 \quad ; \quad z < 0$$

6.3. Function of Multiple Random Variables

Uncertainty analysis in this case is generally involved. The closed form analytical solution can be used only in some special cases. Three such cases which will have major engineering significance are (i) Sum and differences of independent normal variables (ii) Products and quotients of independent lognormal random variables and (iii) Sum of independent random variables with poisson distributions.

6.3.1. <u>Sum and Differences of Independent Normal Variables</u>. Suppose

$$Y = a_1X_1 + a_2X_2 + \ldots\ldots a_nX_n \qquad (20)$$

where X_i's are statistically independent normal variates with mean μ_{x_i} and standard deviation σ_{x_i}. It can be shown that Y is also a normal random variable with mean and variance as shown below:

$$\mu_Y = \sum_{i=1}^{n} a_i \, \mu_{X_i} \tag{21}$$

and

$$\sigma_Y^2 = \sum_{i=1}^{n} a_i^2 \, \sigma_{X_i}^2 \tag{22}$$

It can be stated further that Eqs. 21 and 22 are in fact valid for linear functions of any statistically independent random variables regardless of their distributions. By virtu e of the Central Limit Theorem (6), it can be stated that the sum of a large number of individual random variables, none of which is dominant, tends to the normal distribution as the number of variables (regardless of their initial distributions) increase without limit.

6.3.2. Products and Quotients of Independent Lognormal Variables. In this case assume

$$Y = X_1 \cdot X_2 \cdot \; . \; . \; . \; . \; X_n \tag{23}$$

where X_i's are statistically independent lognormal variables with mean μ_{x_i}'s and standard deviation σ_{x_i}'s. Knowing μ_{x_i} and σ_{x_i} for a random variable X_i, the parameters of the lognormal distribution λ_{x_i} and ζ_{x_i} can be obtained using Eqs. 4 and 5. It can be shown that Y is also a lognormal variable with parameters

$$\lambda_Y = \sum_{i=1}^{n} \lambda_{X_i} \tag{24}$$

and

$$\xi_Y^2 = \sum_{i=1}^{n} \xi_{X_i}^2 \tag{25}$$

6.3.3. Sum of Independent Poisson Random Variables. Assume

$$Y = X_1 + X_2 + \; . \; . \; . \; . \; . \; . \; . \; X_n \tag{26}$$

where X_i's are statistically independent poisson random variables with parameter ν_i's. If X is a poisson random variable with parameter ν, the corresponding density function can be written as

$$P_X(X = x) = \frac{(\nu t)^x \, e^{-\nu t}}{x!} \tag{27}$$

where x is the number of occurrences in time duration t. It can be shown that Y is also a poisson random variable with parameter ν_Y as

$$\nu_Y = \sum_{i=1}^{n} \nu_{X_i} \tag{28}$$

6.4. General Approximate Method

The discussion made in Sections 6.2 and 6.3 are valid when the random variables $X_i^{'}$s are known in terms of statistics (mean, variance, etc.) as well as their distributions. In many geomechanical problems, however, the density function may not be known; information may be limited to the mean and variance of the original variate X_i. Furthermore, even when the density functions are known, the integrations indicated by Eq. 7 may be difficult to perform. Moreover, in case of multiple random variables, the functional forms among variables could be something other than those described in Section 6.3. For these reasons, the approximate mean and variance of the function Y would be practically useful and may be obtained as follows. However, the density functions of Y may not be known. If Y is a function of several random variables, that is

$$Y = g(X_1, X_2, \ldots\ldots X_n) \tag{29}$$

then expanding the function $g(X_1, X_2, \text{---}\ X_n)$ in a Taylor series about the mean values $\mu_{X_1}, \mu_{X_2}, \text{---}\ \mu_{X_n}$, one would obtain

$$Y = g(\mu_{X_1}, \mu_{X_2}, \ldots\ldots \mu_{X_n}) + \sum_{i=1}^{n} (X_i - \mu_{X_i}) \frac{dg}{dX_i}$$

$$+ \frac{1}{2} \sum_{i=1}^{n} \sum_{i=1}^{n} (X_i - \mu_{X_i})(X_j - \mu_{X_j}) \frac{d^2 g}{d_{X_i} d_{X_j}} + \ldots \tag{30}$$

where the derivatives are evaluated at mean values of $X_i^{'}$s.

Truncating the series at the linear terms, the first-order approximate mean and variance of Y can be obtained as

$$E(Y) \simeq g(\mu_{X_1}, \mu_{X_2}, \ldots\ldots \mu_{X_n}) \tag{31}$$

which indicates that the mean of the function is approximately equal to the function of the means; and

$$\mathrm{Var}(Y) \simeq \sum_{i=1}^{n} E_i^2 \,\mathrm{Var}(X_i) + \sum_{i}^{n} \sum_{\neq j}^{n} E_i \, E_j \,\mathrm{COV}(X_i, X_j) \quad (32)$$

where E_i and E_j are constants and are the values of the partial derivatives $\partial g/\partial X_i$ and ∂X_j, respectively, evaluated at the mean values. If X_i and X_j are statistically independent, for all i and j, Eq. 32 reduces to

$$\mathrm{Var}(Y) \simeq \sum_{i=1}^{n} E_i^2 \,\mathrm{Var}(X_i) \quad (33)$$

The above approximate mean and variance may also be improved by including the higher-order terms of the Taylor series expansion of $g(X_1, X_2, \text{---} X_n)$. If X_i and X_j are statistically independent, the second-order mean of Y would be

$$E(Y) \simeq g(\mu_{X_1}, \mu_{X_2}, \ldots \mu_{X_n}) + \frac{1}{2} \sum_{i=1}^{n} \left(\frac{d^2 g}{dX_i^2}\right) \mathrm{Var}(X_i) \quad (34)$$

In many engineering problems, second-order mean and first-order variance would improve the accuracy of the estimation. Although the method is approximate, the method is very powerful. This is explained with the following example on relative density.

Relative density can be defined as

$$D_r = \frac{\gamma_{max}}{\gamma} \quad \frac{\gamma - \gamma_{min}}{\gamma_{max} - \gamma_{min}} \quad (35)$$

in which γ_{max}, γ_{min}, and γ are the maximum, minimum and inplace dry density, respectively, of a cohesionless deposit. Using first-order approximation, the mean and COV of the relative density can be shown to be (12)

$$\bar{D}_r \simeq \frac{\bar{\gamma}_{max}}{\bar{\gamma}} \quad \frac{\bar{\gamma} - \bar{\gamma}_{min}}{\bar{\gamma}_{max} - \bar{\gamma}_{min}} \quad (36)$$

$$\Omega^2_{D_r} \simeq E_1^2 \, \Omega^2_{\gamma_{max}} + \left[\frac{(E_1 + 1)\,(1 - \bar{D}_r)}{\bar{D}_r} \right]^2 \Omega^2_{\gamma_{min}}$$

and

$$+ \left[\frac{E_1 + 1}{\bar{D}_r} - 1 \right]^2 \Omega^2_{\gamma}$$

$$= E_1^2 \, \Omega^2_{\gamma_{max}} + E_2^2 \, \Omega^2_{\gamma_{min}} + E_3^2 \, \Omega^2_{\gamma} \qquad (37)$$

in which $E_1 = \bar{\gamma}_{min}/(\bar{\gamma}_{max} - \bar{\gamma}_{min})$ and γ, γ_{max}, and γ_{min} are assumed to be mutually independent. The coefficients E_1, E_2, and E_3 could be interpreted as the amplification factors for the uncertainties in each of the aforementioned densities. In general, these amplification factors will show the importance of the variables involved in the formulation. In the case of relative density, it is clear that the uncertainty associated with the in situ density measurement is more critical than that in the other densities. Thus, in case of time or money limitations, more attention should be given to the in situ density measurement than the other parameters. This type of probabilistic approach will also help to identify primary factors and secondary factors in geomechanics problems where the number of parameters involved could be very large.

Eq. 37 will also estimate the overall uncertainty in the relative density considering uncertainties in all the other parameters. Although many practicing geotechnical engineers know from their experience that the relative density is an error-prone parameter, it can be proved mathematically using Eq. 37. Haldar and Tang (12) studied this problem as shown in Fig. 3. Theoretically, the plot of calculated versus measured relative densities should lie on a 45° line. However, from Fig. 3, it is clear that there is a considerable spread. The uncertainty in the relative density could be on the order of 0.30 in terms of COV. This type of information needs to be used for all geomechanics formulations.

7. RISK EVALUATION

It can be expected at this stage that all the load and resistance parameters involved in a problem are identified and

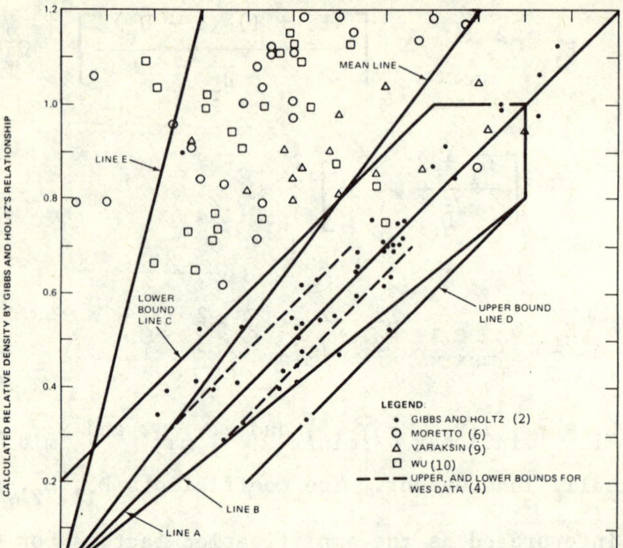

Figure 3. Comparison of Measured Versus Caculated D_r Using Gibbs and Holtz's Relationship

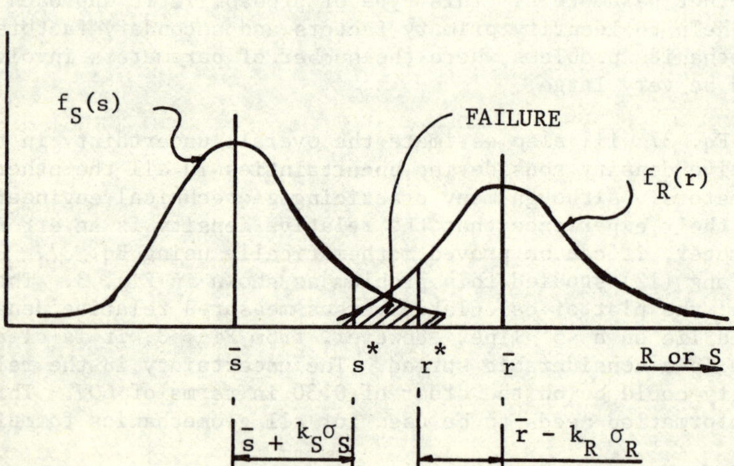

Figure 4. Risk Evaluation

the uncertainties associated with them are quantitied appropriately. The stage is now set to estimate the risk associated with the design or the corresponding safety factor of the design.

Consider first a general case in which the random resistance R and the random load S are known and the uncertainties are quantified as shown in Fig. 4.

Then, for statistically independent R and S,

$$\text{Risk} = P(\text{Failure}) = P(R \leq S{=}s)$$

$$= \int_0^{\infty} \left[\int_0^{s} f_R(r)\,dr \right] f_S(s) \; ds \tag{38}$$

The shaded area in Fig. 4 will represent the risk associated with the design. When the random characteristics of R and S are known, the corresponding risk can be estimated using Eq. 38. The operation could be involved in some cases. However, for other cases, Eq. 38 can be simplified considerably.

Consider first a case where R and S are statistically independent normal variables with mean values μ_R and μ_S, respectively, and corresponding standard deviations σ_R and σ_S. Then

$$P(\text{Failure}) = P(R \leq o) = P(R - S \leq 0)$$

$$= 1 - \Phi\left(\frac{\mu_R - \mu_S}{\sigma_R^2 + \sigma_S^2} \right) \tag{39}$$

where ϕ ( ) is the standard normal probability widely available in the literature.

If R and S are statistically independent lognormal variables with the same means and standard deviations, it can be shown that

$$P(\text{Failure}) = P\left(\frac{R}{S} \leq 1.0\right) = P\left(\ln\left(\frac{R}{S}\right) \leq 0.0\right)$$

$$= 1 - \Phi \left[\frac{\ln \frac{\mu_R}{\mu_S} \sqrt{\frac{1 + \Omega_S^2}{1 + \Omega R}}}{\sqrt{\ln(1 + \Omega_S^2)\,(1 + \Omega_R^2)}} \right] \tag{40}$$

where $\Omega_S = \sigma_S / \mu_S$ and $\Omega_R = \sigma_R / \mu_R$.

For small Ω_R and Ω_S, say < 0.30,

$$P(\text{Failure}) = 1 - \Phi\left[\frac{\ln \frac{\mu_R}{\mu_S}}{\sqrt{\Omega_R^2 + \Omega_S^2}}\right] \tag{41}$$

Using this concept, Haldar and Tang (13) showed how to develop a probabilistic model and estimated risk in liquefaction potential evaluation studies. Similar other studies are also available in the literature (14, 15, 16).

8. RISK AND SAFETY FACTORS

An obvious question that needs to be addressed at this stage is the correlation between the risk and the corresponding safety factor (3, 17, 18, 19, 21). In this respect the concept of safety factor needs some discussion here. The nominal safety factor used in Fig. 4 can be described as

$$\text{Nominal SF} = \frac{r^*}{s^*} \tag{42}$$

The above definition may not have a real physical meaning since in a deterministic study uncertainties are generally not quantified. The nominal resistance r^* is usually a very conservative value (perhaps one, two, or three standard deviations below the mean value). The nominal load s^* is also a very conservative value (several standard deviations above the mean value). Thus, the nominal safety factor fails to convey the real safety factor involved in a design as it fails to include the conservativeness in the load and resistance. In that sense, the mean safety factor, which can be defined as (21)

$$\overline{SF} = \frac{\bar{r}}{\bar{s}} \tag{43}$$

appears to be logical, convenient and consistent with the idea behind the safety factor. Uncertainties in the evaluation in R and S should play a major role in the evaluation of safety factors. Many design failures occur when the critical load and resistance combination is not studied in detail.

The nominal and the mean safety factor can be correlated in terms of uncertainties in R and S and the amount of conservatism necessary for a given problem. The mean safety factor as shown in Eq. 43 can be rewritten as (21)

$$\overline{SF} = \frac{\bar{r}}{\bar{s}} = \frac{1 + 0.75\,\beta\,\Omega_S}{1 - 0.75\,\beta\,\Omega_R} \tag{44}$$

where $\beta = \Phi^{-1}(1-P_f)$ if it is assumed or prescribed that (R-S) or ln (R/S) are normal variates. P_f is the acceptable risk in the design. The nominal safety factor can be rewritten in terms of mean safety factor as

$$SF = \overline{SF}\left(\frac{1 - k_R\,\Omega_R}{1 - k_S\,\Omega_S}\right) \tag{45}$$

k_R and k_S are defined in Fig. 4.

It is clear from this discussion that the probabilistic safety factor conveys more meaning than the nominal safety factor.

The probabilistic safety factor concept was recently used by Vanmarcke (15) in studying the reliability of earth slopes. He observed that failure events involving very long or very short failure zones are highly improbable. This concept is completely new in a deterministic sense. Vanmarcke suggested a probabilistic method which would indicate a fairly distinct most probable failure length for an earth slope.

The safety factor concept is getting new interpretations in many engineering applications. The conventional safety factor concept needs a very major revision. Some countries have already taken a step in this direction or are in the process of doing so.

As the safety factor value (either nominal or mean) is increased, the corresponding risk of failure must decrease. The size of this variation depends on the problem and many other factors. However, it is possible to correlate safety factor and the corresponding risk for a given problem. In structural engineering, considerable progress has been made in this area. In geomechanics progress has been limited thus far.

9. LOADINGS

In many geomechanics problems, the soil deposit attracts the most attention, although the loading aspect of the problem can not be ignored. Most of the common loadings can be categorized as dead load, live load, wind load, snow load, earthquake load, wave actions, etc. The probabilistic aspect of dead load has been studied in great detail (22, 23). The uncertainties associated with the live load must be dealt with on a case by case basis although several studies (24, 25, 26, 27) have been reported. There is a trend in the United States to consider wind load and snow load probabilistically on a regional basis (28). The required building codes could be developed for each city. Although considerable work has been done in the area of seismic loading, a great deal of work still needs to be done, particularly considering the uncertain nature of the future design earthquakes. For any future design earthquake magnitude, acceleration, time-history, duration, three-dimensional aspects of the shaking, etc., are not known in a deterministic sense. Haldar (5) has shown that when earthquake is one of the critical loading conditions to be considered in a design, the uncertainties in the load parameters could be higher than in the soil parameters.

Earthquake acceleration is one of the most critical parameters in many geomechanics problems. If sufficient attention is not given to the selection of the design earthquake acceleration, then no matter how sophisticated the approach is taken to solve the given problem, the final outcome could be misleading. The earthquake acceleration depends on the local seismicity, geologic history of the region, potential sources of earthquakes, direction of faults if any, upper and lower bound earthquake magnitudes for the site under consideration, distance of potential earthquake sources from the site, attenuation characteristics of the region, etc. All these parameters have a considerable amount of uncertainty. This problem needs to be modeled probabilistically. The point source model (29) and line source model (30) are available in the literature and should be used whenever possible. A typical result from this type of study would be acceleration in terms of return period or annual probability of exceedence of the given acceleration as shown in Fig. 5.

In geomechanics problems, uniform intensity cyclic loadings are routinely used in the laboratory instead of irregular-patterned earthquake loadings to investigate the soil properties. Even if it is assumed that the future earthquake time history is known, the intensity and the number of cycles need to be estimated to convert it to uniform cycles. A detailed discussion is made on this area by Haldar and Tang (13). They have shown that to minimize uncertainty in the conversion, the intensity of uniform cycles can be taken as 75% of the maximum intensity and the equivalent number of cycles N_{eq} can be estimated from the following relationship.

$$E(N_{eq} \mid M=m) = 106.1 - 36.4\, m + 3.312\, m^2 \quad ; \quad m \geq 5.0 \qquad (46)$$

and the corresponding $Var(N_{eq}|M) = 29.05$, where M is the earthquake magnitude in Richter's scale.

The solution to a geotechnical problem will be incomplete

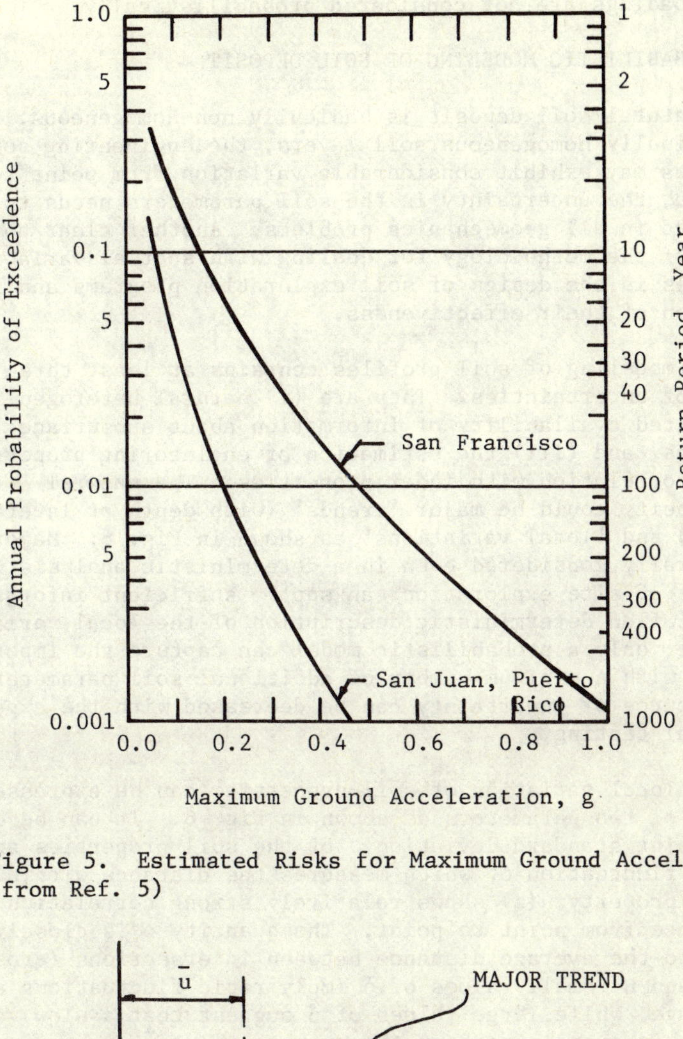

Figure 5. Estimated Risks for Maximum Ground Acceleration (from Ref. 5)

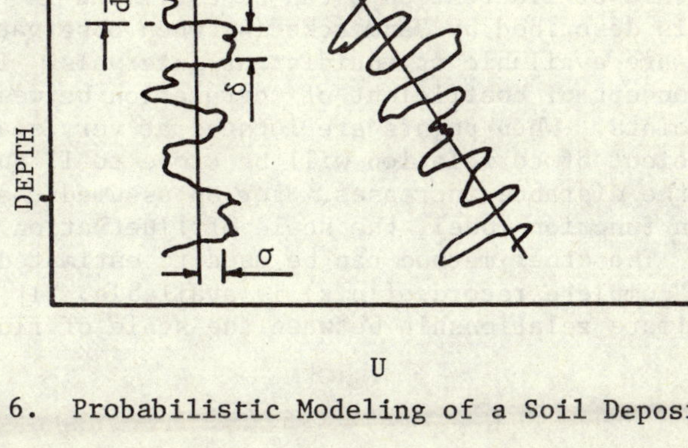

Figure 6. Probabilistic Modeling of a Soil Deposit

if the loadings are not considered probabilistically.

10. PROBABILISTIC MODELING OF SOIL DEPOSIT

A natural soil deposit is basically non-homogeneous. Even with nominally homogeneous soil layers, the engineering soil properties may exhibit considerable variation from point to point. Therefore, the uncertainty in the soil parameters needs to be considered in all geomechanics problems. Another clear potential benefit of the methodology for dealing with spatial variability of soil is in the design of soil exploration programs and in the evaluation of their effectiveness.

The modeling of soil profiles contains at least three major sources of uncertainties. They are (i) natural heterogeneity, (ii) limited availability of information about subsurface conditions, and (iii) the estimation of engineering properties through correlation with index properties. The natural heterogeneity could be major 'trends' (with depth or lateral distance) and 'local variations' as shown in Fig. 6. Major trends are generally considered even in a deterministic analysis. However, no amount of site exploration can supply sufficient information for a detailed deterministic description of the local variations at a site; only a probabilistic model can capture the important features with a minimum number of additional soil parameters. The second source of uncertainty can be decreased with the expense of additional testing.

The local variation of soil properties can be expressed with the help of two parameters as shown in Fig. 6. It can be expressed as the point standard deviation σ of the soil properties and the scale of fluctuation δ, which measures the distance within which the soil property $u(z)$ shows relatively strong correlation of persistence from point to point. The quantity δ is closely related to the average distance between intersections (crossings) of $u(z)$ and $\bar{u}$: small values of δ imply rapid fluctuations about the average, while large values of δ suggest that a slow varying component is superimposed on the average value $\bar{u}$.

The scale of fluctuation δ can be estimated in two ways. A procedure is described by Vanmarcke (4) when observations of soil properties are available at equidistant intervals. This approach used the concept of coefficient of correlation between values of u at two points. When points are located at very small intervals, the coefficient of correlation will be close to 1, and it usually decays as the distance increases. For an assumed theoretical correlation function model, the scale of fluctuation δ can be estimated. The other method can be used to estimated δ if a reasonably complete record of $u(z)$ is available. It is based on the approximate relationship between the scale of fluctuation, δ, and average distance, $\bar{d}$, between the intersections of the fluctuating property, u(z) and its mean $\bar{u}$. The average distance between mean crossing is approximately (4)

$$\bar{d} \simeq \sqrt{\frac{\pi}{2}}\ \delta \simeq 1.25\ \delta \tag{47}$$

The information on the scale of fluctuation is important to a geotechnical engineer. To avoid wasteful redundancy in information gathering, sampling distances should be chosen in such a way that they are large in comparison with δ. On the other hand, when a soil property is being determined by two different tests, the locations of pairs of samples should be well within the correlation distance for maximum effectiveness. Other major information may be obtained on soil deposits when point or local variation is not very important. Consider a site being investigated for liquefaction potential. It is very likely that a very loose pocket of sand is located during the subsurface investigation. It is known that to cause noticeable damage to a structure located on the site, a sufficient volume of sand needs to liquefy. The evaluation of the site considering the loose pocket of sand is thus obviously incomplete. Some sort of spatially averaged soil properties in the critical soil volume needs to be used for this purpose. Consider a soil volume V of size ΔN, ΔY, and ΔZ. For a homogeneous soil deposit, the mean soil property of the volume V would remain the point mean $\bar{u}$, but the average standard deviation over the volume σ_V can be obtained as

$$\sigma_V = \Gamma(\Delta x, \Delta y, \Delta z)\ \sigma = \Gamma(\Delta x)\ \Gamma(\Delta y)\ \Gamma(\Delta z)\ \sigma \tag{48}$$

Each factor, e.g., $\Gamma(\Delta z)$ depends only on the ratio $\Delta z/\delta$. As the name suggests, all these reduction factors will take a maximum value of 1.0 when ΔX, ΔY and ΔZ distances are less than or equal to the corresponding scale of fluctuation δ_x, δ_y, and δ_z, respectively. Otherwise they will always be less than 1.0, suggesting that the average standard deviation over a volume is always smaller than the point standard deviation. For the application of this concept in geotechnical engineering, the readers are referred to reference No. 15.

11. ESTIMATION OF PARAMETERS IN GEOMECHANICAL PROBLEMS

The discussion would be incomplete without consideration of the estimation of parameters in geomechanical problems. Many parameters are expected in a sophisticated theoretical geomechanical problem. The validity of the model depends on how accurately these parameters can be estimated under in situ conditions.

Generally, theoretical geomechanical formulations are verified using case studies. When these verifications are made, it is extremely important to explain the detailed steps involved in estimating the parameters involved in the problem so that in the future anyone interested in the problem can follow the same steps. Statements such as "using triaxial test apparatus" have no real meaning unless other details such as size of the sample, sample preparation method, type of loadings, etc., are clearly mentioned. This will improve the reliability of the procedures. This aspect of the problem is frequently overlooked by researchers but is essential to insure the validity of the methods.

In many geomechanical problems, standard soil index properties are used. Whenever possible, these parameters should be estimated directly from the site without taking any samples. For example, suppose the relative density of a site needs to be determined. It is always preferable to measure the relative density directly from an open shaft in the site. However, this could be very expensive and may not be possible in some cases.

The next best approach would be to collect a sufficient number of undisturbed samples from the site and test them in the laboratory. It is very important to point out that it is very difficult to obtain undisturbed samples, particularly for a saturated cohesionless deposit. Disturbances during collection and transportation of the sample should be kept to a minimum. There are also several sources of uncertainty in the laboratory experiments. Reproducibility of the experiments, calibration error of the instrument, and operator or human error along with random errors can make laboratory investigation ineffective. Even when American Society for Testing and Materials (ASTM) guidelines are used during the studies, care should be taken as these guidelines often not detailed enough. Consider, for example, the standard penetration test (SPT) value estimation according to the ASTM D1586-67 guidelines (32). The guideline fails to mention how to obtain free fall of the hammer systems. There are several techniques used in the industry to produce the free fall condition, and considerable differences in the SPT-values are obtained when different methods are used (33).

In some cases indirect estimation of parameters is the only source of information available for the site. Generally, empirical or semi-empirical relationships are used for this purpose. Consider the estimation of relative density from the SPT-values obtained from a site. Haldar and Tang (12) discussed this problem in great detail. Whenever possible, this type of approach to estimate parameters should be avoided.

11.1 Bayesian Updating Approach

It is always preferable to estimate a parameter by using an adequate number of reliable observations. However, in many practical problems this may not be possible. This does not jeopardize the usefulness and relevance of the probability concepts. The probability concept is a very powerful tool where experience, judgment and observational data can be combined together very effectively. The Bayesian approach can be used for this purpose. It differs from the classical statistical approach in the sense that in the classical approach, the parameters are assumed to be constant but unknown, and sample statistics are used as estimators of these parameters. This requires a relatively large amount of data. In the Bayesian approach the parameters are considered to random variable themselves. This enables one to combine subjective judgments based on intuition, experience or indirect information in a systematic way with observed data to obtain a balanced estimation (34).

Assume that the bearing capacity θ for each pile of a large project needs to be estimated. In a deterministic study a very conservative value can be selected based on the available information. In a classical probabilistic sense, the estimation of the value of θ will be difficult as a sufficient number of observations are not yet available from the site. Of course, an indirect correlation can be made to estimate the value of θ. In this type of problem where θ is basically a random variable, the Bayesian approach would be very helpful as previous experience with similar piles can be combined with limited information from the site as it becomes available. The formulation of the Bayesian updating approach can be expressed as

$$f''(\theta) = KL(\theta)\, f'(\theta) \qquad (49)$$

where $f'(\theta)$ is the prior density function of θ, which may be obtained from prior experience, or judgment; $L(\theta)$ is commonly referred to as the likelihood function of θ, which is basically the likelihood of observing the experimental outcome assuming a given value of θ; k is a normalizing constant required to make $f''(\theta)$ is the posterior or updated density function of θ considering all the available information.

Eq. 49 could be considerably simplified if the mathematical form of the prior distribution and the likelihood function of θ are selected in such a way that they are in the same form. Such pairs of distributions are known in the Bayesian terminology as conjugate pairs or conjugate distributions. It should be emphasized that conjugate distributions are chosen solely for mathemathical convenience and simplicity. The most commonly used pairs of conjugate distributions in engineering problems were summarized by Ang and Tang (6). For a detailed discussion on the topic, the readers are encouraged to read this reference.

12. DECISION ANALYSIS

One of the most significant roles of the probability concept lies in its unified framework for the quantitative analysis on uncertainty and assessment of associated risk, and in the formulation of trade-off studies relative to decision making, planning, and design. The risk as well as the consequences of a design failure needs to be studied to obtain the most desirable solution to a problem. The decision analysis framework will form a logical basis for this considering both the technical and economic aspects of the problem. This area has yet to be explored thoroughly and deserves more attention than it generally receives.

Haldar (35) has developed a decision analysis framework for liquefaction studies. A typical decision tree for a liquefaction problem is shown in Fig. 7. The availability of the probabilistic method to study liquefaction problem makes this type of study possible. A decision can be made with available information or with additional information if additional time and money are available. Collection of additional information may not always be desirable for all projects. It depends on many factors, e.g., types of outcomes, likelihood of each outcome, alternatives under consideration, initial cost of each alternative, cost of damage, etc. If the additional information is desirable, the maximum amount of money that should be spent can also be estimated from this study. Generally this is known as the value of perfect information (VPI). This type of decision analysis framework needs to be developed for other geomechanical problems.

13. REMARKS ON THE USE OF PROBABILISTIC METHODS IN GEOMECHANICAL PROBLEMS

Although the application of probabilistic concepts in geomechanical problems is advocated here, it is necessary to give some cautions. Unjustified probabilistic approaches may not have any practical meaning.

It is very important to consider only the relevant information for the problem under consideration. Increasing the number of observations by some means may not be desirable. Empirical correlation among data obtained from various sites must be done very carefully. The information could be site dependent, and the information may not be correct in all cases. Even information obtained on a soil parameter using the same laboratory equipment but obtained from different laboratories needs to be evaluated very carefully. Combining triaxial test results under cyclic loading obtained from various sources could be meaningless

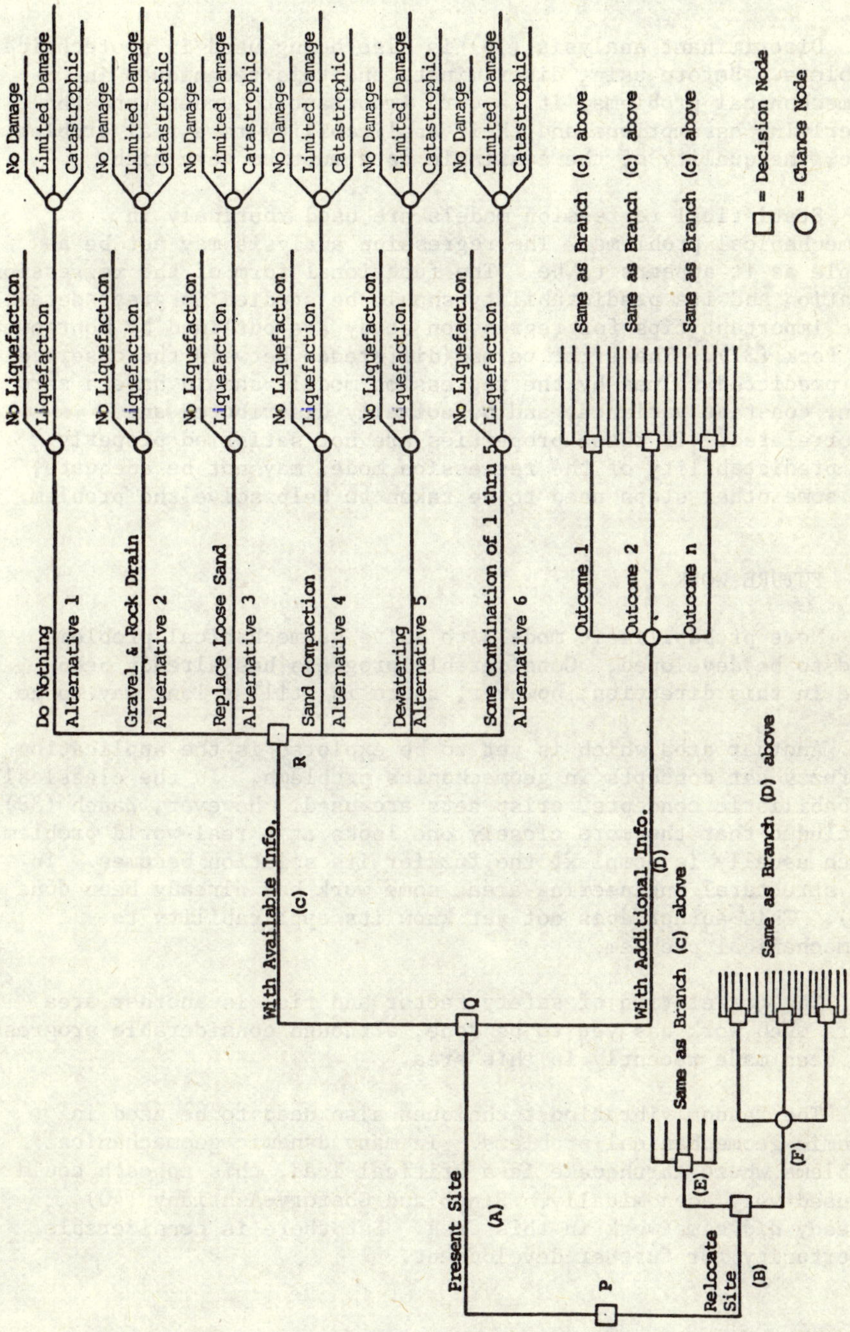

Figure 7. Decision Analysis in Liquefaction Problems

unless information such as type of soil, size of sample, sample preparation methods, disturbance in the sample, etc., is evaluated carefully.

Discriminant analysis (36) is also being used in geotechnical problems. Before using discriminate analysis techniques in geomechanical problems, it is very important to understand the underlying assumptions and their applicability to a real problem. Also, the quality of the available data must be very high.

Statistical regression models are used routinely in geomechanical problems. The regression analysis may not be as simple as it appears to be. The functional form of the regression equation and its predictability should be studied in great detail. Some important tips for regression study are outlined by Montgomery and Peck (37). The error terms (difference between the observed and predicted values by the regression model) should have a zero mean, constant variance, and be normally distributed and uncorrelated. If these properties are not satisfied properly, the predictability of the regression model may not be adequate, and some other steps need to be taken to help solve the problem.

14. FUTURE WORK

More probabilistic models to solve geomechanical problems need to be developed. Considerable progress has already been made in this direction; however, there is still a long way to go.

Another area which is yet to be explored is the application of fuzzy set concepts in geomechanics problems. In the classical probabilistic concepts, crisp sets are used. However, Zadeh (38) concluded that the more closely one looks at a real-world problem, which usually is complex, the fuzzier its solution becomes. In the structural engineering area, some work has already been done (39). This author does not yet know its applicability to geomechanical problem.

The correlation of safety factor and risk is another area where much work has yet to be done, although considerable progress has been made recently in this area.

The random vibration techniques also need to be used in dynamic geomechanical problems. In many dynamic geomechanical problems where earthquake is a critical load, this appoach could be used very economically. Singh and Ghafory-Ashtiany (40) already did some work in this area. But there is considerable opportunity for further development.

15. ACKNOWLEDGEMENT

Some material presented here is partly supported by the National Science Foundation under Grant No. PFR-8006348. Any opinions, findings, and conclusions or recommendations expressed in this publication are those of the writer and do not necessarily reflect the views of the National Science Foundation.

15. REFERENCES

1. Lumb, P., "The Variability of Natural Soils," Canadian Geotechnical Journal, Vol. 3, No. 2, 1966.

2. Singh, A. and Lee, K. L., "Variability in Soil Parameters," Proceedings, Eighth Annual Conf. on Geological and Soil Engineering, Idaho, 1970.

3. Feld, J., "The Factor of Safety in Soil and Rock Mechanics," Proceedings, Sixth Int. Conf. on Soil Mech. and Foundation Engrg. Montreal, Vol. 3, 1965.

4. Vanmarcke, E. H., "Probabilistic Modeling of Soil Profiles," J. Geotechnical Engrg. Div., ASCE, Vol. 103, No. GT11, Nov. 1977.

5. Haldar, A. "Probabilistic Evaluation of Sand Under Earthquake Motions," Ph.D. Thesis, U. of Illinois, Urbana, 1976.

6. Ang, A. H-S, and Tang, W. H., Probability Concepts in Engineering Planning and Design, Vol. 1 - Basic Principles, John Wiley & Sons, Inc., New York, N.Y., 1975.

7. Ang, A. H-S., "Probabilistic Bases of Safety, Performance, and Design," Proceedings, Speciality Conf. on Safety and Reliability of Metal Structures," ASCE, Pittsburgh, Pennsylvania, Nov., 1972.

8. Bolotin, V. V., Statistical Methods in Structural Mechanics, Holden-Day, San Francisco, California, 1969.

9. Benjamin, J. R. and Cornell, C. A., Probability, Statistics, and Decision for Civil Engineers, McGraw-Hill, New York, 1970.

10. Haldar, A., Solution Manual Probability Concepts in Engineering Planning and Design, Vol. 1 - Basic Principles, John Wiley & Sons, Inc., 1975.

11. Hoel, P. G., Introduction to Mathematical Statistics, 3rd Ed., John Wiley & Sons, Inc. 1962.

12. Haldar, A. and Tang, W. H., "Uncertainty Analysis of Relative Density," J. Geotechnical Engrg. Div., ASCE, Vol. 105, No. GT7, July, 1979.

13. Haldar, A. and Tang, W. H., "Probabilistic Evaluation of Liquefaction Potential," J. Geotechnical Engrg. Div., ASCE, Vol. 105, No. GR2, Feb. 1979.

14. Fardis, M. N. and Veneziano, D. "Probabilistic Liquefaction of Sands During Earthquakes," Publication No. R79-14, MIT, March, 1979.

15. Vanmarcke, E. H., "Reliability of Earth Slopes," J. Geotechnical Engrg. Div., ASCE, Vol. 103, No. GT 11, Nov. 1977.

16. Yucemen, M. S., Tang, W. H., and Ang, A. H-S., "A Probabilistic Study of Safety and Design of Earth Slopes," Structural Research Series No. 402, University of Illinois, Urbana, Ill., 1973.

17. Langejan, A., "Some Aspects of the Safety Factor Considered as a Problem of Probability," Proceedings, Sixth Int. Conf. on Soil Mechanics and Foundation Engrg., Montreal, 1965.

18. Lumb, P., "Safety Factors and Probability Distributions of Soil Strength," Canadian Geotechnical Journal, Vol. 7, No. 3, 1970.

19. Meyerhof., G. G., "Safety Factors in Soil Mechanics," Canadian Geotechnical Journal, Vol. 7, No. 4, 1970.

20. Singh, A., "How Reliable is the Factor of Safety in Foundation Engineering," Proceedings, Int. Conf. on Application of Statistics and Probability to Soil and Structural Engrg., Hong Kong, September, 1971.

21. Ang, A., H-S., and Cornell, C. A., "Reliability Bases of Structural Safety and Design," J. Structural Div., ASCE, Vol. 100, No. ST9, Sept., 1974.

22. Borges, J. and Castenheta, M., "Structural Safety," 2nd. ed., National Civil Engineering Laboratory, Lisbon, Portugal, March, 1971.

23. Ellingwood, B. and Ang, A., H-S., "A Probabilistic Study of Safety Criteria for Design," Structural Research Series

No. 387, University of Illinois, Urbana, 1972.

24. Heaney, A., "A Reliability-Based Study Concerning Live Loads and Codified Structural Design," Ph.D. Thesis, U. of Waterloo, 1971.

25. Peir, J. and Cornell, C. A., "Spatial and Temporal Variability of Live Loads," J. Structural Div. ASCE, No. ST5, May, 1973.

26. McGuire, R. and Cornell, C. A., "Live Load Effects in Office Buildings," J. Structural Div., ASCE, No. ST7, July, 1974.

27. Mitchell, G. and Woodgate, R., "Floor Loadings in Office Buildings, the Result of a Survey," CP3/71 Building Design Station, U. K., Jan., 1971.

28. Ellingwood, B., "Wind and Snow Load Statistics for Probabilistic Design," J. Structrual Div., ASCE, Vol. 107, No. ST7, July, 1981.

29. Cornell, C. A., "Engineering Seismic Risk Analysis," Bull. Seis. Soc. of America, Vol. 58, Oct., 1968.

30 Derkiureghian, A. and Ang, A. H-S,. "Fault-Rupture Model for Seismic Risk Analysis," Bull. Seis. Soc. of America, Vol. 67, Ang. 1977.

31. Haldar, A. and Tang, W. H., "Statistical Study of Uniform Cycles in Earthquakes," J. Geotechnical Engrg. Div., ASCE, Vol. 107, No. GT5, May, 1981.

32. ASTM Standard D1586-67," Penetration Test and Split-Barrell Sampling of Soils,".

33. Kovacs, W. D., Evans, J. C., and Griffith, A. H., "Towards a More Standard SPT," Porceedings, 9th Int. Conf. on Soil Mechanics and Foundation Engrg., Vol. 2, 1977.

34. Tang, W. H., "A Bayesian Evaluation of Information for Foundation Engineering," Proceedings, Int. Conf. on Application of Statistics and Probability to Soil and Structural Engineering, Hong Kong, Sept., 1971.

35. Haldar, A., "Liquefaction Study - A Decision Analysis Framework," J. Geotechnical Engrg. Div., ASCE, Vol. 106, No. GT12, Dec., 1980.

36. Christian, J. T., and Swiger, W. F., "Statistics of Liquefaction and SPT Results," J. Geotechnical Engrg. Div., ASCE, Vol. 101, No. GT11, Nov., 1975.

37. Montgomery, D. C. and Peck, E. A., Introduction to Regression Analysis, John Wiley & Sons, Inc., 1981 (in press).

38. Zadeh, L. A., "Fuzzy Sets," Information and Control, Vol. 8, 1965.

39. Yao, J, T, P., "Damage Assessment of Existing Structures," J. Engrg. Mech, Div., ASCE, Vol. 106, No. EM4, Ang. 1980.

40. Singh, M. P. and Ghafory - Ashtiany, M., "Seismic Stability Evaluation of Earth Structures," Virginia Polytechnique Inst. and State Univ., 1980.

APENDICES

LECTURE

LIQUEFACTION OF SOILS DURING EARTHQUAKES

I.M. Idriss

Woodward-Clyde Consultants
San Francisco, California, U.S.A.

DEFINITION OF TERMS

The following definitions pertaining to cyclic loading conditions are based on slight modifications of those originally proposed by Lee and Seed (1967).

FAILURE: When the induced cyclic strains become excessive.

COMPLETE LIQUEFACTION: When a soil exhibits no resistance to deformation over a wide strain range.

PARTIAL LIQUEFACTION: When a soil exhibits no resistance to deformation over a strain range less than that considered to constitute failure.

INITIAL LIQUEFACTION: When a soil first exhibits any degree of partial liquefaction during cyclic loading.

Ordinarily, the first condition reached is INITIAL LIQUEFACTION, followed by PARTIAL, and COMPLETE liquefaction, with FAILURE being reached at some stage during the partial or complete liquefaction stages.

The following definitions are given by Seed, Arango and Chan (1975):

INITIAL LIQUEFACTION: Denotes a condition where, during the course of cyclic stress applications, the residual pore water pressure on completion of any full stress cycle becomes equal to the applied

J. B. Martins (ed.), Numerical Methods in Geomechanics, 507–530.

confining pressure; the development of initial liquefaction has no implications concerning the magnitude of the deformations which the soil might subsequently undergo; however, it defines a condition which is a useful basis for assessing various possible forms of subsequent soil behavior.

INITIAL LIQUEFACTION WITH LIMITED STRAIN POTENTIAL OR CYCLIC MOBILITY: Denotes a condition in which cyclic stress applications develop a condition of initial liquefaction and subsequent cyclic stress applications cause limited strains to develop either because of the remaining resistance of the soil to deformation or because the soil dilates, the pore pressure drops, and the soil stabilizes under the applied loads.

LIQUEFACTION: Denotes a condition where a soil will undergo continued deformation at a constant low residual stress or with no residual resistance, due to the build-up and maintenance of high pore water pressures which reduce the effective confining pressure to a very low value; pore pressure build-up may due either to static or cyclic stress applications.

The following definitions are given by Seed (1979):

LIQUEFACTION: Denotes a condition where a soil will undergo continued deformation at a constant low residual stress or with low residual resistance due to the build-up and maintenance of high pore water pressures, which reduce the effective confining pressure to a very low value; pore pressure build-up leading to liquefaction may be due either to static or cyclic stress applications and the possibility of its occurence will depend on the void ratio or relative density of a sand and the confining pressure; it may also be caused by a critical hydraulic gradient during an upward flow of water in a sand deposit.

PEAK CYCLIC PORE PRESSURE RATIO OF 100%: Denotes a condition where during the course of cyclic stress applications, the residual pore water pressure on completion of any full stress cycle becomes equal to the applied continuing pressure; the development of a peak cyclic pore pressure ratio of 100% has no implications concerning the magnitude of the deformations that the soil might subsequently undergo: however, it defines a condition that is a useful basis for assessing various possible forms of subsequent soil behavior.

PEAK CYCLIC PORE PRESSURE RATIO OF 100% WITH LIMITED STRAIN POTENTIAL OR "CYCLIC MOBILITY": Denotes a condition in which cyclic stress applications develop a peak cyclic pore pressure ratio of 100% and subsequent cyclic stress applications cause limited strains to develop either because of the remaining resistance of the soil to deformation or because the soil dilates, the pore pressure drops and the soil stabilizes under the applied loads. Cyclic mobility may also be used in a broader sense to describe the cyclic straining that may occur even with pore pressure ratios less than 100% in which case the actual peak value of pore pressure ratio may simply be stated.

FAILURE CRITERION

It follows from the definitions summarized in the previous pages that for each situation, a FAILURE CRITERION must be defined and adopted.

The failure criterion should be set based on: (A) desired performance; and (B) consequences of not achieving this performance level

How stringently the failure criterion is to be developed depends on: (A) the naure of the soil; (B) How this soil behaves under cyclic loading conditions; (C) degree of conservatism desired; and (D) degree of conservatism (explicit and/or implicit) that exists in the other parameters that also enter in the evaluation of liquefaction potential. Aspects (A) and (B) are illustrated in Sheets nº 1 and 2.

Table 1 – Estimated susceptibility of sedimentary deposits to liquefaction during strong seismic shaking

Type of Deposit	General distribution of cohesionless sediments in deposits	Likelihood that cohesionless sediments, when saturated, would be susceptible to liquefaction (by age of deposit)			
		<500 Yr.	Holocene	Pleistocene	Pre-Pleistocene
Continental deposits					
River channel	Locally variable	Very high	High	Low	Very low
Flood plain	Locally variable	High	Moderate	Low	Very low
Alluvial fan and plain	Widespread	Moderate	Low	Low	Very low
Marine terraces and plains	Widespread	-	Low	Very low	Very low
Delta and fan-delta	Widespread	High	Moderate	Low	Very low
Lacustrine and playa	Variable	High	Moderate	Low	Very low
Colluvium	Variable	High	Moderate	Low	Very low
Talus	Widespread	Low	Low	Very low	Very low
Dunes	Widespread	High	Moderate	Low	Very low
Loess	Variable	High	High	High	Unknown
Glacial till	Variable	Low	Low	Very low	Very low
Tuff	Rare	Low	Low	Very low	Very low
Tephra	Widespread	High	High	?	?
Residual soils	Rare	Low	Low	Very low	Very low
Sebka	Locally variable	High	Moderate	Low	Very low
Coastal zone					
Delta	Widespread	Very high	High	Low	Very low
Esturine	Locally variable	High	Moderate	Low	Very low
Beach					
High wave energy	Widespread	Moderate	Low	Very low	Very low
Low wave energy	Widespread	High	Moderate	Low	Very low
Lagoonal	Locally variable	High	Moderate	Low	Very low
Fore shore	Locally variable	High	Moderate	Low	Very low
Artificial					
Uncompacted fill	Variable	Very high	-	-	-
Compacted fill	Variable	Low	-	-	-

From Youd and Perkins (1977)

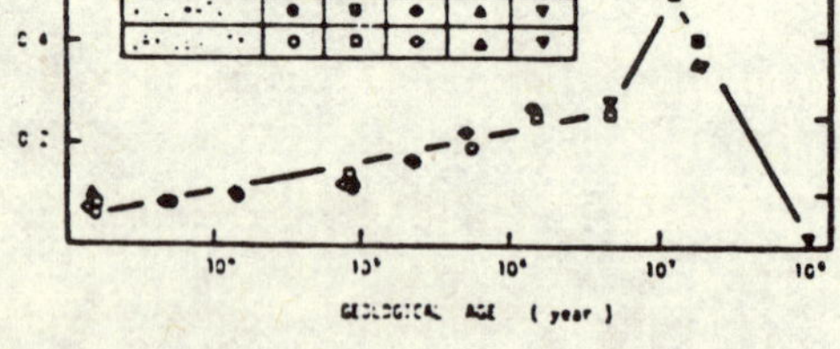

Fig. 7 Effect of Geological Age on the Void Ratio and Dry Density.

FROM TOLNO (1975)

Sheet nº 1

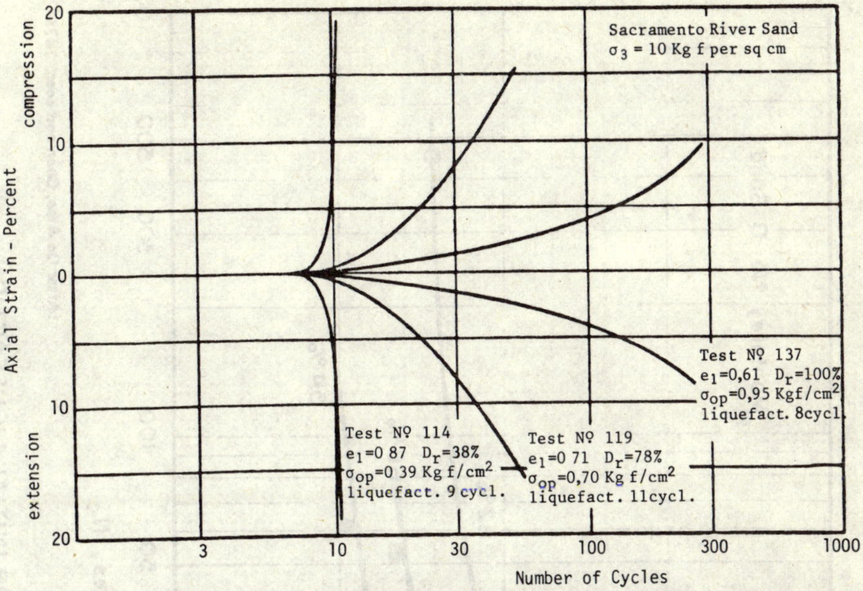

FIG. 9.—AXIAL STRAIN AFTER INITIAL LIQUEFACTION FOR PULSATING LOAD TESTS AT 3 DENSITIES

FROM SEED AND LEE (1966)

FACTORS AFFECTING LIQUEFACTION CHARACTERISTICS OF IN-SITU DEPOSITS

1. MAGNITUDE OF CYCLIC STRESS (I.E., INTENSITY OF SHAKING)
2. NUMBER OF STRESS CYCLES (I.E., DURATION OF SHAKING)
3. INITIAL RELATIVE DENSITY (OR INITIAL VOID RATIO)
4. METHOD OF SOIL PLACEMENT (SOIL STRUCTURE)
5. PERIOD UNDER SUSTAINED LOAD
6. PREVIOUS STRAIN HISTORY
7. INITIAL EFFECTIVE CONFINING PRESSURE
8. LATERAL EARTH PRESSURE COEFFICIENT AND OVER CONSOLIDATION
9. FAILURE CRITERION (INITIAL LIQUEFACTION OR MAXIMUM PERMISSIBLE STRAIN)
10. GRAIN SIZE, GRAIN SHAPE AND GRADATION OF SOIL

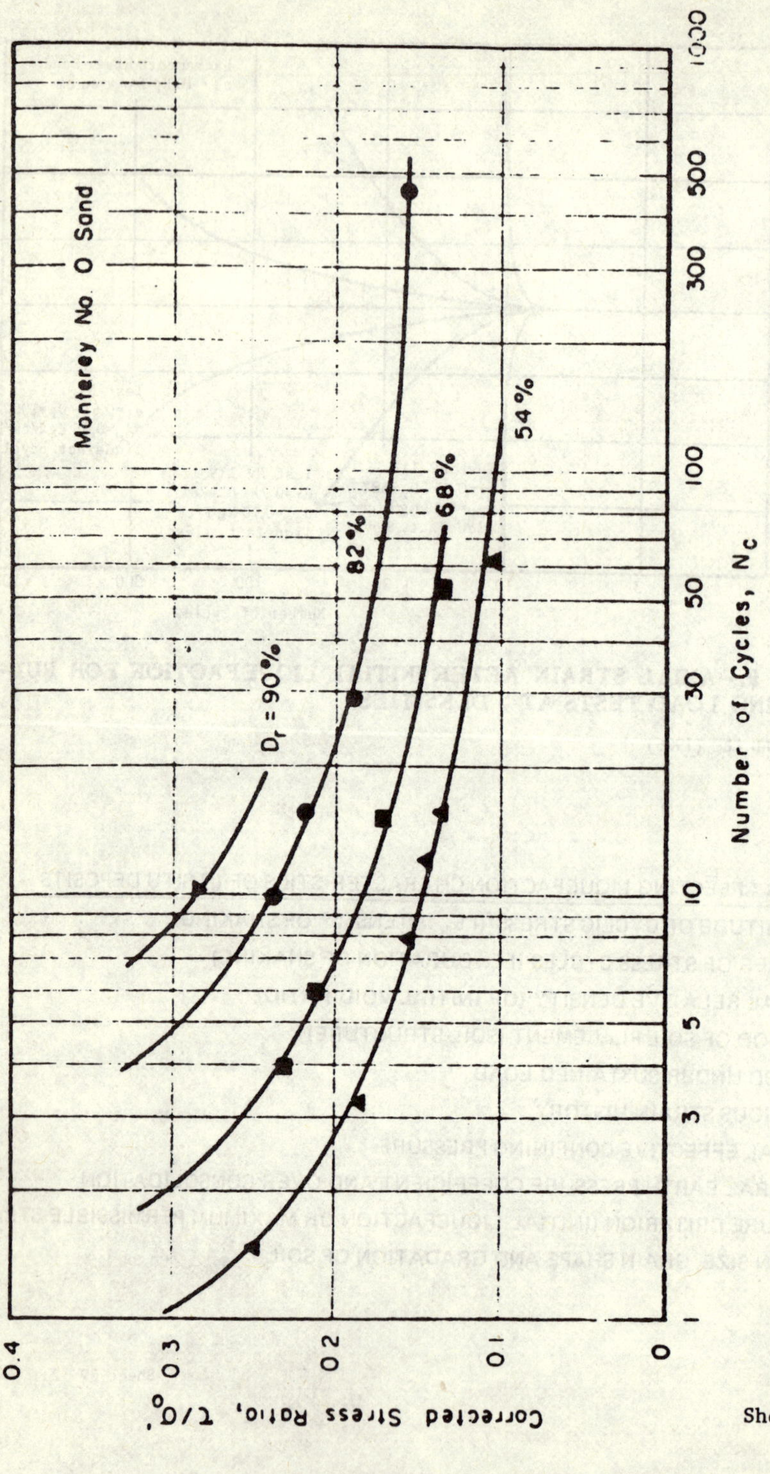

(After De Alba, Chen and Seed, 1975)

Figure 2.3. CORRECTED τ/σ_o VS N_c FOR INITIAL LIQUEFACTION

Sheet nº 3

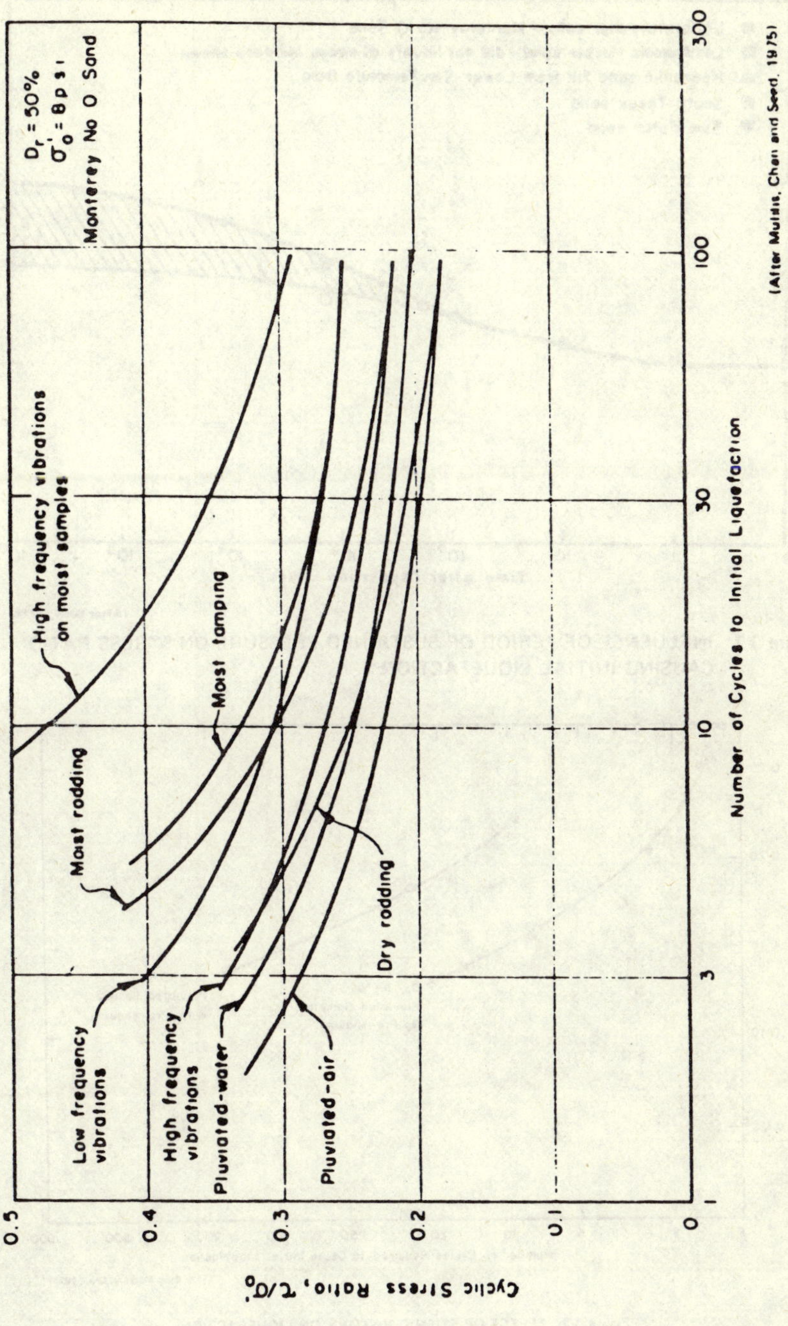

Figure 3-1. CYCLIC STRESS RATIO VS NO. OF CYCLES FOR DIFFERENT COMPACTION PROCEDURES

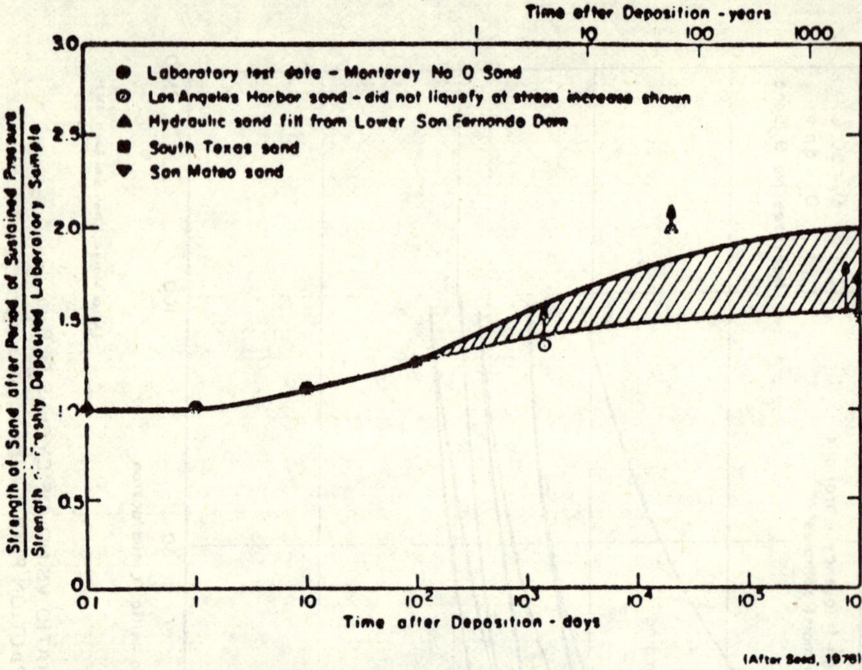

Figure 3-2. INFLUENCE OF PERIOD OF SUSTAINED PRESSURE ON STRESS RATIO CAUSING INITIAL LIQUEFACTION

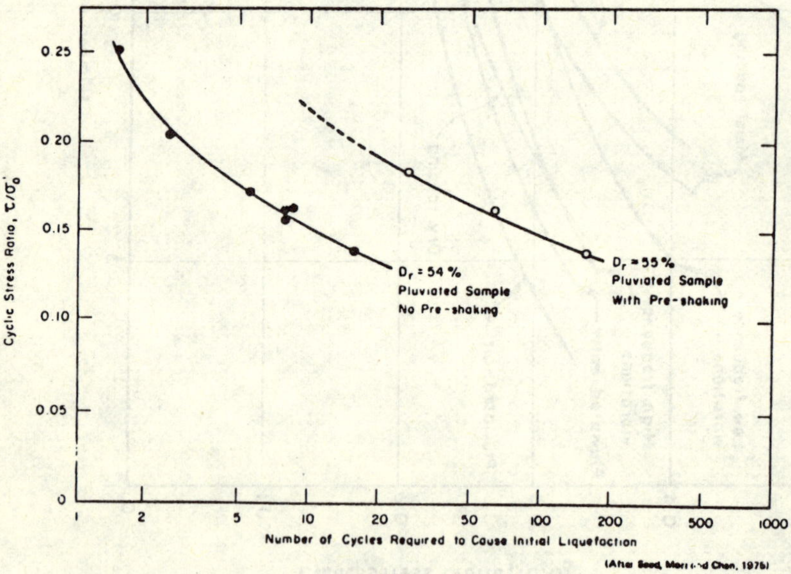

Figure 3-3. EFFECT OF SEISMIC HISTORY ON LIQUEFACTION CHARACTERISTICS OF SAND

Sheet nº 5

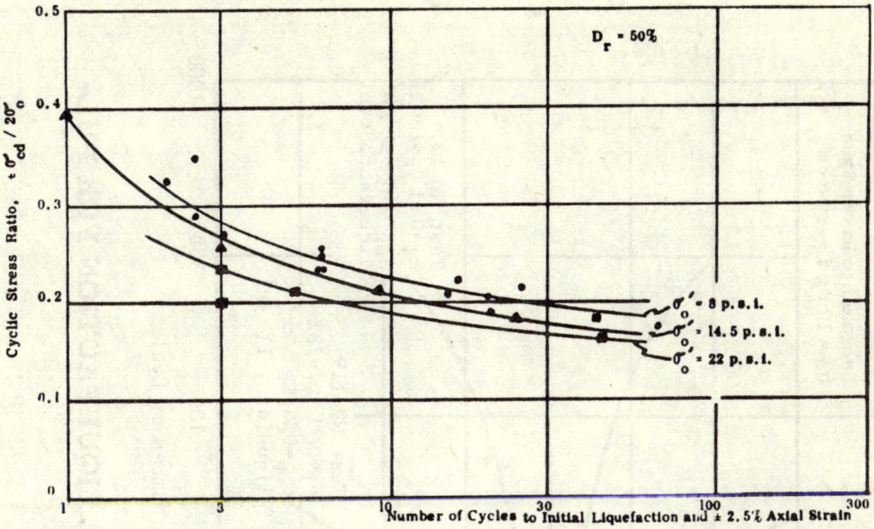

(After Mulilis, Chan and Seed 1975)

EFFECT OF INITIAL EFFECTIVE CONFINING STRESS ON SAMPLES OF MONTEREY NO. 0 SAND FORMED BY PLUVIAL (DRY) COMPACTION

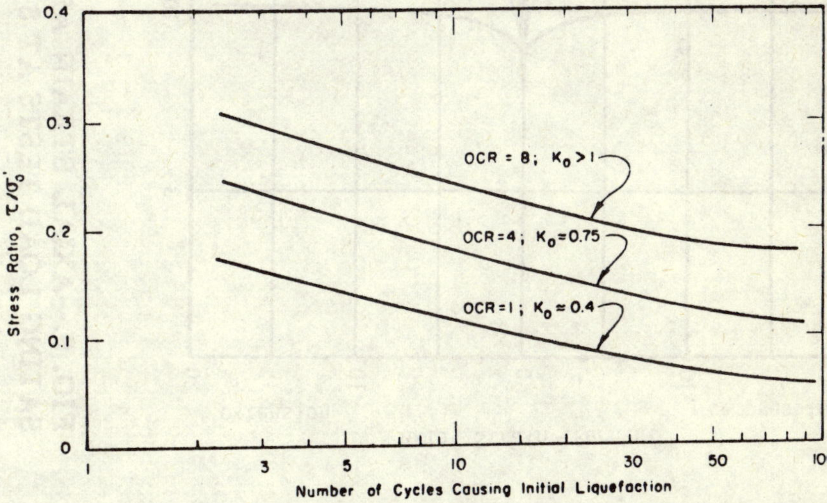

(After Seed and Peacock, 1971)

Figure 3-4. INFLUENCE OF INITIAL PRINCIPAL STRESS RATIO ON STRESSES CAUSING LIQUEFACTION IN SIMPLE SHEAR TESTS

Sheet nº 6

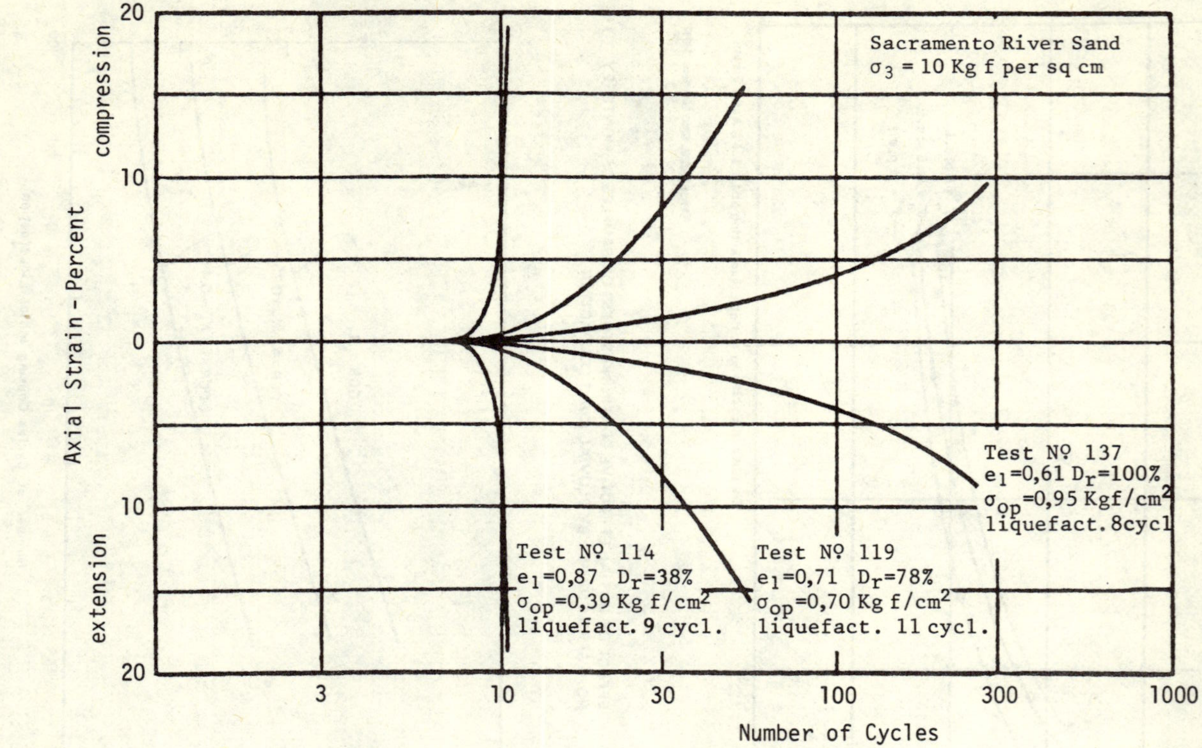

FIG. 9.—AXIAL STRAIN AFTER INITIAL LIQUEFACTION FOR PULSATING LOAD TESTS AT 3 DENSITIES

FROM SEED AND LEE (1966)

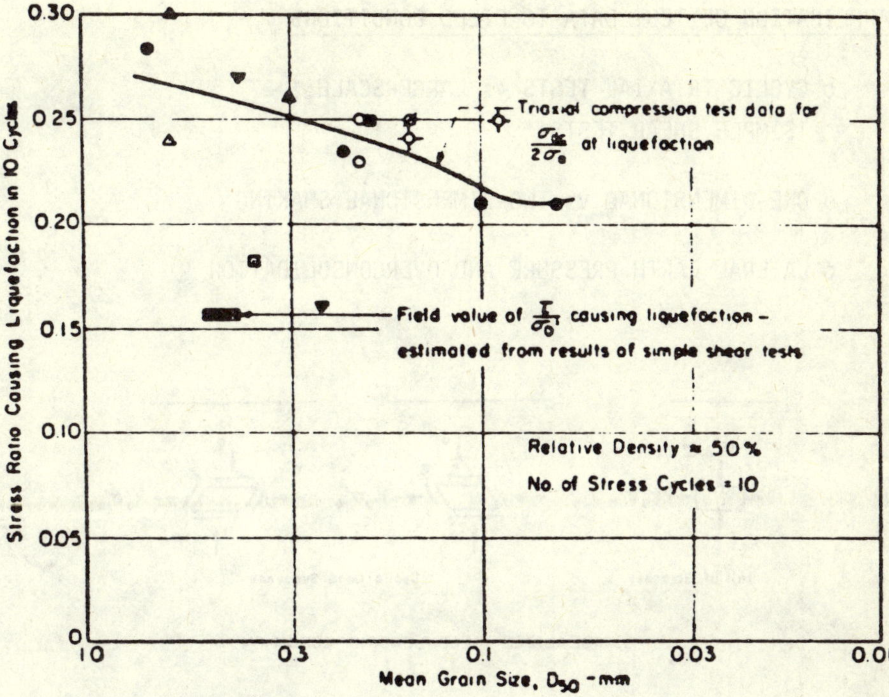

(After Seed and Idriss, 1971b)

STRESS CONDITIONS CAUSING LIQUEFACTION OF SANDS IN 10 CYCLES

<u>LABORATORY TESTS FOR EVALUATING CYCLIC LIQUEFACTION CHARACTERISTICS</u>

CYCLIC LOADING SIMPLE SHEAR TEST
- CYCLIC SIMPLE SHEAR TEST
- LARGE-SCALE SIMPLE SHEAR TEST

TORSIONAL SHEAR TEST

CYCLIC TRIAXIAL COMPRESSION TEST

FAILURE CRITERIA

Sheet n° 8

APPLICATION OF TEST DATA TO FIELD CONDITIONS

- o CYCLIC TRIAXIAL TESTS vs LARGE-SCALE SIMPLE SHEAR TESTS
- o ONE-DIMENSIONAL vs TWO-DIMENSIONAL SHAKING
- o LATERAL EARTH PRESSURE AND OVERCONSOLIDATION

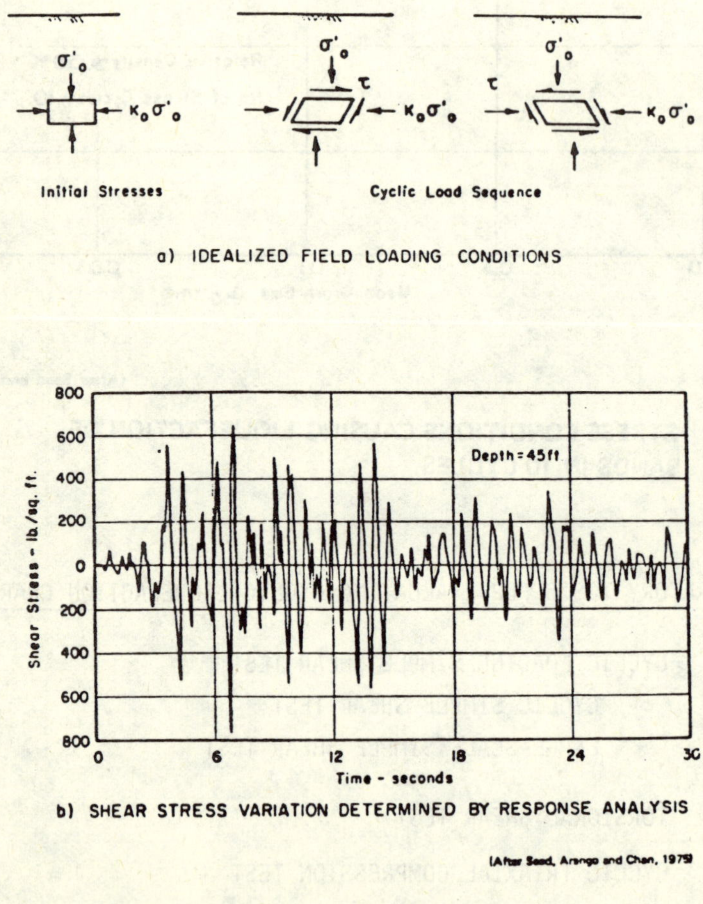

Figure 1-1. CYCLIC SHEAR STRESSES ON A SOIL ELEMENT DURING GROUND SHAKING

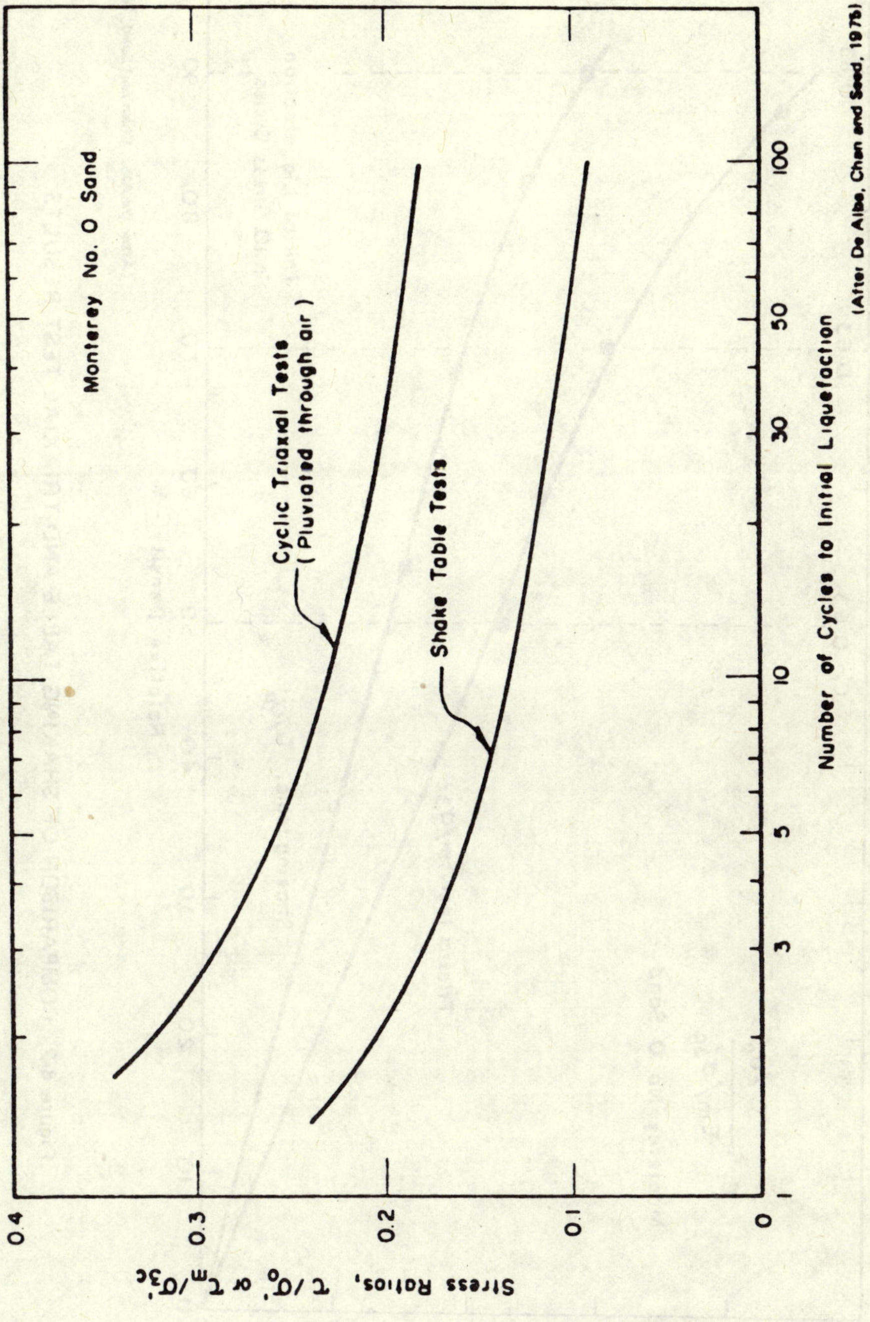

Figure 4-1. COMPARISON OF SHAKING TABLE AND CYCLIC TRIAXIAL TEST RESULTS FOR D_r=50%

Sheet nº 10

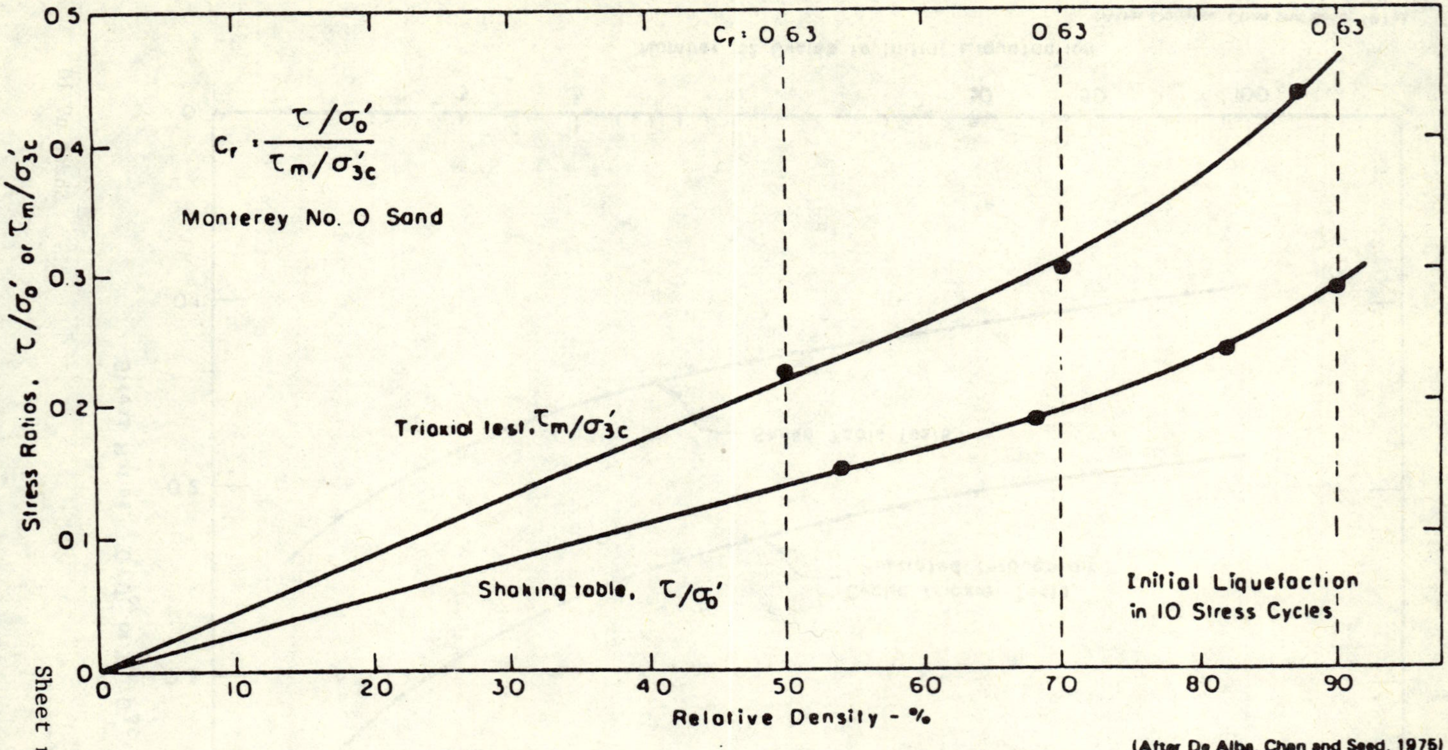

Figure 4-2. COMPARISON OF SHAKING TABLE AND TRIAXIAL TEST RESULTS

Sheet nº 11

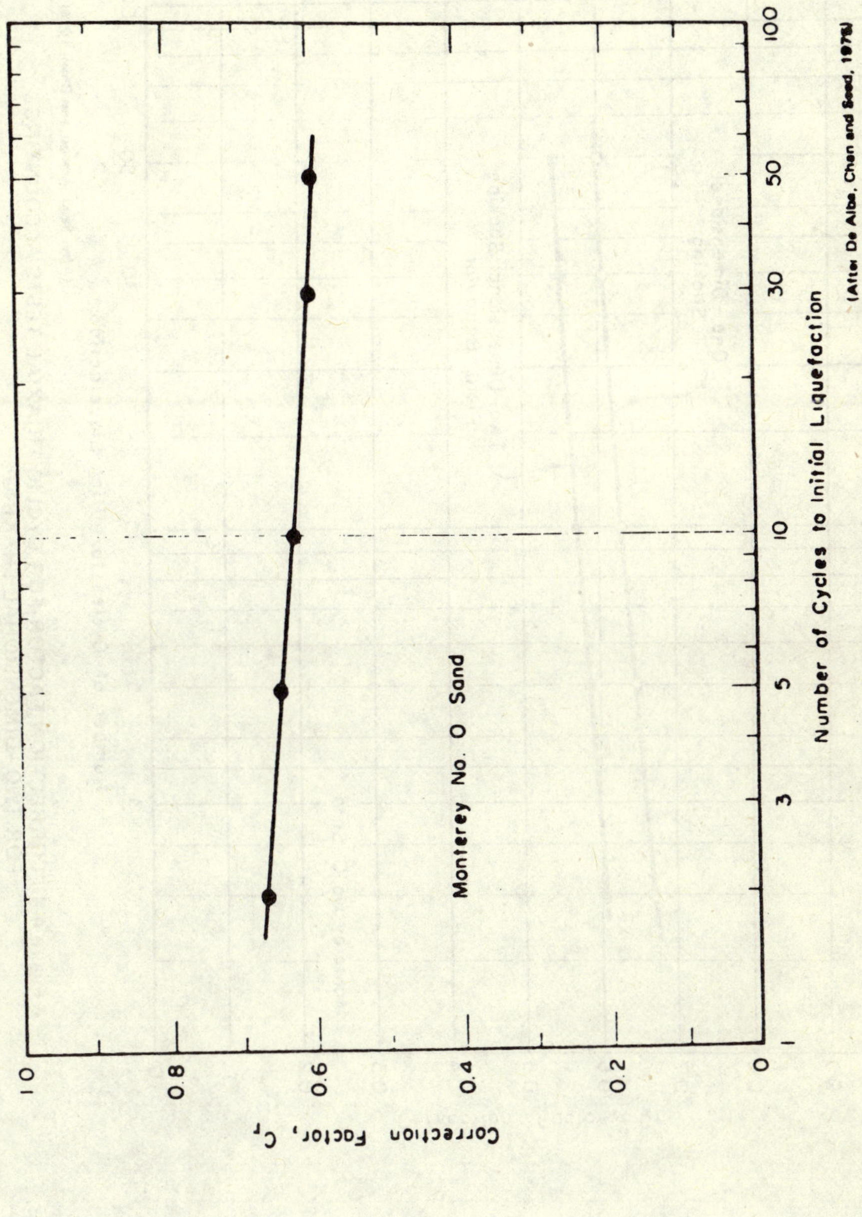

Figure 4-3. CORRECTION FACTORS FOR TRIAXIAL TEST RESULTS

Sheet n°. 12.

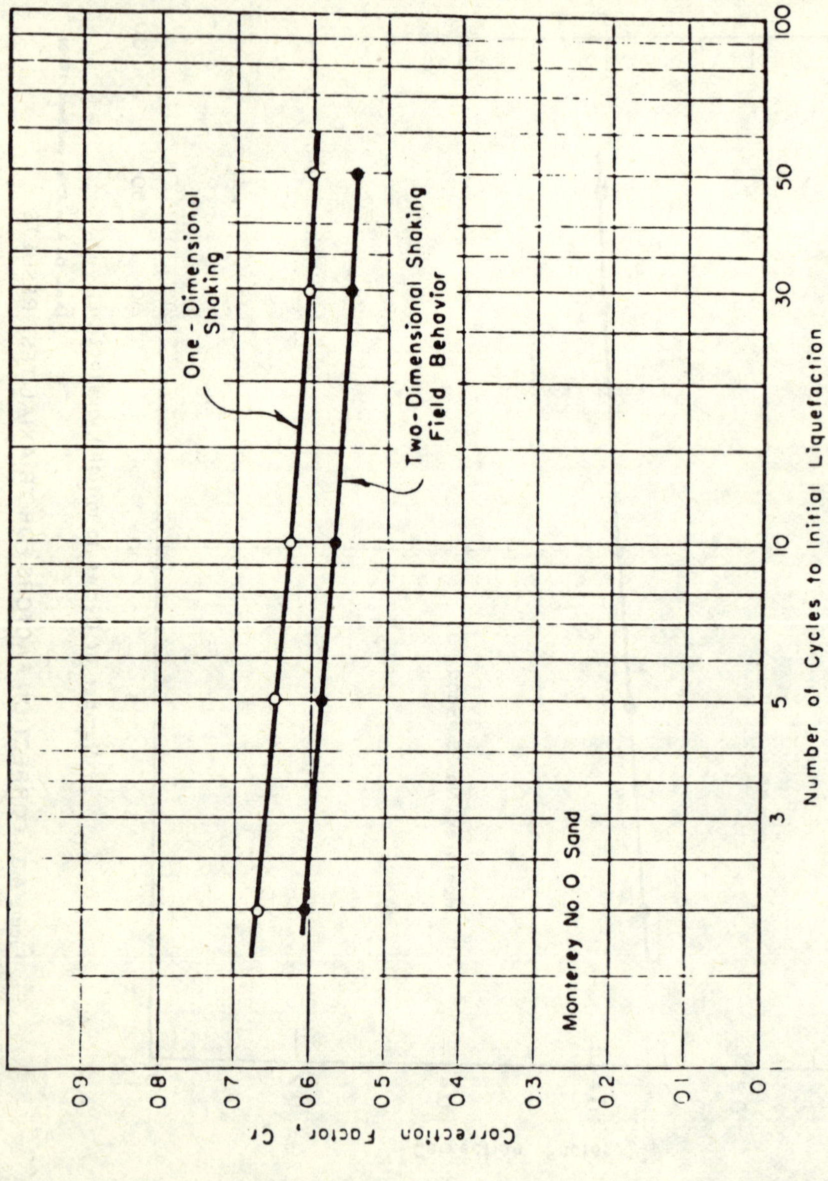

(After Seed, Arango, and Chan, 1975)

Figure 4.4. CORRECTION FACTOR FOR CYCLIC TRIAXIAL TESTS ACCOUNTING FOR TWO-DIMENSIONAL SHAKING

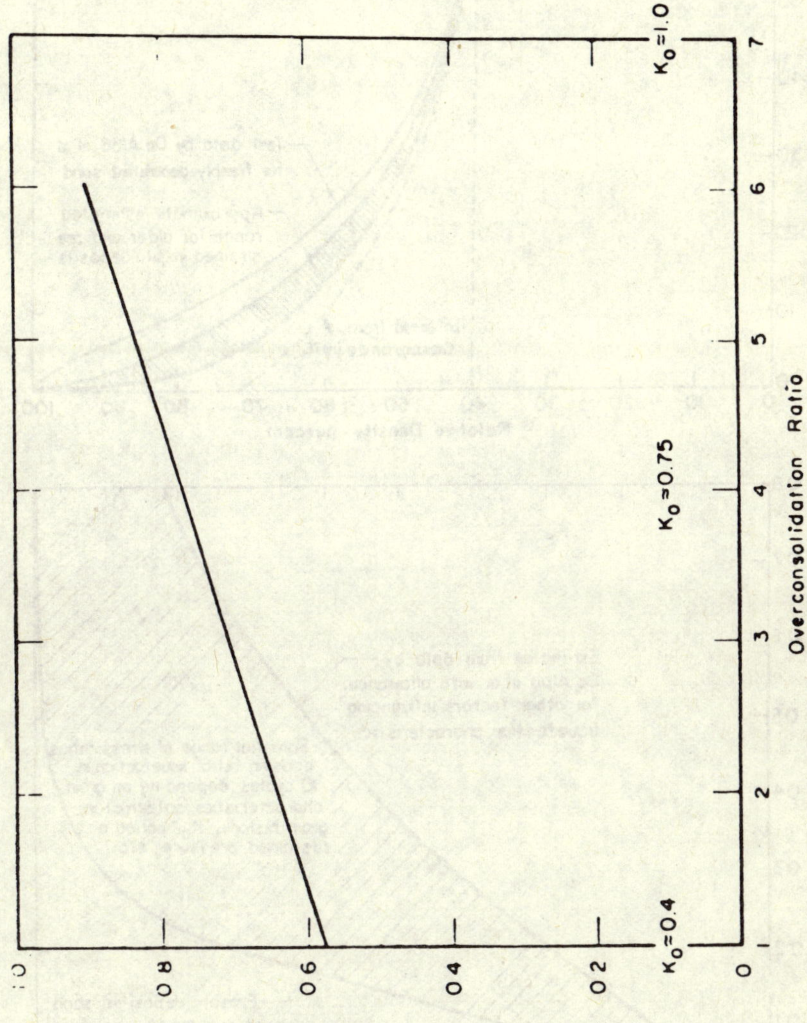

(After Seed, Mori and Chan, 1975)

Figure 4-5. COMBINED CORRECTION FACTOR FOR CYCLIC TRIAXIAL COMPRESSION TESTS ACCOUNTING FOR MULTIDIRECTIONAL SHAKING AND OVER-CONSOLIDATION

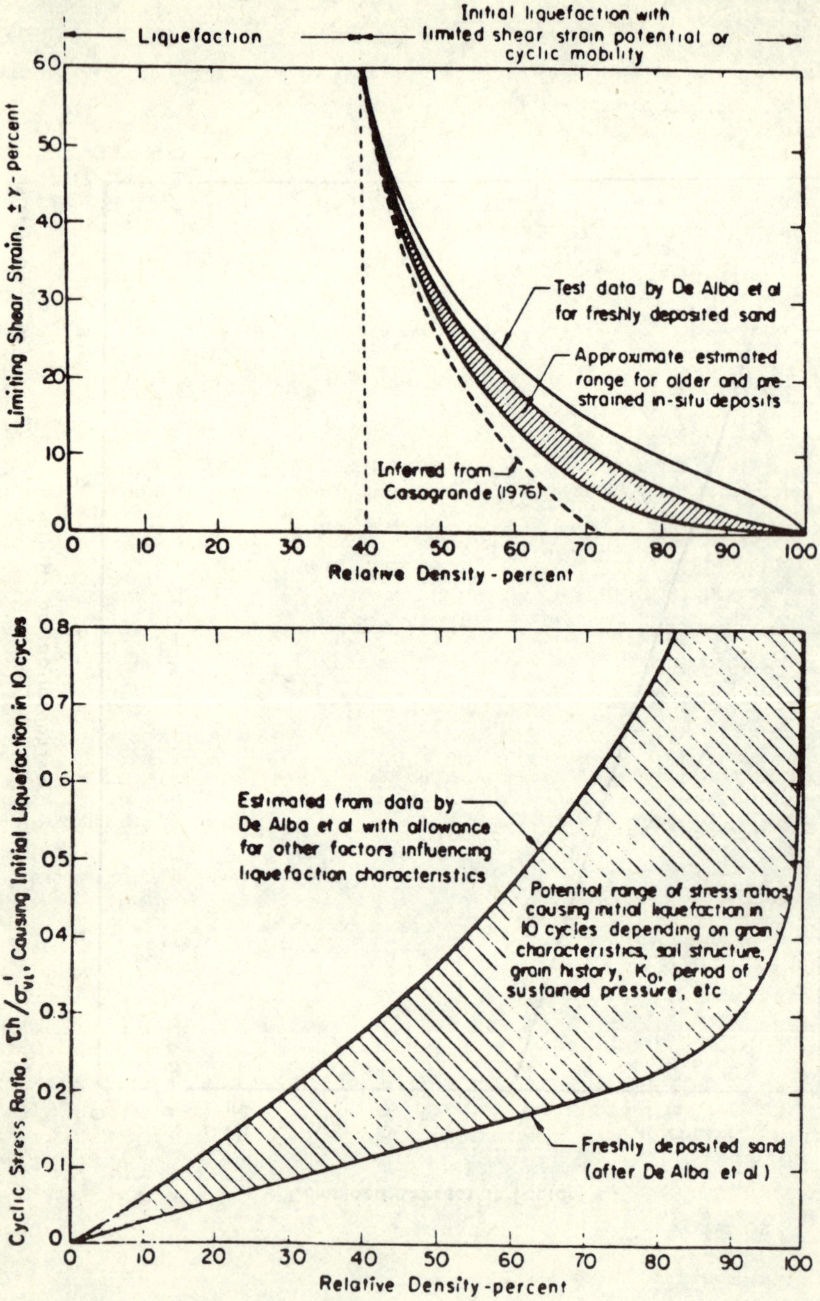

(After Seed, 1976)

Figure 3-5. ESTIMATED RANGE OF LIQUEFACTION CHARACTERISTICS FOR IN-SITU DEPOSITS

Sheet nº 15

METHODS FOR EVALUATING LIQUEFACTION POTENTIAL OF SAND DEPOSITS

The general steps required to evaluate the liquefaction potential in such deposits is outlined in Sheet nº 17. There are two types of methods available for carrying out this evaluation:

1. Methods Based on Observations of Performance of Sand Deposits in Previous Earthquakes

These include both quantitive (e.g., see Sheet nº 1 and quantitive methods. Brief History of evolution is summarized in the paper by Seed (1979) and in the paper by Finn (1981). The quantitative method that is presently most widely used is summarized in Sheet nº 16. It is based on using the SPT blow count N (blows per foot). The measured N is converted to an equivalent N_1 (blow count for an effective overburden of 1 Tsf). See Sheet nº 18 for a chart relating N_1 to N depending on effective overburden pressure and relative density. The value of induced shear stress is estimated using the following equation (Seed and Idriss, 1971):

$$\tau/\sigma_o' = 0.65 \frac{a_{max}}{g} \frac{\sigma_o}{\sigma_o'} r_d$$

a_{max} = maximum acceleration at ground surface

σ_o = total overburden pressure

σ_o' = effective overburden pressure

r_d = stress reduction factor (see Sheet nº 19)

2. Methods Based on Evaluation of Stress Conditions in Field and Laboratory

Analytical procedures for evaluating liquefaction potential were first proposed by Seed and Idriss (1967) and involve two independent determinations: (a) cyclic stresses induced at different levels in the deposit by the earthquake shaking; and (b) a laboratory investigation to evaluate the cyclic stresses required to develop liquefaction (special attention should be paid in this step to select and adopt a failure criterion for use in the evaluation).

This stress analysis-cyclic resistance approach involves the following five basic steps:

1. Development of suitable analytical techniques for calculating stress induced by the earthquake ground motions.

2. Development of suitable procedures for representing the irregular stress history induced by the earthquake by an equivalent number of uniform stress applications.

3. Development of suitable testing procedures to obtain cyclic resistance.

4. Development of an understanding of all the factors having a significant influence on cyclic resistance.

5. Developmentof an understanding of the effects of sample disturbance on the in-situ cyclic resistance of natural deposits.

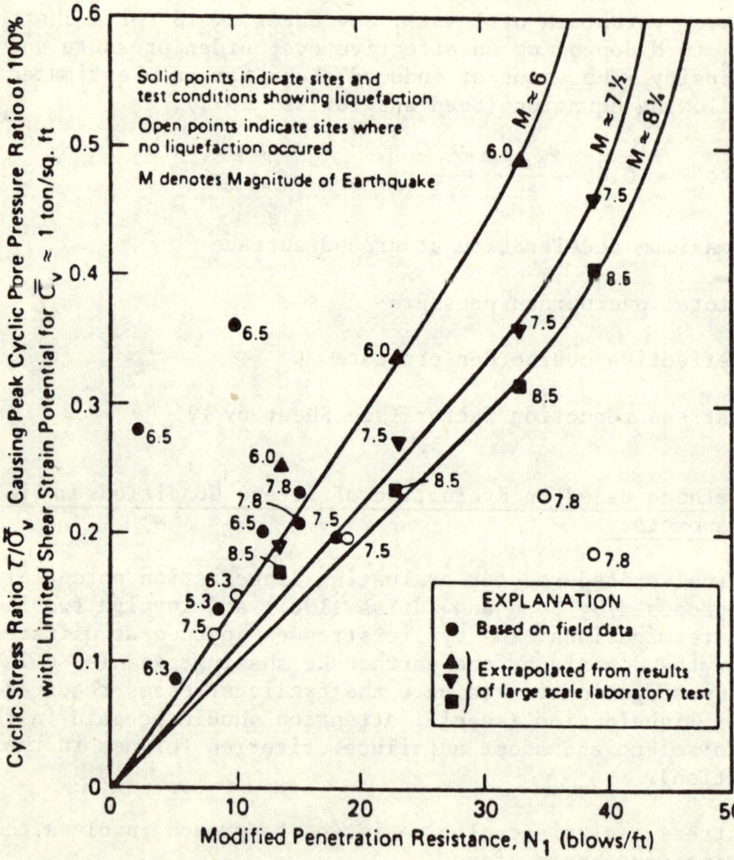

From Seed (1979)

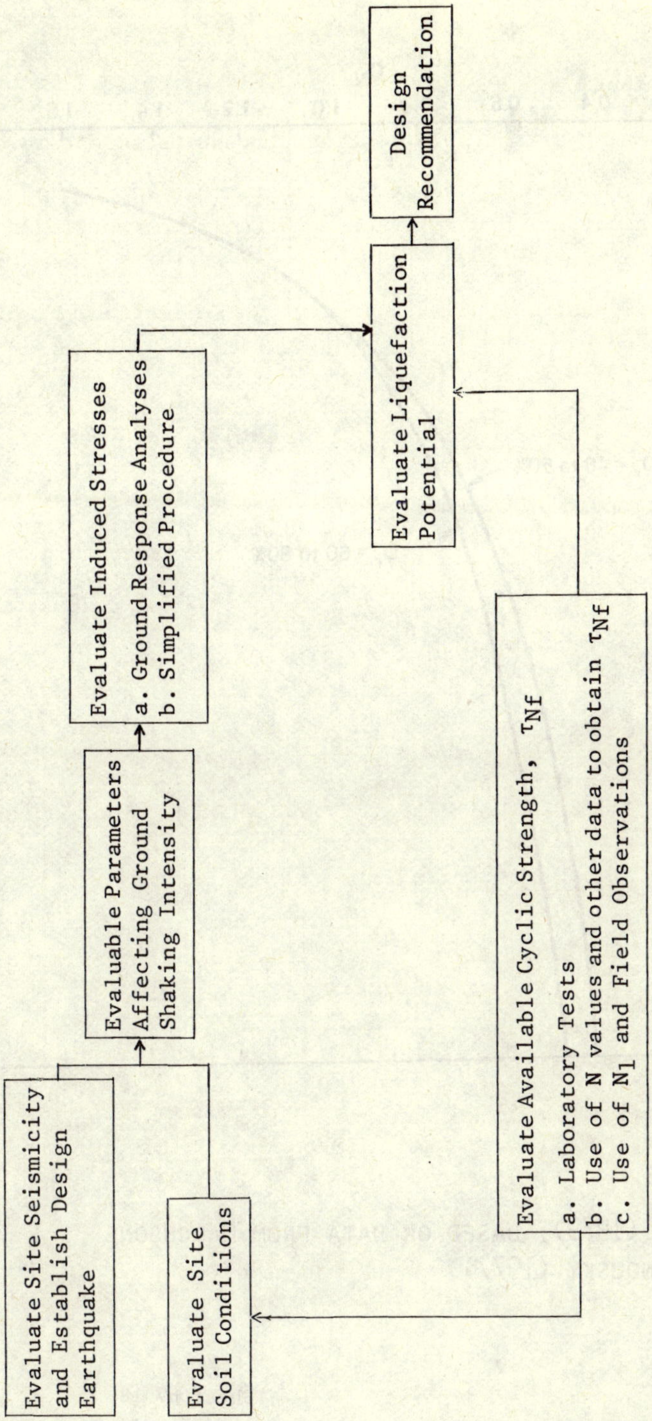

EVALUATION OF LIQUEFACTION POTENTIAL

Sheet nº 17

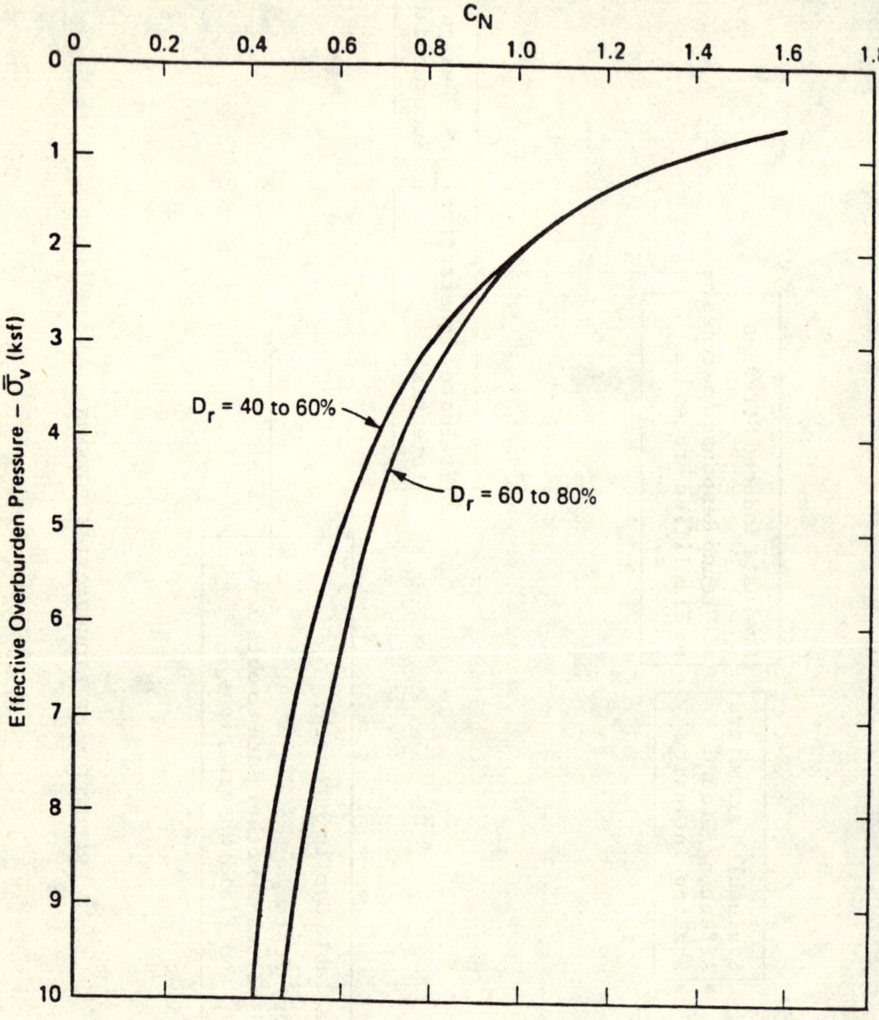

FROM SEED (1979); BASED ON DATA FROM MARCUSON AND BIEGANOUSKY (1977)

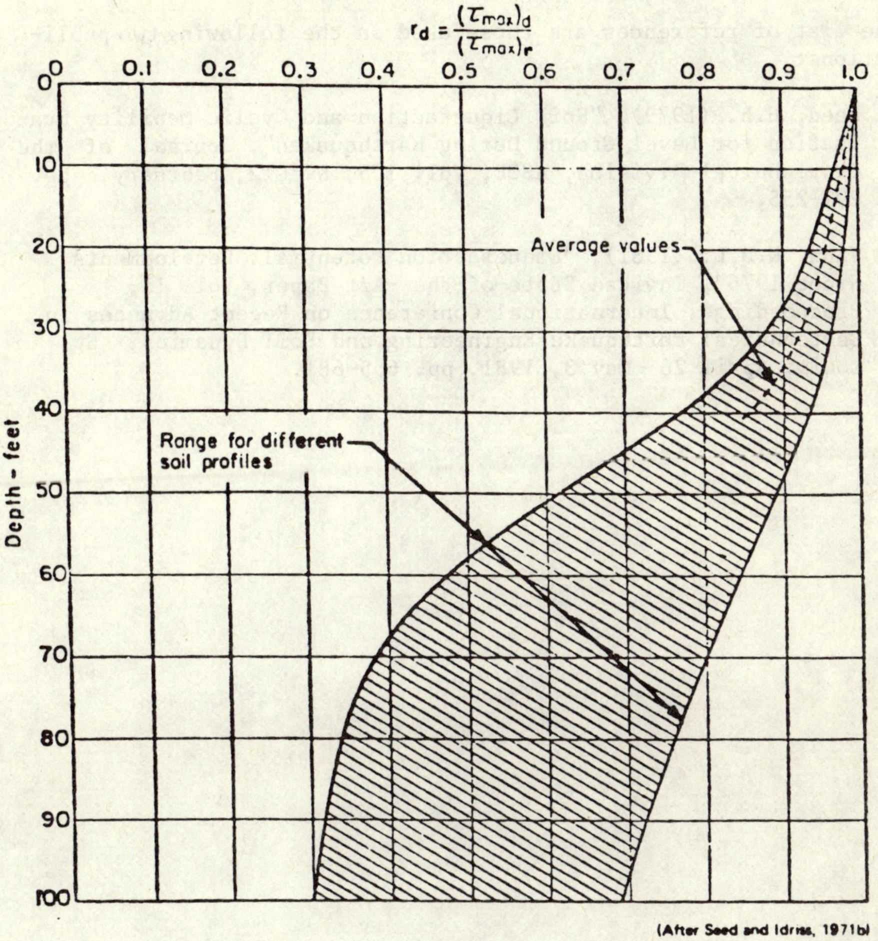

Figure 5-3. RANGE OF VALUES OF r_d FOR DIFFERENT SOIL PROFILES

REFERENCES

The list of references are those used in the following two publications:

1. Seed, H.B. (1979), "Soil Liquefaction and Cyclic Mobility Evaluation for Level Ground During Earthquakes", Journal of the Geotechnical Division, ASCE, Vol. 105, Nº GT2, February, pp. 201-255.

2. Finn, W.D.L. (1981), "Liquefaction Potential: Developments Since 1976", Invited State-of-the -Art Paper, Vol. II, Procceedings, International Conference on Recent Advances in Geotechnical Earthquake Engineering and Soil Dynamics, St. Louis, April 26 - May 3, 1981, pp. 655-681.

RESEARCH PRESENTATION IN PANELS

DYNAMIC, CYCLIC AND TRANSIENT ASPECTS IN SOIL ENGINEERING

M.P. Luong

Lab. de Mécanique des Solides, École Polytechnique, Palaiseau, France

In recent years, ground motion studies have been considerably developed using both constitutive laws for soil dynamics and reduced scale models adequately instrumented in order to analyse ground shock problems of the free field and structure-soil interaction types, for example the case of earthquake loadings on structures.

Constitutive laws for soil dynamics incorporated in large computer codes appear to be a powerful and versatile tool for describing the behaviour of soil foundations subjected to dynamic loading conditions, particularly in the case of parametric studies. The applicability of the dynamic analysis procedure requires realistic understanding and ready determination of the significant factors of the mechanical performance of soil under dynamic, cyclic and transient loading.

It is therefore interesting to discuss the relative advantages or disadvantages of the mathematical models used for dynamic, cyclic and transient computations emphasizing especially:

(a) the ability of the model to fit the laboratory data from a variety of tests;

(b) the theoretical considerations such as uniqueness;

(c) the case of the determination of the material parameters from the experimental data;

(d) the facility with which the model can be integrated into computer calculations; and

J. B. Martins (ed.), Numerical Methods in Geomechanics, 531–539.

(e) their possible use for interpretation of reduced scale models, giving thus more confidence and reliability for answering the question: "how to design against failure in soils subjected to dynamic loadings?".

An experimental approach using the conventional axisymmetric triaxial apparatus suggests the introduction of a characteristic threshold CT as follows (Luong, 1980).

A pore pressure increase characterizes the first stage of any undrained triaxial test irrespective of the initial sand density. As the deviator stress $q = \sigma_1 - \sigma_3$ increases, the pore pressure generation rate $\dot{u}$ decreases, passing through zero to become negative, followed by large irreversible axial strains. This pore pressure generation behaviour parallels phases of contractancy and dilatancy in a drained test.

The stress level corresponding to either the inversion of pore pressure generation rate in an undrained test, or the zero of the volume change rate in a drained one, defines the characteristic threshold CT unambiguously: the granular material is in a "characteristic state" having the following properties:

i - the irreversible volume change rate is zero;

ii - the stress level reached by the material is an intrinsic parameter which defines a characteristic friction angle ϕ_c determining the interlocking capacity of the grain assembly.

The value of ϕ_c is independant of the initial sand density. At any point on the characteristic line CL hereby determined, the rate of the irreversible volume change is strictly zero; this is verified by a cyclic radial loading test along this line, where the obliquity of stresses mobilized by the loading is constant and equals the characteristic value η_c of the material given by the relation

$$\phi_c = \sin^{-1} \frac{3\eta_c}{\eta_c + 6} .$$

The position of the effective loading point relative to the CL line determines its mechanical behaviour. In particular, it can be shown that the material undergoes large deformations after crossing the threshold CT into the surcharacteristic domain.

The characteristic threshold divides the allowable stress space into two regions: (1) subcharacteristic region corresponding to an interlocking of grain structure or contractancy, (2) surcharacteristic region where disaggregation of granular material or dilantacy occurs. Thus a closed load cycle in the subcharacteristic domain exhibits a contracting soil behaviour illustrated by an irreversible volume contraction (or an irreversible increase of pore water pressure)whereas a closed load cycle in the surcharacteristic domain leads to an irreversible volume dilation (or an irreversible decrease of pore water pressure).

A very accurate experimental procedure of determination of the characteristic theshold η_c follows which is readily available under either drained or undrained conditions: η_c is revealed by the appearance of a dilatancy loop (volume change or pore water pressure) during a load cycle crossing the characteristic line CL.

Under either undrained or constant volume conditions, the subcharacteristic region includes all possible effective loading points. As soon as this loading reaches and crosses the characteristic line, it tends to parallel the CL line in the surcharacteristic domain. The length of the section followed determines the degree of memory loss of preceding loading history, reloading being related to a new initial state.

Extensive laboratory tests using the axisymmetric triaxial apparatus on various sand: Fontainebleau sand, Loire sand, carbonate Channel sand (Luong, 1980 - 1981), carbonate marine sediments (Nauroy & Le Tirant, 1981) and Hostun sand (Thanapoulos, 1980) substantiate the following rheological properties:

- Under undrained conditions and cyclic loading:

(1) Sand liquefaction occurs only when load is cycled alternately on both sides of zero deviatoric stress and has reached the characteristic levels. The characteristic friction angle ϕ_c represents the average mobilized angle of interparticle friction.

(2) Cyclic non-alternated deviatoric stress tests show a progressive tendency of the effective stress state to move toward the characteristic level and to stabilize there, i.e. cylic softening is occuring.

(3) Cyclic hardening of sandy soils may be observed when undrained loads are cycled in the surcharacteristic region bounded by the failure line FL and the characteristic line CL. It leads to a stabilization of the granular material on the characteristic threshold. Irreversible strains accumulated during undrained loadings depend on the stress amplitude of cycles.

- Under drained conditions:

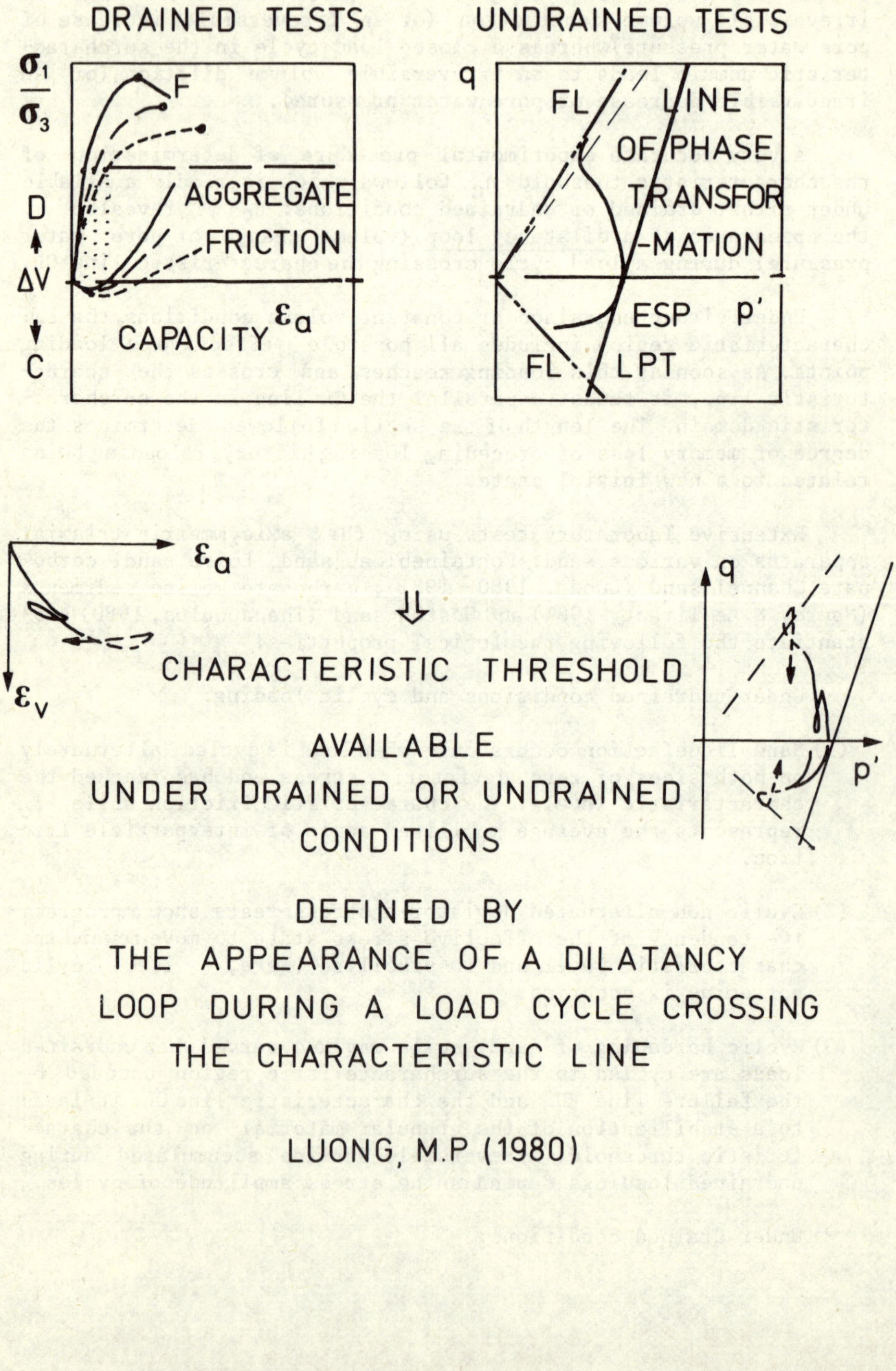
DRAINED TESTS
UNDRAINED TESTS
σ_1 / σ_3
F
AGGREGATE
FRICTION
D
ΔV
C
CAPACITY
ε_a
q
FL
LINE
OF PHASE
TRANSFOR
-MATION
ESP
p'
FL
LPT
ε_a
ε_v
⇓
q
p'
CHARACTERISTIC THRESHOLD
AVAILABLE
UNDER DRAINED OR UNDRAINED
CONDITIONS
DEFINED BY
THE APPEARANCE OF A DILATANCY
LOOP DURING A LOAD CYCLE CROSSING
THE CHARACTERISTIC LINE
LUONG, M.P (1980)

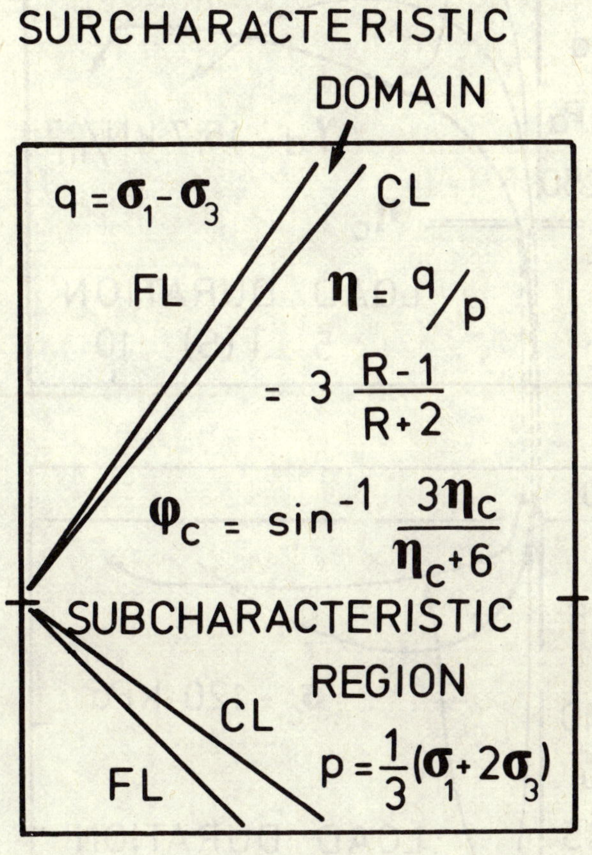
SURCHARACTERISTIC
DOMAIN
q = σ1 - σ3
CL
FL
η = q/p
= 3 (R-1)/(R+2)
φc = sin^-1 3ηc/(ηc+6)
SUBCHARACTERISTIC
REGION
CL
FL
p = 1/3(σ1 + 2σ3)

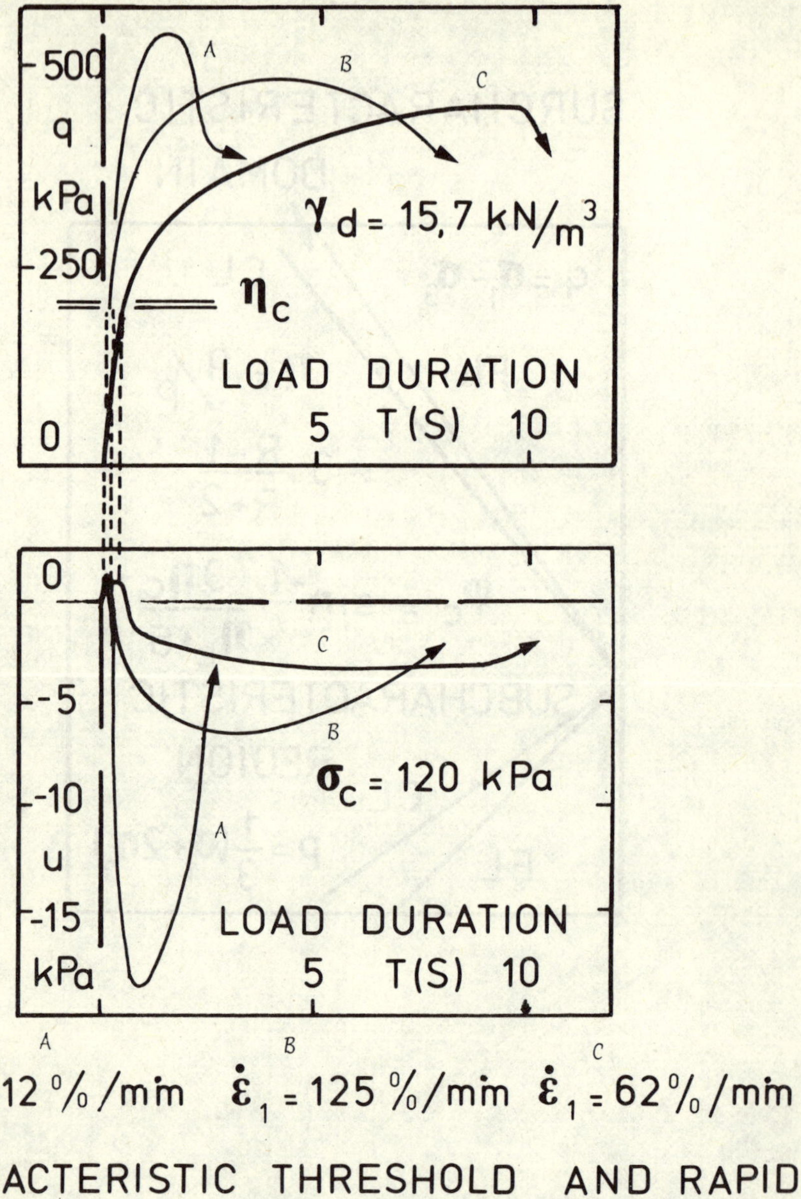

CHARACTERISTIC THRESHOLD AND RAPID LOADINGS WITH PARTIAL DRAINAGE

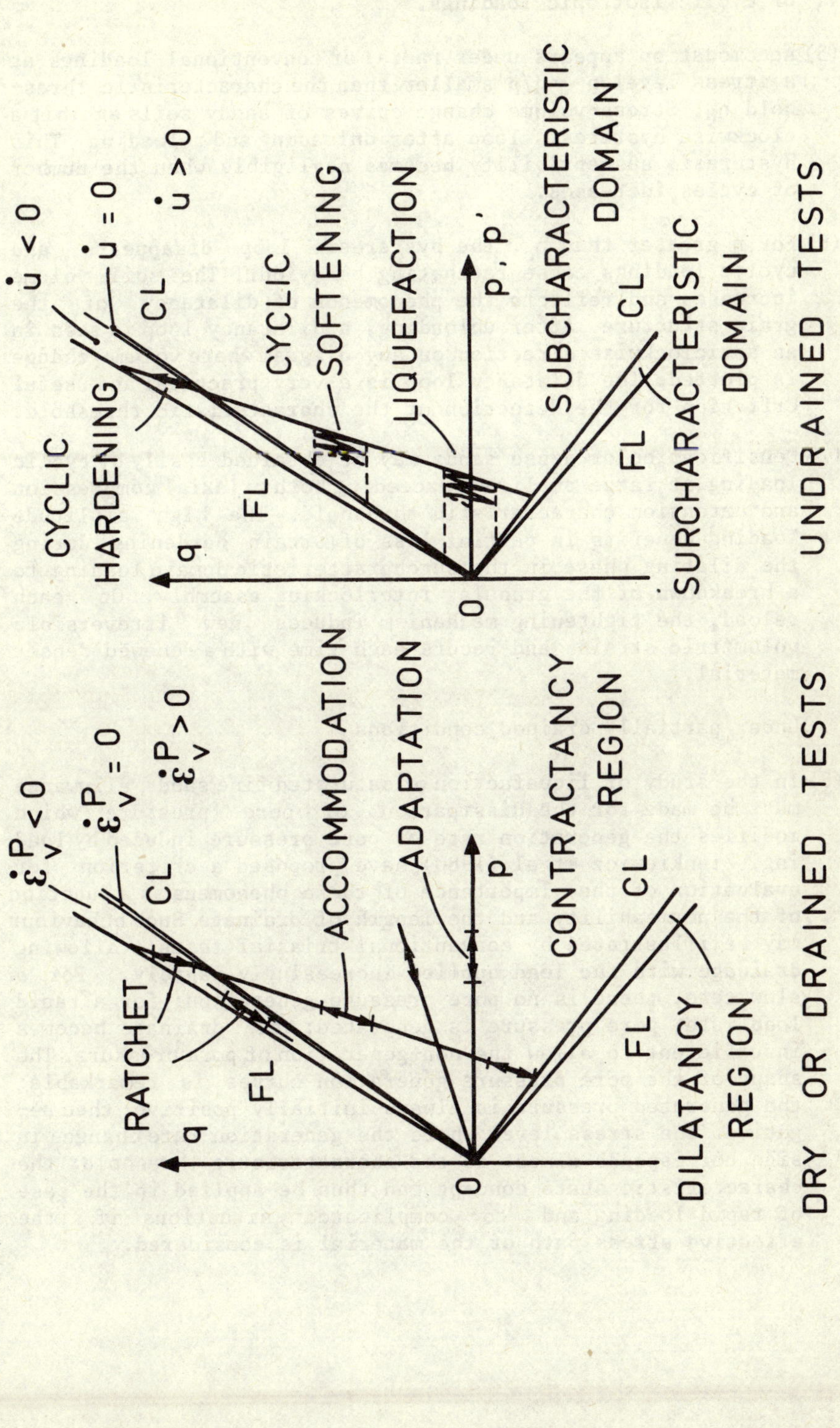

q
RATCHET
FL
CL
$\dot{\varepsilon}^P_v < 0$
$\dot{\varepsilon}^P_v = 0$
$\dot{\varepsilon}^P_v > 0$
ACCOMMODATION
ADAPTATION
p
CONTRACTANCY REGION
FL CL
DILATANCY REGION
0
DRY OR DRAINED TESTS
q
CYCLIC HARDENING
FL
CL
$\dot{u} < 0$
$\dot{u} = 0$
$\dot{u} > 0$
CYCLIC SOFTENING
LIQUEFACTION
p, p'
SUBCHARACTERISTIC DOMAIN
FL CL
SURCHARACTERISTIC DOMAIN
0
UNDRAINED TESTS

(4) Adaptation may be considered as obtained after a finite number of cyclic isotropic loadings.

(5) Accomodation appears under radial or conventional loadings at a stress level $\eta = q/p$ smaller than the characteristic threshold η_c. Stress-volume change curves of sandy soils exhibit a clockwise hysteresis loop after unloading and reloading. This hysteresis susceptibility becomes negligible when the number of cycles increases.

(6) For η greater than η_c, the hysteresis loop disappears and cyclic loadings cause ratcheting behaviour. The soil volume increases and reflects the phenomenon of dilatancy of the grain structure. After unloading, a dilatancy loop is seen in an anticlockwise direction on any diagram where volume change is plotted. The dilatancy loop is a very practical and useful criterion for the detection of the characteristic threshold.

(7) Densification of dense sands may be obtained easily by cyclic loading at large amplitude exceeding both triaxial compression and extension characteristic threshold. The high amplitude loading benefits in partial loss of strain hardening during the dilating phase in the surcharacteristic domain leading to a breakdown of the granular interlocking assembly. On each reload, the tightening mechanism induces new irreversible volumetric strains and recurs each time with a renewed denser material.

Under partially drained conditions

(8) In the study of liquefaction of saturated fine sands, allowance must be made for the dissipation of pore pressure which modifies the generation rate of pore pressure induced by loading. Zienkiewicz et al (1980) have proposed a criterion for evaluation of the importance of these phenomena as a function of the permeability and the length of drainage. Such behaviour may be illustrated by conventional triaxial tests allowing drainage with the load applied increasingly rapidly. For a slow rate, there is no pore pressure generation; for a rapid load rate, pore pressure is generated; the drainage becomes insufficient to allow the non-generation of pore pressure. The shape of the pore pressure generation curves is remarkable: the generated pressure is always initially positive, then negative. The stress level where the generation rate changes in sign corresponds almost to the characteristic threshold: the characteristic state concept can thus be applied in the case of rapid loading and to complicated situations if the effective stress path of the material is considered.

Concluding Remark

The salient features of granular soil behaviour under cyclic loading studied utilizing the conventional triaxial apparatus are readily interpreted within the framework of the characteristic state.

REFERENCES

1. LUONG M.P., (1980), "Stress-strain aspects of cohesionless soils under cyclic and transient loading", Int. Symp. Swansea, Jan. 1980, U.K.

2. NAUROY J.F. & LE TIRANT P., (1981), "Comportement des sédiments marins carbonatés", Xth I C S M F E, Stockholm, June 1981.

3. ZIENKIEWICZ O.C., CHANG C.T., HINTON E. & LEUNG K.H. , (1980) , "Effective stress dynamic modelling for soil structures including drainage and liquefaction", in Soils under Cyclic and Transient Loading, Swansea, Jan. 1980, U.K.

4. THANAPOULOS I., (1980), "Etude théorique et expérimentale des sollicitations cycliques sur sable", Journée de Rhéologie 1980, ENTPE, Vaulx-en-Velin, France.

5. LUONG M.P. & SIDANER J.F., (1981), "Undrained behaviour of cohesionless soils under cyclic and transient loading", Int. Conf. on Recent Advances in Geotech. Earthquake Eng. and Soil Dynamics, St Louis, USA, April 1981.

6. HABIB P. & LUONG M.P., (1978), "Sols pulvérulents sous chargements cycliques", Séminaire Matériaux et Structures sous Chargement Cyclique, Sept. 1978, Ecole Polytechnique, Palaiseau, France.

BURIED STRUCTURES-SURROUNDING SOIL INTERACTION

M.P. Luong

Lab. de Mécanique des Solides, École Polytechnique,
Palaiseau, France

The behaviour of buried structures in earth enbankments is related to the stiffness properties of the composite system comprising both the buried structure and the surrounding soil taken as a basic structural unit in response to loading imposed on the system.

The technical interest of this complicated aspects of soil structure interaction is stimulated and renewed by an increase in the cost and the importance of more and larger structures with greater fill heights. New information has been recently accumulated through research, analysis, models testing and field experience on larger structures, greater loads, new materials and better installation techniques.

The loads on a buried structure depend on the interaction of structure and the surrounding soil because of the "arching-effect" which is considered as the transfer of load to or away from the buried structure as a result of the difference in stiffness properties of the structure, with its adjacent encompassing material, and the surrounding expanse of soil. The stress distribution around the buried structure is therefore different from that which would exist in free field.

If the underground structure deflects more than the adjacent soil; a load reduction is possible in the pressure transmitted to the structure with a corresponding increase in the pressure carried by the adjacent medium. Conversely, under load, the structure may not deform as much as the adjacent soil, and a decrease in the pressure carried by the adjacent medium results. These two opposite conditions are similar to the active and passive earth pressure

J. B. Martins (ed.), Numerical Methods in Geomechanics, 541–546.

conditions.

In order to provide a quantitative definition of flexible and rigid buried structures, both the properties of the soil and the properties of the structure must be considered.

The buried pipe structure is assumed to be an elastic plane strain thin wall cylinder deeply embedded in a soil medium and subject to two-dimensional loading. For soil and pipe materials having elastic constants (Young's modulus and Poisson's ratio) of E_s, ν_s and E_p, ν_p respectively, the following structural parameters can be defined for the system (Gumbel & Wilson, 1981).

- Pipe stiffnesses

Plane strain modulus $E_p^* = E_p/(1-\nu_p^2)$

Flexural stiffness of pipe ring $S_f = E_p^* I/D^3$

Compression stiffness of pipe ring $S_c = E_p^* A/D$

where I and A are respectively the moment of inertia and cross-sectional area of the wall per unit length of pipe.

- Soil stiffness

Plane strain modulus $E_s^* = E_s/(1-\nu_s^2)$

- Pipe-soil interaction parameter

Flexural stiffness ratio $FSR = \dfrac{E_s^*}{S_f} = \dfrac{1-\nu_p^2}{1-\nu_s^2} \dfrac{E_s}{E_p} \dfrac{D^3}{I}$.

A similar nondimensional term has been recommended by Allgood and Takahashi (1972) for a classification as follows:

Flexible system $FSR > 10^4$
Deformable system $10^4 \geq FSR \geq 10$
Rigid system $FSR < 10$.

Structures in the stiff category will experience negative arching, whereas structures in the flexible category will experience positive arching.

Compression stiffness ratio $CSR = \dfrac{E_s^*}{S_c} = \dfrac{1}{1-\nu_s^2} \dfrac{E_s}{E_p} \dfrac{D}{A}$.

A possible classification based on failure mechanism can be

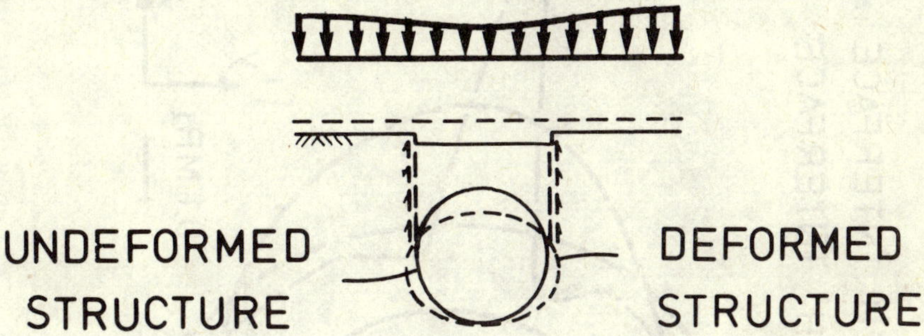

POSITIVE ARCHING EFFECT

NEGATIVE ARCHING EFFECT

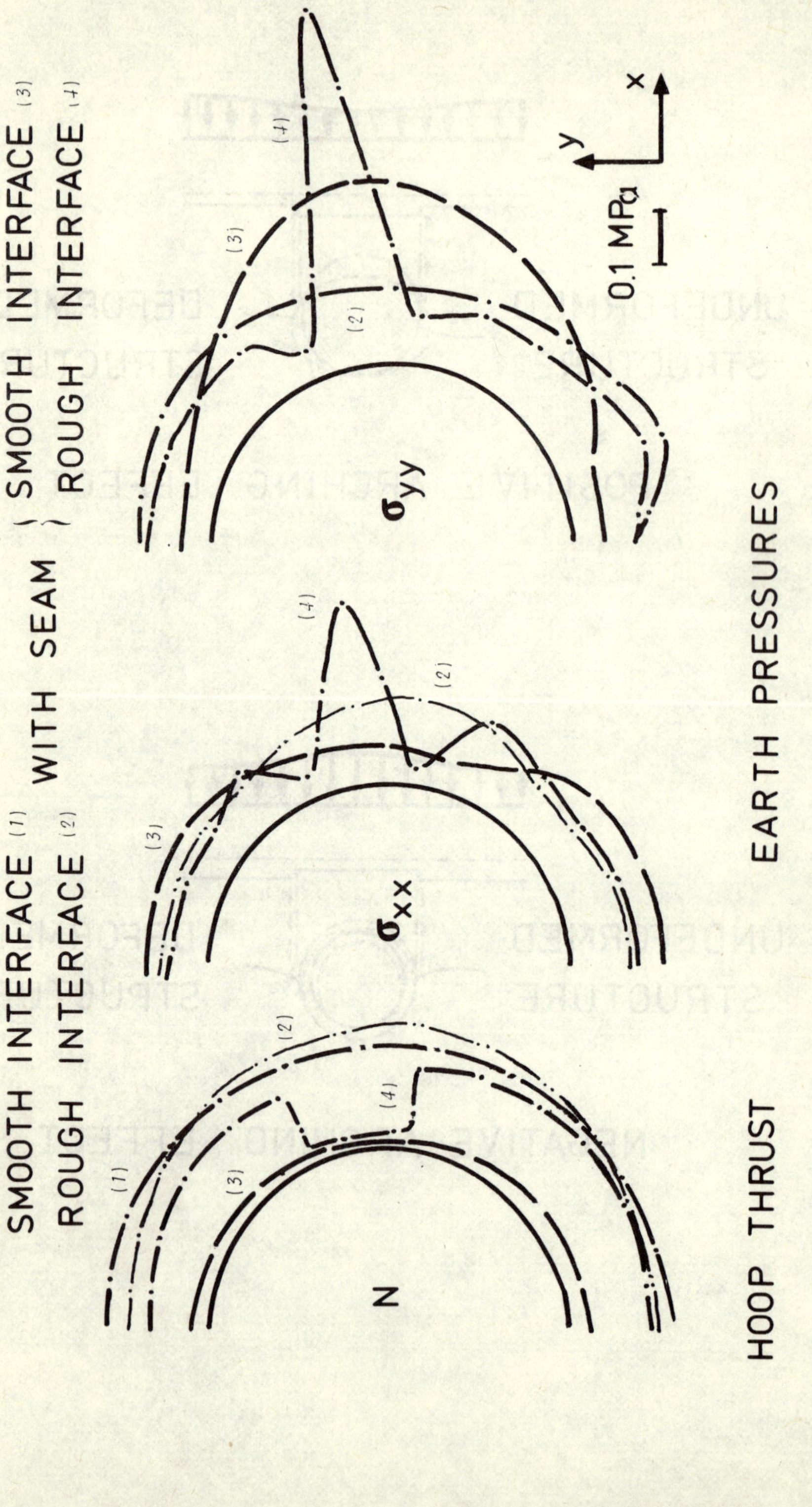
SMOOTH INTERFACE (1)
ROUGH INTERFACE (2)
WITH SEAM
SMOOTH INTERFACE (3)
ROUGH INTERFACE (4)
N
σ_{xx}
σ_{yy}
0.1 MPa
x
y
HOOP THRUST
EARTH PRESSURES

derived from test models:

Extensible system if CSR > 10^{-2}, the buckling occurs by snap-through mechanism;

Inextensible system if CSR < 10^{-2}, the failure results from a buckling by bifurcation mechanism.

For a very thick-walled cylinder ($\frac{D}{A}$ < 20), the wall would be crushed if the ring compression stress reached yield strength. For structures between very thin wall (flexible ring) and very thick wall (rigid ring) structural test results fall within a transition zone with considerable scatter.

Complete definition of the soil-structure system requires specification of the load transfer conditions at the pipe-soil interface. While pipe and soil are assumed to be unbonded in the radial direction (i.e. non-tensile load transfer normal to the pipe wall), tangential load conditions may vary between full slippage, non slippage and non slippage until a prescribed stress has been reached. The upper bound of the prescribed stress would certainly be the shear strength of the soil, and the extent to which the actual srength of the soil, and the extent to which the actual stress at the interface differs from this upper bound would depend on the smoothness of the conduit. In general, it seems quite reasonable to use the shear strength of the weaker soil at the interface between different soil zones.

Research to improve the understanding of soil-structure interaction and flied observations of buried structure performances lead to the conclusion that the fundamental idea behind the use of flexible corrugated metal as conduit wall is to utilize the soil as the principal load bearing and transmitting element of the system. The metal may be viewed as a from that maintains a desired shape of opening in the soil medium and that carries as little load during service as possible.

The finite element method provides the most comprehensive tool available for predicting the detailed performance of buried structures. For example, non linear soil behaviour, bedding details, slippage between the soil and the structure, and any desired geometric shape can be analyzed during construction, service and ultimate conditions.

It also enables the development of new design concepts, for example, a contractable seam recently set up and used in France in order to reduce the excessive value of hoop thrust N and to transfer a significant part of stresses to the surrounding soil.

The numerical results show the importance of smooth interface or rough interface condition on the soil-buried structure behaviour and also on its relative flexibility. In addition, they offer a correct quantitive evaluation of forces in the whole soil-structure system and also of displacements at any point.

An experimental verification by LCPC (Laboratoire Central des Ponts et Chaussées), SETRA (Service d'Etudes Techniques des Routes et Autoroutes) and LMS (Laboratoire de Mécanique des Solides) has been carried out on the site of Dourdan with corrugated metal conduits (Arval) of 2.50 m diameter buried beneath more than 15 m of fill.

REFERENCES

1. **ALLGOOD J.R. & TAKAHASHI S.K.**, (1972), "Balanced design and finite element analysis of culverts". Highway Research Record nº 413 Subsurface Soil-Structure Interaction. Tucson, Arizona.

2. **GUMBEL J.E. & WILSON J.**, (1981), "Interactive design of buried flexible pipes - A fresh approach from basic principles". Ground Engineering, May 1981, vol. 14, nº 4.

3. **LUONG M.P.**, (1981), "Comportement mécanique des ouvrages placés dans les remblais". Proc. Séminaire "Interactions Soil-Structure". CISCO, St Rémy-Lés-Chevreuse, Juin 1981, France.

4. **LUONG M.P.**, (1981), "Déformabilité relative et conditions d'interface sol-structure des buses métalliques enterrées". Discussion, Session nº 5, Sol-Structure Interaction, Proc. X ICMFE, Stockholm 1981, Vol. 4, Juin 1981.

5. **SELIG E.T., ABEL J.F., KULHAWY F.H. & FALBY W.E.**, (1978), "Long-span buried structure design and construction". Proc. ASCE, Jnl of Geot. Eng. Div., vol. 104, nº GT7, July 1978.

BEARING CAPACITY AND SHAPE FACTORS OF SHALLOW FOUNDATION

M.P. Luong

Lab. de Mécanique des Solides, École Polytechnique, Palaiseau, France

The bearing capacity failure of the supporting soils of shallow foundation has been calculated through the theory of yield design developed by Salençon (1978) requiring only the knowledge of the yield criterion and its properties.

The plane-strain and axially symmetric cases are solved for foundations resting over a soil layer of limited or unlimited thickness with a cohesion linearly increasing with depth (Matar & Salençon, 1979 - 1980), (Fig. 1).

A global analysis of the problem leads to the determination of the exact value of the bearing capacity taking into account all the parameters simultaneously. This enables the appreciation of how the bearing capacity value is underestimated through the use of the superposition method by the factor:

$$\mu_c\left(\frac{g+\gamma\,tg\,\phi}{C_o+q\,tg\phi}\,B,\ \frac{B}{h},\ \phi\right)=\frac{q_u-q}{(q_u)_{superp.}-q}$$

which quantifies the under-estimation of the usual method.

The three-entry charts proposed by Matar & Salençon (1979) for several ϕ values allow the interpolation of the values of factors N'_c, $N'_\gamma/\tan\phi$ and μ_c. The exact bearing capacity of strip footing is then given by the following equations.

$$\text{If } \phi>0,\ q_u=q+\mu_c(C_o+q\tan\phi)\left[\frac{1}{2}\,\frac{g+\gamma\tan\phi}{C_o+q\tan\phi}\,\frac{BN'_\gamma}{\tan\phi}+N'_c\right]$$

J. B. Martins (ed.), Numerical Methods in Geomechanics, 547–556.

$$\text{If } \phi = 0, \quad q_u = q + \mu_c C_o \left[\frac{1}{4} \frac{gB}{C_o} + N'_c \right].$$

Similar calculations have been also carried out for the case of circular shallow foundations, axially loaded, within the framework of the theory of yield design. The practical use of the presented in the form of charts is based on the shape factor ν function of the non-dimensional terms of the problem which relates the bearing capacity of the circular footing q_u^o to that of the strip footing q_u with the same width, under the same conditions.

$$\nu \left(\frac{g + \gamma \tan \phi}{C_o + q \tan \phi} B, \frac{B}{h}, \phi\right) = \frac{q_u^o - q}{q_u - q} \ .$$

Several four-entry charts for various values of friction angle ϕ allow the interpolation of factors N'_c, $N'_\gamma/\tan\phi$, μ_c and ν.

$$\text{If } \phi > 0, \quad q_u^o = q + \nu\mu_c (C_o + q \tan \phi) \left[\frac{1}{2} \frac{g + \gamma \tan \phi}{C_o + q \tan \phi} \frac{BN'_\gamma}{\tan \phi} + N'_c \right]$$

$$\text{If } \phi = 0, \quad q_u^o = q + \nu\mu_c C_o \left[\frac{gB}{4C_o} + N'_c \right].$$

The main properties of the shape factor ν are:

(1) For given ϕ and $\frac{g + \gamma \tan \phi}{C_o + q \tan \phi}$, ν is a decreasing function of B/h.
It follows that the shape factor is maximal when the foundation rests on a soil layer of unlimited thickness.

(2) For given ϕ and $\frac{B}{h}$, ν is a decreasing function of $\frac{g + \gamma \tan \phi}{C_o + q \tan \phi} B$.

(3) For $\frac{g + \gamma \tan \phi}{C_o + q \tan \phi} B = 0$, the shape factor ν is the ratio:

$$\nu \left(0, \frac{B}{h}, \phi\right) = N'^o_o \left(\frac{B}{h}, \phi\right) / N'_c \left(\frac{B}{h}, \phi\right).$$

In an unlimited thickness layer $\frac{B}{h} = 0$, we have:

$$\nu \ (0, 0, \phi) = N_c^o \ (\phi) / N_c \ (\phi) = \zeta_c$$

usual shape factor of the cohesion term.

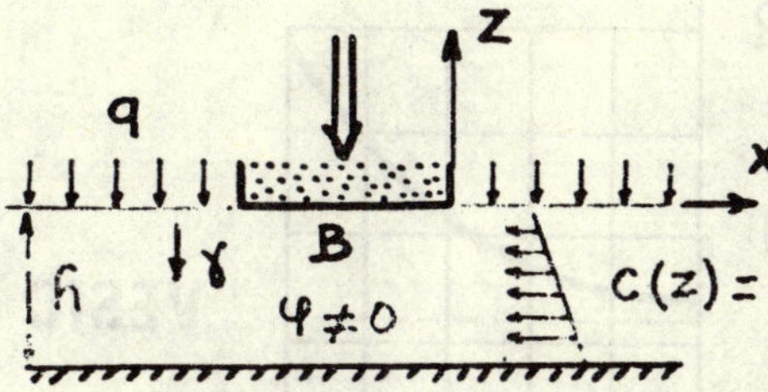

$$q_u \geq (q_u)_{superp.} = \frac{1}{2}\gamma B N'_\gamma\left(\frac{B}{h},\varphi\right) + c N'_c\left(\frac{B}{h},\varphi\right) + q N'_q\left(\frac{B}{h},\varphi\right)$$

$$\mu_c\left(\frac{g+\gamma\tan\varphi}{c_0+q\tan\varphi}B,\ \frac{B}{h},\ \varphi\right) = \frac{q_u - q}{(q_u)_{superp.} - q}$$

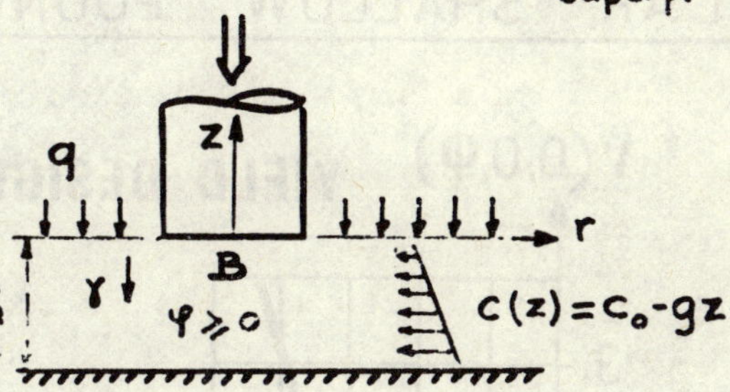

$$\nu\left(\frac{g+\gamma\tan\varphi}{c_0+q\tan\varphi}B,\ \frac{B}{h},\ \varphi\right) = \frac{q_u^0 - q}{q_u - q}$$

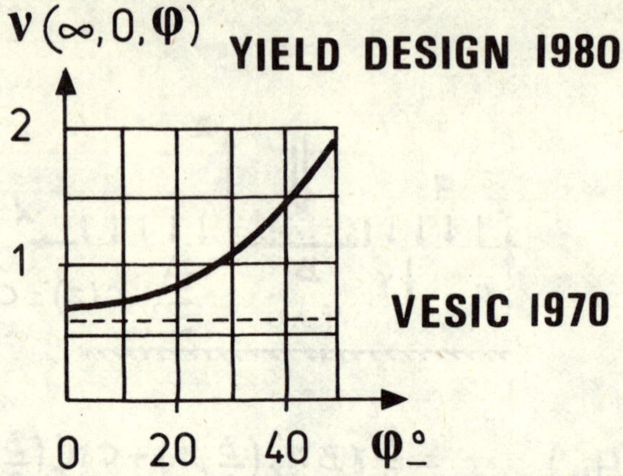

SHAPE FACTOR ζ_γ FOR GRAVITY FORCE

CIRCULAR SHALLOW FOUNDATIONS

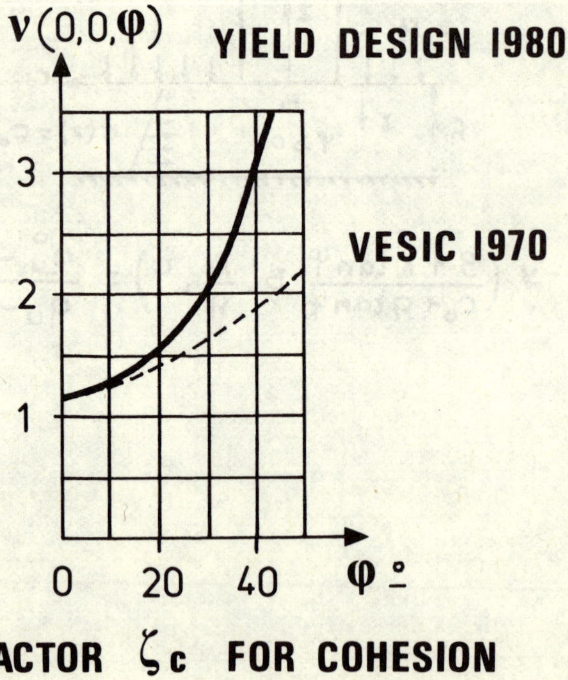

SHAPE FACTOR ζ_c FOR COHESION

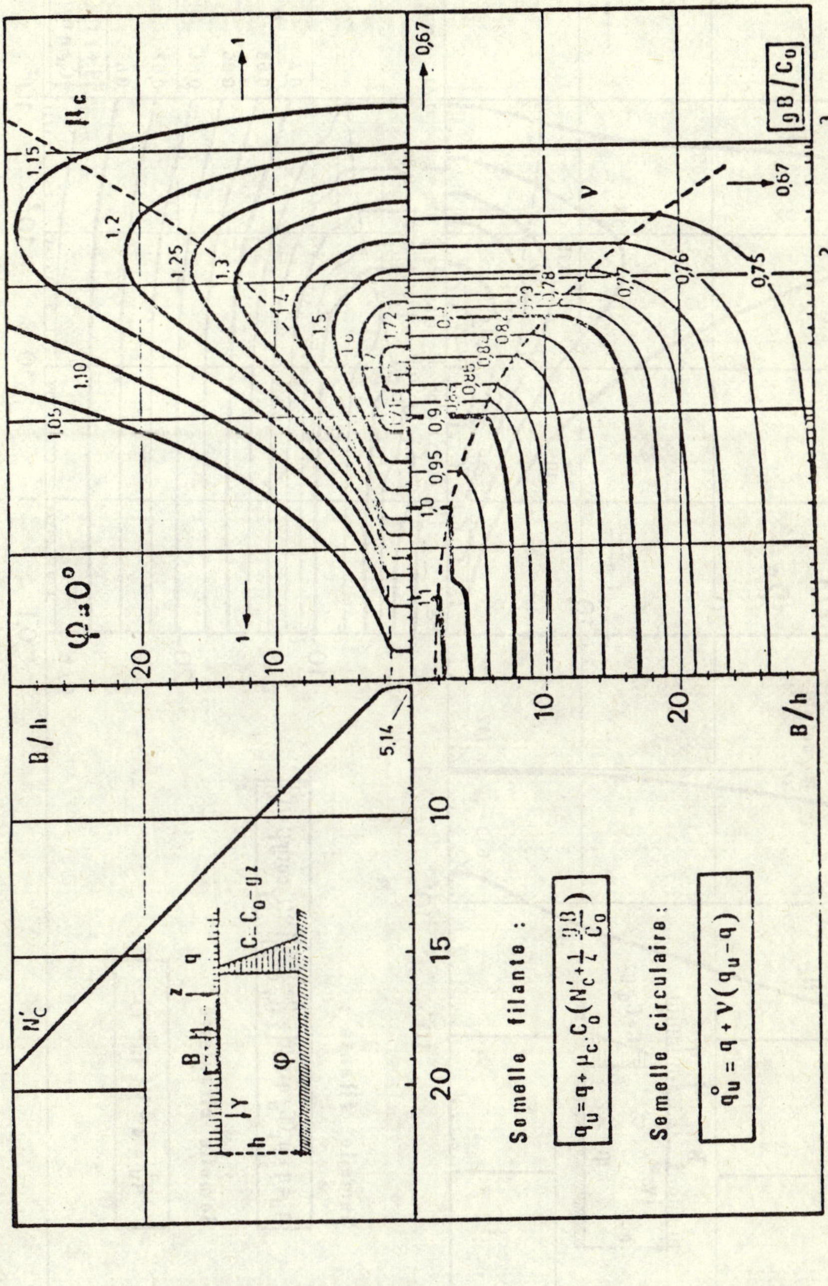
φ = 0°
B/h
gB/C₀
μc
ν
N'c
5,14
Semelle filante :
q_u = q + μc·C₀(N'c + 1/4 gB/C₀)
Semelle circulaire :
q°u = q + ν(q_u − q)

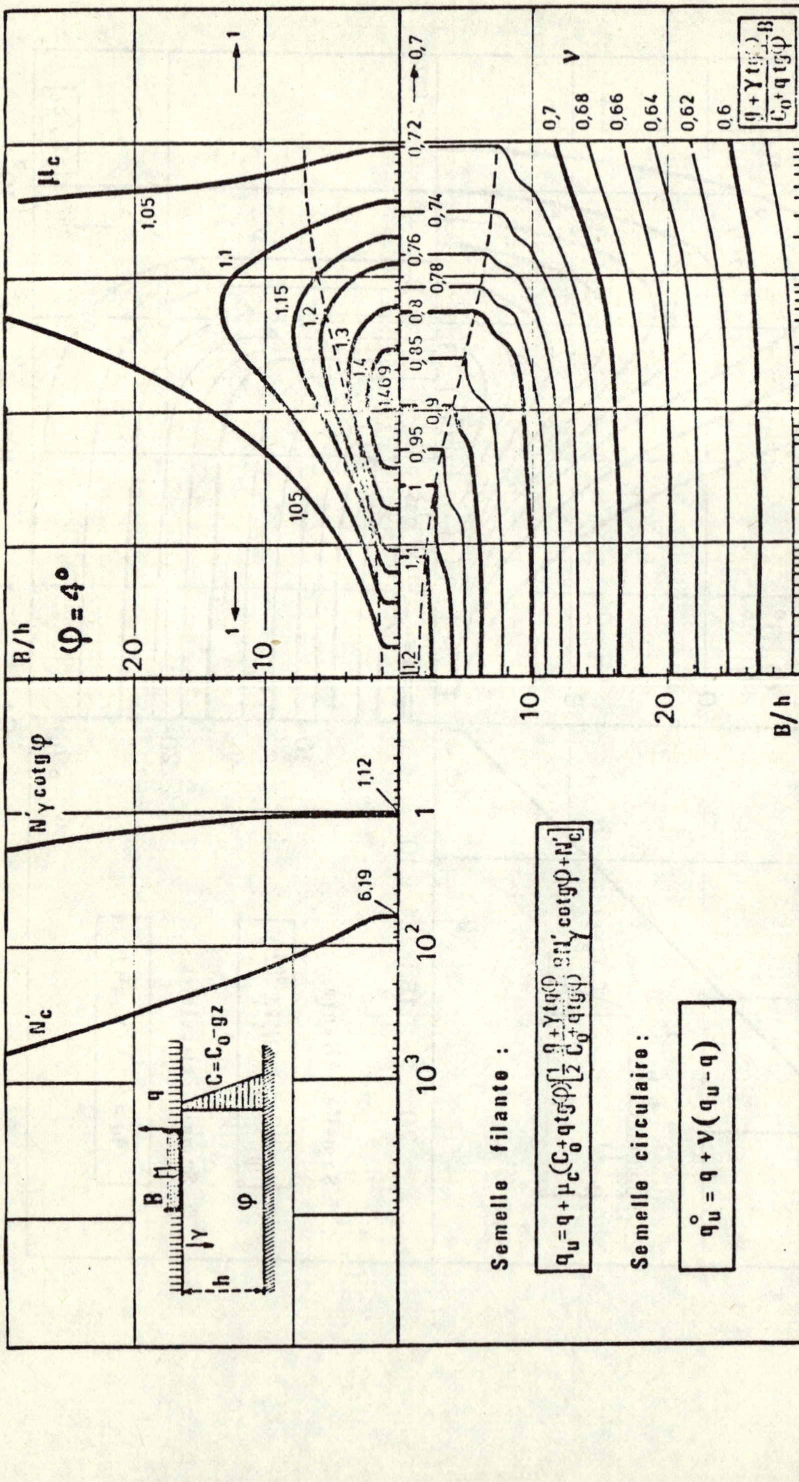
φ = 4°
B/h
N'_c
N'_γ cotg φ
μ_c
ν
C = C₀ − gz
Semelle filante :
Semelle circulaire :
$q^\circ_u = q + \nu(q_u - q)$

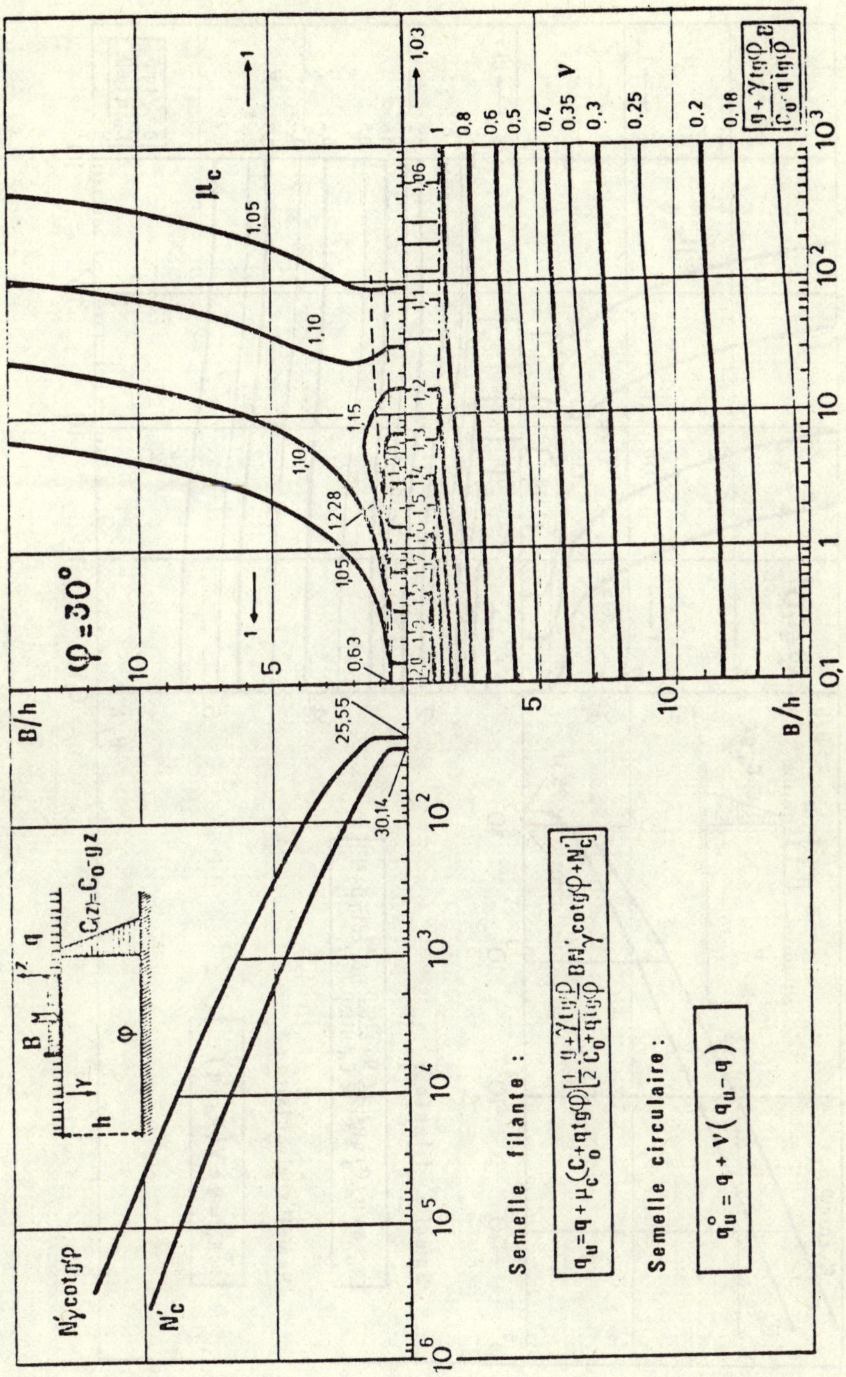
φ = 30°
B/h
μc
ν
N'γ cotgφ
N'c
C(z) = C0 − gz
Semelle filante :
Semelle circulaire :
q°u = q + ν(qu − q)

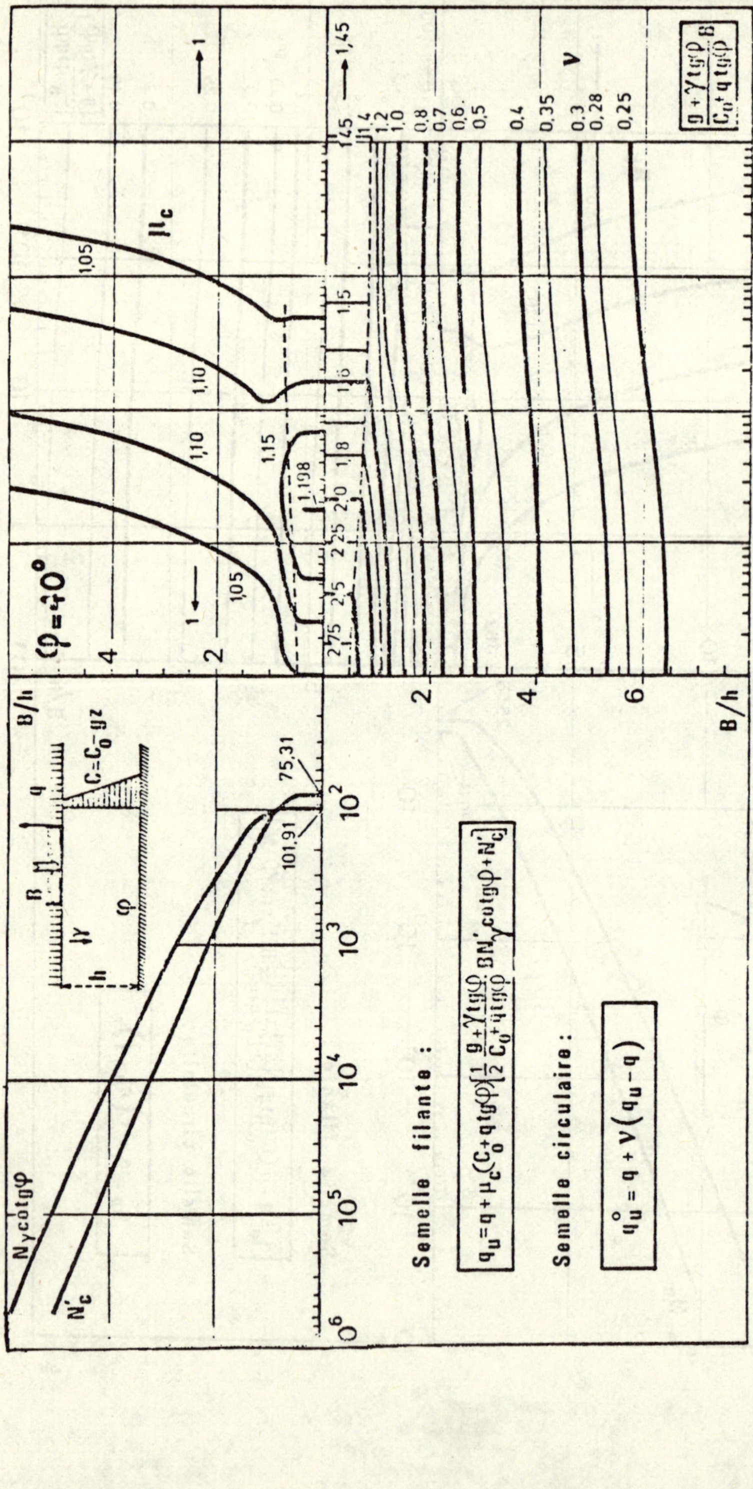
φ = 40°
Semelle filante :
q_u = q + μ_c (C_0 + q tgφ)[1/2 (g + γ tgφ)/(C_0 + q tgφ) B N'_γ cotgφ + N'_c]
Semelle circulaire :
q°_u = q + V (q_u − q)
N_γ cotgφ
N'_c
C = C_0 − gz
μ_c
V
B/h
75,31
101,91
(g + γ tgφ)/(C_0 + q tgφ) B

(4) For $\frac{g + \gamma \tan \phi}{C_o + q \tan \phi} B \uparrow \infty$, we obtain:

$$\nu\left(\infty, \frac{B}{h}, \phi\right) = N'^{o}_{\gamma}\left(\frac{B}{h}, \phi\right) / N'_{\gamma}\left(\frac{B}{h}, \phi\right)$$

- $\phi \to 0 \quad \lim \nu\left(\infty, \frac{B}{h}, \phi\right) = 2/3 \qquad \forall \frac{B}{h}$

- $\frac{B}{h} = 0 \qquad \nu(\infty, 0, \phi) = N^{o}_{\gamma}(\phi) / N_{\gamma}(\phi) = \zeta_{\gamma}$

We recognize the usual shape factor of the gravity term.

The Fig. 2 and 3 shows respectively the comparison of the shape factor of the cohesion term and of the gravity term with semi-empirical values ζ_c and ζ_γ after De Beer (1967) as modified by Vesic (1970) obtained on the basis of comparative loading tests with footing of different shapes.

(5) The shape factor ν is always greater than 1 when the friction angle ϕ is greater than 29°.

(6) There exists a value of $(\frac{B}{h})$ such the shape factor ν is independent of $\frac{B}{h}$ if $\frac{B}{h}$ $(\frac{B}{h})_o$. The representative curve $(\frac{B}{h})$ $\frac{g + \gamma \tan\phi}{C_o + q \tan\phi} B$ is drawn in broken lines on the charts iso - μ_c and iso - ν.

Concluding Remark

The yield design method aims to give to designers a convenient tool for calculating the bearing capacity of shallow foundations, particularly under unusual heterogeneous conditions for classical solutions.

Shape factors for strip footings and circular foundations of the same width and under the same conditions are available by means of four-entry charts corresponding to the usual values of friction angle ϕ for practical needs: $0 \leq \phi \leq 45^o$.

Recent experimental results on several soils from our laboratory seem to be in good agreement with these numerical predictions.

REFERENCES

(1) SALENÇON J., (1978), "Calcul à la rupture". Enseignement d' option Comportement Inélastique des Solides - Ecole Polytechnique, Palaiseau.

(2) MATAR M. & SALENÇON J., (1979), "Capacité portante des semelles filantes". Revue Française de Géotechnique, nº 9, Paris.

(3) SALENÇON J. & MATAR M., (1980), "Capacité portante des fondations superficielles circulaires". Rapport L.M.S., nº 77, Avril 1981, Ecole Polytechnique, Palaiseau.

(4) VESIC A.S., (1975), "Bearing capacity of shallow foundations". Foundation Engineering Handbook, ch.3, Ed. Winterkorn H.F. & Fang H.Y.

COEFFICIENT OF POTENTIAL RUPTURE FOR EARTH STRUCTURES

M.P. Luong

Lab. de Mécanique des Solides, École Polytechnique,
Palaiseau, France

Methods used at the present time for analysing the stability of enbankments and slopes consist of determining the factors of safety associated with a series of sliding lines previously defined by the designer and then obtaining the minimum of these, which is taken as the safety factor...

However a question is raised: does the obtained value correspond to a reliable solution of the problem so-modelled ?

The example described hereafter shows that an analysis by means of the theory of yield design developed by Salençon (1) seems to give better accuracy and more reliable predictions than the classical methods.

This method |1| uses the concept of a convex or star-shaped yield function for the material, |2| does not require a possibly contestable flow rule, |3| is based on two theorems allowing two mathematically dual approaches from inside (static solution) and from outside (kinematic solution) the yield boundary of the system, |4| and define an exact coefficient F of potential rupture for earth structures evaluating the dissipatable work rate of the systems. (2)

$F\ (\frac{\gamma h}{C}, \alpha, \phi) > 1 \Longleftrightarrow$ reliable stability
i.e. potentially stable slope

$F\ (\frac{\gamma h}{C}, \alpha, \phi) < 1 \Longleftrightarrow$ reliable failure
i.e. the equilibrium of the slope

in its initial geometry respecting (C, ϕ) is certainly impossible.

J. B. Martins (ed.), Numerical Methods in Geomechanics, 557–566.

The coefficient F gives a certitude of instability or a stability presumption. No so-called safety factor calculated without complementary knowledge on the behaviour of soil can lead to more precise conclusion. It is important to note that F is quite different - although the comparison is tempting - from the ratio of a dissipated work to a motive force work: the dissipatable work $P(\underline{v})$ represents the maximum of the work mobilizable along the discontinuity line having a given velocity, finite or infinite quantity, and it is precisely the fact that $P(\underline{v})$ must be finite that determines the form of the velocity discontinuity line. Thus it is easy to understand the difficulties of certain "energetic" methods of stability analysis which attempts to give a bound to the value of $P(\underline{v})$ by complementary hypotheses when choosing a priori a curve for which $P(\underline{v})$ is infinite (circular curves with $\phi \neq 0$). That evidently can not give a clear and rigorous interptretation for the so-obtained coefficient (3,4).

In France, the Study Group of Slopes of the Laboratoire des Ponts et Chaussées has listed for the period 1963-1970 more than 300 cases of slope slides often due to an insufficiency of preliminary studies.

Figure 1 describes the cross section of an embankment - studied by Coussy (6) - which serves as a crossing of the valley of the river Dives (20 km east of Caen in Calvados) on the motorway of Normandy. The soil foundation is composed of five layers:

- a layer A of clayey loam
- a layer B of black peat
- a layer C of loam clay
- a layer D of clayey sand
- a layer E of compact loam peat

The weak mechanical properties of these soils require one year for consolidation under an embankment of 3.25 m high. After this first stage, earth-moving operations began. As the height of the embankment reached 5,60 m, the earth structure unhooked on 3 m on its axis whereas a pad of 1 m high was formed at about 30m from this axis. A series of probings determined and characterized the failure: all the layers were strained the layer E, the thickness of which remained constant. Thus it can be said that the failure surface followed the lower part of the layer D. Measurements of cohesion under the embankment in the unstrained zone and in the pad revealed better mechanical characteristics below the embankment, due to consolidation.

The application of the yield design method gives a coeeficient of potential rupture of 0.81 which states with evidence the impossible stability of this embankment. The corresponding weakness line is represented by the full line on the figure. It is noted

DIVES EMBANKMENT

(After O. COUSSY 1978)

--- PROBABLE LINE OF STARTING FAILURE

— CALCULATED WEAKNESS LINE

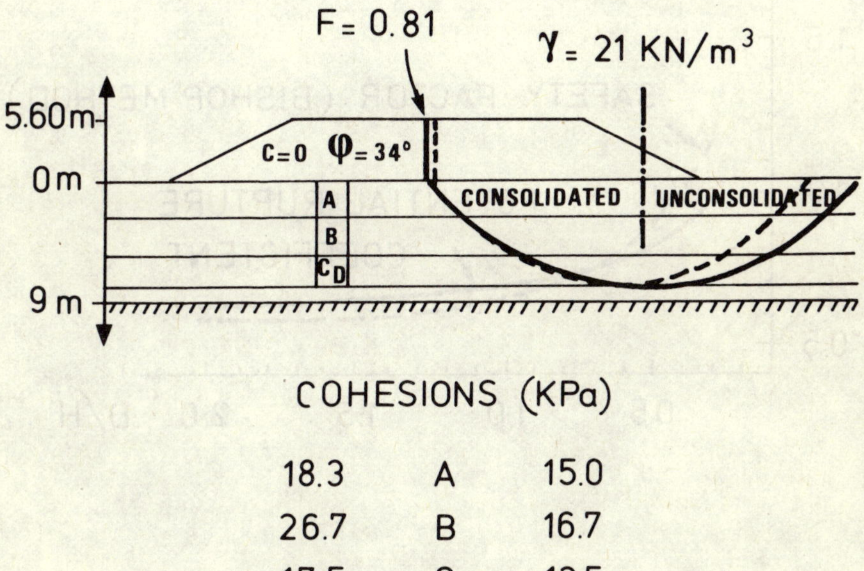

COHESIONS (KPa)

18.3	A	15.0
26.7	B	16.7
17.5	C	12.5
15.0	D	8.4
41.7	E	38.5
CONSOLIDATED		UNCONSOLIDATED

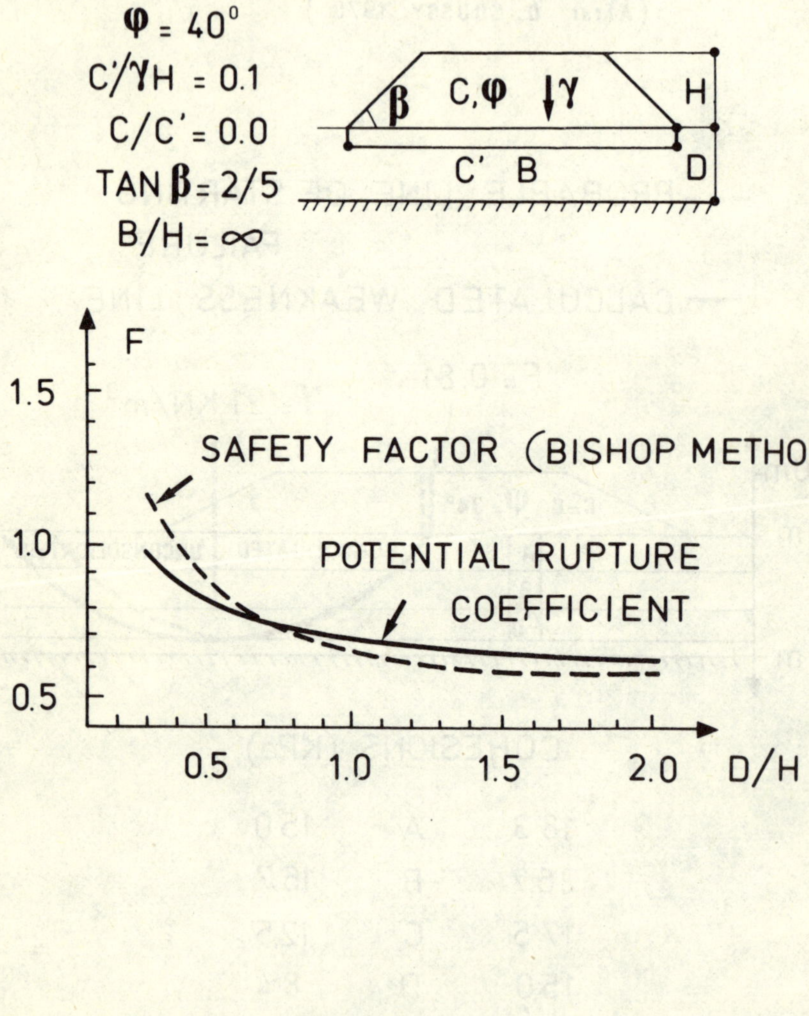

(After O. COUSSY 1978)

Annexe : Méthode du coefficient de rupture : Abaques de calcul de F_M.

$C/C' = 0$

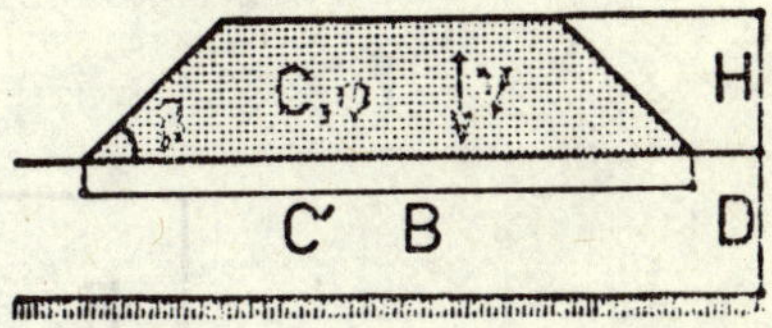

$$F_M = \frac{C'}{\gamma H} K_M$$

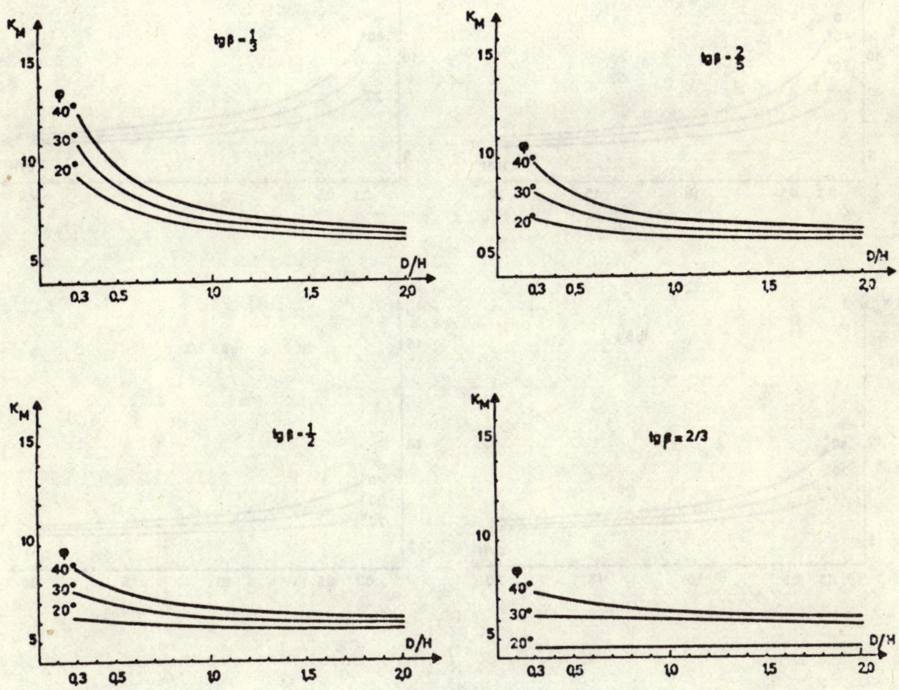

$$C/C' = 0{,}5$$

$$F_M = \frac{C'}{\gamma H} K_M$$

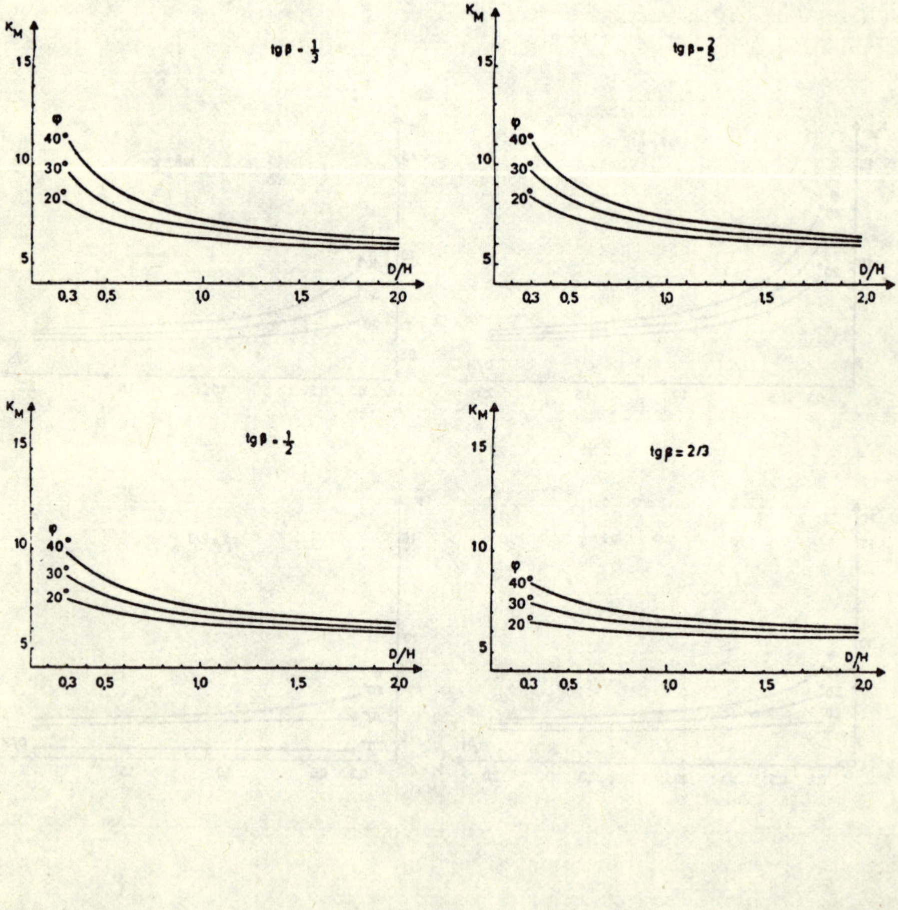

$$C/C' = 0$$

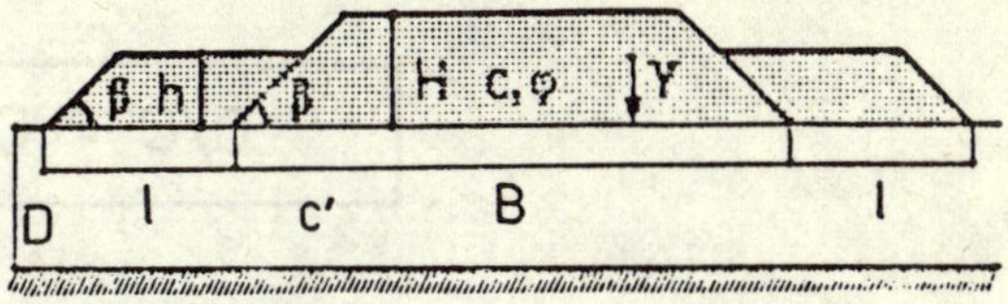

$$F_M = \frac{C'}{\gamma H} K_M$$

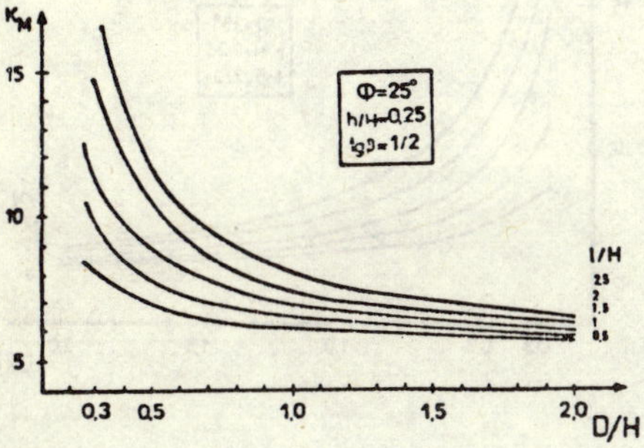

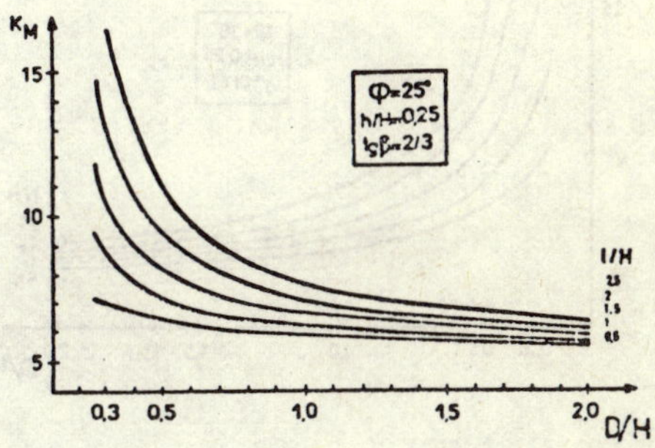

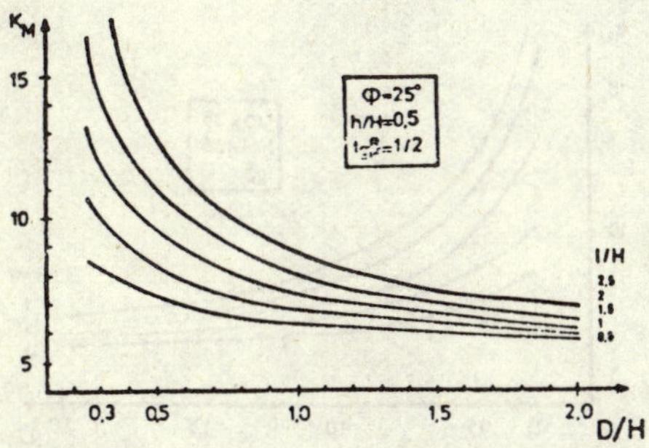

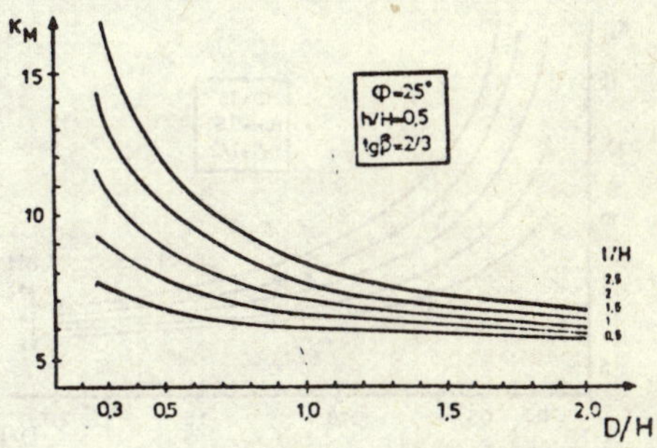

$$C/C' = 0$$

$$F_M = \frac{C'}{\gamma H} K_M$$

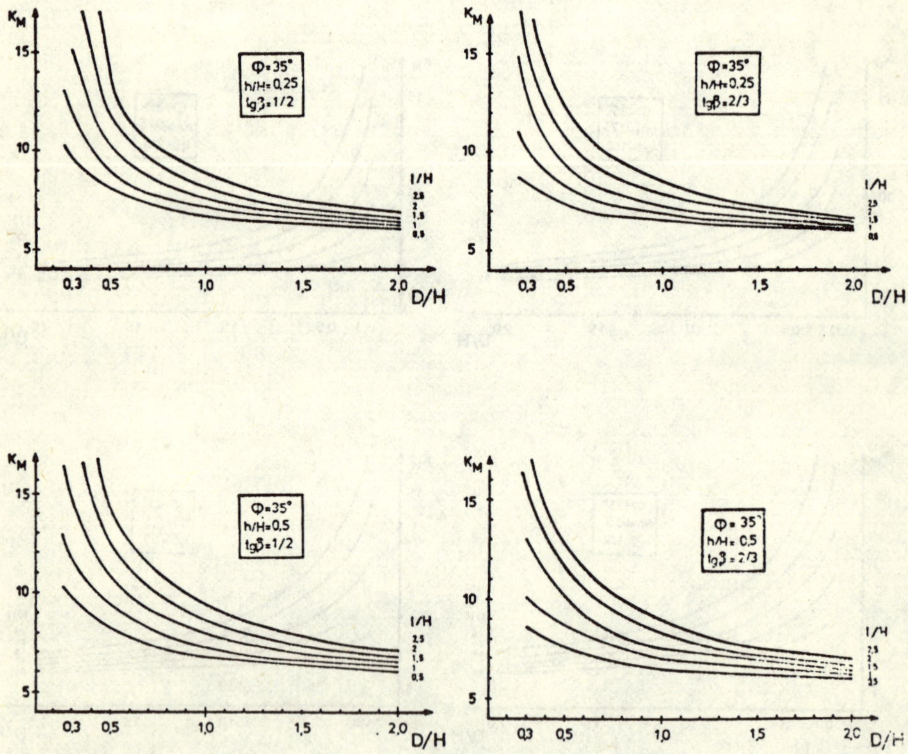

that it is very near to the likely line of starting rupture as proposed in (5):

Despite the purely mathematical aspects of the imposed condition for the velocity discontinuity, the obtained mechanism seems to take account correctly the observed physical phenomenon and particularly the straightening of the likely starting line of rupture in the body the embankment which is practically rectilinear and vertical, and looks like a crack along which occurs the debonding of soil mass.

Concluding Remarks

The theory of yield design employs only a strength criterion defining the allowable stress-state for the constitutive soils and introduces the concept of a potentially stable earth structure in a given geometry, i.e. a structure for which there is compatibility between static equilibrium conditions under the imposed loading and strength condition at each point.

The potential stability on the reliable instability of an earth structure is then characterized by the value of the "yield factor" with respect to 1, based on the utilization of deformation mechanism by blocks in a dualized approach by means of the principle of virtual work and the condition of compatibility between quilibrium and strength which has served as a starting point for the definition of potentially stable earth structures.

The main difference between limit analysis and yield design theory is the use of a flow rule which defines the allowable directions for plastic strain rates $\underline{\underline{d}}(\underline{x})$ at point $\underline{x}$ where the criterion is reached by the stress state $\underline{\underline{\sigma}}(\underline{x})$. The physically allowable directions of $\underline{\underline{d}}(\underline{x})$ lead to a finite values for the dissipatable work which becomes an effectively dissipated work.

REFERENCES

(1) SALENÇON J., (1978), "Calcul à la rupture", Ecole Polytechnique, Enseignement d'Option Comportement Inélastique des Solides, Palaiseau.

(2) COUSSY O. & SALENÇON J., (1979), "Analyse de la stabilité des ouvrages en terre par le calcul à la rupture", Annales des Ponts et Chaussées, 4ème trimestre, 1979.

(3) BISHOP A.W., (1954), "The use of slip circle in the stability analysis of slopes", Géotechnique, vol 5, nº 1, pp. 7 - 17.

(4) PILOT G. & MOREAU M., (1973), "La stabilité des remblais sur sols mous", Abaques de calcul, Eyrolles, Paris.

(5) MAGNAN J.P., PILOT G. & VAUTRAIN J., (1976),"Traversée en remblai de la vallée de la Dives par l'autoroute de Normandie,Bulletin de Liaison des Laboratoires des Ponts et Chaussées, nº 81, LCPC, Paris.

(6) COUSSY O., (1978), "Le calcul à la rupture dans la tenue des ouvrages en terre", Thése de Docteur-Ingénieur, Paris VI.

NAME INDEX

A

Abel,J.F.,546
Abell-Ghaffar,A.M.,447
Alarcon,E.,353
Allgood,J.R.,546
Ang,A.H-S.,501,502,503
Arango,469
Arantes e Oliveira,E.R.,3,37
Aronofsky,J.S.,37
Arulanandan,468
Asoaka,A.,304
Atkin,R.J.,226
Aydinoglu,M.N.,447

B

Backer,R.,141,333
Bancheva,H.,447
Banerjee,P.K.,373
Baron,M.L.,140
Bathe,K.J.,185
Bazant,Z.P.,77,140
Benjamin,J.R.,501
Berezantzev,V.G.,200
Berger,E.,447,449
Best,M.S.,304
Betess,P.,37
Betteness,P.,353
Bhatkar,V.P.,334
Bianchini,G.,448
Bielak,J.,226
Biot,M.A.,37,54,213,226
Bishop,A.W.,185,243,272,333,565
Bjerrum,L.,303
Boehmer,J.W.,227
Bogdanoff,J.L.,447
Bolotin,V.V.,501
Booker,J.R.,77,214,225,243
Boonlaulohr,P.,227
Borges,J.,502
Brebbia,C.A.,337,350,351,352,353, 354,355,373,374,375,386,387, 389,404,405
Brekke,T.L.,142
Brown,C.B.,243
Brune,J.N.,453
Buchmaier,R.,227
Bufler,H.,226
Bui,H.D.,353
Burland,J.B.,78,101,140,243
Butterfield,R.,373

C

Cakiroglu,A.,447
Camargo,W.M.,388
Caquot,A.,168,184
Carslaw,H.S.,374
Carter,J.P.,77,225
Casagrande,468
Castenheta,M.,502
Castillo,E.,333
Castro,468
Cathie,D.N.,404
Chan,468,469
Chandra,
Chang,C.T.,37,55,454,539
Chang,Y.P.,352,374
Chaudonneret,M.,353
Chen,D.J.,352,374
Chen,J.-C.,448
Chen,W.F.,270,404
Cho,468
Chopra,A.K.,448,449
Chow,Y.K.,201
Christian,J.T.,142,224,227,448, 504
Christiensen,N.H.,185
Cividini,A.,304
Cleary,M.P.,185
Cohn,M.Z.,270,303
Collins,J.D.,304
Connor,J.J.,350,387
Cooling,L.F.,243
Cormeau,I.C.,140
Cornell,C.A.,501,503
Corradi,L.,303

Cottle,R.W.,271
Coulomb,C.A.,157,158,168,333
Courant,R.,350
Coussy,O.,565,566
Coyle,H.M.,201
Crochet,M.J.,226
Cruse,T.A.,351,353,353,373,387
Cuny,B.,101

D

Dafalias,Y.F.,61,77,140
Dantzig,G.B.,270
Davis,E.H.,243
Day,S.M.,448
Deresiewicz,H.,271
Derkiureghian,A.,503
Desai,C.S.,103,140,141,225,227,
448
Dikmen,S.U.,448
DiMaggio,F.,140
Dobry,R.,450
Donato,O.D.,270,303,304,333
Doyle,E.H.,450
Drucker,D.C.,77,108,140,157,158,
163,284
Duke,C.M.,451
Duncan,J.M.,77

E

Ellingwood,B.,502,503
Emery,J.J.,451
Ewing,W.M.,448
Evans,J.C.,503
Eykhoff,P.,302

F

Faccioli,E.,448
Falby,W.E.,546
Fardis,M.N.,502
Feld,J.501
Fellenius,W.,333
Felippa,C.A.,54
Feshbach,H.,374
Finn,W.D.L.,448,468,530
Folque,J.,455
Fonseca,A.M.A.,333
Fremond,M.,270

G

Galletley,R.D.,450
Garber,M.,333
Gatti,G.,270
Gazetas,G.,448
Georgiu,P.,353
Ghaboussi,J.,142,225,226,448
Ghafory-Ashtiany,M.,504
Giacomini,S.,271
Giannessi,F.,304
Gibson,G.C.,201
Gibson,R.E.,77
Gioda,G.,270,273,303,304
Gladwell,I.,225
Goble,C.G.,200
Goldberg,J.E.,447
Golder,H.Q.,243
Golubkov,V.M.,200
Gomez-Masso,A.J.,448
Goodman,R.E.,142
Goodwin,G.C.,302
Green,A.E.,226
Grierson,D.E.,270
Griffith,A.A.,163,185
Griffith,A.H.,503
Griffiths,D.V.,184,200
Gumbel,J.E.,546
Gurtin,M.E.,214,226
Gutierrez,J.A.,448,449

H

Habib,P.,539
Hadid,H.A.,351
Hagiwara,T.,452
Hajdin,N.,351
Haldar,A.,473,501,502,503
Hall,J.R.,Jr.,101,452
Hannon,W.J.,449
Hansen,B.,185
Hart,G.C.,304
Haskell,N.A.,449
Hasselmann,T.K.,304

Hayashi,S.,449
Hays,W.W.,449
Heaney,A.,503
Heerema,E.P.,201
Henkel,D.J.,77
Herrera,I.,226
Herrmann,L.R.,140,142
Hess,J.L.,351
Hilbert,D.,350
Hill,R.,100
Himmelblau,D.M.,303
Hinton,E.,55,454,539
Hobbs,R.,227
Hodge,P.G.,405
Hoel,P.G.,502
Hu,G.C.Y.,200
Hughes,T.J.R.,100,101
Humpheson,C.,55,244,334,406
Humphreys,468
Hutchinson,J.,272
Hwang,C.T.,227
Hwang,R.,449,452
Hvorslev,M.J.,100

I

Idriss,I.M.,450,452,469,507
Isenberg,J.,142
Ishiara,468

J

Jaeger,J.C.,374,388
Janardhanam,R.,141
Janbu,N.,272,312,333
Jardetzky,W.S.,448
Jaswon,M.A.,351,373
Jaumann,G.,80,100
Jaynes,E.T.,271
Jenkin,C.F.,163
Johnson,J.J.,450
Johnson,N.E.,450
Josselin de Jong,G.,243
Jowitt,P.W.,271
Joyner,W.B.,450
Jurina,L.,303,304

K

Kanai,K.,450
Kang,C.S.,352,374
Kantorovich,L.V.,350
Katona,M.G.,142
Kausel,E.,430,450
Kavanagh,K.T.,303
Kellogg,O.D.,373
Kellogg,P.D.,351
Kelly,D.W.,353
Kennedy,B.,304
Kerisel,J.,168,184
Khristoforov,V.S.,200
King,I.P.,78,243,405
King,K.W.,449
Kitagawa,468
Koori,Y.,452
Kovacs,W.D.,503
Kovari,K.,145,163
Kramer,J.M.,405
Krause,G.,225
Krelle,W.,270
Krieg,D.B.,100
Krieg,R.D.,77,100
Krylov,V.I.,350
Kuhlemeyer,R.L.,451
Kuhn,H.W.,247,271,285,328
Kulak,R.F.,405
Kulhawy,F.H.,546
Kumar,V.,405
Kunzi,H.P.,270
Kupradze,O.D.,350
Kurata,E.,449

L

Lachat,J.C.,351,373,387
Lade,P.V.,77,140
Langejan,A.,502
Larkin,T.J.,453
Lasdon,L.S.,270
Lau,P.,354
Le Tirant,P.,539
Lee,I.K.,227,468
Lee,K.L.,501
Lee,K.W.,448

Leung,K.H.,55,539
Lewis,R.W.,55,244,406
Lewis,T.O.,304
Liang,G.C.,451
Liggett,J.A.,352
Lightner,J.G.,142
Liou,C.P.,451
Litkouhi,S.,201
Liu,H.,225
Liu,P.L.,352
Lombardi,G.,163
Longworth,T.I.,243
Luco,J.E.,451,454
Luenberger,D.G.,333
Lumb,P.,501,502
Luong,M.P.,531,539,541,546,547,
557
Lysmer,J.,270,409,447,448,449,
451,452

M

Madsen,O.S.,54
Magnan,J.P.,566
Magri,F.,227
Maier,G.,270,271,273,303,304,
333
Malkus,D.S.,100
Mana,A.,143
Mangasarian,O.L.,270
Marcuson,468
Markowitz,H.M.,271
Martin,G.R.,448,468
Martins,J.B.,243,271,303,305,
333,334
Massonnet,C.E.,351
Matar,M.,556
Matos,A.C.,271,333,334
Matsuo,M.,304
McCalla,T.R.,304
McGuire,R.,503
Melan,E.,404
Mendelson,A.,353,404
Meyerhof,G.G.,200,502
Mikhlin,S.G.,214,226,350,404
Minowa,N.,452
Mitchell,G.,468,503
Molenkamp,F.,185
Montgomery,D.C.,504
Moore,J.F.A.,243
Morgenstern,N.R.,185,227,272,310,
334
Mori,468
Morland,L.W.,225
Morse,P.M.,374
Moussa,A.B.,225,226
Mroz,Z.,77,78
Mukherjee,S.,353,404,405
Munro,J.,247,270,271,272
Murarka,R.,452
Murray,D.W.,227
Muskhelishvili,N.I.,350
Mustoe,G.,354

N

Nagaonka,H.,226,227
Naghdi,P.M.,226
Nagtegaal,J.C.,405
Nair,G.P.,451
Nappi,A.,304
Narayan,G.P.,334
Nauroy,J.F.,539
Naylor,D.J.,229,243
Nelson,I.,140
Nicolucci,F.,304
Noll,W.,37
Norris,V.A.,77,78
Nova,R.,78

O

Oden,J.T.,37
Odell,P.L.,304
Ohsaki,Y.,452,468
Oliver,A.A.,450
Ostadan,F.,451
Owen,D.R.J.,163

P

Palmer,A.C.,270,272
Pande,G.N.,142
Park,K.C.,54
Parks,D.M.,405
Pastor,J.,243

Paterlini,F.,271
Payne,R.T.,302
Peaker,K.,184
Peck,E.A.,468,504
Peir,J.,503
Perumpral,J.V.,141
Phan,H.V.,140,141
Pian,T.H.H.,350
Pietruszczak,St.,78
Pilot,G.,565,566
Pister,K.S.,37,226
Podolak,K.,303
Pollard,469
Popov,E.P.,77
Poskitt,T.J.,201
Poulos,468
Prager,W.,108,140,157,158,163, 291,405
Prakash,A.,163
Prandtl,L.,163
Press,F.,448
Prevost,J.H.,79,100,101,140, 225
Price,V.E.,272,310,334

R

Rabicewicz,L.V.,387
Rai,I.S.,227
Ramamurthy,T.C.,334
Randall,M.J.,452
Randolph,M.F.,185
Rankine,W.J.M.,168,334
Rausche,F.,200
Reddy,J.N.,37
Reis,E.B.,271,334
Reissner,E.,350
Revilla,j.,333
Riccardella,P.C.,405
Rice,J.R.,405
Richard,R.E.,101
Richart,F.E.,Jr.,451,452,453
Rizzo,F.J.,351,352,373,386,387, 468
Roesset,J.M.,430,450
Rogers,A.M.,449
Roscoe,K.H.,78,140
Rowe,P.W.,184
Runesson,K.,225

S

Salaam,U.,226
Salençon,J.,270,555,556,565
Sandhu,R.S.,203,213,224,225,226, 227
Sander,I.,140
Sargand,S.M.,143
Sarma,S.K.,272
Scanlan,R.H.,452
Schiff,A.J.,447
Schiffman,R.L.,227
Schnabel,P.B.,452
Schofield,A.N.,78,270
Schreyer,H.L.,405
Scott,R.F.,101
Seed,H.B.,447,448,451,452,468, 469,530
Selig,E.T.,546
Shannon,C.E.,254,271
Sharma,K.G.,142
Shaw,R.P.,373
Shield,R.T.,158,163
Shimizu,N.,452
Shimomura,Y.,453
Shippy,D.J.,352,387
Sidaner,J.F.,539
Siemieniuch,J.L.,225
Silva,W.,453
Silver,468
Singh,A.,501,502
Singh,K.J.,225,227
Singh,M.P.,504
Singh,R.D.,450
Siriwardane,H.J.,140,141,225
Skempton,A.W.,272
Sloan,S.W.,185
Small,J.C.,225
Smirnov,V.J.,351
Smith,A.M.O.,351
Smith,D.L.,271,272
Smith,I.M.,167,185,187,200,201, 225,227
Sodekawa,468
Sokolovsky,V.V.,185
Stippes,M.,387
Streeter,V.L.,451,453
Sture,S.,140,141
Suklje,L.,225
Swedlow,J.L.,353

Swinger,W.F.,504
Symm,G.T.,351,373

T

Tabaddor,F.,226
Tabatabaie-Raissi,M.,451
Tajimi,H.,453
Tajirian,F.,451
Takahashi,S.K.,546
Tanaka,468
Tang,W.H.,501,502,503
Taylor,P.W.,453
Taylor,R.L.,142
Telles,J.C.F.,353,386,389,405
Teng,T.-L.,453
Terzhagi,K.,101
Thanapoulos,I.,539
Thompson,W.T.,453
Toki,K.,453
Tomizawa,M.,304
Tomlin,G.R.,351
Tong,P.,350
Tonti,E.,226,227
Torkamani,M.A.M.,304
Toupin,R.A.,37,226
Townsend,469
Trifunac,M.D.,447,451,453
Truesdell,C.,3,37,226
Truesdell,D.,226
Tsai,C.-F.,451
Tsuchida,H.,449
Tucker,A.W.,247,271,285,328

U

Udaka,T.,451
Ugas,C.,452

V

Vahdani,S.,451
Valliappan,S.,78,227,387,405
Vanduren,W.,353,387
Vanmarcke,E.H.,501,502
Vaughan,D.K.,142
Vautrain,J.,566
Veneziano,D.,502
Venturini,W.S.,375,387
Vermeer,P.A.,225,227
Verruijt,A.,225
Vesic,A.S.,556

W

Waas,G.,450,451,453
Wakugawa,J.M.,101
Walker,S.,351,373,387,404
Warrick,R.E.,450
Washizu,K.350
Watson,J.,351
Whitman,R.V.,54
Wight,L.H.,453
Wilde,P.,78
William,T.,54
Wilson,E.L.,142,224,225,226
Wilson,J.,546
Wolfe,P.,271
Wong,H.L.,451,453,454
Wood,D.M.,78
Wood,W.L.,374
Woodgate,R.,503
Woods,R.D.,101,452
Worth,C.P.,77,78
Wrobel,L.C.,352,355,373,374
Wroth,C.P.,270
Wylie,E.B.,453

Y

Yamagata,K.,226,227
Yanamoto,S.,452
Yao,J.T.P.,504
Yokoo,Y.,213,226,227
Yucemen,M.S.,502

Z

Zadeh,L.A.,504
Zavelani-Rossi,A.,270
Zienkiewicz,O.C.,37,39,46,54,55,
57,77,78,140,163,184,227,244,
334,353,405,406,454,539

SUBJECT INDEX

A

Acceleration,36,413,429
 gravity,36
 maximum,413
 rigid body,429
Accomodation,537
 in cyclic loading,538
Accumulated deviatoric plastic
 strain,72
Activated yield modes,288
Adaptation,537
All-around cell pressure,92
Alternative degradation mode,72
Analogue models,148
Analysis,99
 dynamic,99
 pseudo static,99
Anchored wall,317
Anisotropic,45,71,158,356
 condition,158
 hardening model,71
 media,45
 soil,356
Anisotropy,85
 cross,85
Aquifers,362
 confined,362
Arch dam,146
Arching effect,543
 positive,543
 negative,543
Associated:
 Mohr-Coulomb criterion,399
 Mohr-Coulomb plastic flow,179
 plastic moduli,83
 plasticity,66
Associative:
 plasticity theory,284
 flow rule,82
ASTM guidelines,496
Average values:
 macroscopic,33
Axial and lateral loading,194
 combined,194

B

Back-analysis:
 of field test by FEM,195
Backpressure,92
Backward difference scheme,99,223
Balance principles,20
Ballast-concrete interface,125a
Basic:
 stable system,321
 solution,321
 formulations,3,39
Basis:
 cartesian,5
 orthonormal,5
Bayesian:
 statistics,300
 updating approach,497
Bearing capacity,168,175,188,547
 factors,175
Behaviour:
 dynamic,39
 elastic-perfectly plastic,284
 nonlinear elastic,60
 rock mass,145
 static,39
 undrained,53
Bending moment-rotation law,286
Bessel function,363
 modified,363
Best-fitting,293
Bias removal process,275
Bilinear form,289
Binary mixture:
 of soil and water,207
Biot's:
 assumption,213
 parameters,44
Biquadratic interpolation,215
Black-box,276
Blocks,565
Block characteristics,154
Block-diagonal matrix,284,331
Block formation,150
Body force vector,204

Body waves:
 inclined plane,412
Booker's functional,214
Boolean identification matrix,289
Boundary condition,46
 essential,341
 fluid pressure,208
 natural,341
Boundary elements,337
 in groundwater flow,355
Boundary error functions,345
Bounding ellipse,71
Bounding surface,76
 model with degradation,71
 models, 69
Bounds upper and lower,290
Bulk moduli,280
Broad-band spectrum,413
Buoyant bulk density,204
Buried structures,541

C

Calibration of mathematical models, 301
Cam-clay,182
 modified model,108
Canonical form,321
Cap model,115
Caquot and Kerisel solutions,168
Cauchy's:
 effective stress tensor,80
 stress principle,21
Cell-collocation,342
Characteristic-line method,161, 162,532
Characteristic state concept,538
Characteristic threshold,532
Charts of bearing capacity,547
Chi-square tests,481
Christoffel symbols,11
Circular:
 foundations,191
 piles,191
 surfaces of sliding,306
Classical plasticity,65
Clay kaolinite,91
Coefficient:
 correlation of,494
 variation of,479
Cofferdam,281
Cohesionless soils,86
Cohesive soils,86
Collapse,188,229
 by the direct method,235
 mode of deep cylinder,193
Collocation by subregions,342
Comparison of FE results,173
Compatibility,284
 equations,11
Complementarity:
 constraint nonconvex,289
 problem linear,251,285
 relations,264
 tensors,394
Complete methods,426
Complex moduli,431
Compliance,285
Composite element,215
Compression of:
 solid grains,43
 the fluid,43
Computational model for rock masses,161
Condensed matrix formulation,287
Configurations:
 current,11,24
 reference,9
Confined aquifers,362
Confining pressure,155
Conjugate distributions,497
Conservation of mass,23
Conservativeness,490
Consolidation:
 equation,50
 phenomena,47
Contractancy region,537
Constant volume conditions,92
Constant:
 boundary elements,371
 strain triangle,287
Constitutive coupling symmetric, 213
Constitutive relatios (equations or laws),34,44,53,60,145,209, 284,291
 modeling,103,179
Constrants inequality,278
Continuity equation,207,218

Continuity equation:
macroscopic,36
Continuous yield,68
Continuum:
mechanics,3
two-phase,35
Control:
motion,410
points,329,412
Contravariant:
components,204
matrix,295
Convergence,281
criterion,298
test,298
Convexity conditions,247
Convolutions,214
Correlation:
coefficient of,494
Cost coefficients,251
Coupled:
consolidation equation,47
deformation,203
field equations,99
seepage equation,52
Coupling matrix,216
Covariant components,204
Creep,42
compliance functions,206
Criterion yield,108
Mohr-Coulomb,168
von Mises,168
Critical state,68,182
models,68
Critical void ratio,255
Cross-anisotropy,85
Cumulative probability,481
Current:
geometry,53
state,207
Curves:
stress-strain,87
Cut-off:
tension,157
Cyclic:
aspects,531
densification,69,74
hardening,533
multi-degree-of-freedom,119
pressure,69
Cyclic:
resistance,525
softening,533
strain-controlled tests,85,90
Cyclic loading,74,116,182,538
Cyclic mobility,456,508
Cyclic stresses,456
ratio of,513
Cylinder deeply embedded,542

D

Dam:
coffer:281
earth,281
Damping,196
ratios,415
Darcy's law,36,45,48,356
generalised form of,207
Data experimental,277
Decision analysis,476,498
Decomposions:
cartesian,4
invariant right and left,4
Deconvolution analyses,423
Deep foundations,187
Deep seated movement,230
Deformability modulus,149
Deformation:
gradient,10,80
mechanism by blocks,565
rates,17,260
theory of plasticity,288
Degradation:
bounding surface model with,71
mechanical properties of,90,194
Densities:
fluid,45
force,24
mass,23
Density hardening,68
Description:
material,27
spatial,17,25
Design earthquake,459,492
Design of the experiments,274
Destructive rock excavation,161
Deterministic identification,299
study uncertainties of,490

Deviations standard,293,489
Deviatoric:
 hardening,68
 plastic strain accumulated,72
 stress subspace,82
Devices:
 triaxial,104
Dilantacy:
 loop,533
 region,537
Direction neutral,60
Discrete models,288,291
Discretization,287
 by finite elements,51
Differential-correction,292
Diffusive resistance,205
Dilatancy:
 angle,232
 influence of,168
Dilation,182
Dirac delta function,343
Direct search techniques,276
Directions:
 of plastic deformations,81
 orthogonal principal,7
Dirichlet boundary conditions, 367
Discontinuities,147
Discriminant analysis,500
Displacement:
 at collapse,181
 for pile tip,197
 formulation,222
 measurements,291
 rates,260
Dissipated work,558
Dissipation,538
 fluid pressure of,216,538
Distributions,497
 Beta,477
 Chi-square,477
 Exponential,477
 Extreme value,479
 Gamma,477
 Gaussian,478
 Geometric,477
 Hyperbinomial,477
 Hypergeometric
 Lognormal,293,477
 Normal,477
Distributions:
 Poisson,477
 Standard normal cumulative,478
 Student's,478
Distribution functions,153
Distribution models,480
Disturbances of the experimental data,291
Divergence of a matrix,8
Drainage,538
 full,48
Driving force,45
Drained:
 static solution,48
 tangent matrices,44
Drivability of piles,187,196
Drucker-Prager:
 model,108
 yield criterion,399
Dual problem,325
Duality theory,290
Duration of earthquake,413
Dynamic:
 analysis,99
 aspects,531
 behaviour,39
 consolidation behaviour,53
 equations,47
 methods of analysis,459
 relations,40

E

Earth:
 dams,281
 pressures,169
 slopes,491
 structures,557
Earthquake,507
 magnitude,413
 records stong motion,413
 regular and irregular pulses,459
 response,50
 spectrum,459
Effect of combined loading on axial failure,195
Effects of elastic compressibility, 193
Effective Cauchy stress tensor,80

Effective stress,33,40,42,79
analysis,182
behaviour,66
path,538
tensor,204
Eigenvalue problem,416
Elastic:
isothermal compliance,205
material,31
four-order tensor of moduli,81
regions,61
shear modulus,83
stiffness,280
zone limit of,59
Elastic half-space,419
Elastic perfectly plastic:
behaviour,284
material,156
Elastic predictor-radial corrector method,397
Elasto-plastic:
analysis by boundary elements,389
materials,80
matrix,59,65
rotational springs,285
stresses,329
Element assemblage,284
Ellipse critical state,69
Elliptic integral of the first kind,358
Embankments,229,557
by mathematical programming,305
on stiff clay,233
Embedded structures,412
Embedment partial,160
Endochronic theory,59
Energetic methods,558
Energy:
balance equation,28
balance of,25
function,206
non-mechanical vectors,23
Engineering:
judgement,295
optimisation,248
Enhanced limit methods,237,238
Entropy,254
granular material,254
Envelope failure,115
Morh-Coulomb,66
Epicentral distance,413
Equality-constrained minimality, 249
Equations:
compatibility,11
constitutive,34
energy balance,28
field,3,30
for solids,30
global,25
local,25,27
macroscopic equilibrium,35,36
momentum balance,28
motion of,431
non-linear,3
strain-displacement,31
Equilibrium,284
equation,51
fluid phase of,45
macroscopic equations,35
Equipollent tangential,256
Equivalent linear method,434
Error function boundary,345
Essential boundary condition,341
Estimation of parameters,479,495
Estimation optimal theory,297
Euler's first law of motion,21
Evaluation of liquefaction potential,525
Excavation:
of slopes,229
of tunnel,145
Excess water pressures,179
Excitation level,417,422
Experimental data:
disturbances of the,277,291
Experimental and modelling inaccuracies,289
Expressions for shear and bulk moduli,108
Extraneous energy,196

F

Factor of safety,317
Failure:
criterion,509
envelope,115
pressures,193

Failure:
 progressive,149
 states,168
Fast Fourier transform
 techniques,416
Feasible-conjugate-direction
 algorithm,292
Feasible:
 domain,253,277,320
 solutions,253
Field equations,205
 coupled,99
Finite deformation,80
 effects,92
Finite differences,342
Finite elements:
 discretozation,51
 mesh,175,188,223
 techniques,156
Fittings,293
Flexible polyhedron method
 optimization,276
Flexible volume method,433
Flow resistant viscous,36
Flow rule,557
 associated,63,168
 non associated,63,168
Fluctuaction scale of,495
Fluid:
 density,32
 flow in porous media,203
 incompressibility,32
 interstitial,40
 mechanics,17
 pressure boundary condition,208
Fluids:
 field equation for,31
Footings,231
 of different shapes,555
Force-displacement law,286
Foundation:
 beam,285
 deep,167
 interaction,429
 shallow,167
Fracture in rock deformation,148
Frame-indiference,28
Free-field:
 earthquake motions,428
 site response,410
Frequency phenomena,51
Frequency content,413
Frequency diagram,475
Friction crack element,150
Frictional materials,182
Full drainage,48
Fundamental solution,357
 half-plane,391
Fundamental surface waves,422
Fuzzy-solutions,500

G

Galerkin's method,214,342
Gateaux differential,211
Gaussian probability,293
Genuine rock pressure,159
Gib's relationship,488
Goodness-of-fit tests,480
Gradient:
 components of deformation,10,16
 solid velocity,80
Granular material,532
Gravity:
 acceleration,36,45
 switch-on,173
Green's:
 deformation tensor,11
 functions,338
 third identity,357
Ground shaking,518b
Groundwater flow,355
 round a tunnel,363
 transient,362
Gurtin's method,214

H

Half-plane fundamental solution,
 391
Hardening:
 anisotropic model,71
 deviatoric,68
 linear behaviour,291
 parameters,66,115
 rule,291
Hereditary linear material,206
Higher-order surface waves,418

Histograms,293,475
Historical parameters,60,65
Holonomic,284,288
 elastoplastic analysis,253,288
 response,331
Holonomy,331
Holtz's relationship,488
Hooke's law,157
Horizontally propagation waves,
 412,414
Hybrid finite element,138
Hydrostatic,40
 state of initialstress,161
Hyper-elastic,57,206
Hypo-elasticity,65
Hysteresis loop,72,79,538
Hysteresis susceptibility,538

I

Ideal:
 associated plasticity,66
 non associated plasticity,66
 plasticity,59
Identifiability,280
Identification:
 elastic stiffness,280
 hardening moduli,292
 piecewiselinear elastoplastic
 models,284
 structural,281
 yield limits,288
 young modulus,295
Identification by quadratic
 programming,288
Identification of "numerical"
 methods,276
Identification or "inverse"
 problems,274
Impedance matrix,429
Inclined plane body waves,412
Increment plastic strain,61
Incremental:
 algorithm,99
 equilibrium equations,390
 form,207
 iterative processes,397
 relations,60
Inequality constraints,249,278
Inertial forces,36
Infinite surface model,72
Infinity condition,364
Inflow,45
Influence:
 dilatancy,168
 flow rule,169
 wall roughness,169
Influx density,23
Initial
 condition,46
 state,207
 strain,52
Initial stress,48
 based methods,234
 hydrostatic state of,161
Inner-product,5
Inter-element boundaries,209
Interface,545
Internal:
 boundary conditions,209
 cohesion,157
 continuity constraints,212
 discontinuity conditions,209
 energy,25
 friction,157
Interaction:
 analysis,410
 between tunnel,159
 motions,426
 pressures,281
Interpolation:
 functions,378
 spatial schemes,215
Intersections of element boundaries,
 209
Interstitial fluid,40
Installation of piles,196
Intrinsic density of the fluid,
 204
Invariant,6,25,108
 right and left decompositions,4
 first,19
Irregular pulses earthquake,459
Irreversibility:
 plastic strains,288
 strains accumulated,533
 volumetric strains,538
Irrotational flow,207
Isomorfic operations on matrix,4

Isothermal elastic compliance, 205
Isotropic:
fluid stress,204
loading,114

J

Janbu's method,312
Jaumann derivative,80
Jointed rock slope,316
Jump discontinuities,209

K

Kaolinite clay,91
Kausel-Roesset meyhod,430
Kelvin elastoplastic formulation, 390
Kinematic:
admissibility,266
deformable bodies,9
hardening,71
permeability,50
plasticity,71
solution,557
Koiter's rule,291
Kolmogorov-Smirnov tests,481
Kuhn-Tucker conditions,247,285, 328

L

Lagrange multiplier,249,315
Laplace transforms,214,357
Large time,48
Laterally loaded piles,194
Laws:
bending moment-rotation,286
force-displacement,286
stress-strain,104
Layer interfaces,415
Layered:
regions,362
viscoelastic soil,419
Length of drainage,538
Least square methods,283,342
Limit analysis theory,565
Limit equilibrium:
mechanisms,170
methods,230
Linear complementary,251,285
Linear elastic soils isothermal condition,210
Linear extension of the vector, 12
Linear hardening behaviour,291
Linear hereditary material,206
Linear inequalities,288
Linear programming,316
Linear-Taylor-differential-correction,292
Linear transformation,4
Lined tunnel,383
Lining:
stress in the,376
Liquefaction,72,76,455,507,514b
cone penetration tests,464
during earthquakes,507
effect of horizontal pressure, 466
failure criterion,509
potential,525
rheological considerations,467
sands of,465
SPT results,462
studies,498
susceptibility,464
threshold versus relative density,465
Load:
cyclic,123b
diffusion,194
displacement behaviour,179,188
factor,317
parameter,260
repetitive,123b
unload directions,60
Loading,59,491,538
proportional histories,288
triaxial,66
Local equations,25
seismicity,492
Locations of displacement measurements,291
Logarithmic interpolation,219
Log-normal distributions,293
random variables,484

Long wave excitation,200
Loops hysterysis,72
Loosening pressure,159
Loss function,275,288
Love waves,413
Low frequency wave loading,199
Lower bonds to the collapse,236
Lumped-compliance,285

M

Macrosample,254
Mass:
balance of,23
stiffness matrices,431
Material behaviour:
visco elastic,162
visco plastic,162
Material derivative,80
Material indiference,28
Mathematical modelling,145
Mathematical programming:
embankments by,305
methods,74,245,250
slopes by,305
Matrices:
Boolean identification,289
block-diagonal,284
definite,4
elasto-plastic,65
skew-symmetric,29
Matrix:
equations in consolidation,218
sensitivity,297
Maximum absolute value of the differences, 276
Maximum acceleration,413
Maximum attainable entropy,255
Mean:
second-order,486
pressure,115
stress,115
value,293,479
Measure of uncertainty,254
Measurements:
samples of,295
Media anisotropic,45
Memory,72
models,74
Mendelson's sucessive elastic solutions method,397
Method:
characteristic-line,162
Fellenius and Bishop,306
slices,238
Micromechanical models,148
Microscopic average value,35
continuity equation,36
Microstructure,44
Mikhlin's basic variational theorem,214
Minimization processes,248,283
least square,283
Minimum:
non-global,278
trajectory of the,277
Mixed formulations,210
Mobilisation of resistance,194
Mobilised friction angles,187
Mode alternative degradation,72
Model as input data,273
parameters,85
Modelling inaccuracies,289
for interfaces,116
mathematical,145
Models:
bounding surface,69
cap,115
degrading bounding surface,74
discrete elastoplastic,291
Drucker-Prager,108
elastic-brittle,149
modified Cam-Clay,108
nonlinear material,80
radial rule,73
with residual strength,149
zero-elastic region,72
Modes:
contact,116
no-sliding,116
yielding,285,288
Modified Bessel function,363
Moduli:
associated plastic,83
bulk,42
complex,431
hardening,61
plastic,65
plastic bulk,82

Moduli:
plastic shear,82
tangent,57
Mohr-Coulomb constitutive equation,66,256,314,399
Mohr stress circle,159
Moment of momentum,24
balance of,24,28
Monotonic:
axial,85
cyclic stress,457
drained axial compression tests, 86
non-increasing,253
Motions,9
embedded structures,428
free field,432
interaction,426
rigid body,13
shallow depths,420
Movement of slope crest,174
Multiaxial stress,60
Multiaxial testing devices,104
Multisurface model,69
Multisurface plasticity,82

N

Nanson's formula,22
Natural boundary condition,341
Natural frequencies,417
N_c for smooth strip footings,177
Nested yield surfaces,82
Neutral stresses: see pore pressures
Newmann duality theorem,253
Newmann-Duhamel postulate,206
Newmark procedures,54
Nigata seism,461
Non-associated plastic flow rules,66,92
Non-dimensional groups,48
Non-circular surfaces of sliding, 309
Non-convergence of iterative processes,234
Nonconvex complementary constraint,289
Nonconvex quadratic program,289
Non-dilatant soil,193
Non-global minimum,278
Non-holonomic elastoplasticity, 253
Non-increasing monotonically,253
Non-linear equations,3
Non-linear material,206
models,80
Non-linear soils,210
Non-negative of the probabilities, 254
Non-proportional paths,80
N_q for smooth strip footings,117
Normality condition,159,328
Normally consolidated state,86
No-sliding modes,116
No-tension material,380
problems,375
Numerical methods,276

O

Objective function,250,320
Objectivity:
material,28
Observation-guide design methods, 275
Observational data,497
Observed data,480
Observed and predicted values, 500
Offshore foundations,182,194
Offshore tripod structure,434
Open-ended pile drivability FE model of,199
Operator differential,16
Optimal estimation theory,297
Optimality criterion,274
Optimization:
alternating variable method,276
contraints,289
flexible polydron method,276
methods,248,273
Powell method,277
simplex method,276
Original rock stresses,381
Orthonormal basis,5
Oscillatory error in fluid pressures,216
Ottava sand-concrete interface,122

Outward flux,23
Overburden rock pressure,159
Overconsolidation,68
Over-stress ratio,232

P

Parabolic elements,242
Parallel coupling of rock blocks, 152
Parameters,104
 historical,65
 K and B,150
Parametric identification,273
Parity rule,262
Partial embedment,160
Path dependent:
 plastic behaviour,288
 problem,377
Path strain,90
Path stress,59
Peak cyclic pore pressure ratio, 508
Peak load,179
Penalization,277
Period of sustained pressure, 514a
Periodic force,48
Permanent tunel lining,162
Permeability:
 coefficient,36,45
 kinematic,50
 tensor,207
Pertubations,298
Piecewiselinear:
 constitutive laws,291
 elastoplastic models,284
 representation,90
 yield surfaces,268,288
Piecewiserigid system,285
Pier approximations for collapse, 200
Piles driven offshore,199
 large,199
Piolla-Kirchoff tensor,22
Pipe-soil interaction parameter, 542
Plane body waves,413
Plane-strain consolidation,223
Plane strain:
 problems,170
 strip model,196
Plane stratification,158
Plastic:
 behaviour path-dependent,288
 bulk moduli,82
 collapse,266
 flow,71
 hardening rule,82
 limit analysis,269
 modulus,65,81
 multipliers,284
 potential,65,82
 regime,88
 shear distortions,83
 shear moduli,82
 strains,288
 tangential mechanism deformations, 261
 volume change,83,115,179
Plasticity:
 bounding surface,65
 classical,65
 factor,328
 generalized,57
 multi-surface,82
 rules,69
 theory,284
 two-surface,82
Plate load test,292
Plugged pile,199
Point source model,492
Poisson random variables,484
Polynomial interpolation,215
Polytope feasible,253
Pore pressures,33,40,53,79,179, 538
 built-up,456
 ratio,508
Porosity,45
Potential:
 plastic,65,82
 rupture,557
 yield,232
Powell method optimization,277
Power mechanical,25
Prager's kinematic hardening rule, 291
Prandtl collapse load,236

Predictions and observations,132
Predictor-corrector type
algorithm,99,223,397
Pressure,222
gradient,45
interaction,281
loosening,159
swelling,159
Primal constraints,253
problems,253,325
Primal-dual mathematical programs,
247,252
Principal:
directions,6
invariants,6
stress space,66,157
values,6,115,395
Principle:
material frame-indiference,28
virtual work,214,327,565
virtual work generalization of,
27
Probabilistic:
aspects of dead load,491
design procedure,475
formulation,475
modeling of a soil deposit,493
safety factor concept,491
Probabilty:
distribution,255
functions,475
paper,480
Process models,150
Progressive:
cruching effects,193
failure,149,152,288
Propagation mechanism,418
Proper values:see principal values
Proportional loading histories,288
Pseudo-static analysis,99
Pulsating load tests,511a

Q

Quadratic:
interpolation for displacements,
215
nonlinear programming,284
objective functions,253
Quadratic:
program nonconvex,289,290
programming identification by,
253,288,332
Quadrature formulae,397
Quadrilateral elements,183
Quasi-static,47

R

Radial flux,364
Radial rock deformations,161
Radiation of waves,196
Random:
erros,295
Lognormal,484
Poisson,484
variables,475
vibration techniques,500
Rankine failure state,168
Ratcheting behaviour,538
Rate of tensor deformation,18
Rayleigh-Ritz technique,345
Rayleigh waves,413,428
Reaction principle,20
Redistributions of the stress,
149,182
Reflecting from boundaries,196
Regression models,500
Regular pulses earthquake,459
Relations:
constitutive,40
dynamic and kinematic,40
Relative fluid flux vector,204
Relaxation:
functions,206
procedure,277
Reliability of earth slopes,491
Reloading,59
Representation of piecewise linear,
90
Residual:
angle shearing at the,170
strength,153
Response:
analysis,410
spectra,416,428,443
system,273
Retaining wall strutted,133,168,281

Riemann-Christoffel tensor,11
Rigid base stationary,415
Rigid body accelerations,429
Rigid strip in weightless soil, 190
Rigid-System,285
Risk evaluation,476,487
Risk factors,490
Ritz method,314
Rock block:
elastic-brittle elements forming a,151
coupled in parallel,152
Rock:
bolts,381
confinig pressure p,153
cutting,145
destructive excavation,161,381
failure mechanism,153
friction resistance,155
mass behaviour,145
original stresses,381
properties,147
Rock pressure,162
genuine,159,161
overburden,159
Rock slope,146,163
jointed,316
Rotation infinitesimal tensor, 16
Rotational:
elastoplastic springs,285
symmetry,83
Rule:
plastic hardening,82
Runge-Kutta methods,223

S

Safety factor,229,317,490
Sample space,475
Samples of measurements,295
Sampling distances of probabilistic approach,495
Sandhu's two-field formulation, 213
Sand liquefaction,533
Sandy soils,455
Saturated material,50
Scale of flutuation,495
Scattering,293
Search techniques,277
Seepage coupled equation,52
Seismic:
excitation,419
flexible boundary methods,430
history,514b
problems,408
reflections,427
rigid boundary methods,429
scattering analysis,429
shaking,510
site response,428
soil-structure interaction,426
substructure methods,428
transmitting boundaries,427
viscous boundaries,427
Seismological data,414
Sensitivity matrix in identification,297
Shaft friction,199
Shakedown,182
Shaking table,464,519
Shallow:
foundations,167,175
tunnel,401
Shannon measure of uncertainty, 254
Shape factors of shallow foundation,547
Shear:
between two initially orthogonal vectors,12
devices,119,133
moduli,280
plastic distortions,83
soil test,85,89,170
strength (cyclically weakened), 183
stress (distribution of),182
tests,90
wave theory,422
Shield's representation,158
Significance level,481
Simplex method and algorithm,253, 268,276,321
Simulation:
construction sequences,133
experiments,293

Singular functions,347
Size of the time-step,221
Skeleton,79
 solid,34
Skin friction,187
Slack variables,250,321
Slices:
 method of,238
Slip circle analysis,231,238
Slope crest:
 movement of,174
Slope stability,173,229,269,306
Slopes,557
 by mathematical programming,305
Slow consolidation,54
Small displacement assumption,15
Snow load,492
Softening,68,291
Soil-fluid equations,51
Soil:
 mechanics classical plasticity,66
 skeleton,45
 stiffness,542
 structure interaction,103,119,
 138,409
Soil test:
 compression,89
 extension,86,89
 shear,89
Soils:
 cohesion less,86
 cohesive,86
Solid:
 mixture,45
 phase velocity of,80
 skeleton,34,40
 velocity gradient,80
Solution process,221
Somigliana's identity,347
Source of earthquake,412
Souces:
 density of,25
Spatial:
 coordinates,17
 description,10,25
 discretization,218
 interpolation,214
 variability,494
Spectra:
 response,428
Spectral:
 ratios,437
 shapes,439
Spectrum:
 broad-band,413
 response,416
Spherical and cylindrical waves,
 413
Spin tensor,80
Spread of yield,172
Spring models,194
Stable algorithms,221
Stability:
 analysis slope,306
 cavity,381
 embankments,557
 slopes,316,557
Standard:
 deviation,293,478,489
 penetration test (SPT),496
Staggered procedures,53
State:
 initial stress,161
 critical models,68
 stress/strain,60
Static:
 behaviour,39
 condensation,280
 consolidation behaviour,53
 kinematic duality,256
 kinematic solutions,47,261,557
 soil/fluid equation,51
Static systems:
 stochastic identification of,295
Statically loaded piles,196
Stationary rigid base,415
Statistical:
 independent lognormal variables,
 489
 mechanics of soil,255
 models,500
 probabilistic methods,473
Statistics:
 central limit theorem,484
 general approximate method,485
Steady:
 Darcy flow equation,48
 static drained behaviour,53
Steep valley no-tension solution,
 384

Stieltjes' integral,206
Stipulatons,251
Stochastic identification,287
 problems,275
Storage coefficient,362
Strain,533
 analysis,43,533
 displacements equations,31
 hardening plasticity,59
 history,52
 path,90
 potential,456,508
 softening analysis,59,170
Strain tensor:
 infinitesimal,15
 trace of the,15
Strech of the vector,12
Strength parameters,153
Stress:
 analysis-cyclic resistance approach,525
 concentrations,149
 history,526a
 in the lining,376
 invariants,66
 path,80,538
 ratio,513
 redistributions of the,149
 space,60,109,532
Stress-controlled tests,92
Stress-dilatancy,182
Stress-strain:
 curves,87,104,205
 history,206
Stresses,20
 effective,33,34,42,79
 neutral,33,34
Strip footing,398
Strong motion earthquake records, 413
Structural:
 coefficients,250
 holonomic response,331
 identification,281
 matrix,251
 models for tunnels,146
 three dimensional,146
Structure pushed in soil,127
Subballast,108
Subcharacteristic region,532
Support unilateral,285
Surcharacteristic domain,535
Surface stations,413
Surface waves:
 horizontally propagating,412
Surfaces of sliding,306
 non circular,309
Surrounding rock,159
Susceptibility of sedimentary deposits,510
Sustained pressure,514a
Swelling pressure,159
Symbols:
 Cristoffel,11
Symmetric constitutive coupling, 213

T

Tableau simplex,321
Tangent stiffness,82
Tangential reversible cyclic stresses,456
Tangential stress applied in cycles,464
Temporal discretization,219
Temporary support,160
Tensile cracks,149
Tension cut-off,157
Tensor valued function,80
Tensors,4
 deviatoric,4
 first Piolla-Kirchhoff,22
 Green's deformation,11
 Green's strain,11
 infinitesimal rotation,16
 infinitesimal strain,15
 orthogonal,4
 rate of deformation,18
 Riemann-Christoffel,11
 second order,4
 second Piolla-Kirchhoff,22
 spherical,4
 skew-symmetric,4
 spin,18
 symmetric,4
 trace of the strain,15
 velocity gradient,17
Test compression,128

Test:
extension,128
hydrostatic compression,128
Testing devices multiaxial,104
Tests:
cyclic strain-controlled,85
shear soil,85
Theory:
limit analysis,565
yield design,565
Theory of plasticity,59
Theories of mixtures,209
Thermal-effects,42
Thin interface,119
Three dimensional:
model,197
plastic strain,390
Tightening mechanism,538
Time insensitive,59
Time-marching schemes,363
Time-step,221
scheme,219
procedures,53
selection procedure,221
size,216
Tolerance,298
Torsional shear testing,90
Total rock deformation,162
Total stress:
field,34
tensor,40
Trace of the strain tensor,15
Track support,133
Traction on surface elements, 208
Traction vector,21,34
Trajectory of the minimum loss, 277
Tranfer of load,541
Tranferring stress,381
Transient:
aspects,531
consolidation,48
groundwater flow,362
problems,51
Transmissivity,362
Trefftz method,349
Tresca analysis,232
Triangular finite elements,364
Triaxial compression,155
Triaxial devices,104
Triaxial loading,66
Tunnel:
excavation and lining,145,160
in rock,382
liners,160,281
Tunnels:
structural model for,146
Two-field formulation,212
Two-phase:
continuum,35
medium,39
method in identification,290
Two-rock block model,153
Two-surface plasticity,82

U

Uncertainty,292,487
analysis,476
in the experimental data,292
Unconditional stability,219
of the algorithm,223
Unconstrained optimisation,249
Undeformed state,211
Underground openings,156
Underground structure,541
Undrained:
behaviour,53
capacity,179
conditions,44,76
static solution,47
tangent matrices,44
Uniaxial behaviour,60
Uniaxial compression tests,150
Uniform strain,42
Unilateral support,285
Unloading,59
behaviour,114
Unsteady flow in confined aquifer, 371
Unsteady flows,362
Upper and lower bounds, 231,290

V

Variability spatial,494
Variable moduli model,108

Variables statistically independ
lognormal,489
Variances,479,486
Variational formulations,207,313
Vector space:
N-dimensional,4
Vectors:
acceleration,9
curl of a,9
displacement,9
gradient of a,9
inner product,5
lenghts of,12
linear extension of the,14
non-mechanical energy,23
rotation,16
shear between two initially
orthogonal,12
strech of the,12
traction,21
velocity,9
Velocity of deformation,328
Velocity:
discontinuity line,558
gradient tensor,18
of solid phase,80
Velocities:
rate,17
relative,36
Vertically propagating waves,414
Virtual work:
principles of,327,565
Viscoelastic:
layered soil,419
soil behavior,213
Viscous:
damping,196
flow-resistant,36
force,45
Volume changes,182
Volumetric strain,44,83,538
Von mises yield surface,113

W

Walls retaining,281
Wave:
equation,196
fields,413
interference,413
Waves,413
body plane,413
fundamental surface,422
higher-order surface,418
horizontal propagating,414
inclined SH-,429
multi-layered systems,414
plane harmonic,414
Raleigh,428
spherical and cylindrical,413
vertically propagating,414,428
Weak formulation,344
Weakening,183
Wedge failure,146
Weighted residuals methods,298,
343,357
Weighting function,342
Weightless c - ϕ soil,178
Wind load,492
Winkler soil,286,291

Y

Yield,68,172
design theory,565
limits and conditions,76,81,108,
284,288,319,331
Coulomb,157
Drucker-Prager,158
point,59
Yield surfaces:
nested,82
piecewiselinear,288
Yielded zones,173,232
Yielding modes,285
Yokoo's formulation,213
Young modulus identification,295

Z

Zero-elastic region model,72